OCR
AS/A LEVEL

DESIGN & TECHNOLOGY

John Grundy
(SERIES EDITOR)

Sharon McCarthy
Jacki Piroddi
Chris Walker

Simeon Arnold
(CONTRIBUTOR)

John White
(CONTRIBUTOR)

HODDER
EDUCATION
AN HACHETTE UK COMPANY

Photo credits can be found on page 404 at the back of this book.

Every effort has been made to trace all copyright holders, but if any have been inadvertently overlooked, the Publishers will be pleased to make the necessary arrangements at the first opportunity.

Although every effort has been made to ensure that website addresses are correct at time of going to press, Hodder Education cannot be held responsible for the content of any website mentioned in this book. It is sometimes possible to find a relocated web page by typing in the address of the home page for a website in the URL window of your browser.

Hachette UK's policy is to use papers that are natural, renewable and recyclable products and made from wood grown in well-managed forests and other controlled sources. The logging and manufacturing processes are expected to conform to the environmental regulations of the country of origin.

Orders: please contact Hachette UK Distribution, Hely Hutchinson Centre, Milton Road, Didcot, Oxfordshire, OX11 7HH. Telephone: +44 (0)1235 827827. Email education@hachette.co.uk Lines are open from 9 a.m. to 5 p.m., Monday to Friday. You can also order through our website: www.hoddereducation.co.uk

ISBN: 978 1 5104 0265 2

© Simeon Arnold, John Grundy, Sharon McCarthy, Jacki Piroddi, Chris Walker, John White 2018

First published in 2016 by
Hodder Education,
An Hachette UK Company
Carmelite House
50 Victoria Embankment
London EC4Y 0DZ

www.hoddereducation.co.uk

Impression number   10 9 8 7

Year   2023

Cover photo © Palau / Shutterstock

Illustrations by Integra Software Services

Typeset in India by Integra Software Services

Printed by CPI Group (UK) Ltd, Croydon CR0 4YY

A catalogue record for this title is available from the British Library.

MIX
Paper | Supporting
responsible forestry
FSC™ C104740

Contents

www.hoddereducation.co.uk/ocrdesigntechalevel

Introduction to OCR Design and Technology for AS/A Level

Welcome to *OCR Design and Technology for AS/A Level*. This book has been written to meet the requirements of the following OCR endorsed titles:

- OCR AS Level in Design and Technology: Design Engineering (H004)
- OCR AS Level in Design and Technology: Fashion and Textiles (H005)
- OCR AS Level in Design and Technology: Product Design (H006)
- OCR A Level in Design and Technology: Design Engineering (H404)
- OCR A Level in Design and Technology: Fashion and Textiles (H405)
- OCR A Level in Design and Technology: Product Design (H406)

The textbook is designed to support both AS and A Level learners, with all additional or more in-depth learning required by A Level students clearly signposted.

The book aims to provide clear guidance and support material that will help you acquire a knowledge and understanding of the technical principles and designing and making principles you will need as you progress through your chosen Design and Technology course. It will support you as you develop critical thinking and problem solving skills to identify market needs and opportunities for new products, initiate and develop design solutions and develop and make prototypes that solve real-world problems and that consider a range of needs, wants, aspirations and values.

The textbook is divided into nine chapters that reflect the enquiry nature of the OCR specifications. They are designed to help you make links between topics and to explore, create and evaluate a range of outcomes. The nine topics are:

1. Identifying requirements
2. Learning from existing products and practice
3. Implications of wider issues
4. Design thinking and communication
5. Material considerations
6. Technical understanding
7. Manufacturing processes and techniques
8. Viability of design solutions
9. Health and safety

Each chapter explores these topics in relation to each of the Design Engineering, Fashion and Textiles and Product Design endorsed titles, exploring the core principles of design and technology that are common to all titles, as well as clearly signposting examples and applications that are unique to each of the different disciplines.

Two additional chapters are available online at www.hoddereducation.co.uk/ocrdesigntechalevel

Chapter 10 explores the requirements of the Non-Exam Assessment (NEA) in more detail. It covers both the Product Development NEA for AS students and the Iterative Design Project NEA for A Level learners. This chapter guides you through the key features of the NEA and explains the marking criteria against which you will be assessed. It includes lots of exemplar portfolio content from Design Engineering, Fashion and Textiles and Product Design projects.

Chapter 11, which is also available online, is designed to support you as you prepare for the written exam(s). It provides an overview of the written papers at AS and A Level, and includes general advice on preparing for the exam and types of questions you may encounter.

Features to help you

A range of features appear throughout the book to help you learn and improve your knowledge and understanding.

PRIOR KNOWLEDGE

A summary of prior knowledge you are likely to have acquired in GCSE Design and Technology is included at the start of each chapter.

TEST YOURSELF ON PRIOR KNOWLEDGE:

A series of questions appear at the start of each chapter so you can test your knowledge and understanding of concepts you should already have covered in relation to the topic at GCSE.

LEARNING OUTCOMES

Learning outcomes are included throughout each chapter. They summarise the knowledge and understanding to be covered depending on whether you are studying Design Engineering, Fashion and Textiles or Product Design, and whether you are studying at AS or A Level.

ACTIVITY

Short activities for you to complete appear throughout the book to help you apply your learning.

KEY TERMS

Key terms boxes define all important design and technology terminology.

KEY POINTS

● A summary of key points is included at the end of each section to help you review what you have learnt.

MATHEMATICAL SKILLS

The Design and Technology AS and A Levels require you to apply mathematical skills in a design and technology context. Some questions in the written paper(s) will assess your use of mathematical skills. This box appears at the end of each chapter and summarises the mathematical skills you may need to demonstrate in relation to the topic covered in the chapter.

Further reading

A list of further resources that provide further guidance, examples and extension material on a topic can be found at the end of each chapter.

PRACTICE QUESTIONS

Practice questions for Design Engineering, Fashion and Textiles and Product Design are included at the end of each chapter to test your knowledge and understanding and help you prepare for the written paper(s).

FASHION AND TEXTILES

This box highlights subject content, examples and applications that are unique to the Fashion and Textiles endorsed title.

DESIGN ENGINEERING

This box highlights subject content, examples and applications that are unique to the Design Engineering endorsed title.

PRODUCT DESIGN

This box highlights subject content, examples and applications that are unique to the Product Design endorsed title.

1 Identifying requirements

1.1 What can be learnt by exploring contexts that design solutions are intended for?

LEARNING OUTCOMES

The context any product is used in is the place or setting where it is used, and when and who the product is used by. Designers need to understand the distinctive contexts in which the products they design are used. It is important to understand the knowledge, experience and any mental/physical considerations of the primary users and where the product is used. The context and surroundings are also important as they could have an effect on the product, the materials it could be made from and how it works.

Different people have different needs and wants. What is desirable and helps one person may cause problems and difficulties for someone else. Any new designs of products and systems should improve ease of use or make a task quicker or cheaper for a user. Companies are always looking for new product opportunities to improve users' experiences and lifestyles.

Exploring contexts

Environments and surroundings

The context (surroundings) in which a product is used needs to be understood before any designing takes place. The where, who, when and how a product or system is used can affect many design decisions, from the choice of materials or finish to the size or aesthetic appearance.

- A design solution for use outdoors will need to be suitable for extremes of weather.
- A design solution for use by the public in a town centre will need to be durable for use in all seasons and designed in a way to be understood and able to be used by a variety of people that may include children, adults, the elderly and disabled.
- A design solution for a children's play area needs to be safe and use non-toxic finishes.

The context (setting or surroundings) for a design solution will have an effect on many of the design decisions that will be made, so it is important to consider the surroundings (place) where the product will be used and the users. Purpose (what the product needs to do) and price can also be important factors. When considering the requirements for seating, think of the different contexts a seat might be used in, such as a home, a kitchen, a school, an office, a waiting room in a surgery, a children's ward in a hospital, an outdoor play area, a train station or a town centre. Each distinct context will have an effect on the requirements for the seat in terms of its design, features, function, materials, construction, cost, users, safety factors and maintenance factors and each will have different primary users and **stakeholders**.

User requirements

When creating products and systems, designers always try to create a demand for a product. Consumers may desire products but perhaps due to affordability cannot have that product, but when they desire a product they can afford, it becomes a demand. Consumers have different needs and wants. Basic human needs are for food, shelter, clothing and safety, but people also want products to make things easier and more comfortable and to provide enjoyment or self-expression.

Needs

Human needs are the basic requirements and include food, clothing and shelter. Without these, humans cannot survive. We also consider education and healthcare to be needs. Generally, the products which fall under the needs category of products do not require a **marketing** push. Instead the customer wants to buy it.

Wants

Wants are largely dependent on the needs. For example, you need to take a bath or shower, but you don't need a good smelling soap or bubble bath. You might use it because it is your 'want'.

Example

The smartphone is a relatively recent **invention**. Do we need a smartphone? How did people manage before mobile phones? The first mobile phone used in 1973 weighed 1.1 kg and was 230 mm x 130 mm x 44.5 mm. It made calls and it wasn't until 2000 that text, MMS messages and the internet became common features.

Wearable devices and smartphones have brought more social interaction. We can keep in touch, read the news, shop, listen to music, measure our heart rate and record movement. However, using these devices when driving or walking increases the risk of accidents and interaction through devices can affect our social confidence and make it hard to switch off from the 'connected' world and relax. We don't need smartphones and wearable devices – we want them.

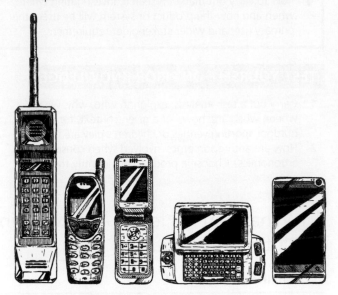

Figure 1.1 The evolution of the mobile phone

Understanding the users' requirements

Most products and systems are used or maintained by humans, so people are important. When we consider users, the primary user is the most obvious but any product will also have other stakeholders with an interest. For example, a company (stakeholder) will have an interest in the equipment they provide in an office environment to ensure their workers (the primary users) can work efficiently and comfortably. The Chartered Institute of Ergonomics and Human Factors and the Health and Safety Executive (HSE) have interests as wider stakeholders to provide guidance for the design of office equipment, from chairs to monitors and keyboards, to minimise back pain and repetitive strain injury (RSI).

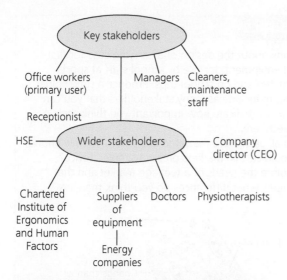

Figure 1.2 **Stakeholder map for office equipment**

Economic and market considerations and product opportunities

Think about the increasingly popular area of wearable devices. Different user groups have differing needs within all the contexts the device will be used. Each of these is distinctive and in turn affects the design in some way. For example, a woman will want a device that can be worn everyday with different outfits that might be casual or smart, yet something that is robust and durable to wear in the gym, out running and possibly when swimming and showering. A child's needs would be different, as would an elderly person's – some of what is important to one user will not be to another.

To explore and appreciate user requirements, it is always important for any designer to investigate the context their product will be used in and the users' needs, observing how people use products and carry out day-to-day tasks.

The makers of Trunki, the famous children's suitcase, extended the range to teenagers and adults. Jurni is a carry-on case for older travellers, which can also be ridden, sat on in queues and pulled along like a conventional travel bag. The Jurni case has a flexible shelf and removable storage pod where travellers can store personal belongings like gadgets and books.

When completing your own design work and identifying new product opportunities, there are many advantages to studying your users in the context where the product will be used.

- At the start of your non-exam assessment (NEA) you should ask your user(s) what they consider to be the important requirements. You can observe them in the design context and find out what the

problems with any existing solutions are, or any problems they have when undertaking a particular task. You should also contact other people who have an interest in, or are affected by, the product/system (stakeholders).
- At the designing stage, it is important to get continuous feedback and carry out user testing with models and prototypes, allowing for further iterations towards an optimum solution.

Figure 1.3 **Although the suitcase was to be used in the same environment, the different primary users and stakeholder needs made the contexts unique and therefore the design requirements and solution were unique**

Whenever you visit your users and stakeholders, you will need to plan your questions, record them and take photographs or videos. Completing a task analysis, thinking about who, why, what, where, when and how, will enable you to develop a list of questions and enable you to plan what you need to find out in advance. The answers obtained will lead to other questions you might not have thought of.
- **Where?** The context, environment, place or places are important – where a product or system is placed, used, stored or moved. Is it indoor or outdoor, will this have an effect on the product, the materials it could be made from, how it works and how often will the product be used? What are the noise and lighting implications?
- **Who?** Both the user and other people (stakeholders) in the context where the product will be used. It is important to understand their knowledge, education, experience and preferences for products,

mental/physical abilities and considerations. The designer needs to understand the primary users and stakeholders and their requirements, to be able to design for them. **Personas** can be created but it is always better to work with real users and stakeholders. The influence and attitudes of individual stakeholders can be important at various stages of designing a solution, but you will always need to judge the significance of these in relation to your project.

- **Why?** The reason a person is doing something. Why the product is needed in the first place, why another method is not preferred.
- **What?** Details of the tasks and activities required and the goals and targets to be met. What does the product that is required need to do?
- **When?** At what times of day or night, regular or irregular use, occasional or frequent, maximum and minimum times?
- **How?** What ways will typical users use the product? How will tasks and activities be carried out? How will you need to market your product and how are similar products marketed?

You can also use **secondary research**, which is other people's observations and findings, sometimes surveys, **standards** and data, which can be accessed online or from libraries or organisations and authorities with an interest in the subject area. This will back up or complement your own **primary research** and observations.

KEY TERMS

Invention – the process of creating something that has never been made or never existed before.

Marketing – the process used by a company to promote and sell its products to consumers.

Persona – often created as a means of representing users or stakeholders. A persona is similar to a user profile. Information about users or stakeholders is used to create a collective persona, a 'typical' person, with their views, attitudes, preferences, lifestyle, skills and so on.

Primary research – the personal collection of research and information. It is carried out through methods such as visits and observations, interviews, testing and surveys.

Secondary research – the collection, collation and editing of readily available information. Such research utilises sources such as published details, company literature and existing test data.

Stakeholders – the stakeholders of a product are all those who may come into contact with it, have some sort of interest or 'stake' in it, or are affected by it in some way. By communicating with your stakeholders, you can ensure that they fully understand what you are doing, understand the benefits of your product and support you in making key decisions.

ACTIVITIES

1. Think about the design of a carrying case for first aid equipment, for use by school staff at sports matches and on school trips. Produce a stakeholder map to include as many stakeholders that you can think of. Indicate how important you think their 'needs' are.
2. Do all users have the same needs for a mobile phone? Would certain markets prefer a simpler device? Explore the needs of a teenage market and the elderly, what differences do you think there are?

Figure 1.4 How grandma sees a remote

KEY POINTS

- The primary user of a product is just one of many stakeholders in the development of a new product.
- The stakeholders of a product are all those who may come into contact with it, have some sort of interest or 'stake' in it, or are affected by it in some way.
- Stakeholders can be:
  - designers, manufacturers, retailers, those involved in marketing the product
  - those who maintain, repair or recycle the product
  - material and component suppliers
  - those related to the location for use of the product
  - energy suppliers
  - various organisations that monitor or regulate designs or manufacturing processes (for example, British Standards, local authorities), experts or specialists in the product area.

1.2 What can be learnt by undertaking stakeholder analysis?

There are a number of methods used to investigate stakeholder requirements that are widely used in the design and manufacturing industries. It is important to understand the requirements of everyone involved in any way with the design, as well as the needs of the primary user. Each of the methods can be valuable. Using them in the right way, for the right reasons and at the right time is important. Which method to use and when or how to use it will differ from project to project.

Some use **qualitative observation** where qualities can be observed without expressing numerical values, so the results could be large, yellow, tubular, soft, etc. **Quantitative observations** collect measurable results that record data, such as measurements, weights and temperatures. Many methods involve designers and manufacturers in direct contact with stakeholders to ensure that all requirements and data are gathered 'first hand' and are therefore accurate and detailed. Specialist market research companies are often commissioned to carry out investigations within the target market, sometimes using **focus groups**.

By communicating with your stakeholders during the design process, you can ensure that they fully understand what you are doing and understand the benefits of your product, supporting you in making key decisions.

User-centred design (UCD)

User-centred design (UCD) is a project approach that puts the intended users at the centre of design and development by ensuring that the designer talks directly to the user at key points in the project. You will be able to apply UCD principles where appropriate in your NEA **iterative design** project.

There are four essential activities in a user-centred design project:
- requirements gathering – understanding and specifying the context of use
- requirements specification – specifying the user and organisational requirements
- design – producing designs and prototypes
- evaluation – carrying out user-based assessment of the designs and prototypes.

These activities are carried out in an iterative fashion, with the cycle being repeated until the project's **usability** objectives have been achieved.

One of the underlying principles of UCD is that the design of a product, or a specific task with a product or system, should match its intended users' behaviour patterns, attitudes and preferences. UCD does not expect a user to adapt their behaviour and attitudes to learn and use a product or system. The result is a product that gives a more efficient, satisfying and user-friendly experience for the user, which in business and commercial terms will translate into increased sales and brand loyalty.

Stakeholder analysis

Completing a detailed stakeholder analysis enables designers to understand the users/stakeholders and the distinctive design context. To develop questions they may need to ask others to help with this, identifying questions to ask their users/stakeholders: 'What do we need to know about who, where, what, and how?'

It is always better to interview users and stakeholders in the context the product will be used, as it might trigger extra questions and observing people helps to understand their needs. Some questions designers might ask are listed below. Note that the reference to 'product' could equally apply to an 'engineered system'.

- What is the specific purpose(s) that the design solution will be used for? How will the product be used?
- What is the main function(s) the design solution needs to have? How often will the product be used and for how long?
- How carefully is the product likely to be handled/treated/worn? Will the product be transported or stored when not in use?
- What physical position will the user be in when they use the product? How and where might the product be cleaned or maintained?
- Who will use the product? Is there anything about the user that affects what the product must be like? Who will purchase the product?
- Does the user have any physical conditions that may cause difficulty using the product, such as strength and range-of-motion, vision or hearing?
- Are there any cultural expectations related to this product?

Stakeholder analysis is used in **project management** to understand and then manage the needs and concerns of stakeholders. The influence and attitudes of individual stakeholders can be important at various stages of designing a solution. A stakeholder analysis focuses on the influence of each stakeholder, which could be an organisation, group or individual and their interest in or attitudes toward the project.

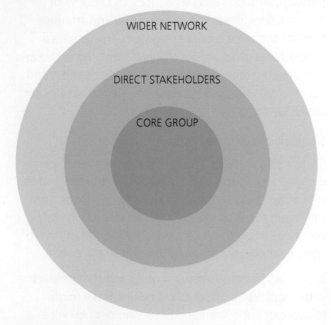

Figure 1.5 Template for stakeholder mapping analysis

In project management, this is often shown using an influence-interest grid and the power-interest grid. The first step in your stakeholder analysis is to identify who your stakeholders are. Think of all the people who will be affected by your design activity, who have influence or power over it, or have an interest in it being successful or unsuccessful. They can then be prioritised by power and interest. You could plot this on an influence-interest or a power-interest grid. There are many templates for these online.

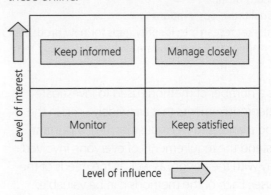

Figure 1.6 Influence-interest matrix

SWOT analysis

Creating any new product involves consideration of its pricing, promotion and place of sale. It is useful to examine the strengths, weaknesses and the opportunities presented to help maximise your resources and be aware of possible pitfalls and problems.

A **SWOT** analysis is one of the methods that is used to evaluate strengths (S), weaknesses (W), opportunities (O) and threats (T) involved in **innovative** ideas, services and strategies. It is widely used in business but can be used when developing new products and systems.

A SWOT analysis sometimes focuses on two factors rather than the four factors. For example, potential opportunities in a specific product can be taken forward by building on the strengths of the product and overcoming the weaknesses identified.

- Strength–Opportunity (S–O) – investigates and analyses the strengths in existing environments/products and opportunities for developing them.
- Weakness–Opportunities (W–O) – identifies and analyses the weaknesses in products or existing solutions to build opportunities for the new product.
- Strength–Threats (S–T) – identifies methods of using a product's strengths to reduce the threats and market risk.
- Weakness–Threats (W–T) – identifies and analyses the weaknesses in a product and builds a strategy

to prevent the product's weakness from being influenced by external threats.

How to carry out a SWOT analysis

Strengths

- What are the advantages of the new product/system?
- What are the product's advantages over similar competitors in market?
- What strength points do people see in the product?
- What are the product's unique selling points (USP)?

Weakness

- What weakness could be improved in the design?
- What issues should be avoided in a new product?
- What are the factors that reduce sales of existing products?
- Does the production process have limited resources?

Opportunities

- What are the opportunities for the new product?
- What are the trends to take advantage of?
- How can we turn strengths into opportunities?
- Are there any changes in the market or legislation which can lead to opportunities?

Threats

- Who are the existing or potential competitors?
- What issues can threaten the product on the market?
- Will there be any shifts in consumer behaviour, legislation or market that can affect the product's success?

Figure 1.7 SWOT analysis

Focus groups

Focus groups are an invited group of intended/actual users who are asked to share their thoughts, feelings, attitudes and ideas on a certain subject, which may be an idea or design for a new product, an engineered system, an advertisement or packaging. The group usually comprises between six to ten people and the informal discussion normally lasts between one and two hours. The group will often represent people from a variety of backgrounds and careers, and a trained moderator ensures that everyone participates as fully as possible. The moderator will pose a series of questions and often use images and objects to gain views and opinions. Focus groups often form part of feasibility studies and market research, and are a good means of getting information from a small sample of people. Trained observers are often situated in an adjacent room with one-way mirror glass and can observe and record participants' reactions. Alternatively, the focus group can be videoed for future observation. Focus groups provide qualitative data – their behaviour is observed and their opinions are obtained.

Usability testing evaluates products, systems and user interfaces by collecting data from people as they use a product or system. A person is invited to attend a session in which they are asked to perform a series of tasks to try out when using a product. Users are usually asked to verbalise what they're doing and why they are doing it. Usability testing can be used as an input to design or at the end of a project. It represents an excellent way of finding out what the usability problems are likely to be.

Participatory design does not just ask users' opinions on design issues, but actively involves them in the design and decision-making processes. An example would be a participatory design workshop in which designers and users work together to design an initial prototype.

Qualitative observations

Qualitative observation deals with data that can be observed with our senses: sight, smell, touch, taste and hearing. Colours, shapes and textures of objects are all **qualitative observations**. Some methods are outlined below.

Interviews usually involve one interviewer speaking to one participant at a time. The advantages of an interview are that a participant's unique point of view can be explored and requirements can be understood in detail. Any misunderstandings between the interviewer and the participant are likely to be quickly identified and addressed.

Card sorting is a method for suggesting intuitive ideas. A participant is presented with an unsorted pack of index cards. Each card has a statement written on it that relates to a product or system being designed. The participant is asked to sort these cards into groups and then to name these groups. The results of individual sorts are combined and analysed statistically. Card sorting is usually used as an input to design. It's an excellent way of determining the most important features to a user or group of users. Card sorting combines qualitative data based on group names and comments from participants, with quantitative data from the groupings of the cards.

Quantitative data can also be collected using surveys. Questionnaires can be used to ask users for their responses to a set of questions and are a good way of generating statistical data. They can be used to obtain a large sample size of opinions and feedback.

Market research

Market research is defined as the process of systematically gathering data on people, goods and services to determine whether a new or updated product or service will satisfy consumers' and stakeholders' needs. Market research can identify market trends, demographics, economic shifts, buying habits and important information on competition. This information can help to define target markets and establish a competitive advantage.

Introducing new or updated products on a constant basis is the best way to get attention and retain market share. New products are usually developed either because of market pull, where consumers demand a particular type of product, or by technology push, where new materials and/or technologies lead to innovative products being released onto the market.

New product development market research can involve:

- **feasibility analysis** – is it possible? Will customers want it? Can it be manufactured at a price that will generate a profit?
- investigation leading to idea/concept generation – from research, technical advisers (for example, engineers), consumers, employees, case studies, design consultancies
- user trialling – testing products with selected consumers, feedback generated
- launch – possible small-scale soft launch, gradual 'roll out' or full national launch of product.

Market research can be used to identify a number of different areas:

- particular markets
- the needs of consumers and stakeholders
- what the competitors are doing
- up-to-the-minute market trends
- customer satisfaction relating to products and services
- how much customers might be prepared to pay
- how the product should be identified
- naming and **branding**.

Market research can be primary, secondary or a mixture of both.

Table 1.1 **Primary and secondary market research**

Primary	Secondary
• Collected first hand using questionnaires, interviews, observations, focus/user groups, surveys and field research • Usually conducted by specialist market research firm • Precise data, meets the exact needs of the organisation • Can be expensive	• Uses widely available data from reference books and publications, government reports (e.g. census), trade/industry journals and online data sources • Provides information such as population trends and regional statistics, legislative trends and industry shifts • Can become quickly out of date or incorrect

Once market research has been gathered and analysed, decisions can be made as to the viability of the new or updated product. The stakeholder requirements identified will form the basis for designing.

Fashion and trend forecasting

This is an increasingly popular method used to identify opportunities for design.

Fashion forecasters gather facts and compile observations from wide-ranging fields such as art and design, science, technology, food and travel to predict the colours, fabrics, textures, materials, prints, graphics, accessories, footwear, street style and other styles that will be popular for the upcoming seasons. The concept applies to not one, but all levels of the fashion industry, including haute couture, ready-to-wear, mass market and street wear.

Trend forecasting takes a broader view, focusing on other industries such as automobiles, medicine, food and beverages, literature and home furnishings.

The fashion forecasting process carried out for a clothing retailer, for example, would include the basic steps of:

- understanding the vision of the business and profile of target customers
- collecting information about available merchandise
- preparing information
- determining trends
- choosing merchandise appropriate for the company and target customer.

Enterprise

In the design context, the term 'enterprise' relates strongly to innovation. It captures brave and courageous decision-making, initiative and resourcefulness, making the most of opportunities to earn money, create or support new businesses, new and energetic undertakings, adventure and challenge. Businesses are often referred to as **enterprises**, particularly those breaking new ground. Enterprise can help drive the development of new product ideas through routes to innovation. These can be by individuals and start-ups or groups of people through **crowdfunding**.

Entrepreneurship

Entrepreneurship has traditionally been defined as the process of designing, launching and running a new business, such as a start-up company, often offering a product or service. Entrepreneurs are the people behind such initiatives or start-ups.

Some famous and successful businesses started in this way. Richard Branson became an entrepreneur at a young age. At the age of 16, his first business venture was a magazine. In 1970, aged just 20, he set up a mail-order record business. In 1972, he opened a chain of record stores, Virgin Records, later known as Virgin Megastores, before also moving into the worlds of trains, aviation and media. An entrepreneur is a person who starts a new business and usually risks their own money to start the venture. Other examples of well-known entrepreneurs include Steve Jobs, Bill Gates and Mark Zuckerberg.

> 'Entrepreneurs are usually confident and self-motivated. They are also willing to fail and start over again, taking the lessons they've learned to create something new and improved.'

Business News Daily, 'What is entrepreneurship' (www.businessnews-daily.com)

Commercial partnerships

As businesses grow and become global, the most forward-thinking companies recognise the need to forge partnerships with others, to share ideas and expertise to grow and gain access to new technologies and markets.

There are several famous examples of business partnerships, including Google, a company founded by Larry Page and Sergey Brin. Page and Brin met at Stanford University as students. Many more successful companies including Apple, eBay and Twitter were built by multiple leaders.

Although partnerships between mature multinational companies and growing companies can be challenging, they can also be mutually beneficial. Multinationals can gain market access and technology, while growing companies can gain a foothold into established markets. There are many examples of companies working together in partnerships.

The 2004 Anglepoise Type75 lamp, designed by Sir Paul Smith, was a wonderful advancement of the classic design we all know, incorporating the same spring mechanism that automotive engineer George Carwardine invented in 1932 for the original Anglepoise. 'Collaborating with existing global brands and designers is a great way to present our brand in new territories', explains Richard Sellwood, the Chief Executive of the Portsmouth-based company.

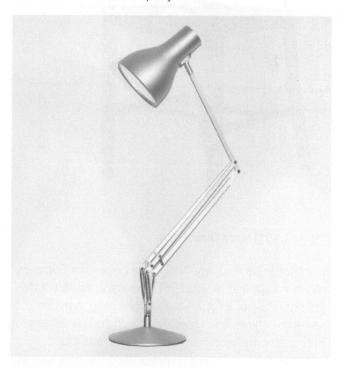

Figure 1.8 Anglepoise's collaboration with Sir Paul Smith enabled the company to sell their products in premium department stores around the world

PlantBottle® packaging is The Coca-Cola Company's way of bringing positive change to the packaging industry using recyclable packaging derived from renewable plant material. Their collaboration with Heinz has helped to generate awareness for PlantBottle™ packaging across industries and consumers. Heinz are using the PlantBottle™ technology for its ketchup bottle. The bottle is 100 per cent recyclable: the resulting by-products can be re-used to manufacture more bottles, or to make other products, such as furniture or clothing.

Coca-Cola and Emeco collaborated to solve an environmental problem, up-cycling consumer waste into a sustainable classic chair. Made of 111 recycled PET bottles, the 111 Navy Chair is a story of innovation. Emeco has estimated that it will process more than 3 million PET bottles.

Figure 1.9 111 Navy Chair

PlantBottle Technology™ from The Coca-Cola Company has been applied beyond PET packaging as part of the interior fabric of a Ford Fusion Energi plug-in hybrid research vehicle. This type of research and development partnership between companies can benefit both companies, even large international companies such as Ford and Coca-Cola.

Nike and Apple have been collaborating since the early 2000s, when the first iPod was released. The partnership has since evolved to become Nike+. Activity tracking technology is incorporated into athletic clothing and gear to sync with Apple iPhone apps to track and record workout data. Tracking transmitters can be built into shoes, armbands and even basketballs to measure time, distance, heart rate and calories burned. This co-branding benefits both companies to provide a better experience for customers.

Venture capitalists

Venture capitalists invest money into small companies that are trying to expand their business or people with great ideas who need finance and funding. In exchange for their investment, venture capitalists become stakeholders (holding a 'stake'), with a strong interest in the businesses they are supporting. Some well-known UK companies were built with venture capital, such as lastminute.com, National Express Group and Trafficmaster. In the USA, companies such as Apple, Federal Express, Compaq (now part of HP), Sun Microsystems, Intel and Microsoft are famous examples of companies that received venture capital early in their development.

Venture capital not only funds new initiatives and innovative new products and companies. A large part of venture money supports a later stage of product development where the innovation is prepared for the market, or 'commercialised'. This involves investment into building the infrastructure needed to grow the business in the expensive areas of manufacturing, marketing and sales. Funding to build the fixed assets of the business and provision of working capital is crucial for success. The best-known example we are all familiar with is TV's *Dragons' Den* where entrepreneurs and inventors pitch for investment from five venture capitalists willing to invest in exchange for equity. Entrepreneurs and inventors often make the mistake of thinking that venture capitalists are just looking for good ideas when, in fact, they are also looking for people with a reputation and skills in managing projects, people and finances, that they can have confidence investing in. They are investing in the person as well as the idea!

Venture funding can be a short-term or a long-term arrangement. Short-term capital investment enables a business to reach self-sufficiency in terms of sales and revenue, or to reach a point of credibility and value where the business can either be sold or become part of a larger company, and the venture capital investor can exit from their involvement with a healthy return/profit on their investment and look for a new venture to support. Long-term funding and investment has the potential to give constant and potentially increasing returns to the capital investor, in a more 'partnership' and 'joint ownership' arrangement.

The funds for venture capital investments are typically very large institutions such as pension funds, financial and insurance companies and university endowments, that put a small percentage of their total funds into what are 'high risk' but potentially 'high return' investments.

Crowdfunding

Crowdfunding is a way of raising finance by asking a large number of people for a small amount of money. Until recently, financing a business, project or venture involved asking a few people for large sums of money. Crowdfunding switches this idea around, using the internet to talk to thousands – if not millions – of potential funders. Typically, those seeking funds will set up a profile of their project on a website (called a crowdfunding platform). They can then use social media, alongside traditional networks of friends, family and work acquaintances, to raise money. There are three different types of crowdfunding: donation, debt and equity.

- Donations are made by people who invest simply because they believe in the cause. They do not expect a payback but might be acknowledged on an album cover, receive tickets to an event or get free gifts and updates.
- Debt funding means investors receive their money back with interest, but also know they contributed to the cause or product's success.
- Equity funding means investors invest in an opportunity in exchange for equity, shares or a small stake in the business, project or venture.

Well-known crowdfunding websites include Kickstarter, Indiegogo and Crowdfunder.

Other successful platforms used to share ideas in the design world include:
- Quirky – this is a community-led invention platform. It aims to partner with inventors in developing and making their ideas successful, empowering everyday problem solvers to share their ideas with the world.
- Behance – this is another online platform to showcase and discover creative work. The creative world updates their work in one place to broadcast it widely and efficiently, allowing companies to explore the work of designers and access talent on a global scale.

ACTIVITY

Explore some of the websites mentioned and find examples of successful new products and design projects that have benefited from crowdfunding, venture capitalist funding or some sort of partnership. Explain the reasons for their success.

KEY TERMS

Branding – the process of creating a product identity and name, logo or design that makes a product identifiable to consumers.

Crowdfunding – an online method of raising finance by asking a large number of people each for a small amount of money.

Enterprise –another name for a business. This refers to the initiative of an individual or organisation and their willingness to take risks to develop new projects and ventures.

Feasibility analysis – considering how possible and realistic it is to proceed with a project, i.e. to manufacture a product.

Focus group – an organised discussion led by a moderator where people are asked about their views and experiences of a product, brand or service.

Innovative – the development of an idea which is new, different and ambitious. The design can be risky and introduce aesthetics or function that hasn't been seen before. Innovation comes from pushing beyond the expected.

Iterative design – a continual and cyclical design development process to refine and perfect the product. Ongoing testing of models and prototypes, incorporating improvements and progressing towards an optimum solution for all stakeholders.

Participatory design – design which actively involves all stakeholders and users in the design process.

Project management – the act of defining, initiating, planning, executing, monitoring and completing the work of a team to meet a specific target. It is the planning and control of everything involved in delivering an end result to meet the stakeholder requirements; getting the job done.

Qualitative observations – these use the senses (sight, smell, touch, taste and hearing) to observe results and explain how we feel.

Quantitative observations – made with instruments such as rulers, scales and balances, graduated cylinders, timers, beakers and thermometers. These results are measurable.

SWOT – an acronym for strengths, weaknesses, opportunities and threats. A SWOT analysis can be carried out on any subject (for example, a product, person, business, industry or place).

User-centred design (UCD) – a design strategy or design approach focused on the user. Its aim is to make products and systems usable.

- A focus group is an organised discussion led by a moderator where people are asked about their views, experiences and perceptions of and attitudes towards a product, brand, service, idea, advertisement or packaging. A focus group is qualitative research because participants give open-ended responses which will convey thoughts or feelings. Focus groups often form part of feasibility studies and market research.
- Sometimes called 'human-centred design', user-centred design (UCD) is a design strategy, or design approach, with the aim of making products and systems usable. It focuses on the user interface – how the user interacts with and relates to the product – creating products with a high level of usability. It draws from many different subject areas and disciplines.

- SWOT is an acronym for strengths, weaknesses, opportunities and threats. A SWOT analysis can be carried out on any subject (for example, a product, person, business, industry or place). The analysis starts with an aim and then identifies the strengths and weaknesses of the subject (product, person, etc.) to achieve the aim, and the opportunities and threats associated with the aim.
- Enterprise captures initiative and resourcefulness, making the most of opportunities to earn money, create or support new businesses. Businesses are often referred to as enterprises, particularly those breaking new ground. Enterprise can help drive the development of new product ideas with individuals (entrepreneurs) and start-ups or groups of people in partnerships or through crowdfunding.

1.3 How can usability be considered when designing prototypes?

LEARNING OUTCOMES

By the end of this section you should have developed a knowledge and understanding of:
- how to analyse and evaluate factors that may need consideration in relation to the user interaction of a design solution, including:
  - the impact of a solution on a user's lifestyle
  - the ease of use and inclusivity of products
  - ergonomic considerations and anthropometric data to support ease of use
  - aesthetic considerations.

- the ergonomic factors that may need considering when developing products, including:
  - anthropometric data to help define design parameters associated with the human body
  - user comfort, layout of controls, software user interface.

The impact of a solution on a user's lifestyle

When developing products and/or systems, designers need to consider how they can meet the needs of a wide variety of users and the way in which they live. Every design decision has the potential to include or exclude some part of the population. **Usability** means making products and systems easier to use and matching them more closely to user needs and requirements.

The international standard ISO 9241-11 that provides guidance on usability defines it as, 'The extent to which a product can be used by specified users to achieve specified goals with effectiveness, efficiency and satisfaction in a specified context of use.'

Usability is about:
- Effectiveness – how easily can users complete tasks and achieve goals with the product (can they do what they want to do)?

- Efficiency – how much effort do users require to complete tasks with the product (often measured in time)?
- Satisfaction – what do users think about the product's ease of use (is it a pleasurable experience)?

This depends on who is using the product, what the user is trying to do with the product and the situation (or 'context of use') – where and how is the product being used?

Usability should not be confused with 'functionality'. Increased functionality does not mean improved usability! Often simple products are the easiest to use. Usability can encompass making things usable for a wider population, like making a handle a shape and size that children, as well as adults, can use. It also includes 'making something do more things'. Just think of the increased usability of your laptop or tablet computer over a desktop computer and how that can improve your lifestyle.

The ease of use and inclusivity of products

Designing for all is known as **inclusive design**; taking into consideration the widest possible audience. This will involve having to consider a number of factors including age, gender, size, weight and disabilities. The vast majority of us will suffer at least temporary disability at some point in our lives as we grow older.

This does not mean you must design for everyone in the world, but it means aiming to exclude as few as possible during the whole design process.

> 'Inclusive design aims to remove the barriers that create undue effort and separation. It enables everyone to participate equally, confidently and independently in everyday activities.'

'The Principles of Inclusive Design' published by CABE – The Commission for Architecture and the Built Environment (the UK Government's advisor on architecture, urban design and public space)

The UK Institute of Inclusive Design actively promotes inclusive design. The Helen Hamlyn Centre for Design and the Design Council provide resources on this topic. See the Design Council website at www.designcouncil.org.uk and search for principles of inclusive design.

Ergonomic considerations and anthropometric data to support ease of use

> 'If an object, an environment or a system is intended for human use, then its design should take into account the characteristics of its human users.'

Stephen Pheasant

Ergonomics

Ergonomics is about the 'fit' between people and the things they do, the objects they use and the environments they work, travel and play in. If a good fit is achieved, the stresses on people are reduced. They are more comfortable, they can do things more quickly and easily and they make fewer mistakes. So, when we talk about 'fit', we don't just mean the physical fit of a person, we are concerned with psychological and other aspects too. Ergonomics involves the layout of controls, use of colour, size of fonts, the software user interface, the weight of products, sound and vibration factors and brightness of display screens, all of which affect how comfortable a product is to handle or use. Good ergonomics often leads to good usability.

Increases in worker productivity over the last century are mainly due to increased education and ergonomics. Increased market share for new products are mainly a result of good user experiences, which owe a lot to ergonomics.

It doesn't take long to get frustrated with a poorly designed product or system. Bad ergonomics not only destroys productivity, but can spoil your mood or ruin your day – bad ergonomics can have a significant psychological impact on users.

Anthropometrics

Anthropometrics are people measurements. Anthropometric data comes in the form of charts and tables. They may provide specific sizes, such as finger lengths and hand spans, but they also offer average group sizes for people of different age ranges. Other sizes to consider are heights, reach, grip and sight lines.

There are three general principles for applying anthropometric data:
- Design for the extreme.
- Design for adjustability.
- Design for the average.

Percentiles

The use of **percentiles** is an important aspect of anthropometrics. The sizes of the human body given in anthropometric data are usually presented in tables and normally include the 5th percentile, the 50th percentile (the average) and the 95th percentile. Look at some anthropometric data and notice the different percentile measurements.

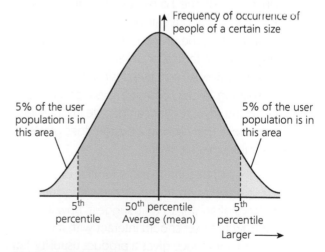

Figure 1.10 Percentiles

The 50th percentile is the most common size, the average. The 5th percentile indicates that 5 per cent of

people (or one person in 20) is smaller than this size. The 95th percentile indicates that 5 per cent of people (or one person in 20) is larger than this size. Very few people are extremely large or very small.

Which percentile range, and therefore which size to be used from anthropometric data tables, will depend on what is being designed and who will be using it. Is the product for all potential users or just the ones of above or below average dimensions?

Aesthetics

Very often, customers are drawn to products because of their visual impact. Aesthetics, in its widest interpretation, is involved with our senses – vision, hearing, taste, touch, smell – and our emotional responses to objects and things.

In *Design for the Real World*, Victor Papanek describes how automotive designers in Detroit, looking at making car dashboards more symmetrical and aesthetically pleasing by relocating ash trays, controls and switches, could have resulted in 20,000 fatalities and 80,000 serious injuries over a five-year period. This happened because of drivers having to over-reach for controls, diverting attention from driving for split seconds longer.

Sometimes a design is considered intuitive. That means that the design works the way the user would expect it to, either adopting the 'standard' or 'stereotype' convention (for example, volume increases through turning a control clockwise) or through design communication and aesthetics such as labelling or the display/interface showing which direction to turn the control to increase volume. Form, colour, control layout or other design elements can all indicate how controls should be used. When the process of using a product moves along a natural, logical or otherwise expected path then it is intuitive. An intuitive design does not need much training. When someone says a product is 'easy to use' or 'intuitive', they mean people instinctively know how to use it, that clicking a button does precisely what they would expect it to do.

Affordance is a word becoming more familiar in industrial design and refers to the aesthetics, features or characteristics of an object that 'suggest' how it functions or how the user should interact with it or operate it. Good affordance gives a product usability. For example, a door with a handle suggests that it should be pulled, so has good affordance.

Figure 1.11 A mains plug naturally plugs into a mains socket – the form and shape of the pins on the plug suggest that they fit in the socket

Much of Apple's success is down to their founder, Steve Jobs, who felt that a core component of design simplicity was making products intuitively easy to use. Sometimes a design can be so sleek and simple that a user finds it intimidating or unfriendly to navigate. His key principle in designing was to 'make things intuitively obvious'. Distinctive design – clean and friendly and fun – became the hallmark of Apple products under Jobs.

Dieter Rams and his team at Braun (1956–78) were very influential, producing hundreds of wonderfully conceived and aesthetically designed objects. The iPhone and iPod Touch were clearly inspired by Rams. Today Dieter Rams is best known for his range of minimalist shelves, made by Vitsoe.

Rams introduced the idea of sustainable development and of obsolescence being a crime in design in the 1970s. Accordingly, he asked himself the question: 'Is my design good design?' The answer formed is now the celebrated ten principles that are important to any designer with usability and good design as their principle.

Good design:
- is innovative – technological development is always offering new opportunities for original designs. But imaginative design always develops in tandem with improving technology and can never be an end in itself.
- makes a product useful – a product is bought to be used. It has to satisfy not only functional, but also psychological and aesthetic criteria. Good design

emphasises the usefulness of a product while disregarding anything that could detract from it.

- is aesthetic – the aesthetic quality of a product is integral to its usefulness because products are used every day and have an effect on people and their well-being. Only well-executed objects can be beautiful.
- makes a product understandable – it clarifies the product's structure. It can make the product clearly express its function by making use of the user's intuition. It is self-explanatory, its appearance helps you to understand its functionality.
- is unobtrusive – products fulfilling a purpose are like tools. They are neither decorative objects nor works of art. Their design should therefore be both neutral and restrained, to leave room for the user's self-expression.
- is honest – it does not make a product appear more innovative, powerful or valuable than it really is. It does not attempt to manipulate the consumer with promises that cannot be kept.
- is long-lasting – it avoids being fashionable and therefore never appears antiquated. Unlike fashionable design, it lasts many years – even in today's **throwaway society**.
- is thorough down to the last detail – nothing must be arbitrary or left to chance. Care and accuracy in the design and its finer detail show respect towards the consumer.
- is environmentally friendly – design makes an important contribution to the preservation of the environment. It conserves resources and minimises pollution throughout the lifecycle of the product.
- is as little design as possible – less, but better – because it concentrates on the essential aspects and the products are not burdened with non-essentials. Back to purity, back to simplicity and simple aesthetics.

These principles are still important to designers of all disciplines today. In the 2009 documentary film *Objectified*, Rams states that Apple Inc. is one of the few companies designing products according to his principles.

Aesthetic appreciation involves shape, colour and form. Shapes are formed as a result of closed lines. Shapes can be visible without lines when a designer establishes a colour area. They may be composed from parts of different objects in an arrangement; they can be gaps, or negative shapes between the objects. Basic shapes include circles, squares, triangles and polygons, all of which appear in nature in some form or another. Form refers to the three-dimensional quality of an object.

Colour is very important in design and can help make a product easier to understand or use. It also creates responses by stimulating emotions and can excite, impress, entertain and persuade. Catching the consumer's attention and conveying information effectively are critical to successful sales. A designer must be aware of how people respond to colour and colour combinations.

Table 1.1 Colour associations in design

Colour		Description
Red	■	Aggressive, passion, strong and heavy, danger, socialism, heat
Blue	■	Comfort, loyalty and security, for boys, sea, sky, peace and tranquillity, cold
Yellow	■	Caution, spring and brightness, joy, cowardice, sunlight
Green	■	Money, health, jealousy, greed, food and nature, inexperience
Brown	■	Nature, aged and eccentric, rustic, soil and earth, heaviness
Orange	■	Warmth, excitement and energy, religion, fire, gaudiness
Pink	■	Soft, healthy, childlike and feminine, gratitude, sympathy
Purple	■	Royalty, sophistication and religion, creativity, wisdom
Black	■	Dramatic, classy and serious, modern, evil, mourning
Grey	■	Business, cold and distinctive, humility, neutrality
White	□	Clean, pure and simple, innocence, elegance

In design, we often refer to 'good taste' or 'bad taste'. We all have our own views and preferences regarding what is 'aesthetically pleasing' and what combinations of shapes, colours and aromas work. Designers and market researchers will aim to predict trends each season and will acknowledge that taste changes quickly, depending on factors such as peer pressure and celebrity endorsement.

Using ergonomic factors when designing and developing products

Ergonomics in action

Figure 1.12 When designing cars, manufacturers have to make sure the controls can be easily reached and the dashboard displays easily seen, by the tallest and the smallest drivers.

Dynamic dimensions or dynamic anthropometry data is used in designing systems that take into account the limits to the reach of body movement. For example, in car design, mock-ups are used to capture key measurements. Car designers build full-scale mock-ups and test them with real people and crash test dummies, full-scale anthropomorphic test devices that represent children to adults. 3D virtual manikins are also used in conjunction with CAD (computer-aided design) software to simulate, analyse and understand the way humans interact with seats, spaces, displays and controls in a

vehicle, as well as lighting and ventilation, discovering ergonomic issues during the design process.

Figure 1.13 Designers can simulate the way humans interact with seats, spaces and controls which makes it possible to predict posture and accessibility to key functions early in the design process

Some useful reference books and websites are given at the end of this chapter. When using ergonomics and anthropometrics in your NEA iterative project it is important to remember:

- Online anthropometric data often uses measurements from military personnel (a sample of 'healthy' people). The data may be dated (the population may be changing year over year). Remember that measurements are taken with nude subjects (normally, people wear clothes and shoes at work).
- Use the data to help design a rough prototype model, then find some testers (perhaps friends or family). Match your primary user's age if possible. Have your testers mimic real-life use of your design and record what difficulties, clearances, reaches, etc. they have. Record what you observe, as well as what they say, and take formal measurements. Make adjustments or tailor the design.
- Record what percentiles are accommodated by the improved design and, explicitly, consider the consequences for users who cannot be accommodated, thinking of safety, discomfort and extra instructions needed. Could adjustments be incorporated into the design to make it more inclusive?
- By adding data about people into the design process, a product or environment can be designed so that all users are accommodated, not just those who resemble the designer!

Anthropometric data can be used to help define design parameters associated with the human body, using sizes to work out the most comfortable fit for a running shoe,

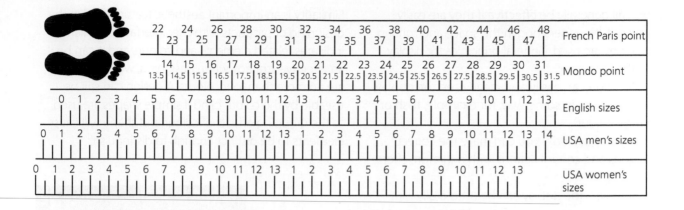

Women's sizes		XS (2)	S (4–6)	M (8–10)	L (12–14)	XL (16–18)	XXL (20–22)
	bust	32.5"	33.5" – 34.5"	35.5" – 36.5"	38" – 39.5"	41" – 43"	44" – 45.5"
	waist	24"	25" – 26"	28" – 30"	29.5" – 31"	32.5" – 34.5"	36.5" – 38"
	inseam	30.5" – 31.5"	30.75" – 31.75"	31" – 32"	31.25" – 32.25"	31.5" – 32.5"	31.5" – 32.5"
	short	28.5" – 29.5"	28.75" – 29.75"	29" – 30"	29.25" – 30.25"	29.5" – 30.5"	29.5" – 30.5"
	long	32.5" – 33.5"	32.75" – 33.75"	33" – 34"	33.25" – 33.25"	33.5" – 34.5"	33.5" – 34.5"
	hips	35"	36" – 37"	38" – 39"	40.5" – 42"	43.5" – 45.5"	46.5" – 48"

Figure 1.14 **Examples of anthropometric data**

the best layout for a control panel or keyboard or the best angle to sit at a workstation.

Ergonomics also provides data on how much force we can comfortably push or pull, what loads we can carry and the best postures for lifting things, what size and type of fonts are most easily readable for displays, what colours are better to use for clarity, what temperatures are comfortable and what noise levels and vibrations are suitable for us to work in.

All devices and displays that we use to form part of a machine interface also have to take usability and ergonomics into consideration. There are displays that are specifically designed to carry changing states to us, whether it's simply that a device is on or off, dials and the screens that we are familiar with from everyday objects such as cars and complex systems such as

aircraft control cabins. Ergonomists need to establish good principles of display design which will transcend different forms of technology.

General principles of good display design can be translated into specific principles within an environment. In the early days of printing, the maximum contrast that could be obtained from the printed word involved using black symbols/letters on white paper. The almost universal use of this principle was not applicable in the early days of computer screen technology. Initially, it was much easier to present the lettering as the bright source against the screen as the dark source. This was the reversal of the normal principle. Current computer screen technology has re-established the principle of dark on light. Many people in designing their web pages choose colours that they think go well together, but often these combinations produce problems for reading text.

Displays will generally be effective if they are easily seen and can be understood in the space and lighting in which they are used. General principles for display include the following:

- visible – you can easily and clearly see the display, it may change in some way, for example, flash or change colour
- easy to understand – you can make the correct decisions and control actions with minimum effort and delay
- compatible – the display can be used easily with others and you are not confused. The movement and layout of displays matches those of their controls.

User interface

The importance of **user interface design** is crucial when we switch on the TV or use our smartphone or washing machine. We need a practical and simple user interface (UI) to prevent common mistakes, speed up our daily activities and make using products more pleasurable. The decisions a designer makes can affect how intuitive a product is to use.

Colour and font style/sizes are all important in interface design. It is recommended that no more than four different font sizes should be used per screen. Serif or sans serif fonts should be used appropriately as the visual task situation demands and generally a mix of uppercase/lowercase text works best. Colours should be used appropriately and make use of expectations, for example using green for 'okay', 'on' or 'go'; yellow or orange for 'caution' and red for 'danger' or 'stop'. Blue text is hard to read on screen, but blue is a good background colour. High-contrast colour combinations are used for ease of understanding/readability and colours should always be used consistently.

8.2 minutes estimated	Tuck your chin into your chest and then lift your chin upward as far as possible. 6–10 repetitions. Lower your left ear toward your left shoulder and then your right ear toward your right shoulder. 6–10 repetitions.
15.1 minutes estimated	*Tuck your chin into your chest and then lift your chin upward as far as possible. 6–10 repetitions. Lower your left ear toward your left shoulder and then your right ear toward your right shoulder. 6–10 repetitions.*

Figure 1.15 Research by Hyunjin Song and Norbert Schwarz shows that the way we perceive information can be affected dramatically by how simple or complex the font is. In particular, their work found that a simple font was more likely to get the readers to make a commitment

Similarity is an important aesthetic feature used in the designing of products with user interfaces. Similarity indicates related elements and colours and it can be an effective strategy when designing, when there are no more than four colours. A design can still be useable by the colour blind if simple shapes are used, as the grouping effects of colour and shape will cause elements to be distinct enough to distinguish.

Figure 1.16 Similarity can be useful when designing control panels. Look at a remote control to see how this has been used

Designing the User Interface by Ben Shneiderman provides some good advice for interface design.

- Visibility of system status. Users should be clear about system operation and status with an easy-to-understand, clear, visible status displayed on the screen.
- Designers should mirror the language and ideas users would find in the real world. Presenting information in a logical order and according to users' expectations derived from their real-world experiences, reduces cognitive strain and makes systems simple and straightforward to use.
- Interface designers should ensure that both the graphic elements and terminology are consistent across similar platforms. For example, icons that represent one category or concept should be similar or the same.
- Because of short-term memory, designers should allow users to recognise rather than recall information.
- With increased use comes the demand for less interaction and faster navigation. Users should be able to customise an interface to suit them and their needs.

- Aesthetic and minimalist design. Clutter on the screen should be kept to a minimum and only show components for the current task.
- Designers should assume users are unable to understand technical terminology – error messages should almost always be in simple language.
- Help buttons should be easily located and worded in a way that will guide simply towards a solution to the issue they are facing.

(www.interaction-design.org/literature/article/shneiderman-s-eight-golden-rules-will-help-you-design-better-interfaces)

Software user interface

Many electronic products feature a microcontroller containing a software program which gives the product a high level of functionality. A microcontroller will also be responsible for the functional part of the user interface and will give rise to the 'smart' operation of the product.

Figure 1.17 All of the basics of good UI (user interface design) can be found in the design of Dieter Rams turntable. For example, the buttons that can be pressed and the dials that can be turned are all immediately apparent

Current electronic products have moved towards minimising the number of buttons (to perhaps a single button) which are then used to navigate through software options or menus, allowing a user to select and adjust parameters should they wish to, while leaving most functions pre-set at a default value which will work well for most users. In some designs, a set of unlabelled buttons (called soft keys) are placed alongside a display screen which indicates the function of the buttons with labels on the screen. The soft key function will change according to the screen display, allowing just a few keys to be used for, perhaps, hundreds of different functions.

The current trend is towards removing buttons altogether and relying on touch screens to act as the output and input for a user interface. Voice recognition and speech technology are also becoming a more common way of humans interacting with electronic systems.

A well-designed software UI should be intuitive to use. Commonly used functions should be quickly accessible and not hidden away in sub-menus.

Figure 1.18 Soft keys on an ATM machine

Some good practice when designing user interfaces

The maximum display-viewing distance should be determined by the size of details shown on a display. The reading distance for displays is usually 300–750 mm, as most displays have to be read at arm's length and must allow you to reach or adjust controls. Displays may have their own internal or back-lighting, but if not, their design should be suited to the lowest expected lighting level.

The preferred angle of view for displays should be 90 degrees. The scale must be legible and you should avoid multiple or non-linear scales. Scale numbers, marking strokes and pointers should contrast well in tone and colour with the display face. Scale numbers should increase clockwise, left to right, or upward, as we would expect them to.

Further guidance for interface design can be found at:
- The International Standards Organisation (www.iso.org) – standard ISO 14915-1:2002 gives guidance on software ergonomics for multimedia user interfaces
- Cornell University's Ergonomics Web (http://ergo.human.cornell.edu/) – look in particular for the ergonomic guidelines for user-interface design.

1 Consider which specific anthropometric data and percentile range you would need to use for the:
   a) the height of a doorway
   b) the size of a handle for a lawn mower
   c) the size of ventilation slots in the casing of a hairdryer
   d) the diameter of a screw-top water bottle.

2 When designing, it is important to try to design tools and hand-held items for operation with both hands. When only right-handed users are considered, left-handed users may be at an increased risk of injury. A simple experiment can demonstrate how the wrong angle can affect strength and grip in your hand. Stand with your arms at your side. Bend your right elbow at a 90-degree angle. Hold your hand out straight with your palm open; your wrist should be straight. Place two fingers from your left hand in your palm; keeping your right wrist straight, make a fist around these fingers and squeeze hard as a test of your grip strength. Now, repeat the procedure, but this time bend your wrist at a sharp angle so your palm faces in towards you. Was your grip affected at this position? Have you ever tried to complete a task that was much more stressful than normal because your body was twisted or extended in an odd way?

3 Data and advice is readily available in publications from the HSE and BSI and online. Try to find advice and data that applies specifically to your area of study.

4 Consider the user interface for a household item you regularly use, such as the microwave, iron, vacuum cleaner. How has the designer considered the user in the design of the product in terms of both usability and UI?

Anthropometrics – the study of the sizes of the human body. (Anthropo- is a prefix from *Anthropos* which means human in Greek; metric should help you remember measurement.)

Culture – the ideas, customs and social behaviour of a particular group of people or society.

Ergonomics – the study of the interaction between people and products, and the application of theory, principles, data and methods to the design of products to ensure maximum usability and efficiency.

Inclusive design – designing products that are accessible to (can be understood and used by) everyone without making changes or adaptations.

Percentile – a means of representing anthropometric data statistically, indicating the sizes of the human body of a specific percentage of people.

Throwaway society – the collective mind-set of consumers to dispose of products before they have reached the end of their life, simply to purchase updated or more fashionable versions.

Usability of a product – involves thinking about different ways in which people interact with the product and how easy it is to understand and use.

User experience design (UX) – the process of enhancing user satisfaction with a product by improving the usability, accessibility and pleasure provided in the interaction with the product.

User interface design or user interface engineering – the design of user interfaces for machines and software, such as computers, home appliances, mobile devices and other electronic devices, with the focus on usability and the user experience.

Interpreting, understanding and presenting data gathered from observations or secondary sources such as anthropometric data.

- Use appropriate methods to present performance data, survey responses and information on design decisions, including the use of frequency tables, graphs and bar charts.
- Understand and apply fractions and percentages when analysing data given in tables such as anthropometric data and charts.
- Interpret and extract appropriate data from technical and graphical sources.

- Interpret and construct charts appropriate to the data type; including frequency tables, bar charts, pie charts and pictograms for categorical data, vertical line charts for ungrouped discrete numerical data. Interpret multiple and composite bar charts.
- Calculate the mean, mode, median and range.
- Compare data sets using 'like for like' summary values.

KEY POINTS

- Assessing the usability of a product involves thinking about the people who will use it (for example, their background and **culture**), different ways in which people interact with the product (for example, opening, closing, operating, carrying, adjusting or switching) and the different situations (for example, bright or dark, outdoors or indoors) in which the product will be used. It can involve both UI and **UX**.
- Inclusive design is designing products that are accessible (can be understood and be used by everyone without making changes or adaptations).
- Ergonomics is the study of how we interact with products and environments, from jewellery, tools, furniture and clothing, to roads, transport, buildings and outdoor spaces.
- Anthropometrics is the study of the sizes of the human body. Data comes in the form of tables and we refer to percentiles. The 50th percentile is the most common size, or the average; the 5th percentile indicates that 5 per cent of people are smaller than this and the 95th percentile indicates that 5 per cent are larger than this size.

Further reading on identifying requirements

- Papanek, V. (2005), *Design for the Real World*, 2nd edition, Thames and Hudson, ISBN 9730500273586
- Klemp, K. and Ueki-Polet, K. (2015), *Less and More: The Design Ethos of Dieter Rams*, ISBN: 9783899553970
- The Design Council, *The Principles of Inclusive Design*, www.designcouncil.org.uk/resources/guide/principles-inclusive-design
- Norman, D.A. (2013), *The Design of Everyday Things*, 2nd edition, MIT press, ISBN 9780262525671
- Pheasant, S. (1987), *Ergonomics-standards and guidelines for designers*, London, BSI Standards, BSI Catalogue Number PP 7317: 1987, ISBN 0580153916
- BSi British Standards, standards relating to ergonomics, human factors, user-centred design, www.bsigroup.com/en-GB/industries-and-sectors/health-and-safety/ergonomics/
- Tilley, A.R. (2002), *The Measure of Man and Woman: Human Factors*, revised edition, Henry Dreyfuss Associates, ISBN: 9780471099550
- The Principles of Design, www.doctordisruption.com/design/principles-of-design-71-accessibility/#more-2014
- Lidwell, W., Holden, K. and Butler, J. (2003, 2011) *Universal Principles of Design (125 Ways to Enhance Usability, Influence Perception, Increase Appeal, Make Better Design Decisions, and Teach through Design)*, Rockport Publishers, Inc, ISBN: 9781592535873
- Core77 (www.core77.com) is an online publisher for design students and professionals. Innovative, entrepreneurial and inspirational design thinking and approaches.
- Shneiderman, B. et al. (2016), *Designing the User Interface: Strategies for Effective Human-Computer Interaction*, 6th edition, Pearson

PRACTICE QUESTIONS: identifying requirements

Design Engineering

1 Carry out a task and stakeholders analysis for a given context, for example, an office space or a restaurant.

2 Describe, using sketches and notes, three examples where anthropometric data has been taken into account in the design of an electrical product of your choice.

3 Describe, using sketches and notes, three examples where ergonomics (other than anthropometrics) have been taken into account in the design of an electrical product or product interface of your choice.

4 Carry out an analysis of the software user interface on an electronic product. Identify how functionality is achieved through the use of buttons, screens or other controls.

Fashion and Textiles

1 Carry out a task and stakeholders analysis for a given context, for example, a retail store or a manufacturing warehouse.

2 Describe, using sketches and notes, three examples where user requirements and sizing have been taken into account in the design of a garment of your choice.

3 Describe, using sketches and notes, three examples where ergonomics (other than anthropometrics) have been taken into account in the design of a fashion accessory of your choice.

Product Design

1 Carry out a task and stakeholders analysis for a given context, for example, an outdoor entertainment venue or public waiting area in a train station.

2 Describe, using sketches and notes, three examples where anthropometric data has been taken into account in the design of a consumer product of your choice.

3 Describe, using sketches and notes, three examples where ergonomics (other than anthropometrics) have been taken into account in the design of a consumer product of your choice.

2 Learning from existing products and practice

2.1 Why is it important to analyse and evaluate products as part of the design and manufacturing process?

The analysis and evaluation of existing products is used by most professional designers as a key stage in the design, manufacture and development of new products. By carrying out product analysis on existing products, the designer has the opportunity to gather useful information as a starting point for further development or redesign. Designers will evaluate fundamental decisions and choices that have been made during the whole process of manufacture. This can help designers develop products further by being able to utilise new technologies and materials or be able to design a product that meets the end purpose or wider considerations more effectively. The completion of a successful analysis and evaluation of an existing product can ultimately result in a competitive product being developed, manufactured and marketed that fulfils a gap in the market.

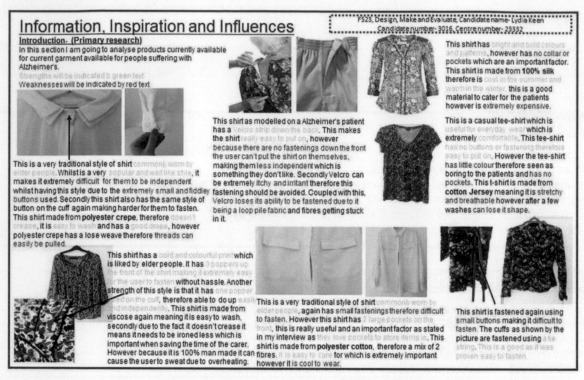

Figure 2.1 **Product analysis of adaptable clothing for consumers with arthritis or Alzheimer's**

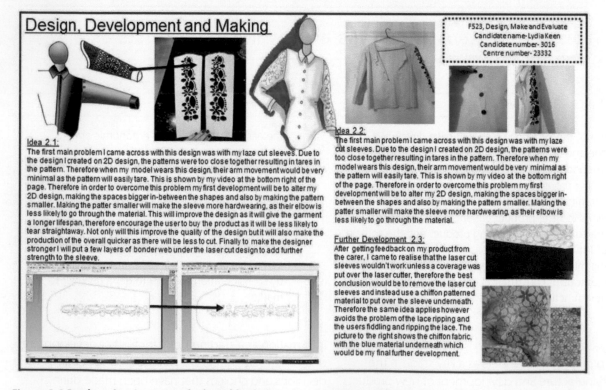

Figure 2.2 **Product development of adaptable clothing for consumers with arthritis or Alzheimer's**

The context of the existing product and the context of future design decisions

The introduction of new and innovative products is essential to any economy. To ensure the successful development of a new product, a thorough analysis of similar existing products is crucial and should be factored into the research stages. A designer will first examine and investigate existing products, taking into consideration all the stages involved, from the **initial concept** through to the **marketing**. This research should help to ensure that a new product design will meet the requirements of its stakeholders and users through any number of developments and will ultimately be commercial and profitable.

Primary and **secondary research** is used to gather important information. Primary research means an in-depth and first-hand exploration of existing products and their users by the researcher and contact with stakeholders affected by the product, whose feedback will be important during the product development. This usually involves direct communication and interaction. For example, products can be evaluated and tested by focus groups to establish their strengths and weaknesses. Secondary research will involve the examination of published data and reviews on products, relating to their performance, features, reliability and returns following product breakage or failure.

However, the crucial factor in both processes is the opinions of the users, as they will be the potential buyers of a new product and therefore determine its success. The objective of product development is to nurture, preserve and increase a company's profit by satisfying a consumer demand. Defining the stakeholders and users of a product, and obtaining their views on existing products, must take place early in the product development process. The following list is not exhaustive and it is likely a variety of these methods will be used during the process of investigating existing products and establishing the requirements for a new product.

1 Interviews and focus groups are used to gain an understanding of stakeholder and user opinions on existing products. This tends to be qualitative research.
2 User surveys can be carried out and analysed to assess measurable data to formulate facts and uncover patterns and trends. This is quantitative research.

3 Feedback from website forums.
4 Social media is now a fundamental method for businesses to access millions of consumers and gather feedback regarding products. However, it can be very difficult to control rumours and speculation about products, which can be disastrous to a brand.
5 Analysis of products returned for repairs. The analysis of reoccurring faults on a product can identify specific problems and lead to a more successful design.
6 Product recalls can be very damaging to a company's reputation as it usually means a product has failed in the same manner on numerous occasions for different users. However, product recalls can also be beneficial to the company, and its competitors when developing similar products, because they can ensure the same mistake isn't made.
7 Analysis of customer service enquiries/complaints can be an excellent basis for further product development.

While the analysis refers more to the examination and investigation of a product, evaluating refers more to assessing its value and suitability. For example, analysing a product to establish the material used but then evaluating its suitability. This can be achieved in a number of ways:

- How easy the product is to use. The most obvious way of assessing this is to study the **ergonomics** of the product. It could also be assessed for how suitable and effective the product is to fulfil the purpose it has been designed for, by observation of the user using the product, possibly carrying out a range of tasks to reveal strengths and weaknesses.
- How well the product and the various component parts perform and function.
- The suitability of the materials that have been listed as part of the analysis. Assessing why the materials have been selected.
- The suitability of the methods of construction, manufacture and assembly. Is it likely that joints could be weak or that fasteners could come loose following repeated use? The materials and construction methods can be evaluated to establish their suitability by carrying out tests. This should give a product developer an insight into the design decisions that were originally made. The results can form a basis to improve the design of the product or to make choices and selections that will improve product function.

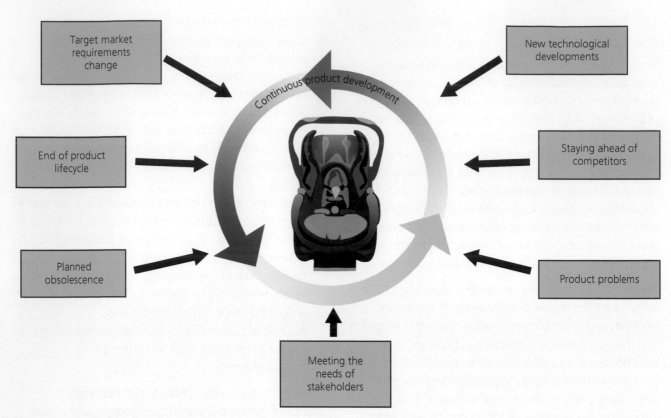

Figure 2.3 **Reasons for product development**

- The impact of the product. When evaluating a product, the designer may have to consider wider issues. For example:
  - What impact will the material selection and manufacture have on the environment?
  - Does the product improve quality of life for the users?
  - Does the manufacture of the product maintain traditional skills?
  - Does the making of the product infringe any human rights?
- Value for money aspects. By comparing and evaluating existing products it is possible to establish the level of market the designer and manufacturer was aiming for and whether the product has been priced to be comparable with similar products. When developing a new product, the manufacturer will want to ensure the product is value for money otherwise they could risk low sales. *Which?* magazine, published by the Consumers' Association, is a leader in product reviews and consumer affairs and is for many a first point of reference when making buying decisions.

The designer will be striving to develop a more innovative but viable product. This can be achieved through the use of new manufacturing techniques, materials or simply by design.

Changing consumer requirements
Consumer and stakeholder requirements are constantly changing, and designers and manufacturers must meet these demands to succeed. This means continual change and updating to prevent their customers switching to a competitor. Many companies employ a research team to source the latest products that are being produced by their competitors. An example of changing consumer requirements can be seen as a result of today's culture to remain active and fit. Designers and companies are developing products to meet this requirement and many fashion designers now produce sportswear ranges alongside their usual products.

Product lifecycle
A product eventually comes to the end of its natural lifecycle. For a company to be successful, it should be able to predict this and develop an existing product or introduce a new one to continue sales. A product goes through various stages:
- Introduction – the new product is introduced to the market.

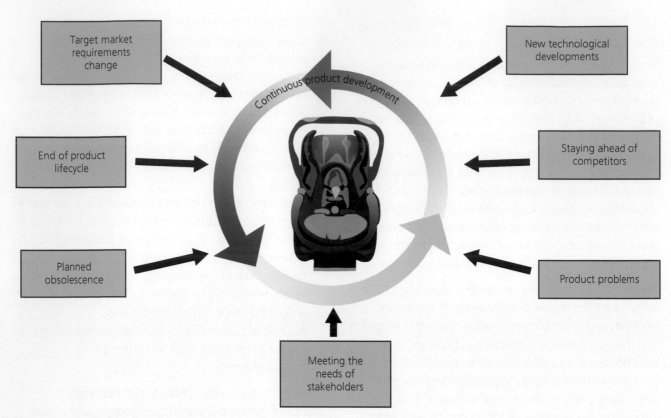

- Growth – the sales of the product continue to grow following successful marketing.
- Maturity – the popularity and sales of the product reach a peak.
- Decline – sales decline as the product reaches its saturation point and as newer versions are introduced following fashion and trends.

Many companies employ people to ensure the **product lifecycle** is assessed and managed to ensure longevity of sales. This management process involves a range of different marketing and promotion strategies, such as advertising to new audiences, price reduction, introducing new features to the product and styling or packaging changes. All these are focused towards making sure the product lifecycle is as long and profitable as possible.

Planned obsolescence

Planned obsolescence, or built-in obsolescence, is a strategy of designing a product so that it has a limited life and will become obsolete. This encourages customers to update or replace what should remain a perfectly serviceable product, usually because the product is no longer fashionable or no longer functional. This gives credence to the idea of a **throwaway society** and can be seen as morally questionable; it could even encourage the target market to purchase products from competitors.

The obvious benefits of planned obsolescence are for the companies that encourage it. They should see continued sales, returning customers and the associated profits. The target market is encouraged to purchase products that feature the latest styles, fashions and technology. However, this also encourages the target market to continually replace rather than repair, which creates more waste and pollution through disposal of the product and increased manufacture.

The multiple materials and components used

Materials used during the manufacturing process must be carefully selected, as they react and perform in different ways in a variety of situations. When examining existing products, it is usually clearly evident that the material choice is dependent on the purpose of the product, enabling the product to perform efficiently and increasing the quality. **Material properties** must be considered when a product is at the planning stage – materials may appear to be appropriate and, from an aesthetic point of view, fit the purpose. However, when in use they might prove to be unsuitable and wear out quickly. Continual advances in materials technology mean that existing products can become outdated before the end of their lifecycle, purely because a more suitable material becomes available to fit the purpose.

When examining existing products, it is crucial to assess what components have been used and why, as with material selection. The designer will have made a decision dependent on the end product use. **Standardised components** are used when the manufacturer needs parts made to guaranteed specifications and of consistent quality. These are usually 'bought in'. Due to the fact that these components are produced in large quantities, the unit costs are low. The use of standardised components does not restrict the designer's or manufacturer's product individuality, as the final product may be unique to the market even when composed from many standardised component parts. Engineered products make extensive use of electronic components and modules such as microcontrollers and **printed circuit boards**, and mechanical components such as motors and gears.

Methods of construction and manufacture

The manufacturing process transforms raw materials into final products or components. It begins with the sourcing or creation of the materials from which the design is made. These materials are then modified to become the required component, part or final product. The processes used depend on a number of factors:
- materials used
- quantity required
- product function
- expected longevity of the product
- wholesale and retail cost of each product
- time allocated for production
- tools and equipment available.

By analysing an existing product we can establish which manufacturing process has been used and, more importantly, why.

Table 2.1 on the next page shows examples of products, the manufacturing process used for each and why the processes were chosen.

Table 2.1 **Examples of manufacturing processes for specific products.**

Product	Manufacturing process	Reason
	Injection moulding has been used to manufacture the main component parts of this child's toy. The main parts are then assembled. • The moulds would have been produced to be placed in the injection moulding machine. • The plastic pellets would have to be manufactured (although these are usually bought in). • Once the individual parts have been moulded they would be assembled and finished.	This toy is mass produced and the age group it is aimed at is one to three years old. The manufacturer will have had to follow strict guidelines to ensure this toy is safe for the child. By using injection moulding the toy can be assembled using as few small parts as possible. Also, while the initial capital outlay for injection moulding is very expensive, if the toy is to be mass produced, this would be the quickest and most effective choice. This method usually means there will not be a lot of assembly once the moulded pieces have been ejected.
	The manufacturing processes used for this garden chair are: • cutting down the trees and splitting the logs to make them small enough to kiln dry • kiln drying, as the wood is to be used outside • splitting and cutting • planing and drilling • sanding • assembly • finishing.	This garden chair has used numerous manufacturing processes from the initial cutting down of the trees to the finished product. It is possible to analyse why each step has been used. For example, the chair has been made out of oak and as it is to be used outside and will be subject to different conditions, the wood needs to be kiln dried to prevent it from warping. The joints used are mortise and tenon joints, which have been used as they are the most suitable for the 90 degree joint angles and are also very strong for frequent use.
	The manufacturing processes used for a standard pair of jeans include: • fabric production • component manufacture, e.g. zips and rivets • cutting out the fabric • assembly of all the pattern pieces • application of rivets and metal components • finishing.	Most jeans are produced in large quantities and are known as hardwearing garments as well as being a staple garment in most wardrobes. They are expected to be durable. For this reason, the fabric selected has been manufactured using a woven structure that is known to be hardwearing. Many denim fabrics are prewashed prior to assembly as this prevents shrinkage once worn and washed. The preferred seam used on jeans is known as the flat fell seam. Again this manufacturing process is used as it is very hardwearing.

Construction methods

Examining existing products enables understanding of the methods of construction that have been used. Carrying out **product disassembly** makes it possible to establish the methods used and also the order of manufacture. Factors that influence construction method choices include the material selection, the end function of the product, the manufacturing equipment available and the cost. For example, it would be inappropriate to select a complex and time-consuming method of construction if the end product will be produced in high volumes and sold at a relatively low price.

Disassembling a product can be a very interesting process for the designer. It is an excellent way to analyse a current product and identify key information about it – how many materials and parts it is made from and how these parts are constructed and assembled. This also allows the designer to predict what will happen when the product reaches the end of its life.

Figure 2.4 **Disassembly of a computer**

By breaking down a computer drive into different components it is easy to see all the manufactured parts that make up the product. You can also see which materials have been used and the methods of construction.

Manufacturing methods

By examining existing products and recognising the end use and the target market, you can also establish and understand the manufacturing systems that have been chosen. The manufacturer will have considered a number of factors that can restrict the method of manufacture, including the available processes and cost restrictions. The systems used in manufacturing are quite diverse and will depend on the complexity of the industry being studied. For example, a **bespoke** dress or piece of furniture is more likely to have used a **manual system** with highly skilled operatives. Wellington boots are more likely to be **batch produced**, as there are certain times of the year when they are in higher demand. Factors to consider when analysing and examining an existing product and the manufacturing system used are:

- the type of product
- demand for the product
- capital funds available
- premises
- equipment and tooling
- labour skills.

How functionality is achieved

A key aspect to consider when analysing and evaluating products is how well they perform and function. This leads on to how suitable and effective the product is and how easy it is to use, which can be established in a number of ways. During the development stage of a new product, focused testing can be carried out on the product to ensure that it meets the required criteria. A simple but effective method is to carry out surveys and user tests to get realistic feedback. For more complex products, the manufacturer will most likely have systems in place to carry out industrial tests. An example of this is would be the drawers found in furniture. Pneumatic tests are carried out to evaluate how many times the drawer can be opened and closed before it starts to malfunction or show signs of strain. As many products consist of a number of component parts that all contribute towards the function, individual tests may need to be carried out on the components as well as the final assembled product.

The end result of industrial tests or user trials can be examined to see if the materials, methods of construction and components have performed as expected. This might result in the manufacturer continuing with production, or it might mean that further research, development and tests must be carried out to ensure the success of the product. Although many tests are carried out before a product is sold on the retail market, there are still occasions when some parts of a product may need to be replaced or updated. Some companies include a warranty with the product, which states that the product will last for a minimum period of time – should any part falter prior to this, the product will be replaced or fixed. Manufacturers want to avoid this, however, as it can be very costly and time consuming and will reflect negatively on the business.

Many products require maintenance checks to ensure that they have not become damaged. Occasionally, the user needs to carry out maintenance to ensure that products remain safe to use, or they might need to be examined by a professional. Many manufacturers include instructions on maintenance within the general instructions or as part of the packaging. In the textile industry, for example, care labels are attached to clothes to ensure correct maintenance following purchase and use. Some garments come with extra buttons, beads and so on, attached within a pack, for future maintenance. A new car comes with advice on the mileage after which a service is necessary. Many products are constructed so that worn-out components can be easily replaced.

The ease of use, including ergonomic and anthropometric considerations

Ergonomics

When designers evaluate and analyse an existing product, a wide range of ergonomic factors should be considered. These could include:

- Height – for example, when designing a table, height is a crucial consideration. This could be in relation to being able to sit without causing strain while working or eating. If a table is too low to work at, it would cause leaning and hunching over.
- Weight – for example, when designing a chair, it needs to be suitable for a rangel of weights. The function of a pneumatic chair mechanism according to different weights could be analysed.
- Hearing – for example, certain noises can signify or trigger certain emotions. The reaction of a group of people to a fire alarm noise could be analysed. The manufacturer would want the noise to signify danger.
- Sight – for example, colour can have different meanings; most people see green as a 'safe' colour. The reaction of users to a product where colour is a key factor in the design or function could be analysed.
- Temperature preferences – for example, varying temperatures within the workplace could be analysed for optimum production and comfort.

See the following websites:

- Health and Safety Executive (www.hse.gov.uk) and search for the publication 'Ergonomics and human factors at work'
- Ergoweb – https://ergoweb.com/
- Bad Designs – www.baddesigns.com/

Figure 2.5 **Using ergonomics to solve work station issues**

Anthropometrics

It is vital for a designer to consider sizes and dimensions when designing a product. When analysing and evaluating existing products, the **anthropometrics** need to be considered to assess how critical dimensions have been used in relation to human form to ensure the product is usable. This ultimately relates to the target market and ease of use. Just because a product is the correct size for one individual within the target market, does not mean it will be suitable for all.

Anthropometric data is used in all fields of design, as all products from buildings to everyday hand tools and clothing relate to human body measurements in one way or another.

There are three general principles for applying anthropometric data:

- Decide who the target market is. When analysing an existing product, think who the product has been designed for. What anthropometric measurements have been taken into consideration?
- Decide which measurements are relevant. Which measurements had been considered to make the product accessible and easy to use?
- Decide if you are designing for the average or the extreme. Has the product been designed for the majority of the population or has it been designed for smaller percentages?

Through the analysis and evaluation of existing products, a designer can assess what improvements are needed to make the product easier and more comfortable to use.

Adjustment

Adjustment is fundamental in the design of many products to address issues relating to anthropometrics, ergonomics and **inclusivity**. In many situations the design of a product could restrict the consumers' use of it. For example, when analysing a work environment it is evident that a wide range of sizes, heights and weights need to be accounted for when organising a productive workplace. The table, chairs, computer screens, lighting and heating should not restrict any employee. One company, Humanscale, has addressed this issue by analysing the products already available and developing the 'Quickstand'. This solves the issue of companies requiring products that can adjust to different workplaces by providing an adjustable system that can attach to the back of any work station.

When you are asked to examine or analyse existing products, taking into account the ergonomic, anthropometric and inclusive design features in relation to the user, you will need to be able to suggest and develop ideas that deal with any of these issues. You will need to understand methods that can be used to test these issues and furthermore, how designers and manufacturers solve the problems raised.

The table below illustrates how existing products can be analysed and evaluated, taking into consideration ergonomic and anthropometric issues.

Table 2.2 **How existing products can be analysed and evaluated, taking into consideration ergonomic and anthropometric issues**

Issue	Method	Example
Product adjustment	Testing the range in different situations with different users	Testing the adjustment on a cycling helmet
Ease of adjustment	Testing how easy it is to adjust the product for a wide range of users	Evaluating how to adjust the height of a chair with a wide range of users to make it suitable for prolonged office use
Auto adjustment	Automatic adjustment according to user	Brightness on a display of a computer screen
Material options	Testing different options of material on a wide variety of users	Testing different materials for handles on a product using a variety of users to evaluate the most suitable or to make the handles interchangeable
Data analysis	Analysing data on the use of products that have resulted in injuries or accidents to address the problem	Repetitive strain injury can be a result of repeated use of a mouse/keyboard. The data could be used to develop a more ergonomic mouse.
Specialist equipment and software	In industry, specialist equipment ergonomic evaluation software can be used to evaluate the function and performance of a product in relation to ergonomics and anthropometrics. The tests will assess human movement, strength and ability to perform tasks over a specified amount of time.	Push/pull force gauge to measure the level of force required to pick up, push or pull an object. This can be used on existing products and on a variety of users to develop a product further and making it more ergonomic.

Percentiles

When analysing and evaluating existing products, it may be an objective of the manufacturer to develop a product to make it more suitable for a larger **percentile** range. This can be achieved by carrying out user trials on a wide range of users and abilities and analysing the results. Some manufacturers specialise in producing products for the extremes.

When designing or developing a new product, the percentile range is a crucial consideration. To put this into context, a car designer would need to ensure that as many potential drivers as possible would be able to reach and depress the foot pedals. To do this, the designer would look at the 5th percentile of 'over 17 year olds', as this would include virtually all drivers with the shortest leg length and they would then assume that everyone else would have longer legs and would therefore be able to reach the pedals. The seat position would also be adjustable to enable people with both extremes of leg length (short and long) to be able to reach the pedals. It would be equally important to assess whether the shorter driver using the driving seat adjusted fully forward could still clearly read the displays on the dashboard, or the taller person within the 95th percentile with the driver's seat adjusted right back could still reach all the switches and levers. Of course, most drivers' seats now have adjustment for seat height, back rake and lumbar support, along with various steering wheel adjustments.

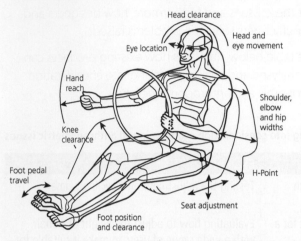

Figure 2.6 Ergonomic considerations of driving positions

Inclusivity of products and appropriate considerations of application to a wide variety of users

It should be possible to see the wider considerations that have influenced the design when analysing and evaluating an existing product. These could include:

- inclusive design principles
- environmental, social and economic implications
- moral issues
- cultural issues.

Inclusive design

When developing an existing product or designing a new one, it is important to consider how it can be made accessible to a wide variety of users. Every design decision has the potential to include or exclude a section of the population. Considering **user diversity** within the target market will inform design decisions throughout the process. It is simply not possible to design for everyone, but best practice is to aim to exclude as few as possible. This is known as inclusive design – taking into consideration the widest possible audience. It will involve consideration of a number of factors, including:

- age
- gender
- weight
- height
- disabilities.

It is particularly useful to test and evaluate the success of an existing product by methods such as trialling it with different age groups, genders and abilities. This will then lead the designer to evaluate how accessible the product is to a wide range of users and whether modifications are needed to make it more inclusive. By having a diverse and wide range of users test a product, the designer is more likely to get an objective point of view to use as a basis to develop a design further.

Look at Figure 2.7 to see how a simple kettle has been developed to include a wider range of users following testing. The manufacturer could see from feedback that holding and tipping the kettle was difficult for certain users and therefore, to make it more inclusive, has developed the design.

Figure 2.7 Development of a kettle to make it more inclusive

The Design Council's website contains some helpful 'inclusive design' case studies. Visit the website at www.designcouncil.org.uk and search for 'inclusive design'.

The UN Convention on the Rights of Persons with Disabilities uses the term 'universal design' and this is defined as, 'the design of products, environments, programmes and services to be usable by all people, to the greatest extent possible, without the need for adaptation or specialised design. Universal design shall not exclude assistive devices for particular groups of persons with disabilities where this is needed.'

Environmental, social and economic implications

As designers, we need to consider the environmental, social and economic implications of our decisions and choices. It affects us now and in the future. By examining and evaluating existing products, we can assess how their manufacture could have an impact on others.

These issues are interlinked: all making involves materials; every material uses energy to produce it; any use of energy causes pollution and affects climate change, damaging people's health. This can result in higher costs and detrimental social issues.

Some products are produced in a way that means workers are subject to poor working conditions and low wages. This is an important social issue. Recycling has environmental benefits but also has associated economic costs and itself uses energy and can cause pollution, again leading to associated social issues. There are no right answers – all we can aim to do is to be informed designers and learn from past mistakes.

Moral and cultural issues

Moral issues occur when a new product could help someone do something that might be considered undesirable or illegal. Moral issues can also relate to the choice of materials and components and the manufacturing techniques used. Whilst these link into environmental issues, it can be immoral to make choices that disregard the negative impact the development of a product could have.

Cultural issues can arise when a new product does not take into account the fact that a particular shape, colour or name can have very different meanings to different groups of people. Designers need to take care not to offend groups of people with different traditions and beliefs. For example, red is the colour for mourning in Africa whereas in China it is considered to be lucky. A careful choice of name, shape and colour can help promote a sense of unity between different global cultures and is particularly relevant if the product is to be sold in a country with a multi-cultural society.

Fitness for purpose

When examining an existing product, it is essential to understand the choices the designer and manufacturer made to ensure that the product is fit for purpose. This can include:

- functionality and usability
- material selection
- component selection
- methods of construction and manufacture
- ergonomic and anthropometric considerations.

Figure 2.8 Factors to consider when analysing an existing product

Examining an actual product allows the designer to get a true 'feel' for the product and its various component parts, and the strengths and weaknesses of the product can be established. This gives a starting point for further design development.

When designing a new product it is essential to consider many **human factors** to ensure there is minimal risk of injury. This includes considering comfort and function, to ensure that repeated interaction with the product will not cause strain to the user.

One of the more obvious examples of this is examining products that have to be held in the hand or gripped to operate. Poor design may result in slower work, discomfort and possible error. The grip of hand tools and level of muscle exertion must be considered. When designing a can opener, for example, the designer must consider the amount of effort necessary to turn the handle and the resulting strain on the wrist, as well as the amount of pressure needed to clasp the handle to the tin. The handles would have been designed for maximum comfort and minimum risk of injury.

The impact on user lifestyles

One of the main aims in product development is to improve the user's lifestyle when interacting with and using the product. This can be through the use of materials, the design or the ease of use. We only have to examine how a product for a specific use has been developed over a number years to understand the aim of the designer in creating a product that will be more appealing to the user. However, some of the developments in technology can have far reaching physical and social consequences.

Mass production has resulted in products being easily available and more affordable for society. As the standard of living has improved, so has the demand for new products for consumers. New methods of advertising and marketing have had a huge impact on user lifestyles, and have led to a competitive consumer society requiring the latest products. Consumers are also driving the need for constant developments to create the latest product with the latest technology. For example, many are determined to have the latest mobile phone, game console or VR (virtual reality) game. This is seen by some as one of the reasons for the high levels of obesity in children: 'today nearly a third of children aged 2 to 15 are overweight or obese,' (Cabinet office, 'Childhood obesity: a plan for action' January 2017, available from www.gov.uk.) Children are encouraged to play on these games through peer pressure and advertising and they see them as part of their everyday leisure activity. Many companies have attempted to counteract this by developing games that involve movement, which are endorsed by sporting celebrities to entice interest.

However, as mentioned earlier in the chapter, many products today do enhance the user's lifestyle and certainly mean they complete a task with greater ease. Now that many products are mass produced there is an increasing requirement for individuality from consumers. Designers and manufacturers respond to this by offering some level of customisation when purchasing. For example, many car manufacturers offer a huge selection of choices to enhance the buying experience and ensure the consumer feels they are getting a product that is customised to their requirements.

Awareness has grown of the impact of mass production and the requirements of a consumer society on the environment. Many designers and manufacturers have recognised the importance of conserving resources when developing a product. By examining an existing product and carrying out a lifecycle analysis, the designer can start to develop new strategies for ensuring greater consideration for the environment and resources in further product developments.

The effect of trends, taste and/or style

Today we buy a huge range of products for many different reasons, including:
- **technical and practical function**
- aesthetics; how the product looks or makes us feel.

In this section you will learn how fashion influences the development, design and success of products by analysing and evaluating past and present-day products.

For any company to be successful, it needs to be market led. This means manufacturing products that are desirable for the target market. For a product to be desirable, it needs to attract the consumer. By following the latest trends, styles and fashions, a company increases its chances of success by staying at the forefront of the market.

Many companies employ **trend forecasters** who can predict future trends and what people will buy. The forecasters examine past and current influences on fashion and trends. This can be done by examining and analysing existing products to establish what influenced the design. It could be that media, culture or celebrities have played a large part in the aesthetics of a product. The impact of fashion on all kinds of products is increasing and is possibly most evident in the textile industry. Fashion is very volatile and can be influenced by many factors.

Throughout design history there have been distinctive styles that can be linked to a period in time, for example:

- Victorian – 1830s–1890
- Art Nouveau – 1890–1905
- Art Deco – 1925–1939
- Pop Art – 1960s.

This list is not exhaustive and you should have knowledge of how various past movements have influenced design.

Each one of the styles above was fashionable during the periods indicated and there are specific features on the products that make each one easily recognisable. Meanwhile, products today are available in many different styles. Designers can choose to emulate styles of the past or follow a new trend. Manufacturers are able to use modern materials and manufacturing processes.

Today, fashion is fast moving and is influenced by any number of factors, but one trend that has become increasingly popular is the desire to protect our environment. Many designers, whether in car design, fashion design or product design, seek to be more **ethical** by creating products that will have a reduced impact on the environment.

Trends and fashions are sometimes company driven; in other words, a company can influence the design and style of a product. This results in other companies of similar or competitive products following that first company's trend. An excellent example of this can be seen in the design and styling of smartphones. The latest features (for example, the size of the camera, the shape and size of the screen, or the methods of security protection) start as trends, which then become considered 'must haves' in order to attract buyers.

Another trend that has become extremely popular and is evident in many products today is the ability of the consumer to customise or personalise their purchase. This is following a trend that has been encouraged by consumers wanting to feel they have had an influence on the design of the product and therefore making it individual. Look at phones with different cases, watches with different faces or straps, trainers that can be customised, or cars that can be customised online before purchase. The list is endless.

It is helpful to examine a range of products and to consider and discuss the trends and fashions that you think have influenced the aesthetics, but also other aspects of the design.

The effect of marketing and branding

Marketing

Marketing is the process undertaken by companies or individuals to promote and sell products or services and it includes market research and advertising. Marketing and **branding** are well established promotional tools that influence customer decisions – a marketing campaign can be the decisive factor in whether or not a product will succeed. Marketing is used to influence potential customers to purchase the products. For marketing to be successful, the consumer needs to feel they 'need' the product. It is therefore vital to understand who the target market is and design the marketing to engage with them.

The manufacturing industry must take into account customer requirements in order to develop new and improved products that will give a competitive edge. New products are usually developed either because of market pull, where consumers demand a particular type of product, or by technology push, where new materials and/or technologies lead to innovative products that are released onto the market.

The success or failure of products brought to market can be due to the marketing strategy that has been used. This is an extremely powerful tool and can literally 'make or break' a product. Imaginative and successful marketing can also ensure longevity in product sales. A successful marketing strategy can keep a company's products, or updated versions of them, at the forefront of a particular market and influence buyers' decisions.

Search the internet for some Coca Cola marketing campaigns to see how the company has remained one of the most popular soft drink manufacturers. This is one of the best known brands in the world and has been successful since it was launched. The company updates its marketing strategy by developing new products.

Branding

Successful marketing usually includes promoting a brand image and developing new markets for the product. A 'brand' can be a name or a logo and it can have a huge influence on the decisions made by the target market. If a particular brand is 'in fashion' it can also prove to be extremely profitable for the company. For example, the product quality may be comparable to a similar non-branded item, but the addition of a brand or logo can influence the customer, perhaps due to celebrity endorsements or simply because it is seen as the 'latest' trend. The cost of the branded product is usually significantly more than the equivalent non-branded product and consumers might buy the chosen brand because it provides a perceived level of reliability, quality and style that makes it seem superior to the consumer. One example where branding is particularly at the forefront of marketing is within the sportswear market. This can be seen as morally questionable, as many young people want to compete with peers and emulate their sporting heroes by wearing the latest brand or logo.

It is possible to examine existing products and evaluate whether it is the design of the product that has made it a success or failure, or the marketing strategy. For example:

● Nike – 'Just do it' – This phrase (coined in 1988) was a phenomenal success. Nike products were originally developed for marathon runners. With the help of a broader increased interest in keeping fit, the launch of the slogan helped more people relate specifically to how they feel when exercising. This, along with the advertising campaign, led to a massive increase in sales.

● Apple – This company is an excellent example of how the products have dictated success rather than massive advertising campaigns. The company's marketing strategy depends on a proactive approach as part of its product strategy. Essentially, the company develops its products based on the existing products of competitors but it improves them by removing undesirable qualities and integrating differentiating features. The results are products that appear new and innovative but are still familiar or recognisable. The Apple logo epitomises the simple, clean, uncluttered designs that characterise Apple products.

The considerations of how to get a product to market

Market research is vital at the beginning of a successful marketing campaign. Using information gathered through market research enables manufacturers to develop and introduce products that consumers will buy. This information can be used to adapt and improve existing products or to find a gap in the market to develop and introduce a new product.

Figure 2.9 **Nike has created an instantly recognisable brand**

Sometimes referred to as **inbound marketing**, market research involves finding out and analysing information about:

● patterns and trends within age groups of the population
● up-to-the-minute market trends and fashion
● consumer needs and requirements
● pricing strategies – how much customers are prepared to pay
● competitors – what they are doing and their strengths and weaknesses
● customer satisfaction relating to existing products and services
● product identification and launch, including naming and branding.

Market research can be either primary or secondary, or a mixture of both. Once market research has been gathered and analysed, a marketing strategy can be put in place with the aim of maximising sales of the product.

This next stage in marketing is sometimes referred to as **outbound marketing** and includes sales, advertising and promotions, customer service and satisfaction.

Marketing mix
The **marketing mix** refers to the set of actions that a company uses to promote its brand or product in the market. The 4Ps make up a typical marketing mix:

● price
● product
● promotion
● place

Table 2.3 **Primary and secondary market research**

Primary research	Secondary research
• Collected first hand using questionnaires, focus/ user groups, surveys and field research, etc. • Trials carried out	• Uses available data from magazines, reference books, government agencies • Provides information such as population trends and regional statistics
Advantages • Precise and up to date • Meets the exact needs of the company and the consumer	**Advantages** • Usually cost effective as already produced
Disadvantages • Expensive • Time consuming	**Disadvantages** • Can be out of date or incorrect • May not be specific to designer needs

Table 2.4 **Marketing mix**

Product	Price
A product is examined on three levels: • The core product relates to the benefit of the product, for example the convenience of a car. • The actual product is the tangible, physical product, for example the car. • The augmented product is about the customer service support offered, for example warranty, guarantee and after-sales service. The quality of a product depends on factors such as its: • aesthetics • performance • maintenance, ease of servicing • durability • range of features • ease/effectiveness of use • brand name. The product's distinctive qualities, features and characteristics that will appeal to customers are summarised as the product's unique selling proposition (USP).	The price of a product can depend on: • demand • costs (need to cover all costs incurred) • government taxes • competition • stage in the lifecycle (price increases in growth and falls in decline). Methods of pricing include: • **Penetration pricing** – price set artificially low to gain market share. Once achieved, the price is increased. • **Price skimming** – if product has a competitive edge, a high price can be set. This will fall with increased supply. • Psychological pricing – charging £1.99 rather than £2. • Predatory pricing – undercutting competitors, creating price wars.
Promotion	**Place/Placement**
Decisions have to be made on how best to promote the product and bring it to the attention of potential customers. The intention is to win new customers or persuade them to change brand loyalty. Methods include: • short-term promotions such as buy one get one free (BOGOF), competitions and coupons • exhibitions and trade fairs • publicity campaigns • personal selling/sales representatives • advertising in the media that specifically targets the intended audience. Promotion can be expensive and risky, so careful budgeting required. The acronym AIDA is used with promotion and advertising: draw attention, create interest, generate desire, invite action.	Placement refers to the location where a customer can purchase a product. It is sometimes known as the distribution channel. It can include any physical store or shop, as well as TV shopping channels, the internet and social media. There are four main channels of distribution: • Manufacturer–Consumer, e.g. mail order, farm shops • Manufacturer–Retailer–Consumer, e.g. high-street stores • Manufacturer–Wholesaler–Consumer, e.g. furniture • Manufacturer–Wholesaler–Retailer–Consumer, e.g. medium-size convenience stores (large supermarkets often cut out the wholesaler). Functions of distribution channels include: contacting prospective buyers, matching the offer to the buyer's needs, negotiating agreement on price and terms and storing and transporting the products.

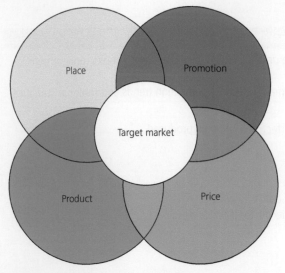

Figure 2.10 **Marketing mix – the 4Ps**

Advertising

Advertising is a strategy used to attract a potential target market to a new or existing product. The most effective advertising will have been planned into the marketing strategy and will utilise the most effective methods and media relevant to the product to reach the widest audience.

The considerations when deciding on the appropriate way to advertise a product are the overall cost, the target market or audience and the most appropriate medium.

Adverts are controlled by the Advertising Standards Agency (ASA), an independent regulatory body set up to ensure that adverts do not mislead or offend potential customers. Adverts are required to be legal, decent, honest and truthful, and to not cause offence to viewers or listeners.

Table 2.5 **Advantages, limitations and costs of advertising**

Advertising medium	Advantages	Limitations	Cost issues
Television	Very high, mass-market coverage; low cost per exposure; can generate powerful emotive response; can use images, sounds and special effects	Quick, fleeting exposure; target markets selected by scheduling, for example between 4 p.m. and 6 p.m. for children	Expensive: key viewing times very expensive
Radio	Very good local impact; high geographic selectivity; national radio effective for consumers on the move (for example in a car)	Audio only; fleeting exposure; fairly low attention, background 'noise'	Relatively low cost for local radio, higher rates for national commercial stations
Newspapers and magazines	Very good local markets, national or geographic selected coverage; broad acceptability; high believability; prestige; magazines often very high-quality images	Very short life; generally poor image quality in newspapers; no guarantees of positioning of advert	High cost in nationals and prestigious magazines; right-hand page often more expensive than left (readability)
Direct mail (post)	Very high audience selection; can be personalised with mail-merge systems	Often discarded as junk mail; poor image	Relatively high cost per exposure
Billboards	Flexibility; high repeat exposure; local targeting; positioned in high traffic areas to catch mobile consumers; some are electronic with several adverts repeating	Little audience selectivity; image only; easily vandalised	Generally low cost; key sites (for example outside airports) can be very expensive
Online (including email)	Used by all industry sectors. High selectivity; instant; can be powerful with moving images and sound; can be interactive; direct access to supplier Pop up advertising on websites can focus on specific buying trends; advertising within smartphone 'Apps' is increasingly popular	Anxiety over invasion of personal space, unsolicited email; linked to spam and spyware 'Ad blocking' software can sometimes be used to block unwanted adverts	Relatively low cost; purchases can be made directly from the advert
Social media	Caters to a very large demographic regardless of age, gender, etc. Through content sharing, it can reach an even wider population. This method encourages and generates interaction and brand loyalty	Time consuming Can attract people not interested in the product, sometimes resulting in false or negative reviews	Low cost method of advertising

The Office of Communications (Ofcom) controls advertising on television and radio.

Product launch

Once a product is ready to launch onto the market, the timing can be crucial to success. The main purpose of a product launch is to make your target market aware of the product. For example, launching a child-friendly product around the school holidays can lead to a large influx of sales due to both the children and parents seeing the product.

The trick is to find the factors that help define what the right timing is.

The four factors of smart timing

At launch time:

- offering unique features that are understandable
- offering features that are currently relevant
- giving reasons as to why the product is superior to those offered by competitors
- product is available, or soon to be available, for purchase.

Companies may be eager to release the product but to ensure it meets the desired requirements and reduce the possibility of a product recall, they must ensure all research and testing has been covered.

It is important to consider the competition, including possible products that they are likely to launch and when. Unless the manufacturer is confident they have carried out sufficient market research to be knowledgeable that their product will propel all other similar products out of the market, it would not be wise to release similar products at the same time.

The launch is extremely reliant on effective communication tactics that may include:

- press release – about the new product and date of launch to create an awareness and 'buzz'
- social media – spreading the word about the new product to the target audience
- product testing – allowing influential people to test the product so they can generate interest.

KEY TERMS

Anthropometrics – the study of the sizes of the human body. (Anthropo- is a prefix from *Anthropos* which means human in Greek; metric should help you remember measurement.)

Batch production – involves the production of a specified quantity of a product. Batches can be repeated as many times as required. This type of production is flexible and is often used to produce batches of similar products with only a small change to the tooling.

Bespoke – the term used to describe a product that has been made specifically for an individual customer.

Branding – the process of creating a product identity and name, logo or design that makes a product easily identifiable to consumers.

Ergonomics – the study of the interaction between people and products, and the application of theory, principles, data and methods to the design of products to ensure maximum usability and efficiency.

Ethical design – considering the impact of a product and whether it is morally correct to produce it.

Human factors – the study of how humans behave physically and psychologically in relation to particular products.

Inbound marketing (also called market research) – the methods a designer or company uses to establish whether there is a need or gap in the market for a proposed idea.

Inclusive design – designing products that are accessible to (can be understood and used by) everyone without making changes or adaptations.

Initial concept – the first design in response to a brief.

Manual system – a product is assembled and constructed by hand or the machinery is controlled or manipulated by a human operator.

Marketing – the process used by a company to promote and sell its products to consumers.

Marketing mix – the best blend of marketing methods, dependent on the type of product and specific target market.

Material property – how well the material performs (often definite). For example, the density or strength of a material.

Outbound marketing – how the customer 'finds' the product. For example, through a search engine.

Penetration pricing – this is when a company sets a low price for its product to attempt to gain market share. The primary objective of penetration pricing is to gain many customers and then use various marketing strategies to retain them.

Percentile – a means of representing anthropometric data statistically, indicating the sizes of the human body of a specific percentage of people.

Planned obsolescence – sometimes known as 'built-in obsolescence', the process of designing products to go out of fashion or no longer function after a specific period of time.

Price skimming – where a company sets its prices high when launching a product to quickly recover expenditures for product production and advertising. The key objective of a price skimming strategy is to achieve a profit quickly.

Primary research – the personal collection of research and information. It is carried out through methods such as visits and observations, interviews, testing and surveys.

Printed circuit boards (PCB) – a non-conductive material with conductive lines printed or etched on it. Electronic components are mounted on the board and the traces connect the components together to form a working circuit or assembly.

Product disassembly – the action of taking apart a product to examine and review the materials and manufacturing processes that have been used.

Product lifecycle – the marketing lifecycle from the launch through to decline of a product (not to be confused with lifecycle assessment).

Secondary research – the collection, collation and editing of readily available information. Such research utilises sources such as published details, company literature and existing test data.

Standardised components – parts that are usually produced in high volumes to the same specification and quality. Bolts, screws and fasteners are common examples.

Technical and practical function – how a product functions and works.

Throwaway society – the collective mind-set of consumers to dispose of products before they have reached the end of their life, simply to purchase updated or more fashionable versions.

Trend forecasters – people employed to predict the mood, behaviour and buying habits of the consumer.

User diversity – considering the different forms and conditions of a target market.

KEY POINTS

- Commercial designers use the analysis of existing products as a key stage in the design, manufacture and development of new products.
- Designers are constantly striving to develop more innovative products through the use of new manufacturing techniques and materials, or simply by design.
- Consumer requirements are constantly changing and designers and manufacturers must meet these demands in order to succeed. This means continual changes and updates to deter consumers in their target market from switching to a competitor.
- Companies should be able to assess a product's lifecycle to recognise when to develop an existing product or introduce a new one to drive continued sales and maintain consumer interest.
- When examining an existing product, it is possible to use a number of criteria to establish which manufacturing processes have been used and why.
- When evaluating a product, it is common to consider how the product has been designed to make it accessible to a wide variety of users. Every design decision has the potential to include or exclude some part of the population.
- Analysing and evaluating existing products recognises how ergonomics have been considered to ensure the 'fit' between people and the things they do, the objects they use and the environments they work, travel and play in, making them easier to understand and safer to use.
- Assessing the anthropometric data that has been considered on an existing product shows how its measurements have been matched to the human body.
- Adjustment in the context of ergonomics or anthropometrics means a small change that improves something or makes it work better. A change that makes it possible for a person to do better or work better in a new situation.
- Designers should consider whether they are supplying the average or extreme users. The average is within the 5th to 95th percentile of the population, whereas the extreme is within the remaining 10 per cent.
- Many companies employ trend forecasters to predict what will be in fashion in the coming seasons and what will influence the market. These forecasters can predict the mood, behaviour and buying habits of the consumer.
- Designers can choose to emulate styles of the past or follow a new trend. Manufacturers produce these products using modern materials and manufacturing processes.
- A marketing campaign can be the decisive factor in whether or not a product will succeed. The aim of all advertising is to influence potential customers to buy the products.
- New products are usually developed as a result either of market pull or technology push.
- Branding products can be incredibly influential on the decisions made by the target market. If a particular brand is in fashion it can prove to be extremely profitable for the company.
- Market research can be used to adapt and improve existing products or to establish a gap in the market to develop and introduce a new product.
- Both inbound and outbound marketing strategies are used to successfully launch a new product.
- The marketing mix refers to the set of actions that a company uses to promote its brand or product in the market. This will include using the 4Ps.
- Companies develop their advertising campaign dependent on the type of product they are selling and, more importantly, the target market they are focusing on.

ACTIVITIES

1 Select a product from your chosen material area and disassemble if possible. Take care when disassembling and assess the risks before you begin.
   - Make sure you use appropriate tools to carefully remove fasteners; never force anything.
   - Electrical equipment must be unplugged from the mains and the plug must be cut off, so there is no chance of plugging it in once you begin disassembly.
   - Remove batteries if possible.
   - Never try to open batteries as they contain harmful chemicals.
   - Never puncture LiPo batteries as they can spontaneously catch fire or explode.
   - Take care with springs as they can exert very large forces and they can fly out unexpectedly, causing injury. Wear safety glasses.
   - Take care not to trap your fingers in moving mechanical parts.
   - If in doubt, seek advice before commencing disassembly of any product.
   a) Take images of the product once separated into all component parts.
   b) What materials have been used for each component part and why? Assess the suitability of the materials used.
   c) For an engineered product, how does the product work? Identify the key subsystems.
   d) Does the product make use of any scientific principles? What are they?
   e) Take measurements so you can carry out a mathematical analysis on key parts, such as gear ratios, battery voltages, weights and surface areas.
   f) Identify ways in which the designer has achieved structural integrity and reinforcement in the product.
   g) What methods of construction have been used and why? Assess the suitability of the methods used by looking for signs of wear, breakage or loose components.
   h) Which manufacturing system has been used and why?
   i) Can you list the stages of construction?
   j) Are there any legal requirements associated with the product?
   k) What cost restriction has the designer had to consider?
   l) What ergonomic and anthropometric measurements have been considered? How easy will the product be to use in a range of situations?
   m) What strengths and weaknesses can you list relating to the product?
   n) How has the manufacturer ensured safe use of the product post sales?

2 Examine another mass-produced product from your material area and discuss how the expected lifecycle of the product can be achieved. Your answer should include:
   a) maintenance factors
   b) standardised parts
   c) cost-related issues.

3 Select two products that have the same function. Compare them to explain how the designers have used anthropometric data in the design of the products.

4 Use examples to explain what is meant by the terms 'anthropometrics' and 'ergonomics'.

5 Select an everyday product and suggest improvements to the design which would make it more ergonomic.

6 Consider and discuss which percentile range you would need to use for the following:
   a) the width of a hallway in a newly built house
   b) the size of the handle on a pair of scissors
   c) the height of a desk for an office.

7 Select a product from your chosen material focus that has been successful over the past ten years. Carry out an investigation into the product and its marketing to answer the following:
   a) How has the product been marketed? What strategies have been used?
   b) What features of the product have contributed to the successful marketing of the product?
   c) What advertising campaigns have been used and what are your opinions on them?
   d) Discuss the reasons for the success of the product. Is it because of the product and its features, or because of the way it has been marketed?

8 Select another product for a specific age group and plan a marketing strategy for it.
   a) Outline the buying behaviour and lifestyle of the target market group.
   b) Describe the key characteristics and features of the selected product.
   c) Design a branding and marketing campaign for the product.

2.2 Why is it important to understand technological developments?

By the end of this section you should have developed knowledge and understanding and:
● be able to critically evaluate how new and emerging technologies influence and inform the evolution and innovation of products in both contemporary and potential future scenarios.

'Like everyone we get frustrated by products that don't work properly. As design engineers we do something about it. We're all about invention and improvement.'

James Dyson – Inventor of cyclonic vacuum technology, www.dyson.co.uk/community/aboutdyson.aspx

New technology

The availability of new technologies which may make a product more appealing could be a factor in encouraging designers or companies to develop a product further to meet a gap in the market. This is part of the **iterative design** process – the new technology means the designer or manufacturer can explore other processes or materials with the aim of improving the existing product. Developing products in this way tends to focus on **innovative** ideas and quality enhancements that will improve functionality or aesthetics. The new technology could depend on materials or components being developed to make a product more successful, or the use of new manufacturing technology. This leads to constant **evolution of products** to meet consumer demand. The increasing use and development of robotics and **additive fabrication** have the potential to define and shape our next era.

When a product evolves or there is a gap in the market for a new product, the material choice underpins the product development. A good example of this is seen in bicycles. In the recent past, the majority of bicycle frames were made from strong but heavy steel. This then progressed to aluminium alloy to enable manufacture of a lighter bicycle and improve performance. More recently, composite metals like carbon fibre have been used, enabling manufacturing of the frame in one piece without the need for welding, which adds to the weight. There is constant development in areas such as bicycle manufacturing to produce materials that perform well and are strong but lightweight.

When considering existing products and practices in the field of design and technology and when using everyday products that seem to fail or malfunction often, it is apparent that there is always room for improvement, to make things better.

Figure 2.11 **James Dyson**

When developing existing products to make use of the latest technologies, whether it is in the design, materials or manufacturing, it is crucial to be able to evaluate the value issues and wider impact of new and **emerging technologies**. There will always be arguments for and against. The development of new technologies can have all kinds of impacts on people, society and economic and environmental issues.

Questions to ask:
1 How will the product impact on lifestyle?
2 What effect does the technology have on the environment?

3 What ethical issues should be considered relating to the new technology or product?

4 What are the economic issues relating to the product or new technology?

One of the latest manufacturing technologies that has been adapted to be beneficial across a wide range of design and manufacturing industries is the 3D printer. This has lent itself to the development of existing products, as prototypes can be trialled and tested as realistic models prior to large-scale production. Already it is possible to 3D print in a wide range of materials, including thermoplastics, metals, ceramics and various forms of food.

Figure 2.12 Rapid prototyping through 3D printing

One person who put imagination into reality and used iterative thinking to solve a real problem is Easton LaChappelle. At the very young age of 14, he decided he wanted to do something extraordinary with his life. Quite simply, he did this by creating a 3D printed prosthetic robotic arm. The first prototype was built from fishing wire, Lego® and a 3D printer. He then developed his idea further by incorporating technology that would allow the user to control movements through the use of brainwaves. To enable the technology and idea to be accessible to all, the files needed to construct the creation have been uploaded to the net, basically meaning the arm can be built by anyone who has access to a 3D printer. His innovative ideas have enabled him to work with NASA and start his own company called Unlimited Tomorrow.

'3D printing allows us to produce something extremely customised for a fraction of the cost. That's really the key, especially within prosthetics – that it is customised to each individual. The customisation process is where the root of the cost

is with traditional prosthetics. It also changes our design philosophy. Because it's printed layer-by-layer, we can have very complex internal structures and essentially fit more into a smaller package. And I think we're just starting to really see the forefront of 3D printing. It's literally changing the world, impacting everything from automotive devices to medicine.'

Easton LaChappelle (www.tonyrobbins.com/leadership-impact/easton-lachappelle/)

LaChappelle's methods in designing and making the prosthetic arm illustrate how fast ideas can become reality. Prototyping has always been used as a method for testing and developing ideas; 3D printing now makes it possible to achieve the whole iterative process, from idea to testing, at a much faster pace.

Not every product that uses new technologies is immediately successful. Some very famous examples are:

● The Sinclair C5, launched in 1985 as a small one-person battery-assisted vehicle and the Segway PT, a two-wheeled, self-balancing scooter launched in 2001. Neither of these products took off as expected but both have influenced future products.

● In 2013, Google allowed several thousand people to test their prototype Google Glass, but bugs and privacy concerns were raised. Google stopped production of the product but is still looking at future iterations.

● Whenever companies create products that fail, they learn from them. Apple released the Apple Newton, a Personal Digital Assistant, in 1993. At this time, the average consumer could see no use for an expensive personal digital assistant, leading Apple to discontinue the Newton in 1998. However, similar ideas were used in later Apple products like the iPad.

Artificial intelligence (AI) is a relatively new technology and is considered by some to be a threat to employment. Professor Stephen Hawking warned that AI will need to be controlled in order to prevent it from destroying the human race. Reports from the University of Oxford and Yale University have found that AI will outperform humans in the near future; translating languages by 2024, driving a lorry by 2027, working in retail by 2031 and working as a surgeon by 2053. Will new technologies, ranging from industrial robots to advances in machine learning, lead to mass unemployment?

Biometrics is another technological development that is being increasingly used in everyday situations. The technology is still expensive but for certain uses it is invaluable, for example, fingerprint scanners are used for a variety of purposes, including recording arrival and departure times at places of work.

Virtual reality (VR) is used increasingly for training purposes. It is still very expensive, but even so it allows training that would otherwise be impossible; for example, allowing pilots to practise emergency routines.

Drones are already used in numerous industries from retail to manufacturing. While this is still a relatively new technology, the use of unmanned flying objects could change the world in which we live. They have

the potential to carry out courier roles delivering goods short distances, but there is still a need for skilled workers to manage a drone delivery network.

Materials such as graphene offer great potential in the design of consumer electronic products.

The use of fabrics that respond to heat, steam and sweat are finding their way into products. 'Bio-skin' fabric that peels back in reaction to sweat and humidity and 'Virus' StayWarm Performance Series of clothing are examples of this.

ACTIVITIES

1 What are the latest material or technological developments within your specialist area?
2 How have these developments benefited the function of a product of your choice?
3 Thinking in the broadest terms possible, discuss the advantages and disadvantages of new technologies being developed.

4 Examine a product of your choice and suggest how it could be developed further by incorporating a modern technology.

KEY TERMS

Additive fabrication – the process of joining materials to make objects from 3D model data, usually layer upon layer.

Emerging technologies – new technologies that are currently being developed, or will be developed within the next five to ten years.

Evolution of products – developments to improve or vary an existing product, or the formulation of an entirely new product that arises from identified customer demand.

Innovative – the development of an idea which is new, different and ambitious. The design can be risky and introduce aesthetics or function that hasn't been seen

before. Innovation comes from pushing beyond the expected.

Iterative design – a continual and cyclical design development process to refine and perfect the product. Ongoing testing of models and prototypes, incorporating improvements and progressing towards an optimum solution for all stakeholders.

KEY POINTS

● By analysing and evaluating existing products and carrying out research on focused groups, designers and manufacturers can develop a product further by introducing new technologies to enhance its aesthetics or performance.

● New technologies and materials are constantly being developed and introduced. It is vital that designers are aware and ready to investigate them so their knowledge base and design thinking is up to date.

2.3 Why is it important to understand both past and present developments?

LEARNING OUTCOMES

By the end of this section you should have developed knowledge and understanding and:
- recognise how past and present design engineers, fashion and textiles designers and product designers, technologies and design thinking have influenced the style and function of products from different perspectives, including:
  - the impact on industry and enterprise
  - the impact on people in relation to lifestyle, culture and society
  - the impact on the environment
  - consideration of sustainability
- understand how key historical movements and figures and their methods have had an influence on future developments.

Designers need inspiration to develop or create new designs and products. A starting point can be to look at how designs have evolved over time. The underlying concept of a product's design can often be linked to influences from the past, which may relate to the actual form of the product or its function. Meanwhile, many products today are purchased not just for their function but also for how they make us feel.

There are many iconic designs – ground-breaking or setting new standards – that have led to the development of a particular product or a selection of products. The designers in the first place would have been influenced by a particular style movement in that era. Examples of these could be the Arts and Crafts Movement, which influenced designers by creating simplified products after years of ornate and overworked designs.

The Bauhaus, a design school founded by Walter Gropius in 1919, was very influential in shaping an understanding of design and taste. Design was considered crucial and integral to the production process rather than merely a visual 'add on'. 'Form follows function' was a phrase often used to counteract the historically prevalent view that beauty was achieved by including additional features, which were not necessarily useful. Architects and industrial designers in the 20th century were beginning to show that the form or shape of an object should be based on its intended purpose or function.

In fashion design, Yves Saint Laurent introduced the tuxedo as part of a woman's wardrobe in 1966, while Chanel introduced the idea of the little black dress (LBD) in the 1920s, a concept which is still copied by many designers today.

The essence of many products from the past remains the same. The basic function may remain unchanged but the range of additional functions and the aesthetics can change dramatically. The best designers manage to combine form and function to create a product that is emulated by others.

It will be necessary for you to carry out further reading and research on key historic movements and designers but some of the most influential ones from the past hundred years will be covered here. You should be able to identify how the styles have influenced designs of today.

By examining iconic and ground-breaking designers' work, we can start to understand the methods they used and what influenced their design decisions. Many elements of designers' work today can be linked to past designers' work. Carry out some research online and look at their work and see how it has influenced different genres of design. You can base your research on the format used in Figure 2.13 to look at the iconic Bauhaus Wassily chair, designed by Marcel Breuer. His familiarity with unforgiving materials led to the design of this chair. The name may not be familiar, but the bent tubular steel chair is still seen in stores today.

Figure 2.13 **Bauhaus Wassily chair**

Impact on industry and enterprise

Every company wants to encourage ideas of creativity, innovation and problem solving. Many companies survive by updating existing products according to trends and technology. Most ideas that are imagined today have probably been thought of before and many designs have been created to solve a real problem, as the existing products available do not perform as the designer requires. The most successful designers generate ideas and technology that is used as inspiration by others or can spark a revolution in manufacturing methods or techniques.

While there are many examples to study, one of the most **iconic designers** of today is James Dyson. He was dissatisfied with the mechanics of the original vacuum cleaners, as the bags had to be replaced and the cleaner lost suction. However, the **invention** of the vacuum cleaner was originally created by Hubert Cecil Booth in the UK and William Henry Hoover in America. Hoover's vacuum cleaners in the UK became such a success that 'to hoover' quickly became synonymous with vacuuming.

How does James Dyson ensure continual product development? It would be very difficult for one designer to continually generate these ideas. To ensure he is at the forefront of the market he nurtures new talent and ideas through an ever-growing team of designers and engineers. The engineers are encouraged to continually refine their ideas; an iterative process. This is carried out at Dyson's research centre in the UK.

Another hugely successful company that encourages and funds research to aid product development is Nike. They are often at the forefront of innovation which enables them to be successful in a very competitive market. They have come a long way since their first running shoe sole was created using a waffle iron. Nike is now synonymous with a huge range of products suitable for various sports. One of their latest developments involved their designers developing the idea of self-lacing shoes and Nike has been securing **patents** over the last ten years, which led to the release of the HyperAdapt 1.0 in 2016. The self-lacing and self-fitting shoes have a sensor in the heel so, as a user steps into the shoes, they tighten and adjust automatically. The shoes also have two buttons, plus and minus, on the outside to allow for self-adjustment, to tighten or loosen the shoes, based on personal preference. The extremely small motors in the shoes tighten and loosen by adjusting the cabling system inside the shoe.

The development or invention of a new design can have both negative and positive effects on industry and **enterprise**. The introduction of a product can result in new methods of manufacture; this could be through material development or the process itself. One good example of this is the introduction of knives with ceramic blades. For many years knife blades have been made from stainless steel but recently ceramic blades have been introduced, which remain extremely sharp and are very hygienic. This has a positive effect on industry in one way, as it creates more jobs for the manufacturers that must create new moulds and manufacturing processes. However, a negative effect is that the introduction of a new product can make another one obsolete.

Figure 2.14 **A ceramic knife**

Impact on people in relation to lifestyle, culture and society

By examining past and present products we can assess the impact they may have had on lifestyle, **culture** or society. This impact could be with reference to the manufacturing techniques that were used, the materials or the use of the product itself.

In the 1970s the punk movement emerged. This was a reaction to society which used many of the same radical tactics used by early avant-garde movements. Punk deliberately cultivated an image of violence, defiance and rebellion at the very beginning of the movement. It wasn't simply a way of dressing or even a style of music; punk was about living the idea of rebellion against authority. There was a definite 'us versus them' attitude. The clothes worn during this time also reflected the punk ideology of wanting to appear careless and against what was considered 'normal' and acceptable by society. It suited the lifestyle of those with limited cash due to high unemployment at this time.

Figure 2.15 **Punk fashion**

If we look at iconic punk fashion items, we can see how they have influenced an image of deconstructed fashion that can still be seen in the work of contemporary fashion designers today. The culture of individual freedom has continued in many forms since the emergence of punk in the 1970s.

The influence of sports and well-being on product design continues to grow. This is not a recent influence but society today has become increasingly conscious about exercise and remaining fit. The idea of 'keeping fit' has been around for years, but has become progressively more popular in many forms and has led to the fitness culture. Many companies now specialise in providing products that help the user to maintain and keep track of their health. This technology has been utilised across fashion and product design. Sport fashion is a product created by the commercialisation of fitness culture and many leading fashion brands have cashed in on its popularity.

With continued developments in sensors and wireless technology, products have now been developed to make it easy for the wearer to assess their fitness level on a regular basis.

One company that has led the field supplying wearable technology is Fitbit. This company started in 2007, when the founders realised that wireless technology had advanced to a stage where products could be developed to help consumers to achieve their fitness goals. This is an excellent example of how products can be technology led. There are now many product variations that make use of this technology to improve the users' lifestyle by making them more aware of their fitness levels.

Figure 2.16 **Wearable technology to monitor and track fitness levels**

The current trend for **fast fashion** and consumers' demand to keep up to date with the latest technologies can have a mixed impact on cultures and economies around the world. Many products are imported and, while buying products that have been manufactured abroad can create jobs and support a local economy, it can also lead to poor working conditions.

People are involved in the designing and making of products, so there are moral implications for us all when we buy them. We may decide to stop buying products because they are considered to be non-ethical, but the families that produce them may no longer be able to support themselves. The current trend for organic and locally produced products across a range of industries has led to local farmers and smaller companies being able to benefit; all producers are able to share in and benefit from the demand for a wide variety and quantity of fresh produce.

Organisations such as Traidcraft only sell ethically produced materials and ingredients, which help both the producers and manufacturers in developing countries. It is the UK's leading fair trade organisation and is dedicated to fighting poverty through trade (www.traidcraft.co.uk).

A recent article from Traidcraft highlights the plight of Bharati, a mother in India and how she struggled with extreme poverty.

'Twelve months ago, Bharati was struggling to feed her four young daughters. Now, after working with Traidcraft on their Sustainable Farms, Sustainable Futures project in Eastern India, she's a source of inspiration in her community. Her farming skills are much improved and she's able to make more of the land she has. As she now grows her cotton organically, she can get a better price for her crop. She also grows food crops which ensure her family always has a nutritious diet.'

(www.traidcraft.co.uk/bharati)

Impact on the environment

Among the main issues relating to product development is the impact on the environment through the consumption of non-renewable resources, as well as waste, pollution and energy use. A few generations ago it seemed as if the world's resources were infinite and people needed only to access them to create businesses and grow.

However, in recent years it has become clear that the impact on the environment of certain manufacturing processes can be irreversible. Concern for the environment is becoming a priority not just for consumers but for governments in countries all over the world. Companies are now encouraged to seek other forms of energy or materials that have less environmental impact, with rewards for being more environmentally friendly.

The influence of advertising, causing people to buy more products, has resulted in the production of excessive quantities of products that flood the market and encourage a throwaway society. This has a knock-on effect on the environment through the production processes used to make the products and their final disposal.

Consideration of sustainability

Some of the techniques that are used by present-day designers to minimise environmental impact and promote sustainable methods of manufacture are outlined below:

- Considering the recyclability of the materials to be used and minimising the range of materials.
- Looking at the materials and components in relation to manufacture. For example, moving from screws to snap clips reduces the amount of time it takes to dismantle the product.
- Using materials that include high levels of recycled content, such as cardboard, metals and most plastics. The Innocent Drinks bottle was one of the first products to be made from 100 per cent recycled PET.
- Using biodegradable packaging that reduces the impact of the packaging at the end of its life.
- Making products that are designed so that more can be transported in one shipment.
- Designing products to interlock or stack in a different way to reduce overall volume and decrease the storage and transport space required.
- Ensuring there is limited or no planned obsolescence. For example, producing a kitchen knife with two blades so that the blade can be swapped and the blunt one sent back to the manufacturer to be professionally sharpened. (This is sometimes referred to as a 'circular economy'.)
- Designing everyday products, such as washing machines, refrigerators and cooking appliances, to meet energy efficiency standards and providing energy labels that help consumers choose energy efficient products. The labelling requirements for individual product groups are created under the EU's Energy Labelling Directive (http://eur-lex.europa.eu/legal-content/EN/ALL/?uri=CELEX:32010L0030t).

One well-known company that has sought to reconsider the manufacture of one of their iconic garments is Levi Strauss & Co. They conducted a lifecycle assessment of a pair of Levi's 501 jeans, and the results were surprising as they discovered a pair of Levi jeans uses nearly 1,000 gallons of water in its lifecycle, from growing the cotton to consumer care. This led the company to find new and innovative ways to design, with sustainability taking the lead.

The company adopted 'Water<Less' finishing techniques which reduced the amount of water used in manufacturing a pair of jeans by up to 96 per cent. Since launching the techniques in 2011, the company has

saved over one billion litres of water. By 2020, the Levi's brand aims to make 80 per cent of its products using Water<Less techniques.

Bourgeois Boheme is a shoe manufacturer based in the UK and was founded in 2005 by Alicia Lai. The company specialises in vegan footwear and ethical production. Due to the impact that clothing and footwear production has on the environment, Bourgeois Boheme tried to improve on material choice and production. Their shoes are completely free from PVC. To avoid 'throwaway fashion' the company designs and manufactures classic shoes that will stand the test of time and uses durable and sustainably sound materials. One of the reasons the company doesn't use leather is because toxic chemicals are used to treat the raw leather, which produces a greater quantity of refuse and has an obvious negative environmental impact. In addition, tanneries consume a lot of energy and are very polluting. Instead the company uses cotton-backere microfiber PU. They use Mycro© PU instead of PVC. The Mycro© is a microfibre with a structure very similar to that of natural leather and suede. Its softness, lightness, breathability, and water and stain resistance provide high-performance, comfortable and very credible leather-looking shoes. The Mycro© holds the EU Ecolabel.

Bourgeois Boheme introduced the use of Pinatec ™ in 2017. This is an innovative, natural and sustainable textile made from pineapple leaf fibres. These fibres are the by-product of the pineapple harvest and therefore no extra land, water, fertilisers or pesticides are required to produce the material.

Figure 2.17 **Bourgeois Boheme, a shoe manufacturer that encourages ethical and environmentally friendly production**

An investigation into a range of products and the historical context in which they were designed and manufactured is likely to reveal the impact on the environment through the use of resources, waste products, pollution from manufacturing processes, end of life disposal, and so on. Today, companies like IKEA ensure they source 50 per cent of their wood from sustainable forests. This sustainable approach is followed throughout the company by having more than 700,000 solar panels powering its stores. In 2012, IKEA announced its goal to be powered by 100 per cent renewables by 2020 – but just four years later, it improved on this further, aiming to be a net energy exporter in the same time.

Case study – Stella McCartney

Stella McCartney is not only a very successful fashion designer but also a supporter of environmental and sustainable methods of productions. Her company achieves this by constantly researching methods to be more sustainable and attempting to carry this methodology through different aspects of the company. The most obvious area is the sourcing and selecting of materials for the collections, but it is also integrated into the manufacturing stage and even in the stores.

Renewable green energy is used to power the Stella McCartney stores. In particular, the UK stores are powered by wind energy. The materials used in the product collections include organic cotton, regenerated cashmere, recycled and forest-friendly fabrics; no leather or fur products are used due to the energy used in production and the impact on forestation. The company is committed to continually sourcing and developing more environmentally friendly alternatives.

Stella McCartney has designed a bag that is made from sustainably certified woods. In places such as Indonesia, Central Africa and the Amazon, the forests are being cleared to make room for livestock and fast-cash crop plantations and are also shrinking as a result of climate change. Stella McCartney is therefore committed to using sustainably certified wood. All of the wood used in her shoes, bags and jewellery comes from sustainably certified sources. Sustainable forestry certifications assure the company that the wood it is using comes from responsibly and sustainably managed forests and plantations. Its products that have ties to the forest therefore do not directly or indirectly contribute to the destruction of the planet's forests.

You will need to carry out regular research to ensure you are aware of new developments in technology to manufacture products that are environmentally friendly and encourage sustainability.

Understand how key historical movements and figures and their methods have had an influence on future developments

Every designer seeks inspiration in some form and design tends to follow patterns and trends time and time again. Inspiration may come from looking into the past to see how historical movements and designers have created styles that have resonated through time. Designers can emulate these styles with the help of technology that offers modern materials and manufacturing processes. Designers need to have an understanding of what has happened before in order to infulence the future.

The following list of design movements looks at how they have influenced future developments:

- The Arts and Crafts Movement – known as the 'reform movement' due to it being associated with 'anti-industrial' manufacture. This promoted the idea of hand-crafted products and created a demand for simplicity of design and fitness for purpose. It is said to have formed the basis of modern design in Europe.

- Art Nouveau – similar to the Arts and Crafts Movement as it encouraged the return to hand-crafted products and is sometimes described as being between art and industry. The philosophy behind this movement can be interpreted as delaying the development of modern industrial design.
- The Bauhaus – known as the centre of modernism and functionalism. The principles laid down by this movement of using modern materials to combine form and function are still meaningful today.
- Art Deco – focused on anti mass-production and one-off expensive products mad from luxurious materials. However, Art Deco influenced the design of many mass-produced products made from new materials such as aluminium, chrome, plywood and Bakelite.

These influential movements and related designers have inspired the creation of many products we see today. The influence may not be immediately apparent as the products are not direct copies but some elements of the philosophy behind the movements are evident, whether it be in the shape, form, colour or material.

DESIGN ENGINEERING

There are some well-known engineers who have made a significant historical impact, such as:

- James Watt, a Scottish engineer and entrepreneur who made significant improvements to steam engine technology which was fundamental to the changes brought by the industrial revolution in the 18th and 19th centuries.
- Isambard Kingdom Brunel, a prolific engineer who made a great impact in the industrial revolution. He was responsible for some of Britain's enduring engineering feats such as the Great Western Railway and the Clifton Suspension Bridge in Bristol.
- James Dyson, a well-know industrial designer and entrepreneur who, among other things, invented the

Dual Cyclone bagless vacuum cleaner and set up his own highly successful business after failing to interest major manufacturers in his design.

However, of more significance to students studying A Level Design Engineering is an understanding of historical key advances in technology which disrupted the established methods and brought about radical changes. Such advances are often brought about by the discovery of a new material, or the invention of a new technology, sometimes linked with a significant scientific discovery. These key advances can sometimes be attributed to a single figure but they are also often associated with groups, businesses or educational establishments, such as universities.

Table 2.8 **Disruptive technological advancements and key figures**

Technological advancement	Key figures	Significance
1895 Invention of radio	Guglielmo Marconi	Marconi proved that radio signals could be used to send messages around the world and to and from ships. Many modern systems rely on radio signals, including Wi-Fi, television, mobile phones, etc.
1906 Invention of the triode valve	Lee De Forest	The triode vacuum valve was the very first amplifying component. It brought about the 'birth' of electronics and resulted in rapid advances in radio technology.

Technological advancement	Key figures	Significance
1925 Invention of television	John Logie Baird	Baird gave the first public demonstration of TV using a mechanical system which was later superseded by a better, fully electronic system. Nonetheless, working alone, this inspirational Scottish engineer successfully built a prototype TV from an old box, darning needles, old bike lamps, wax and glue.
1947 Invention of the transistor	Bell laboratories – William Shockley	The transistor replaced the triode valve. It is a much smaller component, allowing electronic devices to be made smaller and portable, and to be powered from batteries. A transistor is the basic building block of all electronic systems.
1948 Invention of the stored program computer	University of Manchester – Kilburn, Williams and Tootill	This was the first computer to contain all the parts we associate with a modern electronic computer. The computer age developed rapidly with several players contributing significant advancements.
1958 Invention of the integrated circuit (microchip)	Texas Instruments – Jack Kilby	Microchips revolutionised the electronics industry. They allowed thousands (now millions) of transistors to be integrated onto a single tiny chip of silicon, miniaturising electronic products while increasing their functionality. The first digital watches and pocket calculators emerged soon after, and microchips undeniably spawned the space age, allowing flight computers to be built into spacecraft.
1989 Invention of the internet	CERN – Tim Berners-Lee	Networking of computers began in the 1960s. By the 1970s, computer scientists had agreed on a standard network protocol, which spawned the idea of worldwide networking. In the 1980s, Tim Berners-Lee invented the World Wide Web (www.) and wrote the first web browser. The internet has undoubtedly revolutionised the way we communicate, shop, entertain ourselves and conduct business transactions. It has become the main conduit we use for everything from watching TV or ordering a pizza, to controlling our house lights while we are in a different country.
2000s Development of the 3G mobile cellular network	Worldwide players	Mobile phones were first developed in the 1970s, but when the 3rd generation (3G) mobile broadband network was developed, mobile phones became 'always connected' to the data network which resulted in the development of smartphones. For the first time, users had access to the full power of the internet in their pockets.
2004 Isolation of graphene	University of Manchester – Geim and Novoselov	Graphene is one-atom-thick, two-dimensional carbon sheet. It has remarkable properties and is set to revolutionise almost every part of our lives. Currently, only microscopic flakes of graphene have been produced and researchers are racing to produce the material in larger, usable sheets.

ACTIVITIES

1. Take one of the entries in Table 2.8 and research it in detail, identifying key figures and dates.
2. Look at Table 2.8, then research and add three further disruptive technological advancements, identifying key figures and explaining the significance of the technology which followed.
3. Table 2.8 stops at the year 2004. Consider the cutting-edge technological advancements that are currently being made and try to predict the further entries that might appear at the bottom of this table over the next 50 years.

4 Investigate an early product from one of the following categories and create a timeline to show how the product has developed over time. Identify the ways in which changing technologies have impacted on the product's evolution:
   – television
   – radio
   – computer
   – mobile phone.
5 Select a new material that has impacted greatly on the way designers design products.
6 Research a notable historical engineer. Discuss their work and identify their notable achievements, identifying the impact on users at the time.
7 Research a current engineer. Discuss their work.

FASHION AND TEXTILES

The fashion and textiles industry is one sector where it is relatively simple to analyse and recognise how historical figures and movements have influenced future developments. Past trends frequently make a comeback in the fashion world, with slight modifications to the design to reflect a more contemporary style.

Fashion designers are influenced by various factors and research previous styles, techniques and designers when developing their own collections. Any new style that features on the catwalks can usually be linked to a previous trend or movement that has propelled and fuelled the rise of a design feature, colour palette or silhouette.

The UK fashion industry, for example, has a long, proud and diverse history that includes the 'Swinging Sixties', punk and New Romantic movements of the 1970s and 1980s. Each of the eras and designers of that time have continued to influence fashion and textiles over the years and the ideas can still be seen in styles today.

Pedal pushers are an example of a fashionable garment that has re-emerged serveral times from the 1940s to 2018. The original pedal pusher trouser was created by American designer DeDe Johnson in the 1940s as a garment to be worn by ladies for cycling. This was then made into the latest craze by the likes of Audrey Hepburn and Marilyn Monroe in the 1950s. The style has since been worn by Princess Diana and Victoria Beckham in different decades. Each time the garment is reinvented, the style changes slightly to reflect the current trend. This may be in the height of the waist, the length or the width at the hem. For each new version, the fabrics and components are sourced to enhance the style further and ensure it is completely in fashion.

Figure 2.18 **Pedal pushers reinvented**

'The New Look' was created and launched in 1947 by the iconic designer Christian Dior and was a major shift in post-war fashion in the late 1940s. The key elements of the collection were full skirts, tiny waists and soft shoulders. Feminine styles were created through the use of ruffles on a masculine-style jacket or by accenuating the hourglass shape. After wartime rationing, the metres of fabric used in the designs made a refreshing change. The Bar Suit is one of the iconic images from the collection.

Dior then built on this idea of exaggerated shapes and using fabric to excess. He layered materials and used padding around the hip and shoulders to create the desired silhouette. His strength was his evening wear where he could be truly extravagant and he created strapless boned bodices with huge billowing skirts.

The New Look really did change fashion at the time and its influences are still evident today. Even within designer collections seen recently, the basic principles of the Bar Suit can be seen. The hourglass silhouette Dior loved can be seen in many bridal stores in the form of the 'meringue' wedding dress.

Since 1945, mass production has increased due to consumer demand in the fashion and textiles industry. While many features and techniques may be influenced by historical movements and designers, the methods for achieving the same looks will be modified to be produced on a large scale.

Table 2.6 describes the key features of iconic fashion and textile designers and how they have influenced future developments. Carry out your own research into each to see how their ideas have influenced designs seen today.

Figure 2.19 **Christian Dior's 'New Look'**

Table 2.6 **Iconic fashion and textile designers**

Designers	Key features	Influential ideas
Lucienne Day – textile designer	She believed in making textiles affordable and her 1950s designs incorporated modern abstract images influenced by nature. She took inspiration from Paul Klee and created her own style of motifs drawn from nature – flowers, grasses, shoots, the intricate patterns of the landscape – and transformed them into something absolutely new.	Many of Day's printed fabrics were made in long production runs, which kept the price affordable. She made the link between mass production and fine art. Some of her print designs have recently been released and turned into cushions by the department store John Lewis.
Charles Worth	Considered to be the master of haute couture. He changed the idea of the designer visiting the clients by opening the House of Worth where the clients would visit him for consultations.	He was the first designer to add labels to his designs and the first to use real models to promote his clothing to clients. He is also credited as the first designer to use fashion shows to promote his designs. He introduced the princess line in dresses which is a style still used today.
Coco Chanel	Chanel popularised the little black dress; (LBD). Before Chanel, the colour black was reserved for funerals and women in mourning. She pioneered the androgynous look and was one of the first women to wear trousers as a fashion item. She was the first designer to use jersey fabric. Her emphasis was on making clothing that was more comfortable for women to wear.	The skirt suit with a collarless, cardigan-like jacket has influenced many look-alikes, but the original design is solely attributed to Chanel. She was also the first designer to encourage the use of costume jewellery, in particular pearls.

Designers	Key features	Influential ideas
Madeleine Vionet	She created lingerie-inspired dresses on a mannequin, rather than on paper. This was an unusual method but allowed her to see how the fabric moved as she made a dress. She was also the first to cut her fabric on the bias, which allows for sensual, sculptural, draped dresses that hug the figure.	Similar versions of her bias-cut dresses can be seen on many catwalks today, in particular, those of Ossie Clark, Halston, John Galliano, Comme des Garçons, Azzedine Alaia, Issey Miyake and Marchesa.
Claire McCardell	The American designer introduced mass-produced, affordable and stylish women's sportswear. Her designs embraced comfort, functionality and everyday fabrics.	Her casual approach to dressing can be seen in collections of designers such as Tommy Hilfiger and Ralph Lauren today.
Yves Saint Laurent	A modern designer who conceptualised men's clothing, like the tuxedo (Le Smoking) and the safari jacket, into chic women's wear. He wanted women to look comfortable and elegant.	Many versions of the tuxedo for women can be seen today in varying formats and price brackets.
Mary Quant	She became popular in the 1960s and has been credited with creating the miniskirt. She was also one of the first designers to use polymer fabrics.	Her innovative use of new fabrics paved the way for future designers to experiment with less conventional fabrics. Many versions of the mini skirt have appeared since she introduced it.
Roy Halston Frowick	Halston redefined American sportswear with fabrics such as cashmere and Ultrasuede in minimalist yet sophisticated silhouettes.	He influenced designers such as Michael Kors, Calvin Klein and Donna Karan

PRODUCT DESIGN

You will need to research other designers in order to refer to them in your work and exam. You need to understand how past designers and movements have inspired and influenced designers today, and how designers have also been influenced by a number of other factors. For example, the machine age of the 1920s influenced designers, as they responded by producing products that used the relatively new invention of mass-produced materials. Following the two World Wars, there was a demand for natural and handmade products that reflected a humanist, individual approach. It may be that styles that past designers have been famous for introducing can be seen in products now.

Table 2.7 **Key designers and their ideas**

Designers	Key features	Influential ideas
Ettore Sottsass	Italian designer who designed innovative products that express the look of the particular period. He is known for combining materials not normally used together, such as polymer laminates with expensive wood veneers. He was influenced by 1960s pop culture and art.	His ideas feature bold, angular, geometric shapes and bright colours for furniture and many other products. One of the most iconic designs is the Valentine typewriter for Olivetti. He has influenced many designers since and elements of his ideas can be seen in stores today where modern materials are combined with those that are more traditional.
Mario Bellini	He originally trained as an architect but is more famous for product design. His work is distinguished by its clean lines and innovative styling. He has collaborated with many companies to create a wide range of products.	One of the most influential is the concept car he designed, named the Kar-A-Sutra. This never went into production but is thought to have inspired the people carrier.

Designers	Key features	Influential ideas
Ron Arad	Ron Arad started designing furniture by recycling a Rover car seat and building a frame for it. He is known for experimenting with the latest technologies to create innovative designs and has worked with a range of materials. He says he is influenced by everything around him to create something completely new.	His designs are innovative and individual and, as such, inspire designers to be creative and not be restricted by everyday perceptions of design or by the materials they use.
Dieter Rams	A German designer, known as one of the most influential designers of the late 20th century. He made electronic gadgets and furniture. He worked as head of design for the electrical company Braun. Rams uses hand drawing to portray his ideas and has never been a fan of using digital technology to present his designs. He defined his design philosophy into ten distinct points that he believed reflected good design.	Dieter Rams has been an inspiration for younger designers, notably Jonathan Ive and Jasper Morrison. Jonathan Ive is the chief designer at Apple and he has noted Dieter Rams as the inspiration behind Apple's iconic products. Jasper Morrison works for Rowenta and has acknowledged Rams in his work.
Michael Graves	One of the Memphis Group, he was originally known as an architect. He then became a product designer focusing on domestic products that featured original forms. His most iconic designs were made by Alessi, the Italian houseware manufacturer.	In 1985 Graves designed a stainless steel kettle featuring a red whistle shaped like a bird. This has become Alessi's bestselling product.
Philippe Starck	He uses inspiration from the past. He is known for combining unusual materials such as glass with stone. He was responsible for designing the three-legged chair seen in many bars and restaurants. He favours manufacturing commercial, relatively inexpensive products for mass production.	Starck's design practice has produced many products, including furniture, domestic appliances, staplers, toothbrushes, lemon juicers, tableware and even clothing, food and architecture.
James Dyson	James Dyson is a British designer famous for his design of a cyclonic, bagless vacuum cleaner. He has also designed hair dryers and fans as well as batteries. He studied furniture and interior design at the Royal College of Art (1966–1970) before moving into engineering. His first original invention was a modified version of a wheelbarrow – using a ball instead of a wheel – known as the Ballbarrow.	James Dyson aims to design products that are lean and efficient in the way they work and use materials. He discovered the flaw with vacuum cleaner bags clogging and not picking up dust, and his aim was to invent something more effective. It is function, not form, that excites Dyson and his team of design engineers, who create machines that are lightweight but robust. They also use fewer raw materials and are less of a drain on resources and energy to process, resulting in lighter, more user-friendly designs for their users.
Jonathan Ive	Jonathan Ive is Apple's Chief Design Officer, who is responsible for all design, hardware, user interface, packaging and major architectural projects at Apple, as well as new ideas and initiatives. Born in Britain, he studied industrial design at Northumbria University, before being asked by Apple to create a design for their new laptop. Since 1996 Jonathan has led the Apple design team, which is regarded as one of the world's best.	He is influenced by the Dieter Rams and Bauhaus/Ulm design philosophies of form follows function and 'less is more'. Ive's first design assignment at Apple was the iMac, which helped Apple's success with other designs such as the iPod, iPad and iPhone, thus establishing Apple's leading position with a series of functional and aesthetically pleasing products that offer users an intuitive experience.

ACTIVITIES

1   Select a past design movement and discuss how the style has influenced products that are designed and manufactured today.
    a)   Produce sketches to show the style of products made by designers in that movement.
    b)   Explain how the materials used in the product influenced its design.
2   Discuss what you think is the most important design or product that has been introduced in the last ten years within your specialism and why.
3   Investigate domestic products that have been inspired by the Arts and Crafts Movement. Explain the similarities and differences between the products.
4   Take an iconic product from a designer of your choice from a past design movement. Explain how the product was manufactured when first introduced and how it would be manufactured today. What changes would need to be made and why?
5   Select a product or material that has advanced in design or function dramatically over the last 20 years. How has this impacted on consumer lifestyle?
6   Select a designer or manufacturer that has not compromised on the function or design of their products by selecting materials or manufacturing methods that are eco-friendly. Explain how they have achieved this.

KEY TERMS

Culture – the ideas, customs and social behaviour of a particular group of people or society.

Enterprise – another name for a business. This refers to the initiative of an individual or organisation and their willingness to take risks to develop new projects and ventures.

Fast fashion – a term used to explain that designs move from catwalk to high-street stores quickly to capture current fashion trends.

Iconic design – a design that is 'ground breaking' and sets new standards in its field.

Invention – the process of creating something that has never been made or never existed before.

Patents – granted by the government to protect designs and inventions, providing strong protection against copying of the technical and functional aspects of an invention.

KEY POINTS

● Examining and researching how designs have developed over time and following advances in technology or consumer requirements can help the iterative design process.
● New developments in the design industry can have both positive and negative effects on companies.

Positive because of new products being introduced, new processes being used and the resulting new streams of income; negative as it can make existing products and industries obsolete.

2.4 What can be learnt by examining lifecycles of products?

LEARNING OUTCOMES

By the end of this section you should have developed knowledge and understanding of:
● a product's marketing lifecycle from initial launch to decline in popularity, including:
    –   consideration of initial demand, growth in popularity and decline over time
    –   methods used to create more demand and maintain a longer product popularity
    –   [A Level only] new models of marketing and the influence of social media.

A product begins its marketing lifecycle when it is first thought of and it continues until (in most but not all cases) the demand for the product drops to a low level and it is phased out. Older products can become unfashionable and therefore less popular, whereas the release of a new product can increase its demand rapidly, depending on popularity.

Consideration of initial demand, growth in popularity and decline over time

Lifecycle of products

When we assess a product's lifecycle there are four factors to consider:

1 Products have a limited life.
2 Product sales pass through distinct stages, each posing different challenges, opportunities and problems to the seller.
3 Profits rise and fall at different stages of the **product lifecycle**.
4 Products require different marketing, manufacturing, purchasing and human resource strategies at each stage of the product's marketing lifecycle.

Product lifecycle stages

There are four basic product lifecycle strategies.

1 **Introduction** – this stage relates to new products being launched on the market for the first time. The crucial factor at this stage is to determine the stakeholders that will be involved in the whole lifecycle process. One key aspect is being able to determine the target market and identifying how to reach them effectively. Carrying out surveys at the prototype stage with the target market can give the manufacturer considerable and constructive feedback that will be invaluable to achieve **market penetration** with the product. This is typically the stage where the highest monetary investment is required as there are the costs of product design and development, investment in tooling and machinery and marketing costs. In many cases, products are not profitable at a market introduction stage due to low sales and lack of product awareness. Some retailers may be apprehensive about stocking a new product until they are confident it will bring profits and high sales. It is crucial to ensure the **pricing strategy** is correct to attract buyers. The manufacturer will need to decide on the pricing strategy that they feel will be the most appropriate to attract sales. This can be either:

- **Price skimming**, where the highest price possible is charged initially and then the price is slowly reduced. One example of this is seen in the ongoing developments with televisions. The latest developments of OLED (organic light-emitting diodes) TVs are introduced at a very high cost. This is because the manufacturer or retailer is aware that the product may have a certain term of popularity and they want to recover the introduction costs.
- **Penetration pricing**, where the lowest price is charged and then the price is gradually increased. This pricing is typically used when the market is already saturated with the product being introduced. **Penetration pricing** gives an edge to the company because many customers are attracted to the price and could decide to switch brands. A good example of a company that has used penetration pricing to tempt buyers to switch brands is Samsung. The company did manage to take a large share of the smartphone market away from Apple. However, Apple used skimming pricing to introduce its products and has remained very successful using this strategy due to customer brand loyalty and innovative product development.

2 **Growth** – this is the stage where the popularity of the product is increasing and sales are steadily growing. The product has been accepted by the target market and profit should be positive. Due to the fact sales have increased, the total manufacturing costs per unit are reduced due to economies of scale. This is also the stage where competitors may introduce similar products. It is crucial at this stage to penetrate the market by not only maintaining sales to the target market but also by attracting competitors' customers. This can be achieved by further investment in advertising, marketing and continued promotional spending. The growth stage is also the ideal phase in which to make improvements to the product to maintain consumer interest, for which further investment may be required.

3 **Maturity** – during the maturity stage, sales reduce and the profit margins start to decline. This can be due to a number of factors, such as market saturation or the introduction of other competitive products. However, this stage can often present opportunities for companies through a number of solutions, if the company has the funds and

creativity to maintain sales and attract new interest. This can be done by expanding to international markets or introducing new features. For example, Apple managed to extend the life of the iPod by introducing the iPod Touch which included new features, applications and a touch screen. The company continually updates products to maintain interest and sales. It is at the maturity stage where products and sales are most vulnerable and companies should plan and research for a next generation product or technology before the decline stage.

4 **Decline** – this stage happens when the market is saturated with the product or competitors have taken a larger portion of the sales. The company sees a decline in profits and cash flow is reduced; sales decrease. It is very important that a company is able to predict this stage and have a strategy to exit the product from the market to prevent profit loss. An exit strategy normally coincides with the introduction of a new product. Companies must decide in advance what strategies to implement when their products enter the decline stage. Some products decline slowly, whereas others can have a rapid decline. Many fashions and trends for the younger generation tend to have a short lifecycle. Technical products such as digital cameras, cell phones and video games often have limited lifecycles.

Methods used to create demand and maintain popularity

The main goal of every company is to maintain or increase profits and stay ahead of the market. There are many different approaches a company can take to create demand and maintain popularity in a product once it gets to the maturity or decline stages. One method of extending sales is to generate further interest in the product by making it appeal to a different target market. This is an extension strategy. Further methods include:

- increasing the uses of the product; this could be through an add-on to the product or accessories to maintain interest and remain competitive
- reducing the price
- adapting the original product to ensure it has a **unique selling point or proposition (USP)**. For example, car manufacturers modify their vehicles slightly each year to offer new styles and new safety features. Every three to five years, they do more extensive modifications
- introducing promotional offers

- introducing new advertising campaigns
- changing the image or branding of the product to appeal to another target market
- staying up to date with technology so customers will think they have the latest version
- attracting new markets; with the growth in the number of online shoppers, more organisations sell their products through the internet.

Fifty years after the launch of the original Fiat 500, Fiat Auto decided to reintroduce the Fiat 500 to the market in 2006, with great success. Fiat Auto carried out extensive research to collect a huge amount of information about their potential customers and about their preferences on colours and accessories. For example, Fiat realised that some colours were more appreciated than others and that the price of the car could be increased. Consumer demand for customisation led to Fiat introducing several new accessories and models, such as the open-roofed Fiat 500 and the Italian flag sticker. Originally the Fiat 500 met consumer needs for a trendy retro small car. However, Fiat then developed their Fiat 500 range to include the 500X to meet the ever-growing market for larger family cars.

Figure 2.20 **The new model of the Fiat 500**

Another good example to examine is the development of nylon, which was invented by a chemist and inventor working for DuPont. The sales and popularity of nylon fibre have been repeatedly and systematically extended since its introduction. The first nylon uses were primarily military, for parachutes, thread and rope. This was followed by use in circular knit hosiery, which has enabled it to dominate the women's hosiery business. After some years the profit began to level but DuPont had the foresight and planning to ensure they had a strategy to maintain sales and interest in the fibre. Du Pont took certain steps which pushed hosiery sales upward. They adopted strategies that expanded sales via four different routes:

1 Promoting more frequent usage of the product among current users – their research showed that there was an ongoing trend to have bare legs. At this point it could have been very difficult for DuPont to maintain sales. However, they introduced tinted and patterned hosiery in an attempt to attract more users who would see the hosiery as a fashion staple and a required accessory.

2 Developing more varied usage of the product among current users – by introducing a wider range it attracted a wider audience.

3 Creating new users for the product by expanding the market – to attract new users to the range and the idea that hosiery was a fashion staple, further advertising and marketing using well-known fashion leaders was called for.

4 Finding new uses for the basic material – for example, socks, carpets and tyres. Recent developments have led to nylon being used for artificial muscles.

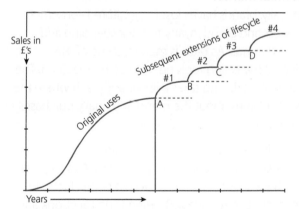

A, B, C and D = sales of nylon products plateau

#1 = More frequent use of the product is promoted to existing users
#2 = Developing more varied usage of the product among current users
#3 = Creating new users for the product by expanding the market
#4 = Finding new uses for the basic material

Figure 2.21 Sales of nylon and methods used to extend its lifecycle

It is vital for a company to decide whether the product has further possibilities or whether it is time to remove it from the market. Extending a product's life means that the company will normally have to invest more time and money. It is a fine balancing act to decide whether it is worthwhile. Another option for the company is divesting the product from its offerings. The company might choose to sell the brand to another firm or simply reduce the price drastically in order to get rid of all remaining stock. Many companies decide the best

strategy is to modify the product in the maturity stage to avoid entering the decline stage.

New models of marketing and the influence of social media

Social networks provide advertisers with information about the likes and dislikes of their consumers. The majority of social media platforms have their own built-in analysis tools which give companies the ability to track the success of advertising campaigns. Companies are able to target potential stakeholders to gain interest in their product through **social media marketing**. This is a very effective market research technique as it provides businesses with a specific target audience. The methods used to attract customers fall into two categories:

● Passive approach – this is where individuals use social media platforms to share their reviews and recommendations of brands and products. This information can be analysed by marketers to track problems and identify market opportunities.

● Active approach – this refers to how social media can be used as a direct marketing tool. Companies may initiate some form of online dialogue with the public to build a rapport.

Social media marketing techniques

The following techniques can be used by companies for effective marketing of their products and services.

● Social media marketing (SMM) uses social networking websites as a marketing tool. The goal of SMM is to produce content that people will then share with their social network to help a company increase exposure and broaden customer reach.

● COBRA (Consumer's Online Brand Related Activities) facilitates spreading awareness of a particular product on the social media platform with the aim of gaining responses and demand for the product. An example of this is when a consumer uploads a photo of their latest purchase on a social media site, which builds awareness of the product. The company could then analyse this information to see which products are popular and specifically target this audience when advertising future products.

● eWOM (Electronic Word of Mouth) is a very powerful marketing method and as such can influence consumers' purchasing decisions. Information on a product or service can be passed on through written text, images or even movies. A consumer can recommend, like or dislike a product and the message can reach a multitude of people at the same time so it has a greater potential of becoming viral. Again, as with COBRA, the information is analysed by the manufacturers to assess a product's success. An example of eWOM would be an online review.

- Search engine optimisation (SEO) is a strategy for drawing new visitors to a website. SEO can be done two ways: adding social media links to content, such as feeds and sharing buttons, or promoting activity through social media by tweets or status updates.

Companies could encourage credible eWOM by including a review site and links to social media on their web page. This ensures that consumers can easily find links to credible sources. Examples of these links on a web page are shown here with Tesco's Hudl. They illustrate that the product has been awarded 4.7 stars, gained from 10,553 customer reviews.

What makes eWOM marketing so powerful?

- People like to share – our need to share explains the success of consumer review sites, discussion forums and social media sites. One person's comment reaches several others and each of them can share the message with their network of friends and so on. This explains the viral potential of eWOM.
- People seek advice online – whenever we plan to make a purchase, we search for information online. Online reviews are used as important information sources, shape consumer attitudes towards a product and influence sales.
- People trust other people – we trust the opinion of a real person more than we trust advertising.

There are negative and positive aspects to social media marketing. The obvious positive is the fact that the companies and brands are getting free publicity that can reach a huge audience and influence the success of a product. However, on the negative side, if the consumer comments on a faulty product or poor service when giving a review, this can harm the company's reputation.

Many companies today use influential people on social media to send messages to their target audiences. These people can have a huge following through any number of social media sites and their opinion is trusted. They can review products and services for their followers, which can be positive or negative towards a particular brand. These people have a social status and are influential because of their personality, beliefs, values, etc. They are known as **opinion leaders** and those who follow their lead are known as **opinion followers**.

Marques Brownlee is one such person who has become extremely popular through YouTube. He has 1.8 million subscribers who are interested in his reviews on the latest hi-tech products. Under the username MKBHD, he tests a wide variety of products that have just been released. One of the reviews he carried out was on the sapphire crystal display for mobile phones. In the video, which received about 8 million views, he took keys, his foot, and even a blade to the screen to prove its hardness and toughness.

New models of marketing

Millennial market

Historically, brands have closely monitored consumer awareness and brand loyalty to track spending and shopping trends. As brands mature, some of the traditional marketing tools need to be re-imagined. The majority of marketing takes a linear approach where the consumer learns about the brand and finally purchases the product.

Millennials were born between 1977 and 2000 and make 21 per cent of consumer purchases. As such, they have a huge influence on the success of products. 46 per cent state they have 200+ friends on Facebook and therefore their ability to share shopping experiences and purchases is wide. Over half of millennials use Instagram every day, so an Instagram account is a must for companies aiming products at this age range.

The aim of a millennial model remains to get the customer to make a purchase. But millennials will ask friends, family and their online networks for recommendations and opinions and will then share the experience on the same network they used to find out about the product.

Table 2.9 Examples of social media platforms

Social media platform	Marketing method
Twitter	Twitter allows companies to promote their products through short messages limited to 140 characters. These messages can contain links, videos and photos.
	Twitter incorporates a range of tools and independent applications are also available to monitor and manage tweets in the marketing of products.
Facebook	Facebook allows companies to market their products in much more detail than Twitter. They can include videos, photos and links. It allows others to comment on the pages for others to see.
	Facebook incorporates a range of business and marketing tools for companies to manage their posts and advertisements.
Instagram	Instagram provides a platform where user and company can communicate publicly and directly, making it an ideal platform for companies to connect with their current and potential customers.
	Instagram is a highly visual medium, where photos and videos are shared, showing products in action. This is free advertising and increases brand awareness and exposure.
YouTube	YouTube allows companies to promote a product in a way that reflects the audience's style and taste. It also allows companies to sponsor certain videos, etc.
	People can subscribe to YouTube channels that they are interested in, giving companies the opportunity to introduce new products using videos.

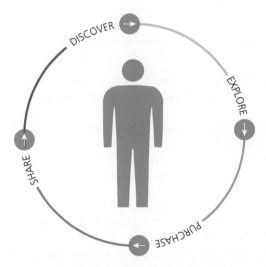

Figure 2.22 Millennial marketing circular approach

SAVE approach
The SAVE marketing framework can replace the traditional 4Ps marketing mix. SAVE focuses on the Solution, Access, Value and Education in relation to marketing:
- Focus on Solution instead of Product – many companies are moving their business model from product oriented to service and solution oriented.
- Focus on Access instead of Place – many companies are moving their business model from ownership to 'access to'. Nowadays, many businesses operate around continual high-speed internet access.

Traditionally 'place' (where you sell your product) has been crucial to attract the most interest; today that 'place' is usually online and therefore it is vital to have access.
- Focus on Value instead of Price – customers still have concerns about price, but only after their concerns about value. They want value for money, whether this is the budget market or high end.
- Focus on Education instead of Promotion – companies today can provide current and potential customers with information relevant to their interests to create a sense of familiarity and trust long before a purchase is even made.

Digital marketing
Digital marketing can offer detailed data and analysis of consumer behaviour, as well as accurate and detailed data on the effectiveness of marketing strategies. Consumers increasingly demand marketing messages and offers that are highly personalised, relevant and targeted.

Four digital marketing models:
- Digital branders are companies that are investing in more immersive digital multimedia experiences that can connect consumers to the brand in new ways. They are reimagining how they engage consumers, with the primary goal of recruiting new consumers to the brand and promoting loyalty through multiple experiences.

- Customer experience designers use customer data to create superior brand experiences for their customers. For example, airlines, hotels and retailers build their business models around customer service.
- Demand generators focus on driving online traffic and converting as many sales as possible across channels to maximise marketing efficiency. All elements of the digital marketing strategy; website design, search engine optimisation, mobile connected apps and engagement in social communities, are tailored to boost sales and increase loyalty.
- Product innovators use digital marketing to identify, develop and launch new digital products and services.

OCR Design & Technology for AS/A Level

ACTIVITIES

1 Take an everyday product of your choice.
   a) Identify the key features that contribute to its success.
   b) Suggest methods for extending the product life and therefore increasing sales.
2 Take an existing well-known product
   a) Investigate the different methods that have been used to market and advertise it since product launch.
   b) Produce a timeline poster to explain how the marketing has developed and why.

3 Give an example of a company that has launched a new product to the market. Explain the pricing strategy used during the launch and explain why you think the manufacturer chose this method.
4 Explain why the marketing costs related to a product are typically higher during the introduction stage.
5 Explain why and when penetration and skimming pricing are used in the introduction stage of the marketing lifecycle.
6 What different strategies are used by companies to extend the lifecycles of their products throughout the maturity stage?

KEY TERMS

Market penetration – the activity of increasing the market share of an existing product, or promoting a new product, through various marketing and advertising strategies.

Pricing strategy – the method companies use to price their products. This is usually dependant on production, labour and advertising expenses. It also includes a certain percentage so they can make a profit.

Opinion followers – this is the majority of the public, who are consumers searching for information or guidance from sources such as the media.

Opinion leaders – consumers that exert considerable influence; the decisions of others are influenced by those of the opinion leaders.

Penetration pricing – this is when a company sets a low price for its product to attempt to gain market share. The primary objective of penetration pricing is to gain many customers and then use various marketing strategies to retain them.

Price skimming – this is where a company sets its prices high when launching a product to quickly recover expenditures for product production and advertising. The key objective of a price skimming strategy is to achieve a profit quickly.

Product lifecycle – the marketing lifecycle from the launch through to decline of a product (not to be confused with lifecycle assessment).

Social media marketing (SMM) – a form of internet marketing that utilises social networking websites as a marketing tool.

Unique selling point (USP) – a factor that differentiates a product from that of its competitors.

KEY POINTS

- All products are at some stage within the product lifecycle. Where a product is within this cycle is usually dependent on product age, popularity and functionality.
- For a company to be successful it needs to maintain a competitive edge and manufacture products that are desirable. To achieve this they have to update or improve existing products, or create new products to meet target market requirements.
- The product lifecycle helps a company understand the stages a product will go through once it is launched into the marketplace.

MATHEMATICAL SKILLS

Mathematical skills will be used to analyse data when examining existing products and carrying out research (for example, surveys, focus groups) to assess how they could be developed further. This could be:

- Qualitative data – this is non-numerical and descriptive, often giving people's reasons or opinions. This is primarily exploratory data and is harder to analyse than quantitative data. It is used to gain and uncover an understanding of trends, reasons and opinions. The collection methods vary, including focus groups, individual interviews and observations.
- Quantitative data – this is numeric, or information that can be converted into numbers and then used in a statistical way. Quantitative investigations gather data which can be put into categories, in rank order or measured in units of measurement. This type of data can be used to construct graphs and tables.
- Anthropometric data for scale and proportion to determine product scale and dimensions.
- Statistical data to determine user needs and preferences.

Mathematical skills will be used in analysing data obtained from social media sites as part of the development of marketing strategies.

Mathematical skills will be used in the production of timelines, graphs and charts, such as product marketing lifecycles and pricing strategies.

APPLICATION AND EXAMPLES FOR DESIGN ENGINEERING

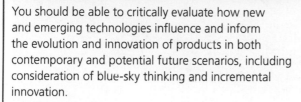

You should be able to critically evaluate how new and emerging technologies influence and inform the evolution and innovation of products in both contemporary and potential future scenarios, including consideration of blue-sky thinking and incremental innovation.

Blue-sky thinking

This is a term that describes the method of being extremely creative when generating ideas. It may lead to unconventional, innovative and unusual ways to solve a problem. It's called blue-sky because the sky's the limit and innovative ideas should form out of nothing (there should be no preconceptions), like a solitary cloud appearing in an otherwise blue sky. You are free to explore any ideas and the possibilities are endless. You think outside of the box to solve design challenges. One of the benefits of approaching projects in this way is that it provides the opportunity to change normal methods of designing. This could be by carrying out research to examine how current competitors have solved problems by designing and manufacturing products. They may have utilised the latest technologies and materials. This approach can assist the designer by using all the information gathered to solve the challenges they have to address. Innovative and fresh ideas give a business a competitive advantage where their products and technologies stand out alongside similar products.

Blue-sky thinking can involve a group of people looking at a design problem with fresh eyes. As many ideas as possible are generated in an ideas generation session where no ideas are rejected.

Incremental innovation

Incremental innovation is a series of small improvements or upgrades made to a company's existing products or technologies. The changes implemented through incremental innovation are usually focused on improving an existing product's development, efficiency and productivity. Many companies use incremental innovation to help maintain or improve a product's market position.

An example of a company that has used incremental innovation to ensure its products are widely used is Nest. Traditionally, this company supplies thermostats and fire alarms for use in homes but neither of these products has developed in a significant way.

Nest advanced these products with sensors and intelligent algorithms which enable these devices to understand user preferences, interact in a more human manner and talk to similar devices to ensure incidents are reported in a more informative way. Nest has created a new market in an area that has traditionally been based on cost and identical products.

You might not think of Gillette as one of the great innovation leaders but, in fact, the brand is a great example of a company that has used incremental innovation to stay ahead of the competition. Gillette razors started life with a single blade but their product has evolved, adding different features and more blades as the company has continued to meet customer needs.

There are continual developments and emerging technologies throughout the fashion and textile industry and it is important to be familiar with the latest developments to stay up to date. Technical requirements usually relate to the end-product use and how the fabric or product has to perform. This could mean that the development focuses on the function rather than the aesthetics.

Military textiles

The military have expressed an interest in using the latest developments in colour-changing technology. This involves a membrane of tiny crystals being used within the structure of a fabric. These crystals react when exposed to light and can change their formation. This could be used for camouflage, similar to how a chameleon changes colour and adapts its appearance according to its environment.

Governments have invested in the development of military uniforms to improve a soldier's performance and comfort and offer the potential to save lives during combat. The main aims of military research into nanotechnology are to improve medical and casualty care for soldiers and to produce lightweight, strong and multi-functional materials for use in clothing, both for protection and to provide enhanced connectivity.

The Ministry of Defence in the UK has predicted that technologies such as medical nanobots and nano-enhanced reconnaissance and communication devices will begin to be used from 2030 onwards.

The military are working with leading textile organisations in a range of ways. For example, they are working with an outdoor clothing specialist that produces clothing and textile products for extreme conditions. They are also developing tents that capture solar energy to make its user energy autonomous.

There is continual development of electronics embedded within textile fibres. The possibilities within this area are massive. The fabric can sense if weapons are being targeted at the clothing and send warning signals. This has been further developed to create heat sensors to indicate if a soldier has been injured.

Nanofibres

The 'nano' in nanofibre comes from the Greek word *nanos*, which means very small objects.

A nanofibre is a polymer membrane formed by electrospinning. The polymer is dissolved in a solution, which then evaporates as the fibre forms. The fibres are measured in nanometres. The size of nanofibres are just three to four atoms thick, with diameters of 50–500 nm (nanometres). Polymer-based nanofibres are a new class of material used in a wide variety of applications, such as filtration, hand wipes, personal care, material reinforcements, garments, insulation and medical. Their unique properties make them a perfect solution applicable across a wide range of industries.

Nanofibres

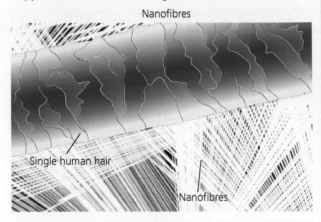

Single human hair

Nanofibres

Figure 2.23 The comparison between a single human hair and nanofibres

When compared to conventional fibres, nanofibres offer numerous advantages, including:

- high surface-area-to-volume ratio
- wide variety of polymers and materials can be used to form nanofibres
- ease of fibre functionalisation – this can be achieved by blending the polymer solution either pre- or post-spinning
- ease of material combination – different materials can be easily mixed together for spinning into fibres
- relatively low start-up cost
- variety of nanofibrous structures can be constructed
- mass production is easily achievable, making it suitable for industrial production and commercial applications.

Table 2.10 **Nanofibres: applications and benefits**

Application	Role	How?
Medical	Drug delivery Wound healing Tissue engineering	• The use of nanotechnology in medical applications has improved the process of healing wounds. The nanofibres can allow liquids and gases to pass through the dressing but at the same time keep bacteria away. This is particularly useful when treating severe wounds, such as burns. This helps to prevent infections. Antibacterial bandages are made using multiple layers of nanofibres with polymers to dress ulcers and chronic wounds. • Nanofibres are particularly effective as scaffolds for tissue reengineering and repair as most human organs are made up of nanofibre-like structures. The nanofibres can be used to create a 3D model so cells can grow and attach to it. • Nanofibres with anti-adhesion properties are used in post-surgical tissue adhesion of internal organs. The nanofibres used have the ability to dissolve into the body without causing any toxicity. Nanofibres used for tissue re-engineering are completely biodegradable.
Sportswear	Waterproof Breathable Insulation Protection	• Nanofibres are becoming increasingly popular for replacing breathable membranes in high-end performance sportswear. They have the potential to create a waterproof fabric with a high level of breathability. • Carbon nanofibres are used to increase the thermal comfort of garments. This is particularly useful for skiwear. Its high tensile strength and insulating properties, combined with the fact it is lightweight, make it both comfortable and protective for sportswear. • Nano socks are treated with silver nanoparticles. The silver acts against infection and odour. • Nanofibre shark-skin is used for swimwear. This incorporates a plasma layer enhanced by nanotechnology to repel water molecules and is designed to help the swimmer glide through the water. • Nano sensors provide a personalised healthcare system, monitoring your vital signs as you run up a hill or responding to changes in the weather.
Fashion/textiles	Aesthetics Protection Self-cleaning	• Stain repellent and wrinkle-resistant threads are woven into textiles. • Fabric treated with UV nanoabsorbers ensures that textiles can deflect the harmful ultraviolet rays of the sun, reducing a person's UVR (Ultraviolet radiation) exposure and protecting the skin from potential damage. • Self-cleaning fabrics have been used for a while. They are created by wrapping the original fibre in minute whiskers using nanotechniques. The whiskers trap air which creates a super **hydrophobic** surface when in contact with water. This allows dirt to be captured in the beads of water, which then rolls off the fabric. • Silver can be applied in the form of nanoparticles on the surface of the fibres. It reduces bacterial growth on the textile (clothing, household and furnishing) by releasing silver ions.
Geotextiles	Filtration	• A nanofibre membrane is used for water/soil filtration.
Military	Protection against chemicals using a mask Water filter membrane Protection	• Nanofibre membranes are very sensitive to chemicals and as such can be used as protection against gas and chemical weapons. They can absorb and decompose the chemical agents. • The nanofibres can be used as a water filter in combat zones. • Nanofibres can switch reversibly from a highly breathable state to a protective one in response to an environmental threat. In the protective state, the uniform material will block the chemical threat while maintaining a good breathability level. • Silver nanoparticles are used to minimise bacterial growth and decrease odour. • In bullet-proof vests, nanotube fibres are used to make a material 17 times tougher than the Kevlar. • Body warmers use Phase Change Materials (PCMs), which respond to changing body temperatures.

APPLICATION AND EXAMPLES FOR PRODUCT DESIGN

The design and manufacturing industry is diverse and there are excellent examples to be studied to understand how products have developed over time, depending on consumer requirements and technological developments. The essence or part of a product's design may stay the same but many products now have the benefit of the latest developments in materials and components or have extra features that relate to new technology to make them more efficient, ergonomic or safer to use.

An interesting way to research how products have developed according to various influences is to complete a timeline. It is then possible to understand the factors that determine their development.

Many developments are consumer led and the following website illustrates how bicycle designers worked from set criteria to make a product more suitable for its purpose:

http://bicycledesign.net/2014/11/design-and-development-of-the-cylo-bike/

Further reading on learning from existing products and practice

- www.core77.com/ – shows the latest innovative designs in all design categories.
- http://productlifecyclestages.com/new-product-development-stages/ – explains product lifecycle stages, each stage in detail, with examples.
- http://wonderfulengineering.com/30-innovative-products-you-did-not-know-exist-but-are-too-awesome-to-miss/ – excellent examples of innovative products. Some are novelty and some solve real problems.
- www.dyson.co.uk/community/aboutdyson.aspx – explains how James Dyson developed the original vacuum cleaner and how products continue to develop.
- www.baddesigns.com – a collection of illustrated examples of things that are hard to use because they do not follow human factors principles.
- www.b2binternational.com/publications/product-development-research/ – explains how research is used for product development.
- www.traidcraft.co.uk/bharati – the story of Bharati in Section 2.3, plus more.
- www.gore.com/en_gb/industries/ – see the development of textile technology for a wide range of applications.
- www.oxgadgets.com/2015/03/gillette-rd-a-journey-through-innovation-and-implementation.html – see how a product is developed by using the process of prototyping, testing and redesigning: the iterative process.

PRACTICE QUESTIONS: learning from existing products and practice

Design Engineering

AS Level

1 a) Explain what is meant by 'blue-sky' thinking in relation to the creation of innovative ideas.

  b) Explain what is meant by 'incremental innovation' in relation to the evolution of engineered products.

2 Explain the difference between ergonomics and anthropometrics relating to the design of engineered products.

3 Select an engineered product and discuss how technology has influenced its development. Use examples of the product through its development to support your discussion.

4 Social media is a huge influence on consumer choices. Discuss the disadvantages and advantages to both the manufacturer and consumer of using this marketing strategy. Use examples to support your answer.

5 What factors need to be considered when evaluating an existing product to assess its suitability in informing the development of a new product?

A Level

1 Describe a disruptive technological advancement and explain how it impacted on the design of engineered products which followed.

2 Describe two reasons why the invention of the microchip was necessary before computers could significantly develop in complexity and power.

3 Give five applications which rely on radio waves for communication, other than entertainment broadcasting.

4 Explain the impact on users of being permanently connected to the internet through their smartphones. Refer specifically to lifestyle, culture and society.

5 Explain the wider implications of planned obsolescence within the technology industry. Give examples to support your answer.

Fashion and Textiles

AS Level

1 Explain how the emergence of nanofibres and conductive dyes have been used to enhance fashion and textile products. Use industrial examples to support your answer.

2 Some fashion and textile designs may become iconic, while others are known as fast fashion. Compare the two terms and discuss the advantages and disadvantages of each. Give an example in each category.

3 Select a fashion or textiles product and discuss how technology has influenced its development. Use examples of the product through its development to support your discussion.

4 Social media is a huge influence on consumer choices. Discuss the disadvantages and advantages to both the manufacturer and consumer of using this marketing strategy. Use examples to support your answer.

5 What factors need to be considered when evaluating an existing product to assess its suitability?

A Level

1 Select a past or current trend and explain how it has influenced fashion and textile products used today. Make reference to materials, technological developments and aesthetics.

2 New technologies have had a major impact on different areas of the textile industry in the last 50 years. Identify some key changes and use products to support your answer.

3 Examine the product lifecycle of polyester using the four stages. Explain how its use has been maintained over the last 50 years. Include references to fashion and trends.

4 Explain the wider implications of planned obsolescence within the fashion and textiles industry. Give examples to support your answer.

Product Design

AS Level

1 Explain how the emergence of new technologies or materials have been used to enhance consumer products. Use examples to support your answer.

3 Select a product and discuss how technology has influenced its development. Use examples of the product through its development to support your discussion.

4 Social media is a huge influence on consumer choices. Discuss the disadvantages and advantages to both the manufacturer and consumer of using this marketing strategy. Use examples to support your answer.

5 What factors need to be considered when evaluating an existing product to assess its suitability?

A Level

1 Select a past or current trend and explain how it has influenced products used today. Make reference to materials, technological developments and aesthetics.

2 New technologies have had a major impact on different areas of the design industry in the last 50 years. Identify some key changes and use products to support your answer.

3 Examine the product lifecycle of a consumer product.

4 Explain the wider implications of planned obsolescence within the design industry. Give examples to support your answer.

3   Implications of wider issues

3.1 What factors need to be considered while investigating design possibilities?

Any new product that is created will have an impact on the environment through the use of raw materials and energy. Historically, little thought was given to the impact on the environment when we used materials and manufactured products – it was more important to keep costs down; and advertisers encouraged consumers to buy more, resulting in more products being produced. It is only recently that designers, consumers and manufacturers have realised the impact that comes with the creation of new products and systems and the social responsibility we have. New legislation and regulations arising from the global consumption of natural resources and creation of waste and pollution have also had an influence on the design and manufacture of products. As some materials and resources become scarcer we have begun to think about disposal and designers are exploring and developing ways in which we can adopt a cradle-to-cradle approach in designing.

Many designers encourage the use of environmentally friendly and sustainable methods of manufacture and materials. Products that can be easily disassembled at the end of their life are being developed, as we move towards adopting a circular economy where products can be put back into the system to be regenerated, then reused or recycled to reduce waste.

Consideration of lifecycle assessment (LCA) at all stages of a product's life (A Level)

Lifecycle analysis is a raising of awareness of environmental issues through a detailed examination of the lifecycle of products and was introduced in the late 1960s and early 1970s. Current practice involves a lifecycle inventory (LCI) where detailed data is compiled followed by a **lifecycle assessment (LCA)**, which interprets and evaluates the environmental impact of a product from 'cradle to grave' – from the extraction of raw materials required to manufacture the product to end of use and disposal.

Products, services and processes all have a lifecycle. For products, the lifecycle begins when raw materials are extracted or harvested. Raw materials then go through a number of manufacturing steps until the product is delivered to a customer. The product is used, then disposed of or recycled. Traditionally, product designers have been concerned primarily with product lifecycles up to and including the manufacturing step. Increasingly, product designers must consider how their products might be recycled, or components re-used. They must consider how consumers will use their products and what environmental issues might arise.

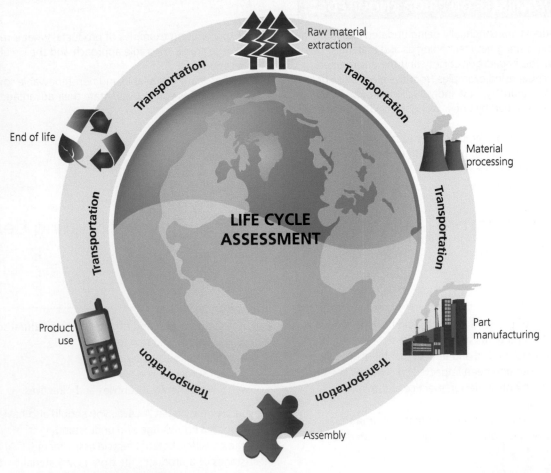

Figure 3.1 **Lifecycle assessment**

In order to make an assessment, the energy needs in the life of a product or system are analysed. The table below covers stages that need consideration. The assessment can be used to decide whether an idea is viable enough for it to be made into a product by thinking about the amount of energy needed and the possible environmental implications.

Table 3.1 **Analysing the energy needs in the lifecycle of a product or system**

Stage	Considerations
Acquisition of raw materials	All products or systems are created from raw materials. Consider the energy needed to extract oil, ores and timber. Look at the environmental impact of mining, **deforestation** and other issues related to the extraction of raw materials.
Transporting raw materials	Consider how raw materials are transported nationally and internationally and examine the environmental impact of, for example, oil tanker disasters and pollution of the air by fuel emissions. Using electric vehicles is cleaner for road users but the generation of electricity to recharge vehicle batteries impacts on the environment.
Processing raw materials	Consider the energy requirements and environmental effects of transforming raw materials by chemical or physical processing methods, for example, smelting and converting ores into usable materials, making polymers from oil.
Manufacturing the product	Most products require machine processing. The manufacturing industry requires energy for machines, lighting, heating, etc. Textile products are often dyed during manufacture and the chemicals used may have an environmental impact. Often manufacturing doesn't take place in the same area as material processing. Transporting materials, components and completed products for distribution involves considerable energy use and impacts on the environment.
Using the product	Some products require no further energy in usage. Many products, such as cars, washing machines and electrical items use significant amounts of energy. Some products, such as milk bottles, are reused; energy is used for cleaning before refilling. Detergents used may have an environmental impact.
Disposal and recycling	The collection of waste requires energy. Incineration centres use energy to dispose of waste, although many reclaim the energy created by incineration for useful purposes. Landfill systems may impact on the environment. Often recycling materials can use significant amounts of energy, but this will use less raw materials and conserve valuable natural resources.

Nearly all types of polymers can be recycled, however the extent to which they are recycled depends upon technical, economic and logistic factors. The UK uses over 5 million tonnes of polymers each year, of which it is estimated 29 per cent is currently being recovered or recycled. The growth in polymer bottle recycling has been incredible. Back in 2000, only around 12,000 tonnes of polymer bottles were recycled in the UK – by 2017 it was nearly 20 times that amount. We are also seeing rapid growth in infrastructure to reprocess polymer bottles – around half of the polymer bottles collected for recycling are now reprocessed in the UK.

Lifecycle of a polymer bottle:
- Most polymer bottles are made from polyethylene terephthalate (PET) polymer, produced from crude oil. Oil extraction releases greenhouse gases, and harms habitats and the environment. After the oil has been extracted it is transported to a refinery.
- Distilled oil is shipped to a manufacturer, who creates polymer pellets. Bottle producers then melt down these pellets into 'pre-forms' (parisons) that resemble polymer test tubes.
- In the bottling plant, polymer pre-forms are expanded to size by blow moulding, sterilised and filled with water (or other drinks) before being capped, labelled and packed into cases for shipping to supermarkets and shops.
- Consumers purchase bottled drinks. Sometimes they reuse/refill a bottle but it will end up either being recycled or disposed of in landfill. In recycling, polymer bottles are shredded and washed, then melted down and reformed into pellets.

Figure 3.2 **Polymer bottles pressed and packed for recycling**

It is estimated that around 60 million polymer bottles were recycled in Europe in 2014 – just over half of all bottles used. Polymers can take hundreds of years to decompose. Polymer waste in oceans is a major environmental problem, polluting the water and threatening sea creatures and birds.

Many companies have realised the importance of LCA. Cars are manufactured at production plants. Raw materials, recycled materials and components enter the factory at one end of the enormous building. Completed cars are driven out of the opposite end and transported to their owners. During their useful life, most cars consume petrol or diesel and this pollutes the atmosphere. From time to time they need servicing and repairing. Eventually, after many years, a car cannot be repaired any further and it will break down for the last time. The car is taken to the scrap yard where useful parts are stripped from it. This may include a wide range of materials and components, including the alloy wheels, recyclable polymers and metals. The remainder of the car is crushed and melted down. The recycled metal can be used to manufacture new cars or other products.

Audi is an example of a manufacturer that has used LCA in the development of its cars. The Audi A6 model uses lighter metals, meaning that the car weighs less and uses less fuel than other models. Some of the materials used require higher CO_2 emissions during manufacturing, but by using recycled aluminium instead of steel, Audi is able to begin recycling materials from the very start of the production process. You can read more about Audi's LCA process at www.audi.com/content/dam/com/EN/ corporate-responsibility/product/audi_a6_life_cycle_ assessment.pdf.

Electrically propelled vehicles cause no local emissions. Of course, responsibility does not end at the electric power socket. In the use phase, CO_2 emissions caused by the generation and supply of electric power have to be included in the LCA, so no ecological benefits are made unless renewable energy is used to create this electricity.

ACTIVITIES

1 Produce a lifecycle assessment for the following products:
 a) denim jeans
 b) disposable coffee cup and lid
 c) fizzy drinks can.
2 Consider how choosing different materials or methods of providing energy can improve the lifecycle.

The source and origin of materials; and the ecological and social footprint of materials

Human activities (including using materials to manufacture products) consume natural resources such as energy and materials and produce waste and pollution. These activities produce what is known as a footprint, a mark that humans make on the planet. Sourcing and extracting materials can produce both an ecological and a **social footprint**.

- **Ecological footprint** refers to the impact we have on the environment and natural resources.
- Social footprint refers to the impact we have on other people.

Another common term used is carbon footprint, which is the total amount of greenhouse gases produced to directly and indirectly support human activity. This can be thought of as part of our ecological and social footprint. A carbon footprint is made up of two parts – the primary footprint and the secondary footprint.

- The primary footprint measures our direct emissions of CO_2 from the burning of fossil fuels, including domestic energy consumption and transportation.
- The secondary footprint is a measure of the indirect CO_2 emissions from the products we use.

You can find more information at the websites of the following organisations:

- Few Resources (www.fewresources.org)
- Earth Day Network (www.earthday.org)

Figure 3.3 The carbon footprint of human activity is calculated by considering all of the materials consumed and the wastes generated over a year

The World Wide Fund for Nature has estimated that in the UK, the average ecological footprint of a person is 5.6 hectares, three times what is considered to be sustainable. In other words, if everyone in the world consumed resources at this level we would need three planets to support us. We need to aim for a sustainable planet, one which replenishes all the natural resources used and is able to absorb all the waste produced. Designers and manufacturers are increasingly considering the sources of the materials we use with sustainability in mind as they design and develop products. The choices we make when we choose and source materials for products can have an effect on our ecological footprint. An understanding of the ecological and social footprint of materials is needed for the written examination. As part of your NEA project, you will also need to investigate and show your consideration of relevant factors, reflecting this in your designing. Be careful not to confuse a social footprint with a digital footprint.

Materials can be categorised into two types: **natural materials** and man-made or **synthetic materials**.

Natural materials
A natural material comes from plants, animals or the ground. Natural materials include wood, bamboo, wool and cotton, leather, stone, glass, metals such as copper, iron and gold, clay and porcelain and synthetic materials made from chemicals. Examples of synthetic materials include polymers or synthetic fabrics such as polyester, nylon and Kevlar.

Metals
Metals are found in ores that naturally occur in the Earth's crust, often formed millions of years ago. Metal production occurs all over the world, including the United Kingdom. The mining process used to extract metal ores can have a harmful impact on the environment, affecting the landscape, destroying wildlife habitats and creating waste. It also consumes large amounts of energy. Mining affects the lives of those who live and work near sites as dust and noise are created by the machinery and transport associated with mining activities. Mining affects lives and can even cause subsidence.

Extraction of metals can have further ecological, environmental and social consequences as it produces carbon monoxide and carbon dioxide. Aluminium is the most widely distributed metal. Its ore is called bauxite and is mainly found in Africa, the West Indies, South America and Australia. Bauxite is purified to make aluminium oxide and aluminium can be extracted through electrolysis, which uses a large amount of energy. Silver is extracted from its ore by smelting or chemical leaching. The smelting process emits sulphur dioxide gas, which reacts in the atmosphere to produce acid rain, affecting soil, rivers and lakes and harming fish and other wildlife.

Figure 3.4 Wood and gold nuggets are natural materials

Much of the world's gold supply is mined using open pit or underground mining tunnels. Gold tends to be found in rock that contains acid-generating sulphides and its mining produces more unwanted rock than the mining of other minerals. The process can cause chemical reactions that produce acids. Gold has been mined commercially in both Wales and Scotland. In the UK, gold prospecting can only take place with the explicit permission of the land or river owner and any activities that cause pollution can result in a criminal prosecution for environmental reasons.

Timber

Much of our furniture is made from wood, mainly oak or pine. But even furniture that is made from manufactured board comes from timber sources. The use of timber without replacement of trees that are cut down has led to deforestation, soil erosion, flooding, landslides and loss of habitats for wildlife. Natural timbers are classified into two groups: hardwoods and softwoods (see Chapter 5 for more detail on these categories of timber). Hardwoods such as mahogany and teak are commonly used in the production of garden furniture. However, use of these woods has a lasting environmental impact. More than 50 per cent of mahogany species in Asia, for example, are now endangered, as the time taken to replenish the trees is so great. Coniferous trees grow more quickly than deciduous trees, which means that softwoods are usually less expensive than hardwoods. It also means that commercial production of softwood timber is more sustainable, as trees that are harvested can be replaced more quickly. There are many different species of tree that are grown commercially to produce softwoods and designers can reduce the environmental impact of their products by considering softwoods rather than hardwoods when making material choices. For example, Douglas Fir is used for building, panelling and furniture.

Choosing timber from well-managed forests helps to minimise the ecological and social footprint of timber use and reduce the risks of destroying forests. The Forest Stewardship Council (FSC) ensures that harvesting of timber does not have an impact on the biodiversity or ecological processes of the world's forests. Timber that is FSC certified has been harvested in an environmentally responsible way.

Bamboo is becoming more popular as a construction material. It grows rapidly, up to 60 cm every day, and as such it can be harvested every four to five years, as opposed to 25–70 years for many other tree species.

As a grass, it continuously creates new shoots after harvesting without the need for replanting.

Man-made boards such as plywood or MDF (Medium Density Fibreboard) also offer environmental benefits because they can be made from lower grade timbers and sometimes waste wood.

Natural fibres/fabrics

Cotton

Most of the world's cotton is grown in the USA, China and India. Other cotton-growing countries are Brazil, Pakistan and Turkey. According to CottonConnect (an organisation that works with leading clothing manufacturers and retailers to promote and implement large-scale improvements in the environmental and sustainability of cotton production and processing), in 2017 only 12 per cent of the world's cotton was classed as sustainable. Cotton tends to be intensively farmed and uses large volumes of water (10,000 litres per kilogram of cotton), which puts pressure on resources. Cotton production also has **social impacts** – cotton-producing communities tend to offer little in the way of workers' rights.

Organisations like CottonConnect are introducing steps to reduce the impact of cotton production on both the environment and the people that work in the industry. For example, they are:

- identifying risks in the cotton supply chain and developing sustainability strategies
- creating a traceability system to enable tracing of cotton from farm to spinner
- working towards UN Sustainable Development Goals (for example, setting up farmer training programmes and support for water management)
- educating farmers and cotton-producing communities on fair working conditions, gender equality and human rights.

Organic cotton production, which uses non-genetically modified plants grown without the use of any synthetic agricultural chemicals such as fertilisers or pesticides, can also reduce the environmental impact of cotton production.

ACTIVITY

Visit the website for CottonConnect (www.cottonconnect.org) and read the article entitled 'The environmental and social footprint of cotton' which gives details of the impact of cotton production and steps being taken to reduce that impact.

Leather

Leather is now one of the world's most widely traded commodities. More than half the world's supply of leather raw material comes from developing countries, with China as the world leader.

Leather is a durable, long-lasting and biodegradable material commonly used in fashion products such as footwear and clothing and in many furnishings. Many people see leather as a by-product of the meat industry but in some instances animals are slaughtered for their skin alone. The leather-tanning process (that turns the animals' skin into usable leather material) also produces toxins that cause pollution. People for the Ethical Treatment of Animals (PETA) is an animal rights organisation focusing on the food industry, clothing trade, laboratory testing and the entertainment industry. It carries out various activities to raise awareness of and monitor animal rights issues. It estimates that from the millions of cows (cowhide is the most common material used for leather) slaughtered for meat each year, 20 per cent of the income from each cow is for the hide but the softest most luxurious leather comes from the skin of the newborn calf. We need to consider the ethical issues surrounding all materials we use. PETA promotes non-leather alternatives to shoes, clothing, bags, belts and wallets.

Papers and boards

Papers and boards are made from wood pulp, which comes from softwood trees such as spruce or pine and hardwoods such as eucalyptus and birch This means, all of the issues associated with the environmental and social impacts of harvesting timber can also apply to paper and board products. It requires approximately 12 mature trees to make one tonne of newspaper and around 24 trees to make one tonne of white copier paper. Designers and manufacturers can consider making greater use of recycled paper (which uses fewer trees) as opposed to virgin wood to reduce the environmental impact of paper production. Reducing the amount of paper and board used in the packaging of products can also help. Most consumers recycle paper and cardboard and understand the ethical reasons for this.

Synthetic materials

Materials can also be made from chemicals. These are called synthetic materials and they can also have ecological and social footprints similar to those associated with natural materials. Examples of synthetic materials include polymers such as PVC, acrylic and polypropylene or synthetic fabrics such as polyester, nylon, acetate and Kevlar.

Polymers

Polymers are everywhere in our world and can be synthetic or organic materials. They're usually made of long-chain molecules called polymers that are created from materials such as oil.

Polymer material needs to be sourced and go through several processes before it is 'workable' and can be made into the different stock forms, such as pellets, tubes or sheets ready to be manufactured. The main source of synthetic polymer is crude oil, although coal and natural gas can also be used. To make polymers we need hydrogen and carbon and the most convenient way of accessing them is by taking them out of oil. But hydrocarbons can also be made from methane, coal and biomass. Other possible raw materials for polymers are starch, cellulose, sugars and organic waste. Currently fossil fuels represent 99 per cent of the polymers raw material we use, but there is a growing interest in the use of biomass as supplies of oil diminish. Most polymers can be recycled up to six times.

In August 2008, a British waste expert predicted that due to high raw material prices, UK landfills may be mined and oceans scoured for polymers in the future. The futurologist Ray Hammond in his book *The World in 2030*, predicts that in the future oil will not be burnt away and wasted in energy and transport but reserved for 'high value processes and products such as polymers manufacturing … and energy trapped within the polymers can either be recycled or recovered and used for heat generation.'

Polymers can be found naturally and occur in such things as plants and trees from which cellulose, latex, amber and resin can be extracted, and animals and insects from which horn and shellac is obtained.

Figure 3.5 Nylon is a synthetic material

Biopolymers are polymers in which all carbon is derived from renewable feedstocks including corn, potatoes, rice, palm and wood cellulose. They are often used for disposable items, such as crockery, cutlery and packaging. They are also used in non-disposable applications including mobile phone casings and carpet fibres. New electroactive biopolymers are being developed that can be used to carry current. These polymers are not biodegradable at the end of their life but are made from sustainable resources.

Biopolymers can be divided into three main groups:
- polymers that are both bio-based and biodegradable
- bio-based or partly bio-based non-biodegradable polymers
- polymers that are based on fossil fuel resources and are biodegradable.

Material choice

You can see that the material choice you make can reduce your ecological footprint. You may like to ask yourself the following questions:
- Where did the material come from? Is it local?
- Did it require a lot of energy to extract it or to get it to you?
- Can it be replaced or replenished easily at the source?
- Was it recycled? Can it be recycled or reused easily at the end of its life?
- Is the material durable? Will it last a long time or will it need to be replaced?
- Is there an alternative material that you could choose which would be better?

The depletion of and effects of using natural sources of energy and raw materials

When we use natural sources of energy and raw materials we have an effect on the environment, which can lead to deforestation, rising temperatures, loss of biodiversity and extinction of animal and plant species.

The main sources of natural energy are the sun, wind and water, and we also can get energy from the earth from fossil fuels and natural gas.

Fossil fuels are concentrated organic compounds found in the Earth's crust. They are created from the remains of plants and animals that lived millions of years ago. Fossil fuels currently meet about 81 per cent of world energy needs. (http://data.worldbank.org/)

Although estimates of available reserves vary, it is considered by some that, at current annual rates of production, we have left, worldwide, about:
- 50 years of oil. Because we use oil to manufacture many materials, including polymers, we use oil at a much faster rate than either natural gas or coal. People have been expecting oil to run out within the next few years since at least the 1990s. No doubt it is currently getting scarcer, but current estimates by British Petroleum (BP) suggest we will not actually run out for 50 years.
- 50 years of natural gas with current usage levels. Some experts believe that current natural gas deposits fill around 6,000 trillion cubic feet.

Shale gas and oil are natural sources of energy trapped in rocks (formed millions of years ago from decaying plants trapped in clay) and are extracted by a controversial process called hydraulic fracturing (or fracking).

Most of the coal deposits have not been tapped. The decline of the coal mining industry in the UK means that many coal seams are lying undisturbed. If we carry on using coal at the same rate as we do today, we could have enough coal to last well over a thousand years. However, if other fossil fuels run out, particularly oil, the use of coal to provide energy may increase, reducing that time span.

Using fossil fuels to generate energy has a significant environmental impact. Burning fuels produces waste products. Gases such as sulphur dioxide, nitrogen oxide and other volatile organic compounds can have a harmful effect on the environment. The burning of fossil fuels creates carbon dioxide, which contributes towards global warming.

Figure 3.6 **Fossil fuels are an energy source**

The generation of energy, particularly the production of electricity, has contributed greatly to atmospheric pollution. Coal-, gas- and oil-fired systems emit large amounts of carbon dioxide, which many scientists agree is a significant contributory factor to global warming, but have benefits along with an impact on the environment. The table below compares the main natural energy sources, including renewable and non-renewable energy sources.

Renewable energy systems that use water, solar power and wind power are being increasingly used to provide large-and small-scale, effective, environmentally acceptable power. Over the past decade, world wind power capacity grew more than 20 per cent a year. In 2015, Denmark produced 42 per cent of its electricity from wind turbines, which is the highest figure recorded worldwide. Between 2009 and 2014, solar panel prices dropped by three-quarters, helping global photovoltaic (PV) installations grow 50 per cent per year. Contrary to popular belief, solar panels will work when it's cloudy and raining. Transportation will move away from oil as electric vehicle fleets expand rapidly and bike- and car-sharing spreads. Bike-sharing programmes have sprung up worldwide in recent years. More than 800 cities in 56 countries now have fully operational bike-share programs, with over 1 million bikes.

Table 3.2 **The advantages and impact of various energy sources**

Method	Description	Advantages	Impact
Gas/Coal/Oil	Fuel is burnt to generate heat, heats water to generate steam, steam turns turbines, turbines turn generators and electricity is distributed.	Readily available, ease of transport of fuel, gas-fired stations very efficient	Air pollution including carbon dioxide and sulphur dioxide, finite resources, visual impact of extraction, large stocks needed (coal), will eventually run out as non-renewable
Hydro-electric	Dam is used to trap water, water released turns turbines, turbines turn generators and electricity is distributed.	Very low cost once dam built, no air pollution, reliable, up to full power very quickly	Can impact on environment (flooded area, reduced flow at base), initially expensive
Wind	Blades designed to catch wind, blades turn turbines using gears, turbines turn generators and electricity is distributed.	No fuel needed, no waste or greenhouse gases, can be used in remote areas or offshore to minimise impact	Unreliable, unsightly effect on the landscape, some designs are noisy, can harm flocks of birds, a large space is needed
Solar photovoltaic	Photovoltaic cells convert light to electricity.	Low cost after initial outlay, no pollutants or waste, used on small or large scale in remote areas	Expensive initial cost, unreliable, storage system needed
Tidal barrages	Barrage built across river estuary, turbines turn as tide enters (and when tide leaves), turbines turn generators and electricity is distributed.	Low cost after initial outlay, no pollutants or waste, predictable	Very expensive initial cost, environmental costs – can damage habitats, only generates power at set times during the day
Wave	Motion of waves forces air up cylinder to turn turbines, turbines turn generators and electricity is distributed.	Low cost after initial outlay, no pollutants or waste, suitable for remote coastal areas	Unreliable, hostile environment, high maintenance, not a large power output
Geothermal	Cold water pumped underground through heated rocks, steam turns turbines, turbines turn generators and electricity is distributed.	No pollutants or waste, minor cost of pumping, resource 'free', very small stations, no negative visual impact	Only work in certain locations, possibility of gas emissions
Biomass	Fuel (wood, sugar cane, etc.) is burnt to generate heat, heats water to generate steam, steam turns turbines, turbines turn generators and electricity is distributed.	Readily available fuel, can use waste materials, low-cost process	Air pollutants, requires large amounts of fuel, can be seasonal

Raw materials

Timber

In recent years, deforestation has become an issue which has affected our forests and jungles but its effect is wider, increasing the effects of climate change and affecting biodiversity and resulting in species becoming extinct. When forests are cleared, they affect the levels of carbon dioxide in the air. Deforestation accounts for about 11 per cent of global greenhouse gas emissions caused by humans – comparable to the emissions from all of the cars and lorries on Earth combined. (The World Carfree Network (WCN) states that cars and trucks account for about 14 per cent of global carbon emissions, while most analysts attribute upwards of 15 per cent to deforestation.) Illegal logging occurs when timber is harvested, transported, processed, bought or sold in violation of the law. It exists because of the demand for timber and paper-based products. At the Earth Summit in Rio de Janeiro in 1992, a Statement of Forest Principles was adopted which sets out a global framework for the management, conservation and sustainable development of forests. The main criteria are:

- maintenance and appropriate enhancement of forest resources
- maintenance of forest ecosystems for health and vitality
- maintenance and encouragement of productive functions of forests
- maintenance, conservation and appropriate enhancement of biological diversity in forest ecosystems.

Figure 3.7 The Forest Stewardship Council® (FSC®) and the Sustainable Forestry Initiative (SFI) both have certification schemes for wood from sustainably managed forests

In the UK, the Forestry Standard (UKFS) and its associated guidelines set out the criteria and standards for the sustainable management of its forests based on this. This criteria can be found at the Forestry Commission website (www.forestry.gov.uk).

The Forest Stewardship Council® (FSC®) and the Sustainable Forestry Initiative (SFI) both have certification schemes for wood from sustainably managed forests. In 2012, only 7 per cent of the world's forest areas were certified to FSC standards. It can make it hard to be sure you are sourcing wood that has been ethically sourced. However, companies like IKEA have been working to improve the growth of FSC-certified wood supply and promote better forest management, aiming to contribute to ending deforestation. Swedwood is a subsidiary of IKEA which manages forests and operating sawmills, components, including man-made boards, and furniture production for IKEA.

Figure 3.8 One of the largest furniture retailers, IKEA, used 13.56 million cubic metres of solid wood and wood-based board materials, not including paper and packaging, meaning IKEA alone uses almost 1 per cent of all wood used commercially around the world

Fabrics

Bamboo as a fabric isn't always as natural as it sounds, as sustainable bamboo needs to be harvested correctly. Once Guada bamboo is more than seven years old, it starts to dry and lose its **mechanical properties**, which means it is suitable only for fuel pellets and charcoal. The fibres of bamboo are very short (less than 3 mm). To make fabric, the fibres are broken down with chemicals and extruded through mechanical spinnerets. Pandas and mountain gorillas depend on bamboo shoots for their diet, so bamboo also needs to be from managed plantations to avoid endangering species.

An alternative option to bamboo is eucalyptus, which provides good quality fibres with less waste, making it a better option for the environment than bamboo. Eucalyptus as a fabric material is made from the pulp of eucalyptus trees.

Cotton farming relies on pesticides derived from petrochemicals. According to the Organic Cotton website, 'Only 2.5% of the world's farmland is used to grow cotton, yet 10% of all chemical pesticides and 22% of all insecticides are sprayed on conventional cotton.' The use of pesticides reduces biodiversity and affects ecosystems and also contaminates water supplies. When pests exposed to synthetic pesticides build up a resistance to them, farmers have to use more pesticides to grow the same amount of cotton. Organic cotton is grown using methods and materials that have a lower impact on the environment by maintaining the fertility of the soil and reducing the use of synthetic pesticides. To source cotton that uses less pesticides look for the 100 per cent organic cotton on the label.

For more details, visit the Organic Cotton website (https://organiccotton.org) and look for information on the risks of cotton farming.

Figure 3.9 Asian bamboo forest and ripe cotton ready for harvesting

Metals

The mining sector is responsible for some of the largest releases of heavy metals and other air pollutants, including sulphur dioxide, into the environment. Heavy metals are hazardous to human health in several ways. For example, lead, cadmium and mercury have all been associated with neurological issues and silica dust is associated with lung cancer and other health problems.

Different types of mining can create different environmental issues. Open pit mining for iron ore requires many pits to be dug, as the ore is found in small concentrated amounts; creating the mines involves moving vegetation and rock. Underground mining can cause subsidence and release toxic compounds into the local environment. Gold mining can pollute rivers with sulphuric acid and causes mercury pollution.

Some types of mining have less of an environmental impact. For example Bauxite, the ore used to make aluminium, is found near the surface, which means that there is less damage to land, and topsoil cleared before mining begins can be stored and replaced. The Aluminium Association (www.aluminum.org/) believes bauxite reserves will last for centuries, although it has been suggested elsewhere that resources may run out within 70 years if global demand continues at its current rate. Aluminium is widely used as it is both lightweight and corrosion resistant. It is being used in ways to benefit both consumers/users and the environment (for example, Jaguar car bodies). The material's high strength-to-weight ratio creates lighter car bodies, which in turn lowers fuel consumption. It also benefits from lower corrosion of car bodies and lower costs of manufacturing and, as a result, is more likely to be considered as an option by designers. Aluminium also has the advantage of being recyclable. Aluminium scrap is melted down, requiring only 5 per cent of the energy needed to make new aluminium using ore, although some of the aluminium will be lost during the process.

Our smartphones and other technological equipment rely on metal elements such as copper, aluminium and steel, but also less well-known materials such as indium, which is an important part of touch screens, flat screen TVs and solar panels. It has been suggested that the Earth's known supply of indium will have been used up within the next ten years.

As consumers, we need to all find responsible ways of recycling our used electronic products, and designers and manufacturers need to make sure they create products that allow for easy recycling. Supply and demand will affect us – as material sources deplete, the price and availability of the materials used in many products in our lives will increase. So, while no material is so scarce that it is about to disappear completely, we should worry that many could soon become out of reach and should look to find alternatives or create ways of using less, and allowing for ease of disassembly and recycling.

Electronic waste or e-waste refers to discarded electrical or electronic devices. According to the United Nations University Institute for the Advanced Study of Sustainability (UNU-IAS), 20–50 million metric tons of e-waste are created every year. The majority of e-waste in 2014 (almost 60 per cent) was discarded kitchen and laundry equipment. Devices such as mobile phones, personal computers and printers accounted for 7 per cent of e-waste in 2016. For every one million smartphones recycled, the EPA (Environmental Protection Agency USA) states that 16,000 kg of copper, 350 kg of silver, 34 kg of gold, and 15 kg of palladium can be recovered.

Polymers

Polymers are part of the solution to halting climate change by replacing traditional materials in cars, saving weight and cutting fuel consumption. Fuel consumption of cars is reduced by 12 million tonnes in Europe and CO_2 emissions by 30 million tonnes.

But production and disposal of conventional polymers can cause major environmental problems. Polymer pollution ends up in our streams and oceans, resulting in the deaths of millions of animals annually. Globally nearly 300 million tons of polymer is produced every year, but more than 8 million tons of polymer is thrown away and ends up in our oceans every year. It is estimated that around 50 per cent of polymer is used just once and thrown away. Polymer debris in the oceans pollutes rivers, beaches and oceans. As time passes, we know that polymers will eventually photo-degrade (break down into smaller and smaller fragments by exposure to the sun) but even then they are still polymers. At sea, the polymer fragmentation process occurs as well, due to wave, sand action and **oxidation**. Estimates for polymer degradation at sea ranges from 450 to 1,000 years. For more information see www.plasticoceans.org/the-facts/.

Figure 3.10 Electronic waste

Polymers are accumulating in the world's oceans in areas of concentrated litter from humans which has made its way into the seas. This litter is known as marine debris. One of the beskt known areas of marine debris is the Great Pacific Garbage Patch, which is thought to have formed gradually as a result of ocean pollution gathered by the currents. A similar patch has also been discovered

in the Atlantic. Toxins in polymers affect the ocean, as the chemicals are ingested by animals which live there. It is thought they could be passed on from contaminated fish to humans through the food chain.

Figure 3.11 Waste and pollution washing on the shores of the beach in the city of Colon in Panama, April 2015

A 19-year-old student recently unveiled plans to create an Ocean Clean up Array that could remove 7,250,000 tons of polymer waste from the world's oceans. Boyan Slat devised a system through which, driven by the ocean currents, the polymers would build up in a specific area, reducing the theoretical clean-up time from millennia to mere years. In February 2013, he dropped out of his aerospace engineering course to start The Ocean Cleanup. The first prototype system was deployed in June 2016 and The Ocean Cleanup now prepares to launch the first working pilot system in late 2017. Read more at www.theoceancleanup.com/.

Biopolymers are part of the solution to the polymer problem. The production and use of biopolymers is seen as sustainable when compared with polymer production from petroleum, because it requires less fossil fuel for production and creates less emissions and air pollution. The use of biopolymers can also result in less waste than oil-derived polymers, which can take hundreds of years to degrade. But petroleum is often still used as a source of materials and energy in the production of biopolymers. Biopolymers result in a 42 per cent reduction in carbon footprint, but producing biopolymers in large quantities could contribute to deforestation, so managed plantations and forests would need to be set up. Biopolymers are therefore not the only solution to the problem of polymer pollution in our environment. Changing the way we think about, use and dispose of polymers is essential in effecting longer-term change. Recycled polymers can also reduce greenhouse gases and mean fewer polymers end up in landfill or polluting the environment.

We are all aware of social and ethical issues with reports of workers in factories and mines around the world with no real safety equipment, handling chemicals and machinery, working long hours with low pay and child workers being discovered working in factories.

Planned obsolescence

Planned obsolescence is planning or designing a product with a limited useful life, so it will become obsolete (that is, unfashionable or no longer functional) after a period of time. It is sometimes known as 'built-in obsolescence'.

Products that are obsolete often end up as landfill and create pollution issues, as chemicals from the products can escape and damage the ground or atmosphere. Electrical products are often received for recycling by people in low-wage economies, where they burn and melt the polymer to access the components and/or precious metals. This can create gases and pollution which harm the environment and are a health hazard. The processes to recycle the products are often more expensive than landfill but it's important to use less raw materials and move towards a cradle-to-cradle approach where we reuse materials.

The morals and ethics of an economic system and a consumer society that encourage high levels of consumption are increasingly under question, given the environmental implications. Continually replacing instead of being able to repair or upgrade items contributes to waste and pollution and this makes the practice of planned obsolescence difficult to support.

Light bulbs are one of the best known, earliest examples of planned obsolescence. In the 1920s, light bulbs became mass-market, as companies made bulbs disposable, putting replacement costs on to their customers and ensuring continued need for new bulbs. This practice cropped up in other industries, although the term 'planned obsolescence' wasn't widely used until the 1950s, with competition between car companies enticing consumers to buy new cars. Many white goods such as refrigerators and washing machines have parts that wear out after about five years. Planned obsolescence is evident in examples relating to durability, where key parts wear out or fail, and aesthetics

where upgrades make older product versions seem less stylish or fashionable. Clothing and accessories are often subject to planned obsolescence as new styles, trends and patterns are introduced to entice consumers, making last year's clothing no longer as desirable to wear. This has become known as **fast fashion**. Children grow out of their clothes quickly so the highest quality materials aren't always suitable or required.

Medical equipment such as sterile surgical gloves or needles are disposable and cannot be reused. They have a planned life for reasons of health and hygiene. Other products have a limited working life to ensure that safety factors can be incorporated into later versions of the product, so sometimes planned obsolescence can benefit consumers.

The design of technological items such as smartphones is frequently improved with new features, and many users want the latest design with the latest features to replace their old model. These handsets often get discarded after a couple of years' use. Screens or buttons break, batteries die, or their operating systems and chargers can no longer be upgraded. This can also benefit consumers, with innovation and improved technologies in smartphones, faster processors, better cameras and functionality. Many customers opt to pay more for products that have better craftsmanship, stronger materials and a greater durability and resale value.

Figure 3.12 **The circular economy**

A related issue is that the price of repairing a product can sometimes be more than that of replacement, due to the products being cheaply produced for a mass market and the cost of repair being high due to the need for replacement parts. Manufacturers may choose to make spare parts so expensive that consumers choose to buy a complete new product instead. This is part of the circular economy debate, which suggests that products can be designed in a way that enables parts to be replaced more easily and cheaply. (See www. ellenmacarthurfoundation.org.)

Buying trends

The way we buy is changing. Technology and our lifestyles are enabling us to use smart and synchronised devices to save time and money. Mintel's research highlights that consumers are already thinking about how their devices work together as part of their research and purchasing process. In the UK, as many as 76 per cent of potential TV buyers say they are interested in a TV with the ability to wirelessly stream content from other devices and a further 28 per cent say they would pay more for this feature. Meanwhile, over a third (34 per cent) of UK fridge shoppers expect or would pay more for a bar code reader synced to online shopping. (See 2015 buying trends at www.mintel.com/.)

The subscription model, where people pay to use a service or product, is changing consumers' approach to ownership. We may see an increase in renting items such as clothes and other products from high street brands; we already lease many technological devices and cars.

The internet has enabled widespread reviews of companies, products and services. Online reviews and comments by users on their buying experiences and product reliability has become a valuable source of information, which many customers check before making a purchase.

Crowdfunding (see Chapter 1) is helping to make deliveries cheaper and can tap into a freelance workforce or make use of journeys consumers or delivery companies are already making. Social media has continued to become more visual, with pictures and video becoming more important, encouraging us to purchase and starting to provide consumers with ways to buy directly via social channels. Burberry live streamed its catwalk shows over the internet in 2016, allowing some customers to click and buy certain garments as soon as they see them on the catwalk. Television is also set to play a big role in the future with more retailers starting to use tools like live-streaming to engage consumers.

Technology has changed the way many retailers and brands have operated over the last decade. Some of the key technologies that will have an effect on buying trends are:

- The **Internet of Things (IoT)** could affect retailers in many ways, allowing them to sell connected products or develop automated services. Some retailers are already using their stores to show customers what a smart home could look like, with the aim of inspiring and educating. We could all get used to the idea of our fridge ordering milk for us.
- Virtual reality is now commonly used in marketing. Fashion retailers have created virtual catwalks and travel companies are creating virtual brochures.
- Robots are used throughout the retail supply chain. They are an increasingly common sight in warehouses and several retailers are working on building robots that can assist with and speed up the purchasing process. Driverless courier vehicles and drones are being explored.
- The first devices are starting to come on the market that seek to bring artificial intelligence (AI) technology into the home. This can be used to answer questions, turn on lights and order a taxi. AI-powered shopping assistants are being explored and specialised earphones with an AI personal trainer are now available to order. These will help shoppers navigate stores, improve their fitness and lifestyle and ask their phones for advice on shopping and health. The use of chatbots (virtual assistants) are also on the rise and retailers are using them for customer service and ordering.

Figure 3.13 **Virtual reality internet shopping**

Environmental incentives and directives

At a meeting in Paris in 2016, the world's governments agreed to limit global temperature rises to no more than 2°C and preferably less than 1.5°C above pre-industrial levels. That agreement has now come into force. A UN review of the pledges to cut carbon emissions published in 2016 found that countries had fallen short of even keeping future temperature increases below 2°C. The Paris agreement binds both rich and poor nations to take action to combat climate change. World leaders are becoming more aware of the need for us to unite globally to tackle climate change

There has been concern in 'green' circles over Brexit, because 70 per cent of green regulation in the UK stems from the EU. However, most of the EU directives relating to a sustainable built environment have been transposed into UK law and UK climate change targets are likely to remain unchanged.

The EU Renewable Energy Directive (RED)

This directive requires the UK to generate 15 per cent of its energy from renewable sources by 2020 – up from just 3 per cent when the directive was adopted in 2009. This includes both electricity and energy used in heating and transport.

To hit the overall target, the UK is expected to need to generate 30 per cent of its electricity and 12 per cent of its heating energy from renewable sources. It is also bound by a sub-target for transport, requiring 10 per cent to come from renewable sources. In the electricity sector, the RED is one of the reasons why the Government has backed the rapid expansion of renewable power sources such as wind and solar farms and biomass power plants, paid for through subsidies on energy bills.

In 2010, just 7 per cent of UK electricity came from renewable sources; by 2014 that had risen about 18 or 19 per cent. In transport, about 5 per cent of road fuels came from biofuels in 2017. Ministers are looking to increase that level and are pursuing plans to ban all new petrol and diesel cars and vans from 2040.

In the longer term, the 2030 emissions reductions targets will require a continued switch to low-carbon energy sources and greater energy efficiency. Withdrawing from the EU might allow the UK to take an easier route on heat and transport policies in the short term, if Britain was no longer obliged to hit the RED target for 2020. Post-2020, Brexit appears unlikely to make a difference to UK energy policy, because Britain's own unilateral Climate Change Act actually imposes even tougher requirements for cutting carbon emissions. Under the Act, the UK must cut its carbon emissions by 80 per cent on 1990 levels by 2050.

Wood from controlled sources

By checking compliance of wood from Controlled Sources with the regulations on the exploitation of forests it is possible to ensure that the EU Timber Regulation (EUTR) is being fulfilled; both the PEFC and FSC mark this wood as coming from Controlled Sources. The FSC (Forest Stewardship Council) indicates reliably identified forest management when it comes to producing wood by-products. The PEFC (Programme for the Endorsement of Forest Certification) is a professional and international non-profit organisation which promotes sustainable forest management and the certification of forest-origin raw materials. At present it is made up of 38 national organisations around the globe and it came into existence in 1998 as a voluntary initiative of the forest private sector.

Other environmental targets

The UK must hit a target of recycling 50 per cent of its household waste by 2020. The EU is also considering imposing targets requiring the UK and most other EU nations to recycle 65 per cent of their rubbish by 2030. Seven EU nations which are significantly behind on their 2020 targets could be given more lenient 2030 goals. Households around the UK are now asked to separate out recyclable items from their rubbish. Recycling rates in England have increased from just over 10 per cent in 2000 to about 44 per cent in 2017. However, the growth rate has stagnated in recent years and some councils are expected to have to step up their efforts to hit the targets, asking households to separate out different types of recycling, such as paper, glass, metal and polymer, as well as separate collections of food waste. Recycling rates are similar in Scotland to England, while in Wales they are already in excess of 50 per cent. In England, leaving the EU could give ministers leeway to set more lenient targets.

The EU's Ambient Air Quality Directive has set a series of targets to limit the levels of dangerous pollutants in the air. One of the most challenging has proven to be for nitrogen dioxide. The Department for Environment, Food and Rural Affairs brought out a new strategy in 2016 to try to improve air quality with new clean air zones in five UK cities, as well as London which has a similar plan. Old diesel lorries, vans and taxis will face charges for driving in the zones.

While the UK will continue to be bound by international agreements on climate change, air pollution and wildlife protection, these treaties are normally expressed in less prescriptive terms and tend to be a lot less enforceable than EU law.

Any designer and manufacturer needs to take into consideration the environmental directives when choosing materials or energy sources. The flower is the symbol of the European Ecolabel scheme and identifies products and services that are a genuinely better choice for the environment. It is designed to help manufacturers, retailers and service providers to get recognition for good standards and for purchasers to make reliable and informed choices.

Figure 3.14 Recognised throughout Europe, the EU Ecolabel is a voluntary label promoting environmental excellence which can be trusted. Backed by all EU Member States and the European Commission, it is Europe's and the UK's official environmental label

The RoHS (Restriction of Hazardous Substances) directive sets out to reduce the use of certain hazardous materials, including lead, mercury and cadmium, in the manufacture of electronic equipment. Hazardous chemicals can be found in electronic components, solder, batteries, paints and polymers, and so on. The aim is to reduce the problem of environmental pollution caused by the unethical disposal of unwanted products, such as throwing away used batteries in normal household rubbish.

The average UK citizen will dispose of over 3 tonnes of electrical and electronic products in their lifetime. The WEEE (Waste Electrical and Electronic Equipment) directive requires all manufacturers and producers to take responsibility for what happens to the products they sell at the end of their lives. In practical terms, this means that retailers of electronic products must now provide a free take-back service for customers to hand in the unwanted product they are replacing. The retailer must then dispose of the product at an approved treatment facility. You may have spotted in your local supermarket the collection bin for old batteries – all battery retailers must now provide this take-back facility.

In January 2018 the UK Prime Minister set out the government's strategy on plastic with plans to tax single-use plastic packaging and assist supermarkets to create plastic-free aisles. (https://www.packagingnews. co.uk/top-story/ government-targets-plastic-part-25-year-plan-11-01-2018)

You can find more details of regulations that must be followed in design and manufacture in Chapter 9.

KEY TERMS

Biopolymers – polymers in which all carbon is derived from renewable feedstocks, including corn, potatoes, rice and wood cellulose.

Deforestation – the removal of a forest or trees or stand of trees where the land is then not used for forests, but for urban use or farming. The most concentrated deforestation is in the rain forest.

Ecological footprint – the impact we have on the environment and natural resources.

Fast fashion – a term used to explain that designs move from catwalk to high-street stores quickly to capture current fashion trend.

Internet of Things (IoT) – the internet of physical devices, buildings and other items embedded with electronics and software that allow objects to collect and exchange data. Cars, lights, refrigerators and more appliances can all be connected to the IoT.

Life cycle assessment (LCA) – an assessment of all stages of a product's life from raw material to disposal.

Mechanical properties – characteristics that indicate the behaviour of a material under pressure (force), such as brittleness, flexibility, ductility, toughness and tensile strength. Mechanical properties determine the range of usefulness of a material.

Natural material – any product or physical matter that comes from plants, animals or the ground.

Planned obsolescence – sometimes known as 'built-in obsolescence', the process of designing products to go out of fashion or no longer function after a specific period of time.

Social footprint – the impact we have on other people.

Social Impact – how the use of a matenal or manufacturing method could impact on people's lives and the community

Synthetic material – a material made from chemicals.

KEY POINTS

- As materials and resources become scarcer, designers are adopting a cradle-to-cradle approach in designing, with the use of environmentally friendly and sustainable methods of manufacture and disassembly.
- A circular economy is one where products can be put back into the system to be regenerated, then reused or recycled to reduce waste.
- Lifecycle analysis looks at the extraction or harvesting of raw materials, the manufacturing and delivery of the product to a customer and the product's use and how it is disposed of or recycled.
- Sourcing and extracting materials can produce both an ecological and a social footprint.
- A natural material comes from plants, animals or the ground; synthetic materials are made from chemicals.

- When we use natural sources of energy and raw materials we have an effect on the environment. It can lead to deforestation, rising temperatures, loss of biodiversity and extinction of animal and plant species.
- Renewable energy systems that use water, solar power and wind power are being increasingly used to provide large and small-scale, effective, environmentally acceptable power. Over the past decade, world wind power capacity grew more than 20 per cent a year.
- Planned obsolescence is planning or designing a product with a limited useful life, so it will become obsolete (that is, unfashionable or no longer functional) after a period of time. It is sometimes known as 'built-in obsolescence'.
- The way we buy is changing and technology and our lifestyles are enabling us to use smart and synchronised devices to save time and money.

3.2 What factors need to be considered when developing design solutions for manufacture? (A Level)

LEARNING OUTCOMES

By the end of this section you should have developed a knowledge and understanding of:
- The responsibilities and principles of Design for Manufacturing (DFM)/Total Quality Management (TQM), including:
  - planning for accuracy and efficiency through testing and prototyping
  - being aware of issues in relation to different scales of production
  - designing for repair and maintenance
  - designing with consideration of product life.

Design Engineering

By the end of this section you should have developed a knowledge and understanding of:
- product lifecycle management and engineered lifespans, considering system compatibility, the need for maintenance of machinery, product support and end of life (EOL).

Fashion and Textiles

By the end of this section you should have developed a knowledge and understanding of:
- issues related to product lifecycles that extend useful product life, such as:
  - products standing the test of time in terms of durability and style

  - maintenance and aftercare
  - re-working and recycling systems.

Product Design

By the end of this section you should have developed a knowledge and understanding of:
  - product lifecycles that extend useful product life through planning for and consideration of maintenance, repair, upgrades, remanufacture and recycling systems.

By the end of this section you should have developed a knowledge and understanding of:
- how environmental factors impact on:
  - sourcing and processing raw materials into a workable form
  - the disposal of waste, surplus materials and components, by-products of production including pollution related to energy
  - cost implications related to materials and process.
- sustainability issues relating to industrial manufacture, including:
  - fair trade and the Ethical Trade Initiative (ETI)
  - economic issues and globalisation
  - material sustainability and optimisation, availability, recycling and conservation schemes, such as exploring:
    - the impact and use of eco-materials
    - how materials can be up-cycled.

Responsibilities and principles of Design for Manufacturing

Design for Manufacturing (DFM) is the method of designing products to be easy to manufacture at the lowest cost. It considers the choice of raw material and the manufacturing of all the parts that will form the product after assembly. It is concerned with reducing overall part production cost and minimising the complexity of manufacturing and assembly operations. The goal is to design a product that is easily and economically manufactured. Approximately 70 per cent of the manufacturing costs of a product (cost of materials, processing and assembly) are determined by design decisions.

Every aspect is considered – raw material selection, dimensional requirements and secondary processes such as finishing, and even how the product is packaged. Two important Design for Manufacturing questions are 'What needs to be added?' and 'What needs to be removed?' DFM has been around for many years and is also known as value analysis or cost reduction systems.

Some key points DFM aims for are:

- minimise the number of parts
- standardise the parts and materials, make maximum use of purchased parts, modular design and standard design features
- reduce the number of manufacturing operations
- create modular parts assemblies
- create methods for efficient joining
- minimise any re-orientation of parts during assembly. This can be done by making parts symmetrical and avoiding the use of left- and right-handed parts.

FASHION AND TEXTILES

Total Quality Management (TQM)

TQM aims to achieve quality in everything a company does. TQM involves everyone in an organisation. In a manufacturing company, TQM usually involves sampling a random selection of the products made.

Product quality is usually determined by physical, chemical and environmental reasons but can also relate to comfort and usability. It is easy to measure and collect data relating to these factors but some characteristics (such as form, shape and colour) are not easily measured and quantified.

Products of the fashion and textile industries are valued by consumers in their integrity – both quality and fashion are considered. Quality of the construction or structure can be apparent in both manufacture and use (for example, grain of leather goods or the soles of shoes or water resistance of shoes and gloves). Quality of workmanship and precision are key if products are to be absent of faults, for example, evenness of seams and overlaps, finishes and symmetry of garments and left and right shoes, gloves and other items.

Quality plays an important role in the marketing of fashion and textile products. Quality arguments may arise around a product specification or sample, functionality and fitness for purpose and consistency within a batch.

Companies that implement TQM are always aiming to improve the performance of its organisation and its products and services. TQM involves the whole manufacturing process, reviewing and monitoring every stage of management and manufacture across the company. Checks are made at every stage from the delivery of resources through to the final delivery of the product to the customer. TQM is also covered in Section 7.5.

Planning for accuracy and efficiency through testing and prototyping

When manufacturing components, the correct detailing of tolerances and surface finish is important to ensure parts can be manufactured as accurately, efficiently and economically as possible, meaning the different parts of a product fit together and perform exactly as required.

By creating parts to be multi-functional or have multi-use within different products, manufacturers can focus on creating fewer parts with greater accuracy. By minimising assembly directions and handling, greater accuracy will be achieved. All parts should be assembled from one direction and, to facilitate orientation, symmetrical parts should be used whenever possible, as parts that fit either way round create less chance of error. Using standard fittings, including bolt sizes and the spacing of holes, will also make accuracy and efficiency greater, as standard components are less expensive than custom-made items and the availability of these components and reliability of quality and supply can reduce the manufacturer's concern of meeting production schedules.

Being aware of issues in relation to different scales of production

The reduction of the number of parts in a product will reduce manufacturing costs. Fewer parts means less handling, processing time and assembly difficulty. One-piece structures or manufacturing processes such as injection moulding and casting are often used, which can be part of batch and mass production.

Modular designs can simplify manufacturing activities and assembly. They allow the use of standard components to minimise product variations and the use of common parts across components and products. This creates easier fabrication, avoiding painting, polishing, finish machining and so on, and can allow for faster production in mass and batch production.

The use of fasteners increases the cost and time taken to manufacture a part due to the handling and feeding operations that have to be performed. In general, fasteners should be avoided and replaced, for example, by using tabs or snap fits. If fasteners have to be used then standard components should be used whenever possible.

Designing for repair and maintenance

To facilitate greater repair and reuse of materials and to improve the lifespan of a product, it is important that designers maximise the opportunity for re-use of as much of a product as possible at the end of its life. They can do this by:

- using threaded/bolted connections in preference to welded or permanent joints to allow products to be dismantled
- where feasible, ensuring that the materials are free from coatings or coverings
- allowing easy identification of the origin and properties of the component by bar-coding, e-tagging or stamping parts
- using multi-functional parts that can reduce the total number of parts needed in a design, for example, a part could be designed to act as both an electric conductor and as a structural member. Different products can share parts that have been designed

for multi-use. These parts can have the same or different functions when used in different products.

Designing a product to allow maintenance means including features that allow parts to be easily replaced, such as access panels and standard screws. Alternatively, the product might be designed to be made from a series of standard modules, meaning that if something went wrong, only the faulty module would have to be repaired or replaced.

Designing with consideration of product life

In addition to estimating cost and other traditional manufacturing performance measures, manufacturing firms are increasingly interested in evaluating and reducing the environmental impact of their products and how to reduce the energy that the manufacturing processes consume and the hazardous waste materials that are produced. They aim to create products that use less energy during their life and require less maintenance. Design for Environment (DFE) and lifecycle assessment (LCA) are also tools for all stages of the design process.

Most products are only expected to last for so long before they stop working, are worn out or are thrown away. How long a product will last is an important design consideration. The longer the product's life, the fewer new materials will be needed for replacements. However, a longer life also means that the manufacturer will sell fewer replacement products.

The life of a product can be extended by using materials with better properties, for example, by using stronger materials or materials that resist corrosion. Dyson uses Polycarbonate-ABS, a strong and flexible polymer used to make riot shields and police helmets which also produces over 50 per cent less CO_2 emissions during production than the aluminium equivalent. Another way is through design that allows product life to be extended by maintenance. This allows the product to have a longer life and can include anything from repairing worn-out parts to replacing batteries.

DESIGN ENGINEERING

System compatibility

One factor that limits the usable life of engineered products is their technical compatibility with newer systems. Recent history is full of examples of technologies which became obsolete because they were incompatible with newer systems. One such example is the development of home music playback systems, which have progressed from vinyl disk, through several types of magnetic tape (none of which were compatible with each other), optical compact disk (CD), magnetic disk (Minidisk), downloadable files (MP3) through to our current method of streaming music on demand through the internet.

The emergence of a disruptive technology tends to cause incompatibility between old and new systems because the technologies in use are completely different. It is generally beneficial to a manufacturer to produce new products which maintain at least a degree of compatibility with existing systems, as this broadens the products' application and encourages customer loyalty and confidence with a brand. Users can invest large amounts of money developing their equipment and accessories trusting that, when they upgrade, their investment does not become obsolete. The principle of making newer products compatible with older versions is called backwards compatibility.

Product upgrades, such as software upgrades, are used to keep a product current for as long as possible, but as the new software demands more processing power, the older products will run slower and eventually reach a point where they are no longer compatible with the latest software and cannot be upgraded any further. System incompatibility can arise from many causes, including different connector systems, different communication protocols, wireless systems operating on different frequencies and so on.

The need for maintenance of machinery

Mechanical systems will inevitably wear during use. A well-planned product lifecycle management will include a planned service schedule to ensure that mechanical systems continue to work reliably and efficiently throughout the product's life. Planned maintenance is intended to eliminate breakdowns and the need for repairs by eliminating problems before they occur. This ensures that machinery continues to work safely and is unlikely to break down unexpectedly.

A car service schedule is an example of preventative maintenance. The oil and filter in a car engine should be replaced after a specified time. Oil is a vital component in an engine, providing lubrication, cooling and cleaning functions. The oil will degrade if it is not changed and its lubrication efficiency will be reduced. This then leads to excessive wear of critical moving parts which will eventually cause engine failure leading to very expensive repair bills. Routinely changing the oil prevents this.

Certain parts in mechanical systems are intended to wear out and be replaced over time. Brake pads in a car are such items. Serviceable parts in mechanical systems should be designed to be easily accessible and able to be replaced quickly.

A machine manufacturer should produce a service manual which provides instructions for routine maintenance. They may also sell a service contract which reassures the customer and guarantees extra revenue for the manufacturer.

Product support and end of life (EOL)

Manufacturers should consider the level of support they will put in place at the launch of a new product. Aside from the legal requirement to offer product guarantees, a great deal of investment can be made to support a product over its lifetime. This could include software upgrades, spare parts availability, downloadable support and technical helpdesks to assist users. Strong product support boosts a brand image and sustains the **marketability** of the product.

A manufacturer may decide to remove product support either because it finds that it is no longer economical to sustain it, or because of a planned obsolescence strategy. Once support for a product is removed, the product will eventually become obsolete.

A product's end of life, from a seller's point of view, occurs when the seller stops marketing, selling and supporting the product. The end of life strategy may simply be concerned with discontinuing the product, but consideration would be given to addressing the marketing needs that the product fulfils, which may result in the development of a new product.

A successful EOL strategy will feature early in the product development stages and will consider the destination of the discarded product. This ensures that the product is designed with a view to supporting an environmentally sustainable EOL scenario that is economically viable and follows environmental directives.

1   Research the development and obsolescence of the following technologies:
    a)   home video playback (watching films on demand at home)
    b)   photography and cameras.
2   Research the development of the connecting leads used over the last 40 years to carry video/audio between parts of your home TV system.

FASHION AND TEXTILES

Products standing the test of time in terms of durability and style

Many products stand the test of time both in terms of durability of materials and parts and of style. Products such as leather jackets and denim jeans are manufactured from materials that are long lasting and constructed in a way that lasts. Designs that stand the test of time often become known as iconic designs. Sometimes these are everyday objects that are used by a vast majority of the population. Sometimes they can be so user-friendly that they disappear into our daily routine.

Maintenance and aftercare

Many products require maintenance during their life, which can involve cleaning or washing, oiling, greasing, descaling or replacement of batteries. Many products come with advice about maintenance and aftercare.

Care labels are often a deciding factor when choosing fashion and textile products. While some consumers look for washable garments, some look for items that are dry-cleanable. Care instructions are provided by manufacturers, such as 'For best results, dry clean.' This tells consumers that the garment can be washed but that dry cleaning may be better for appearance and durability. Leather and suede products can be specialist cleaned but

products to improve durability can be applied and life can be prolonged by polishing. Wood-based products sometimes require regular oiling and metal products polishing; again the manufacturer provides instructions for after care.

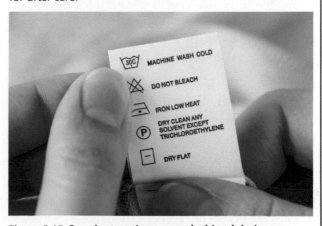

Figure 3.15 **Care instructions on a clothing label**

Re-working and recycling systems

Re-working and recycling materials is increasingly important. You can read more about these in the sections on material sustainability below.

PRODUCT DESIGN

A business-minded approach to smarter recycling, reuse and repurposing is beginning to change the way we think. Environmental consciousness of the waste generated by a throwaway/disposable culture has risen and a move to make some consumer goods less disposable has begun.

Designers and companies are starting by thinking more carefully about the choice of materials used, choosing easily recyclable materials or using recycled materials in products, such as carpets that are returned and recycled into new carpets after use. They are designing with the circular economy in mind (cradle-to-cradle) and designing products to have a longer life span, for example, the Tripp Trapp highchair, which grows with the child to last

Figure 3.16 **Tesla electric car**

longer. Smartphone providers now provide upgrades and recycling schemes and many companies are designing products that are easy to disassemble and choosing materials that are recyclable or easily broken down to limit the effect on the environment or the need to dispose.

Tesla, the electric car manufacturer, has plans to take back the spent batteries in its clients' cars and repurpose them for home energy storage. The company also auto-downloads and upgrades the software in its cars as the vehicles charge overnight. These upgrades 'future-proof' the sensors and hardware in the vehicle.

Impact of environmental factors on design solutions

The sourcing and processing of raw materials into a workable form

Once materials have been sourced, mined or harvested they need further processing into a workable form we can use for manufacturing. Some materials require processing that can use energy and create pollution; others have less environmental impact.

Metals

Metals are mined and are usually in the form of ore which then requires extraction from their oxide form or through electrolysis.

Iron is extracted from iron ore in a huge container called a blast furnace. Iron ores such as haematite contain iron oxide. The oxygen must be removed from the iron oxide to leave the iron behind. Steel is then made by reducing the amount of carbon in molten iron by blowing oxygen into it. This process produces carbon monoxide and carbon dioxide, which escape from the molten metal. These by-products are greenhouse gases that contribute to air pollution.

Most copper ores have a high concentration of copper compounds. Copper can be extracted from these ores by heating them in a furnace, a process called smelting. The copper is then purified using a process called electrolysis. Electricity is passed through solutions containing copper compounds, such as copper sulphate. During electrolysis, positively charged copper ions move towards the negative electrode and are deposited as copper metal.

Silver is extracted from its ore by smelting or chemical leaching. The smelting process emits sulphur dioxide gas, which reacts in the atmosphere to produce acid rain, which affects soil, rivers and lakes, harming fish and other wildlife.

Gold doesn't require processing as it is mined in its pure form but it tends to be found in rock that contains acid-generating sulphides and its mining produces more unwanted rock than in the mining of other minerals.

Aluminium is the most widely distributed metal. Its ore is called bauxite and is mainly found in Africa, the West Indies, South America and Australia. Bauxite is purified to make aluminium oxide and aluminium can be extracted through electrolysis, using a large amount of energy in the extraction process. Although bauxite is readily available, aluminium is expensive. This is because the processes to extract it has many stages and requires large amounts of energy. Aluminium also has the advantage of being recyclable – aluminium scrap is melted down, requiring only 5 per cent of the energy needed to make new aluminium using ore, although some of the aluminium will be lost during the process.

Iron and steel are the world's most recycled materials and are easier to recycle than many other materials. They can also be reused to produce metal that does not downgrade with each reuse. Recycling preserves limited resources and requires less energy, so it causes less damage to the environment.

Timbers, papers and boards

Trees are converted from timber logs into a usable plank of wood, other timber products or into pulp to be made into paper or boards. This takes place at a sawmill, where logs are cut into boards using equipment such as circular saws and bandsaws and is called 'conversion'. Natural wood contains moisture and needs to be seasoned before it can be used in furniture and construction, which sometimes takes place in a kiln (which uses energy). The moisture content in wood can be 50 per cent, but during the seasoning process a tree loses its free water and a high proportion of its cell water and, as a result, is less likely to warp or deform. This can also be done naturally but is more time consuming.

The market value of timber can be further increased through manufacturing sawn-timber products – called secondary processing. This involves the wood being made (either by man or machine) into a more refined product, such as boards of a specific size and dimension. At this stage any preferred treatments to timber such as fire or rot resistance is added. Treated timber in sawn form is used either directly in construction or to prepare construction components, such as timber frame panels. Pulp mills convert the wood chips or other plant fibre source into thick fibreboards which are shipped to a paper mill for processing into papers and boards.

Polymers and synthetic fabrics

During the refining of crude oil petrol, paraffin, lubricating oil and petrol are the by-products. These create monomers, which when linked together are called polymers. Ethylene and propylene, for example, come from crude oil, which contains the hydrocarbons that make up monomers. The hydrocarbon raw materials are created by 'cracking' (the process used in the refining oil and natural gas) and these hydrocarbons are then chemically processed to make hydrocarbon monomers and other carbon monomers (like styrene, vinyl chloride and acrylonitrile) used in polymers.

Transporting petroleum or oil can result in oil spills which have a major effect on the environment and sea life. The refining process also uses energy. Further processing then takes place to create polymers in stock forms such as pellets, rolls and sheets.

Natural fabrics

Cotton tends to be intensively farmed. It uses large volumes of water (10,000 litres per kilogram of cotton), which puts pressure on resources. Seed cotton is then processed by ginning to separate the lint and seed. This process creates dust which can cause health issues because of pesticides used on the cotton crops. The cotton first goes through dryers to reduce moisture content and then through cleaning equipment. Once cotton has been baled it is sent to a mill for spinning or weaving into fabric. At this point, dyes, prints or special finishes are added.

Wools and silks are also harvested and prepared to be made into fabrics. Wool fleece needs washing before it can be turned into wool yarn, due to the lanolin and grease in the shorn fleece. Making silk is a labour intensive process that involves treating cocoons created by the caterpillar of the silk moth with hot air, steam or boiling water.

Leather is another natural fabric which requires work to get it to a usable state. Chrome tanning of leather is a toxic and water-intensive process. It produces waste that has a high potential for becoming carcinogenic chromium (VI), which has been known to contaminate water supplies. Vegetable tanning avoids the harsh chemicals that chrome tanning uses and creates leather that is safer to recycle or dispose of.

The disposal of waste, surplus materials and components, by-products of production including pollution related to energy

The generation of energy, particularly the production of electricity, has contributed greatly to atmospheric pollution. Coal gas and oil-fired systems emit large amounts of carbon dioxide, which many scientists agree is a significant contributory factor to global warming. Many of the processes outlined above use high amounts of energy and cause pollution to the air, which contributes to global warming and acid rain.

The Earth is surrounded by a layer of gases including carbon dioxide (CO_2), methane (CH_4) and nitrogen oxide (NO). This layer allows the sun's rays to penetrate. Around 30 per cent of the rays are deflected back by ice caps and clouds, but the majority are absorbed by the earth and oceans and released back towards space as infrared radiation.

The layer of gases prevents all of the radiation from leaving the planet and traps enough to heat the lower atmosphere. Without this layer, temperatures would be at least 30°C cooler. Increased amounts of carbon dioxide, methane and nitrogen oxide are effectively making the blanket thicker and creating global warming.

Electricity production emits other pollutants. Over 80 tons of mercury, the most toxic heavy metal, is released into the atmosphere every year. Waste incineration plants add to this total.

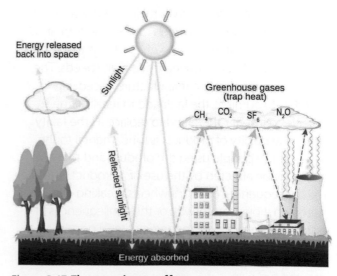

Greenhouse effect

Figure 3.17 **The greenhouse effect**

Coal-fired power stations emit sulphur dioxide (SO_2), nitrogen oxide (NO) and polycyclic aromatic hydrocarbons (PHA) into the atmosphere. Regulations were introduced to limit emissions when the destruction of plant life and pollution of rivers was directly linked with power production plants.

Governments have reacted with legislation to impose targets for reduction of the key pollutants. Companies are required to develop and use better technologies to filter and significantly reduce the emissions of SO_2, NO and mercury.

> ### ACTIVITY
> Choose a raw material and research the environmental issues that arise from sourcing this material and processing it into a workable form.

Cost implications related to materials and process

These days, many large businesses either move from place to place or country to country, wherever the labour force and premises costs are cheapest. They might out-source the work to manufacturers anywhere in the world where the costs are the lowest. As two of the most significant costs for any business are labour costs and premises costs, the out-sourcing of work to industrially developing countries makes economic sense to businesses. This can be to be closer to the source of raw material used for manufacture.

The quality of products must be suitable for their intended purpose. When companies make cost savings using lower cost materials or manufacturing methods, they need to ensure quality is maintained. For example, a manufacturer of alloy wheels for high performance cars found tiny faults in some of its wheels that could result in the wheels breaking up at high speeds. The moral dilemma that the manufacturer faced was whether to admit to the fault and to recall all the wheels, to inspect them and to replace all the faulty ones. The quality of products is fundamental to the safety of the people using the product and to those who may be affected by the use of a product. Imagine the consequences of a car's wheels breaking up at high speed on a busy motorway, or the consequences of poor quality smoke alarms and electrical goods.

Sustainability issues relating to industrial manufacture

Products are made by people. There are sustainable and moral implications for all of us when we buy things.

Manufacturers have to ensure that the workers they employ are treated fairly and materials are sustainably and ethically sourced. The situation is complex; organisations like ActionAid work alongside their partners, such as the Labour Education Foundation, to slowly improve working conditions and health and education in developing countries where some of the raw materials that are used to make products are sourced. **Ethical trade** means that retailers, brands and their suppliers take responsibility for improving the working conditions of the people who make the products they sell. **Fairtrade** and the ETI (Ethical Trade Initiative) are two well-known initiatives that try to ensure manufacturers follow guidelines and best practice to ensure fairness for everyone.

Fairtrade

Fairtrade as a concept is about better prices, decent working conditions and fair terms of trade for farmers and workers. It exists to make sure that farmers and workers are treated fairly and safe working practices are followed. Environmental protection is also a key element of Fairtrade's view of sustainability. There are a set of Fairtrade Standards that require smallholder farmer and larger labour production set-ups to comply in key areas. This has been shown to lead to good agricultural practices, which have encouraged environmentally sustainable production. Fairtrade works to benefit small-scale farmers and workers, who are among most marginalised groups globally, to enable them to maintain their livelihoods and reach their potential through trade rather than aid.

For certain products, such as cocoa, cotton and rice, Fairtrade only certifies small organisations offering rural families the stability of income which enables them to plan for their future. But Fairtrade also certifies plantations for bananas, tea, and so on. Standards for such large-scale production units differ and protect workers' basic rights; keeping them safe and healthy, preventing discrimination and ensuring there is no illegal child labour.

The Ethical Trade Initiative (ETI)

The Ethical Trading Initiative (ETI) is an alliance of companies, trade unions and non-governmental organisations (NGOs) that promotes respect for workers' rights around the globe. The ETI Base Code is an internationally agreed code of labour practice, specifying in detail the requirements applicable to all areas of employment for which members are responsible. (www.ethicaltrade.org/eti-base-code)

The ETI and Fairtrade have distinct origins, but their approaches are complementary. Both focus on helping make international trade work better for poor and otherwise disadvantaged people.

An example of a company committed to fair trade is People Tree, an ethical and sustainable fashion brand. They aim to be 100 per cent fair trade throughout their supply chain and purchase products from marginalised producer groups in the developing world. Nearly all of their range is produced using organic cotton, which is both better for the environment, as it uses natural pesticides rather than chemical ones, and provides better income for farmers. Read more at www.peopletree.co.uk/.

For a number of years, companies such as Primark, Zara and the Arcadia Group, owner of outlets such as Topshop, Miss Selfridge and Dorothy Perkins, have operated programmes to monitor and manage the social and environmental impacts of their business. Arcadia's is called 'Fashion Footprint' and commits to ensuring that the workers in their supply chains are treated fairly. When customers buy Arcadia goods they should be reassured that they have been produced under acceptable conditions. Every factory that manufactures products for Primark must commit to meeting internationally recognised standards before an order is placed. M&S also recently committed to trace the source of all its cotton. But child labour has been reported in all the major cotton-growing countries – China, India, Pakistan, Brazil, Uzbekistan and Turkey. In 2005, The Better Cotton Initiative was set up, supported by a collective of major organisations including Adidas, Gap Inc., H&M, ICCO, IFAP, IFC, IKEA, Organic Exchange, Oxfam, PAN UK and WWF. See http://bettercotton.org/ for more information.

The children's charity World Vision has information about forced labour in the cotton industry. You can find more information on their website at www.worldvision.com.au.

It may not be beneficial to suddenly stop buying non-ethical products – the families that make them might not be able to support themselves otherwise. Customers, however, need to ask questions and challenge the ethics of those who design, manufacture and sell products, and support those organisations that are trying to change the lives of families for the better.

ACTIVITY

Research these issues and find examples of where individuals or groups have been disadvantaged by unfair working conditions or trading practices.

Economic issues and globalisation

Globalisation can be summed up by the term 'world trade'. We are now familiar with products that are made in countries from around the world. The internet, improved transportation links and free trade agreements have made it possible for us to buy things from anywhere in the world and electronic payment systems allow transactions to occur across borders. Advances in transportation, such as the development of large container ships, have made transporting goods around the world easier and more cost effective. Trade between different countries has been facilitated by new free trade agreements.

Another feature of globalisation is the internationalisation of design, manufacturing and the market place. In simple terms, a product may be designed in one country, with the manufacture of its component parts and final assembly taking place in a number of countries across the world. The market place for the product may also be international.

Globalisation has benefits for those in developed countries. We have access to more foreign goods, we are able to eat fresh fruits and vegetables in winter and there is more diversity of products available to us at lower costs. It can help developing countries. When manufacturers set up in developing countries they create jobs for people, but this can then cause human rights problems such as child labour. Many nations have benefited from the globalisation of the economy over the years. Most countries are dependent on international trade and business.

Multi-national corporations work globally, taking advantage of interest rates, currency rates, cheap labour and labour skills across the world. IKEA is an example of a company that has taken advantage of globalisation. In part, this success can be credited to the company's universally appealing brand attributes of pricing, simplicity and sustainability, but IKEA also understands its international audiences and room sets (displays) will vary from store to store to suit local customs. For example, in Japan, they will often feature tatami mats, a traditional Japanese floor covering. IKEA has 43 manufacturing facilities in 12 countries as of 2017.

Some of the political action around the world in recent years could be linked to a backlash against globalisation – the Brexit vote or the rise of certain politicians. Globalisation has exposed where domestic companies have become inefficient or complacent; they lose market share to cheaper or more innovative companies.

Transportation of goods globally has also put a strain on the non-renewable sources of energy, such as petroleum. Many companies and governments are looking to strengthen local manufacture. Advances in technologies like 3D printing will 'digitise' some manufacturing, allowing production to take place closer to the customer, with potential for local and even mobile manufacturing.

Material sustainability and optimisation

Sustainable development

Sustainable development was a term originally coined by the Brundtland Commission, which was set up by the United Nations with the aim of uniting countries to work together in ways that did not deplete the environment and natural resources in an unsustainable way, as development that 'meets the needs of the present generation without compromising the ability of future generations to meet their own needs'.

It has been suggested that if everyone in the world lived as we do in the UK, we would need at least three planets to sustain us. We can all make a difference by being aware of sustainable issues and making informed choices when we design or buy products or source ethically and sustainably produced materials. As future designers, we need to consider the environmental,

social and economic implications of our decisions and choices. These are all interlinked: all making involves materials wherever in the world it occurs; every material uses energy to produce it; any use of energy causes pollution and affects climate change, damaging people's health.

This can result in economic costs and social issues. As designers we can aim to reduce, reuse and recycle. But the real message is perhaps to try to use less – less products, less materials, less energy, less resources. Less packaging results in less waste and less impact on the environment. By using less material we can reduce our products' ecological impact. A few simple questions can be asked:

- How far are the materials you are using travelling to get to the factory? Could they be substituted for more locally sourced materials that will result in less impact?
- Would it be possible to use materials that are easily recycled? If so, could you make it obvious to the consumer that it is recyclable?
- Could your product and its packaging be designed so that it is easily recycled, with the materials easy to separate?

The 6Rs are a checklist of the best ways of managing waste:

Table 3.3 The 6Rs

The 6Rs	Questions
RETHINK	How can it do the job better? Is it energy efficient? Has it been designed for disassembly? Are the materials chosen environmentally friendly? Could they be replaced?
REUSE	Which parts can I use again? Has the product got another valuable use without processing it?
RECYCLE	How easy is it to take apart? Are the materials recyclable? How can the parts be used again? How much energy is required to reprocess parts?
REPAIR	Which parts can be replaced or fixed? Which parts are going to fail? How easy is it to replace parts?
REDUCE	Which parts are not needed? Do we need as much material? Can we simplify the product or remove parts?
REFUSE	Is the product really necessary? Do we need it? Is it going to last? Is it fair trade? Is it too unfashionable to be trendy and too costly to be stylish?

One way to reduce the impact on the environment is to use less material in the product. This might mean asking questions about what materials are needed, or if the product could be made smaller (or have thinner walls or parts) and still do the same job. It could also mean using an alternative material with better properties, so that less of the material is required. Some materials have less impact on the environment than others. A renewable resource is one which can be replaced naturally in a relatively short time, such as wood from a managed forest. Most

manufacturing processes produce various kinds of waste as a by-product. Sometimes these by-products contain toxic substances harmful to people or the environment.

The 6Rs are a common approach to considering sustainability and asking questions to help you think about the environmental impact when looking at existing products or creating your own. More information and useful resources can be found at the website of the NGO Practical Action, who are committed to using technology to challenge poverty (https://practicalaction.org).

One way in which manufacturers are reducing the environmental impact of products and materials is to adopt **Design for Disassembly** (DfD). This is a strategy which looks at designing products with a view to how they might be repaired, refurbished or recycled later in their lifespan to avoid using more materials and allow easy recycling of material within products. Some of the ways in which designers might think about disassembling a product are:

● The fewer parts your product has, the fewer parts there are to take apart.
● As with parts, the fewer fasteners you use, the better.
● Standard fasteners that require only simple tools will help to simplify and speed disassembly.
● Screws are faster to unfasten than nuts and bolts.
● Glues and surface finishes should be avoided.
● Building disassembly instructions into the product will help consumers understand how to take it apart.

Optimisation

In our highly competitive world with complex and highly refined products and systems, it is essential to design the best products and systems. The smartest, most reliable and most efficient products succeed. Design **optimisation** is a means of identifying the best choices from design alternatives in terms of optimum use of materials, manufacturability or ease of assembly, quality, performance, size, weight, design features, sustainability and so on. Creating an optimal or optimum design is the ultimate aim of an iterative design process. The optimal design will most efficiently and effectively meet the needs of users and stakeholders, while being good for our environment and resources.

Optimisation software and methods have been developed by companies including Altair and Dassault. Systems have been created that allow engineers and designers to automatically and quickly determine the solution of identified design optimisation problems, such as topology (shape) optimisation. Design optimisation can be used to modify a design. Its goal is to automate the design process by using a rational, mathematical approach to improve designs, yet keep it functional within a set of parameters.

The design optimisation process can reduce the **lead time** on products. Instead of requiring months of prototyping, the performance of parts can be simulated using optimisation software, allowing the engineering

team to move quickly to the optimal design. Companies can get the design right, modifying size or structure and satisfying any parameters set. There are many different types of optimisation tools available for designers and engineers to use:

● Structural optimisation methods use algorithms to solve structural problems using finite element analysis.
● Size optimisation explores modifying size-related structural elements, such as shell thickness, beam cross-sectional properties, spring stiffness and mass.
● Topology optimisation is perhaps the best known to us. It optimises material layout, sometimes using biomorphic (naturally occurring) shapes within a given physical volume that will still achieve performance targets.

Traditionally, engineering has been performed by teams, each with expertise in a specific discipline, such as aerodynamics or structures. Each team would use its members' experience and judgement to develop a workable design, usually sequentially, which takes time. Optimisation software allows a much faster time to market. Optimisation often means less materials are used, saving both costs and raw materials, which improves sustainability.

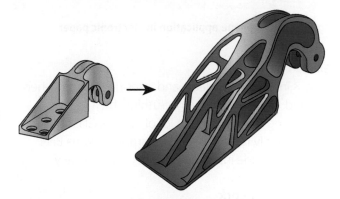

Figure 3.18 **An example of design optimisation**

The impact and use of eco-materials

Eco-materials are defined as, 'those materials that enhance the environmental improvement throughout the whole lifecycle, while maintaining accountable performance' (Halada and Yamamoto, 2001). Using eco-materials minimises the environmental impact of materials and can improve the recyclability of materials. Eco-materials are sometimes known as environmentally friendly or eco-friendly materials.

Papers and boards

Recycled card is commonly used to make all sorts of products from furniture to packaging and is a well-known eco-material. The two types of cardboard that can be recycled are the flat cardboard used in cereal and shoe boxes and corrugated cardboard, which has a fluted layer between the two flat pieces of cardboard and is often used in packing boxes. Corrugated cardboard can be a very strong material. There is even a range of affordable cardboard furniture available from the company Eco Floots (http://ecofloots.com/).

China has developed the world's first graphene electronic paper that is set to revolutionise the screen displays on electronic gadgets such as wearable devices and e-readers. The material is also currently the world's lightest and strongest material. In terms of eco-materials, graphene-based e-papers can easily be produced cost-effectively and are derived from carbon.

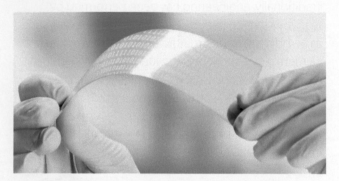

Figure 3.19 **Graphene application in electronic paper**

Glass
Glass is infinitely recyclable and loses no strength when recycled. However, washing, crushing, melting and remoulding the glass uses a lot of energy. Glass can also be commonly repurposed, for example, Corona Beer bottles are made into work surfaces and jewellery.

Bamboo and cork
Bamboo and cork are both quick growing so can be rapidly regrown once cut down. Although impermeable, the surface of cork can be waterproofed to extend the life further. It can be applied to walls or floors or used as a textile material for handbags. Bamboo can be grown and harvested in three years. It is now widely used in furniture, building and construction and, as a fibre, bamboo can be used for all types of clothing.

Wood
The use of reclaimed wood is common. It can be re-finished and cut to make new wood flooring, kitchen cabinets or cladding. Railway sleepers and pallets are commonly repurposed. Sustainably sourced wood is commonly available and often sourced locally and as forests and trees are cut down, they are replanted.

Biopolymers
Bio-compostables are made of sugar cane fibre, corn and potato starches and can be used in products that might otherwise be made from petroleum-based polymers such as utensils and coffee cups.

Fabrics
Hemp can be used to replace materials used for clothing and around the home. It is one of the fastest growing plants in the world and requires little use of pesticides.

Soy fabric is made from leftover pulp created by producing tofu and soy milk and is extremely soft, making it similar to cashmere.

Exploring how materials can be upcycled
Upcycling is taking an item that is no longer needed or wanted and giving it new life as something that is either useful or creative by refashioning it into something different. For example, you could cut a T-shirt into strips of material to use to create a new product, but it will still be the same materials as when you started.

There are many great examples of upcycled ideas and products on the internet. For instance, Kevin McCloud from the TV show *Grand Designs* challenged three designers to find a new use for every single piece of an Airbus A320. He said, 'Like so many beautifully designed machines we use them and then we chuck them away. But what if instead of putting them into landfill or melting them down we could make them into something new?' The programme showed that the designers were able to sell their upcycled products to consumers for a profit. McCloud concluded, 'The most exciting part of this is we have shown the potential of upcycling.'

> **ACTIVITY**
>
> Find examples of upcycled products and eco-materials and consider how you could use them in your NEA project.

Figure 3.20 An upcycled fruit bowl from an old record

KEY POINTS

● Design optimisation is a means of identifying the best choices from design alternatives in terms of optimum use of materials, manufacturability or ease of assembly, quality, performance, size, weight, design features, sustainability and so on.
● Design for Disassembly is a strategy that considers the future need to disassemble a product for it to be repaired, refurbished or recycled.
● Fairtrade and ethical trade is about better prices, decent working conditions, local sustainability and fair terms of trade and rights such as working hours, health and safety and wages for farmers and workers in the developing world.

KEY TERMS

Design for Disassembly – a strategy that considers the future need to disassemble a product for it to be repaired, refurbished or recycled.

Eco-materials – often called environmentally friendly materials, those materials that enhance the environmental improvement throughout the whole lifecycle.

Ethical trade – this is about having confidence that the products and services we buy have not been made at the expense of workers in global supply chains. It encompasses a breadth of international labour rights such as working hours, health and safety, freedom of association and wages.

Fairtrade – this is about better prices, decent working conditions, local sustainability and fair terms of trade for farmers and workers in the developing world.

Globalisation – the process by which the world is becoming increasingly interconnected as a result of increased trade and cultural exchange.

Lead time – total time required for item manufacture.

Marketability – the potential of a product or service to be successful in the real world. A product's innovative features, its pricing, branding and packaging are some of the aspects that give it commercial potential and value. Identifying a potential Unique Selling Proposition (USP) is a good starting point.

Optimisation – a means of identifying the best choices from design alternatives in terms of optimum use of materials, manufacturability or ease of assembly, quality, performance, size, weight, design features, sustainability and so on.

Upcycling – taking an item that is no longer needed or wanted and giving it new life as something that is either useful or creative.

3.3 What factors need to be considered when manufacturing products? (A Level)

LEARNING OUTCOMES

By the end of this section you should have developed a knowledge and understanding of:
● how to achieve an optimum use of materials and components, including:

– the cost of materials and/or components
– stock sizes and forms available
– sustainable production.

Achieving optimum use of materials and components

When we talk about choosing materials for a component, we need to consider different factors:
● material properties – the expected level of performance from the material
● material cost and availability

● processing – how to make the part
● the effect that the environment in which the product will be used has on the part
● the effect the part has on the environment
● the effect that processing has on the environment.

The designer of any product must consider material selection. Only occasionally will the exact grade of material be specified by the consumer or stakeholder, but the designer must still understand the material to be able to design the product.

When designing, it is worth analysing what the product needs to do and to be like.

The following questions can help:
- What does the product or component do?
- Where will the product be used?
- Will it need to be easily maintained?
- What should it look like (colours, etc.)?
- How will any electrical or mechanical parts work and interact?
- Who will be using it?
- How much will the final product cost?
- How many are going to be made?
- What manufacturing methods might need to be used?
- What material properties are needed?
  - Mechanical – strength, modulus, etc.
  - Physical – density, melting point, etc.
  - Electrical – conductivity, resistivity
  - Aesthetics – appearance, texture, colour
  - Processability, ease of working, ductility, mouldability.
- How sustainable are the choices?

The requirements of any product will include the following:
- performance
- reliability
- size, shape and mass
- cost
- manufacturing and assembly
- industry standards
- Government regulations
- intellectual property
- sustainability.

Identifying as many of the requirements as possible will help to determine the best material. For many products, some of these requirements are not applicable, making the process easier.

The material selection criteria are specific properties derived from the requirements; for example, for a component that must support a specific weight. Materials will be ruled out that will not satisfy all the material selection criteria. Trials and experiments with different materials can help to find the optimum choice. A suitable process that is also economic should be chosen.

Cost of materials and/or components

Materials will be considered that satisfy all the material selection criteria at the lowest cost. Cost includes the cost of the material and the cost to fabricate a component or form a joint between components. Thinking about the design from an ergonomic and functional viewpoint is relevant. Is the product performance driven or cost driven?

Stock sizes and forms available

Although materials are often chosen first, sometimes it is the shape and process which is the limiting factor. The availability and stock forms of materials also affect price, as commonly available forms are more cost effective than special sizes. They are made in quantity, so bulk purchasing can mean less transportation costs and this can also benefit the environment because they are processed and transported in larger quantities. For these reasons, most designers and engineers try to work with stock sizes.

Material forms tend to be sheets, rolls, bars, tubes and sections. Tubes are sometimes used for frames as they are lighter and use less material than a solid bar and they can also be more cost effective to use. Working out the best way to buy materials for cost efficiency and minimising waste is key. If the material is bought in sheet form, then it is necessary to calculate the size of sheet that will most efficiently provide the number and shape of pieces required.

Sustainable production

The impact of the choice of material and processes on the environment must be considered. The less material we use, the less of a drain on resources and the less energy intensive processing is required. It also means a lighter, more user-friendly design. During the manufacturing stage, efforts can be made to use fewer resources (material, energy, water, etc.) or produce less pollution and waste. The concept of material efficiency can also be extended to reducing the impact of distribution and sales as using less materials makes products weigh less, which in turn will use less fuel for transportation.

Industrial facilities must follow internationally established guidelines as to how they should manage energy. These are known as ISO 50001 Energy management and include all aspects of energy use, aiming to reduce energy costs and greenhouse gas emissions. In manufacturing, the link between energy use and CO_2 emissions is very strong. The use of renewable energy is a key part of the 'Sustainable Plant' initiative being taken up by some companies, such as Toyota Manufacturing UK (TMUK). In 2011, TMUK worked in partnership with British Gas to install the UK's largest Solar PV (Photovoltaic) array connected to an industrial site. The glass tower has an annual energy output of 9,632 kilowatt hours and provides an annual CO_2 saving of 4,400 kg. IKEA aims to be 100 per cent renewable by 2020 – producing as much renewable energy as they consume, using renewable sources, such as the wind and sun.

The American car manufacturer Tesla is building a 'Gigafactory' in Nevada which will be used to make batteries for their electric cars. They will use heat pump technology and solar power to provide all the energy needed for manufacture. Once it is complete, Tesla expects the Gigafactory to be the biggest building in the world and for it to be entirely powered by renewable energy sources, with the goal of achieving net zero energy.

Selecting materials for a project

The following sources will be useful when you make material selection for your NEA project:

- Textbooks are good for general information and provide tables of properties. Although they won't have detailed specifications and properties, they can be a useful starting point.
- Data books are one of the quickest sources of detailed information of materials. They contain grades and specifications as well as properties, but they can be expensive.
- Manufacturers' literature varies in quality and usefulness and can be **biased**. They usually do not compare materials but can be good for final selection to check stock sizes and availability.
- The internet can be a good source of information, but searches can bring up superfluous data and it can be hard to find technical information.
- Material selection charts allow easy visualisation of properties and show balances of properties, such as strength versus cost.

ACTIVITY

Using the link below, look at these material selection charts and complete the tasks suggested. This will help you understand how to use these charts in your own NEA work.

www-g.eng.cam.ac.uk/millennium/now/mfs/tutorial/non_IE/charts.html

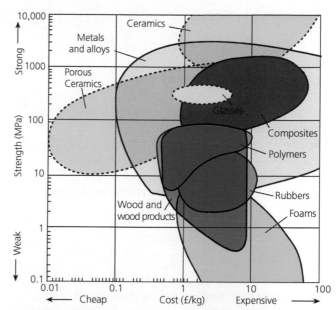

Figure 3.21 Material selection charts published by Cambridge University online and available from TEP

KEY TERM

Bias – a personal prejudice or inclination against a concept or approach.

MATHEMATICAL SKILLS

You will need mathematical skills to read data and interpret charts. The material data charts have logarithmic scales – each division is a multiple of 10. Material properties often cover such huge ranges that logarithmic scales are essential. You may need to consider tessellation and nesting of parts to minimise material use when calculating quantities for ordering.

KEY POINTS

- When selecting materials for components, we need to consider different factors:
  - properties
  - the expected level of performance from the material
  - cost and availability
  - processing – how to make the part
  - the environment in which the product will be used
  - the impact of the material choice on the environment.

3.4 What factors need to be considered when distributing products to markets? (A Level)

LEARNING OUTCOMES

By the end of this section you should have developed a knowledge and understanding of:
- the issues related to the effective and responsible distribution of products to market, including:
  - cost-effective distribution
  - environmental issues and energy requirements

- social media and mobile technology
- global production and delivery.
- the implications of intellectual property (IP), registered designs, registered trademarks, copyright, design rights and patents, in relation to ethics in design practice and consumer rights.

The effective and responsible distribution of products to market

Products should be transported safely and in a way that ensures that they will not be damaged in transit and that people, wildlife and the environment will not be harmed by the transportation process. Deciding the best distribution channel to sell a product and the best way to market it to customers is important.

Distribution refers to the location where a customer can purchase a product. It is sometimes known as the distribution channel. It can include any physical store or shop as well as TV shopping channels and the internet – anything that helps provide convenience for the consumer.

There are four main channels of distribution:
- Manufacturer – Consumer, for example, mail order, farm shops, bespoke made items
- Manufacturer – Retailer – Consumer, for example, high-street stores
- Manufacturer – Wholesaler – Consumer, for example, furniture or white goods
- Manufacturer – Wholesaler – Retailer – Consumer, for example, medium-size convenience stores (large supermarkets often cut out the wholesaler).

Cost-effective distribution

In the world of business and industry, all products are made to be sold for a profit. Cost-effective distribution is a factor in ensuring the profitability of a product. When considering how to distribute new products or services, companies and manufacturers need to consider a number of distribution models. The traditional distribution model has three levels: the manufacturer, the wholesaler and the retailer. Some manufacturers might go straight to retailers or even the end customer.

Going straight to the consumer/end customer is known as direct distribution. Think of a farmer setting up a shop on their farm and selling direct to their customers rather than going through a wholesaler or retailer such as a supermarket or local store.

Direct distribution cuts significant costs from the system because manufacturers don't have to share their profits with wholesalers and retailers, as companies sell and deliver their product themselves, using their own salespeople and warehouses.

Another consideration when selecting a distribution method is the way the product is purchased by consumers, as this can create further costs. Customers may wish to try a product before purchasing, for instance garments or footwear. This means the method of distribution is going to have to include a changing room, provide a live demonstration of the product or offer a free returns system if sold online. The distance the customer is willing to travel to purchase the product or service is another key consideration as to whether the product will be in high demand and unique.

Environmental issues and energy requirements

Any business or organisation that produces or uses packaging, or sells packaged goods, has an obligation to follow rules which help to:
- reduce the amount of packaging produced in the first place
- reduce how much packaging waste goes to landfill
- increase the amount of packaging waste that is recycled and recovered.

The regulations can be found on the Government information website, www.gov.uk.

Some examples of ways in which the environmental impact of product distribution can be reduced include:
- having a central distribution centre rather than multiple distribution centres, which can reduce energy consumption and is also likely to be more cost effective
- using rail or water as alternatives to road or air distribution

- using fuel efficient engines or electrical vehicles when road distribution is used
- using intelligent transport management systems
- using local suppliers for materials and services to reduce the carbon footprint of products
- outsourcing storage and deliveries, and using local networks or wholesalers, reducing journeys made
- optimising packaging materials and design, which can significantly help **logistics** by improving vehicle load. A fully loaded lorry can increase the volume of goods being transported at once and reduce CO_2 emissions, time and cost.

One of the most sustainable trends in storage solutions is JIT (just in time) production, which means that products are delivered directly from supplier to producer without warehousing. This form of direct transportation is more efficient when large products or quantities are to be moved, but transportation through regional distribution warehouses is more efficient in the case of small batches of products.

Many large companies are looking at the lifecycle assessment (LCA) of their products and in particular the transportation of supplies and goods. Mars have identified methods to make fewer journeys by fully loading trucks and containers and are wasting less fuel by training their staff to drive more efficiently and using new vehicles that allow pallets to be double-stacked. They are also exploring new technologies like hybrid electric engines and biofuels.

Electronic distribution of goods and products offers an environmentally friendly way to deliver particular products and services to the end-consumer. This can only be applied to products that can be available for download instead of delivering them on media such as DVD or CDs. Technical manuals are no longer printed and shipped with products, but instead available as electronic print-on-demand documents. The future possibilities of 3D printing within our homes or in local centres might mean in the future we can download all sorts of products, print them and avoid the need for transportation. Another way to reduce environmental impacts of distribution processes is to move customers' expectations towards the availability of fresh and seasonal food or items that can be sourced locally.

Social media and mobile technology
Social media marketing (SMM) utilises social networking websites as a marketing tool and as a way to sell directly to consumers, who can click and purchase. Users are encouraged to share with their social network to help a company increase brand exposure and broaden its customer reach. Search engine optimisation (SEO) is a strategy for drawing new visitors to a website and then promoting further social media activity, such as tweets or status updates. Websites use cookies which mean users leave a digital trace, which can allow companies to contact a registered customer and offer incentives to purchase. Technology and apps also allow customers to review products, services and delivery services quickly, allowing potential purchasers easy access to reviews before making purchases.

Mobile technology is used by companies and consumers to track deliveries, allowing customers to have real-time, up-to-date information on delivery times. With shortened shipping times becoming the norm, within two hours is offered in some cities. Real-time access to tracking packages improves customer satisfaction.

Large stores and warehouses use RFID (radio-frequency identification) tags to track items because of the high quantities of items shipped between stores and from warehouses. This technology gives their logistic centres visibility through the shipping process. It is proving to be a good way to reduce mistakes in shipping, packing and customer service. Mobile technology and apps are enabling warehouse managers to collect and act upon information gathered. Wearables, along with smaller, more accurate handheld scanners and devices are making it easier to track items and for stock levels to be kept up to date easily. Mobile technology is also improving road safety as it enables drivers to clock in and out remotely, providing managers with more accurate insight into how long the drivers are actually working for.

Global production and delivery
Globalisation means goods and products are transported worldwide by air, sea and road. Many large businesses either move from place to place or country to country, wherever the labour force and premises costs are cheapest. They may also outsource the work to manufacturers anywhere in the world where costs are the lowest. As two of the most significant costs for any business are labour and premises, the outsourcing of work to industrially developing countries makes economic sense to businesses, but with this comes transportation costs that aren't just financial.

Unilever distributes a huge number of goods worldwide and have been looking to lessen their impact. They have transport management systems looking to improve efficiency of their logistics operations and aim to reduce the number of kilometres products need to travel by strategically locating factories and distribution centres. Pallet heights are standardised for different products to ensure optimal truck fill, which helps lower emissions and reduce costs. China has recently launched a Smart Transportation Programme, which uses more rail than road, reducing CO_2 emissions and costs while transporting larger numbers of goods just as fast as by road. Over 40 per cent of products were moved by rail in 2016.

Long-distance freight transport is specifically being addressed by governments worldwide and the European Commission has set a target of a shift of 30 per cent of road freight transport over 300 km to other modes, such as rail or waterborne transport by 2030 and more than a 50 per cent shift by 2050 (EU 2011a).

A new class of container vessel is being developed for overseas transport and in Europe research projects are analysing the maximum length of trains. Having identified that rail freight transport would benefit from longer trains, research also points towards automated/autonomous transportation being a future solution in industry.

An innovative concept called CargoCap (www.cargocap.com/), an alternative to the conventional systems of road, rail, air and water, is being researched and explored. It would provide a safe and economical way to carry goods quickly and on time in congested urban areas by underground transportation pipelines. Interdisciplinary collaboration in research and development at the Ruhr University of Bochum under the direction of Professor Dr.-Ing. Dietrich Stein is supported by the current Ministry of Innovation, Science, Research and Technology of the State of North Rhine-Westphalia.

The implications of intellectual property

In the design world a number of laws exist to protect ideas and inventions. They are known as **intellectual property (IP)** laws and relate to all kinds of intangible (not physical) types of property that people can own, for example, the creative outcomes from the mind such as design ideas, written material, artistic and musical composition.

IP law allows you to own the things that you create in a similar way to owning physical property. You can control the use of your IP and use it to gain reward or protect it from copying. IP rules ensure that the financial gain from the design goes to the creator.

The main types of IP protection are **registered designs, trademarks** and **copyright, design rights** and **patents**.

In Britain, the UK Intellectual Property Office (IPO) is responsible for registering intellectual property rights. The IPO website (part of www.gov.uk) has a search facility where details of all IP registered in the UK can be found.

A registered design will give you ownership rights for the appearance of a product, protecting both the shape and the pattern or decoration. A registered design also covers visual features such as lines, contours, colours, shape, texture and materials of the product. To be registered, a design must:
- be new and original
- have an individual, unique character
- not resemble, in the opinion of an informed person, an existing design.

Registered designs must be renewed every five years (for up to 25 years).

A trademark is used by a company or an individual to identify their brand and distinguish their products from those of others. The trademark could be in the form of a word, name, song or a symbol.

A name, logo, slogan, song, domain name, shape or sound can be registered as a trademark. A trademark must be:
- unique and distinctive
- fair and accurate, not deceptive
- morally acceptable.

Trademarks must be renewed every ten years.

One special category of trademark is known as trade dress protection. It was initially used for product packaging, for instance Tiffany's little blue boxes or Louboutin's red soles. Even without looking at the inside of the shoe and seeing the brand name, you see the red sole and know it is a Louboutin shoe.

A copyright is a set of exclusive rights or protection given to creators of original ideas, information or other intellectual works. It is often seen as 'the right to copy' an original creation.

Copyrighted material can only be copied, used or recreated with the owner's permission.

Copyright protection is automatic and registration is not needed. Most works are marked with the copyright symbol ©, the creator's name and the date, but even without the mark, the work is protected. Copyright lasts 70 years for most types of written, dramatic and artistic works, and at least 25 years for photographs. The copyright on other types of work last for different lengths of time.

Copyright does not protect the ideas for a piece of work. When the idea is confirmed, for example in the form of a logo, copyright automatically protects it. A copyrighted work may have another IP right connected to it, for example, a logo may be registered as a trademark.

Design rights concern the rights of the creator of a design or designs, unless a third party commissions the work. Unregistered design rights will protect the configuration or shape of a product and are used to prevent copying of an original design without permission.

Design rights do not give protection for any of the 2D aspects, for example patterns, but 2D designs can be protected as registered designs or by copyright law.

Design rights can be bought, sold or licensed. They stay in force for either ten years after the first marketing of the product(s) that use the design or 15 years after creation of the design, whichever is earlier. Anyone can be prevented from copying the design for the first five years. For the rest of the time, the design is subject to a licence of right. Design rights only give you protection in the UK.

Patents are the most common type of IP protection. Although they are difficult to obtain, they hold the strongest protection. Patents are granted by the government and protect designs and inventions for 20 years, providing strong protection against copying of the technical and functional aspects of a person's invention without permission. It covers details such as how the item works, is made and what it is made of. Obtaining patents can include long, expensive, technical processes and inventors must make all the details of their product known to the public. For example, James Dyson obtained a patent for his bagless vacuum cleaner and took Hoover to the High Court and won when they developed a similar vacuum cleaner in 2000. Nike is one of the top ten holders of US design patents for use of technologies in clothing and footwear.

To be considered for a patent an invention must:
● be new
● have an inventive step (a non-obvious feature) that is not obvious to someone with knowledge and experience in the subject
● be capable of being made or used.

Case studies

In 2016, the Supreme Court ruled that a rival product to the Trunki (a ride-on suitcase for children), called the Kiddee Case, was different enough in appearance for Trunki maker Magmatic not to be able to prevent its sale. The creator of the Trunki, Rob Law, tweeted to say he was 'devastated' with the ruling. In giving the ruling, Justice Lord Neuberger said the visual design of the Trunki was 'significantly different from the impression made by the Kiddee Case' and therefore Magmatic's rights were not infringed.

A similar case between Dyson and Vax has been rejected by the courts and ongoing disputes between Apple and Samsung have been well documented in the news.

ACTIVITY

Use these examples and find examples of your own to discuss how legislation can protect designers and consumers and how companies have an ethical responsibility to follow IP laws.

KEY TERMS

Copyright – a set of exclusive rights or protection given to creators of original ideas, information or other intellectual works.

Design rights – the rights of the creator of a design or designs. Design rights do not, however, give protection for any of the 2D aspects of the design, for example patterns.

Intellectual property (IP) – all kinds of intangible (not physical) types of property that people can own, for example, the creative outcomes from the mind such as design ideas, written material, artistic and musical composition.

Logistics – the careful organisation of a complicated activity so that it happens in a successful and effective way.

Patents – granted by the government to protect designs and inventions, providing strong protection against copying of the technical and functional aspects of an invention.

Registered design – gives ownership rights for the appearance of a product, protecting both the shape and the pattern or decoration.

Social media marketing (SMM) – a form of internet marketing that utilises social networking websites as a marketing tool.

Trademark – used by a company or an individual to identify and distinguish its products from those of others.

- Distribution is the action or process of supplying goods to retailers/consumers, often referred to as the distribution channel. It can include any physical store or shop as well as TV shopping channels and the internet.
- Social media marketing (SMM) utilises social networking websites as a marketing tool and is often are used to directly sell to consumers, allowing them to click and purchase.
- Search engine optimisation (SEO) is a strategy for drawing new visitors to a website and can then promote further social media activity such as tweets or status updates.
- Globalisation means goods and products are transported worldwide by air, sea and road.
- In the design world a number of laws exist to protect ideas and inventions. These are known as intellectual property (IP) laws and relate to all kinds of intangible (not physical) types of property that people can own, for example, the creative outcomes from the mind such as design ideas, written material, artistic and musical composition.

3.5 How can skills and knowledge from other subject areas, including mathematics and science, inform decisions?

LEARNING OUTCOMES

By the end of this section you should have developed a knowledge and understanding of:
- the need to incorporate knowledge from other experts and subjects to inform design and manufacturing decisions, including the areas of science and mathematics
- how undertaking primary and secondary research and being able to interpret technical data and information from specialist websites and publications supports design development.

Knowledge from other experts and subject areas

Other people are an extremely valuable source of ideas that often gets overlooked. It is truly amazing the viable ideas that can come from subject experts in your own and other fields. Other people may be able to provide knowledge to which you would not otherwise have access.

Designers should never be afraid to ask others for ideas or opinions. Even if someone does not have any solutions to offer, they may know where to locate information or a person who can provide information. This is called networking and can be one of the most powerful tools in any business.

As we have learnt in other chapters, you will need regular consultations with users and stakeholders regarding their requirements.

Designers are increasingly consulting scientists and technologists in order to incorporate technological developments into their products and to ensure that these aspects of their designs are successful. Experts can provide advice on human body functioning, physiological factors, investigative skills, toxicology, ecology, psychological information relating to ergonomics, social sciences and cognitive processes.

During your NEA project you will find opportunities to research and learn from the work of others. You will also draw upon knowledge gained in other subjects. You should record and document this exploration as it occurs.

You can read more on how designers collaborate with experts in different areas, such as manufacturers and other specialists, in Chapter 4.

Undertaking primary and secondary research and interpreting technical data and information

Throughout the process of design and manufacture of a product it is necessary to gather and interpret information and data. This could be in the initial stages of design to gather, analyse and relate anthropometric data; in the development of a product to find suitable materials or components for the product; or during the planning of the manufacturing process to organise suitable tooling and machinery for successful production.

Sources of information that could be useful when designing include:
- information collected first hand (primary research) – for example, carrying out questionnaires or surveys,

interviewing users and stakeholders, using focus groups and first-hand observations

- looking at and analysing solutions to similar problems, other products with similar features and other products used in the context you are designing for
- sizes and other technical details of items that your product or system will need to function with or the space it needs to function within
- data from magazines, reference books, government agencies. This could take the form of data about material properties or component characteristics, or be tables of anthropometric data, information on forces or strengths required for certain tasks.

The precise source and nature of the data and its significance to each stage of the work of a designer will depend enormously on the field of work. Whatever the field, information is often available from:

- trade organisations, for example:
  - The Timber Research and Development Association (TRADA)
  - The Society of Motor Manufacturers and Traders (SMMT)
  - The Building Research Establishment (BRE)
  - The British Polymers Federation (BPF)
- manufacturers' data sheets
- Engineering Village (www.engineeringvillage.com) the most comprehensive database for engineering
- Granta Design (www.grantadesign.com), which as supports materials education and research across a wide variety of subjects in engineering, science, and design

- EBSCO, a large multidisciplinary database that is good for product design and ergonomics
- ProQuest, a multidisciplinary database which includes trade publications, news and peer reviewed journal titles
- British Standards Online, which has access to all current British Standards in full text. These have to be purchased, but copies of commonly used ones are found in some libraries
- Espacenet (https://worldwide.espacenet.com) and the Intellectual Property Office (www.ipo.gov.uk), which provide information about patents
- Keynote Reports, which publish the latest market research information, market size, competition analysis, SWOT analysis and market and trend forecasts
- regulatory bodies that provide information on materials or components you may wish to use
- other published sources, such as directories, books and periodicals.

Although traditional printed media should never be forgotten as a source, information is increasingly available electronically.

There are many specialist internet search engines that support the design process by collating information sources in particular fields of design, for example, the engineering search engine IEEE Globalspec (www.globalspec.com). Some of these search engines operate as free sources, while others operate a charging policy as commercial sources for business users.

You should always state the source of information you have found and explain how it has helped further your design iterations.

3.6 What energy factors need to be considered when developing design solutions?

LEARNING OUTCOMES

Many electrical and mechanical engineered systems need a source of energy to operate.

When exploring energy sources for a particular application, a design engineer would need to consider some of the following factors:

- cost of the energy
- reliability of the energy source
- how the energy can be stored until needed
- method of refuelling or recharging

- power requirements (**voltage** and current for electrical sources)
- portability of energy source
- how the energy can be transferred to where it is needed
- environmental impact
- energy density (the physical dimensions of the source and how long it will last).

Electrical energy sources for products

A Level design engineers will mostly be interested in a supply of electricity for their projects, which is normally derived from low voltage batteries or 230 V mains electricity. For reasons of safety, you should never try to use mains electricity in your own projects, except for a commercial low voltage **power supply unit (PSU)** which plugs in to the mains and produces a low voltage output.

Chapter 6 describes the use of PSUs and batteries as energy sources for electronic systems.

Table 3.4 **Comparison of electrical energy sources for products**

	Advantages	Disadvantages
Mains electricity	• Reliable • Able to provide continuous high power • Low cost	• Requires a wired connection to a supply • Not portable • Dangers of high voltage
Non-rechargeable batteries	• Readily available • Simple solution • Achieves product portability	• Expensive source of energy • Significant environmental impact • Can run out unexpectedly • Need to be repeatedly changed • Limited shelf life • Requires product to be opened, compromising case seals
Rechargeable batteries	• Achieves product portability • Can be sealed inside product – no need to open case • LiPo batteries can be very slim, with a very high energy density	• Limited charged shelf life • Limited number of charge/discharge cycles • Need for appropriate charging circuit • Time taken for battery to recharge • Dangerous if not used properly
Wind generator	• Free, clean energy • Ideal in remote locations	• Output not consistent • Need for servicing – limited operating life • Noisy • Expensive to purchase
Solar photovoltaic (PV) panel	• Free, clean energy • Ideal in remote locations • Long lasting – no moving parts	• Low power output • Output not consistent • No output at night • Need to keep panel clean

Small wind turbines or solar panels are a realistic option for powering small products and systems. They are relatively expensive to purchase, but they supply a free, low environmental impact energy which may be the only option in remote areas.

Electronic systems require a steady **power** supply voltage. Fluctuations in the **power supply** can cause microcontrollers to crash or behave erratically. Consequently, solar panels and wind generators are always used in conjunction with a rechargeable battery; the battery is kept charged by the solar panel or wind generator but, should the sunlight level or wind speed reduce, the battery will maintain a steady voltage for the electronics. Electronic systems frequently feature a 'brown-out' detector circuit which monitors the battery voltage and safely shuts down the system if the battery voltage drops to a very low level.

Other energy sources

Electrical energy is not the only source for engineered systems. Pneumatic systems use compressed air stored in a pressure vessel, transport vehicles use energy stored in liquid fuels such as petrol and diesel and some mechanical systems use stored elastic energy, such as a wind-up radio or torch.

Hydrogen fuel cells are likely to feature more in future energy scenarios. A fuel cell combines hydrogen with oxygen (from the air) to produce electricity and heat, and water as a waste product. Current fuel cells are used in highly specialised applications, such as in space craft. They are not yet cheap or efficient enough to replace conventional ways of producing or storing electricity. As the demand increases, however, you can expect fuel cells to become more popular for certain applications. It is worth remembering that the fuel, hydrogen gas, does not exist naturally on Earth, but only combined with other substances. Therefore, energy is required to extract hydrogen for use as a fuel and this energy would need to come from a conventional electricity power station.

Figure 3.22 **Energy can be generated by solar panels and wind turbines**

ACTIVITIES

1 As a group, think about other subject experts and specialists that you might contact in your NEA project.
2 Compile a list of websites and publications that could be useful for your specific material area.

KEY TERMS

Power – energy produced per second.
Power supply – the voltage supply which powers the electronic system.
Power supply unit (PSU) – provides a steady voltage while pushing current through the system.
Voltage – a measure of electrical 'pressure'.

MATHEMATICAL SKILLS

● Use appropriate methods to present any performance data, survey responses and information on your design decisions, including the use of frequency tables or graphs and bar charts.
● Interpret and extract appropriate data from technical and graphical sources.
● Interpret and construct charts appropriate to the data type, including frequency tables, bar charts, pie charts and pictograms for categorical data and vertical line charts for ungrouped discrete numerical data. Interpret multiple and composite bar charts.
● Calculate the mean, mode, median and range to help you work out sizes, etc.

KEY POINTS

● Throughout the process of designing and manufacturing a product it will be necessary to gather and interpret information and data. This could be in the initial stages of design, in the development of a product or for the product's manufacture.

● Use appropriate methods to present any performance data, survey responses and information on your design decisions and always acknowledge sources of information used.

Further reading on implications of wider issues

● Chock, A and Micklethwaite, P. (2011) *Design for Sustainable Change: How Design and Designers Can Drive the Sustainability Agenda*, Ava Publishing
● The Sustainability Handbook for D&T Teachers: www.practicalaction.org
● Fuad-Luke, A. (2009), *The Eco-design Handbook: A Complete Sourcebook for the Home and Office*, 3rd ed. Thames and Hudson Ltd

● Wallman. J. (2015), *Stuffocation: Living More with Less*, Penguin
● Barbe, S., Cozzo, B., and Tamborrini, P. (2009), *EcoDesign*, hf ULLMANN
● McDonough W. and Braungart, M. (2009), *Cradle to Cradle: Remaking the Way We Make Things*, Vintage
● Klantenm, R. (2012), *Cause and Effect: Visualizing Sustainability*, Die Gestalten Verlag

PRACTICE QUESTIONS: implications of wider issues

Product Design

1 Discuss how a designer's responsibility extends beyond meeting the needs of the consumer and manufacturer.

2 Discuss the environmental implications of increased energy use.

3 Discuss ways in which consumers can help to conserve the environment.

4 Many UK-based companies manufacture their products in other countries. Discuss the moral and/or ethical considerations of the globalisation of product manufacture.

5 Discuss the environmental implications of using metals for food and drink cans.

6 Discuss the implications of using non-sustainable resources in disposable products.

7 Discuss the implications for the design of packaging to enable a reduction in the volume of disposable waste.

8 Discuss the implications of using recycled materials in the manufacture of products.

9 Many products are designed to have a limited life expectancy. Discuss the implications of the increased availability and use of 'throw-away products'.

10 Discuss the benefits and drawbacks of the use of legislation to protect IP.

Fashion and Textiles

1 Discuss how a clothing manufacturer's responsibility extends beyond meeting the needs of the consumer and manufacturer.

2 Discuss the environmental implications of cotton picking and ginning.

3 Many UK-based companies manufacture their textile products in other countries. Discuss the moral and/or ethical considerations of the globalisation of product manufacture.

4 Discuss the environmental implications of using synthetic fabrics for fashion clothing and accessories.

5 Discuss the implications of using recycled materials in the manufacture of textile products.

6 Many textile products are designed to have a limited life expectancy. Discuss the implications of the increased availability and use of 'disposable fashion items'.

7 Discuss the benefits and drawbacks of the use of legislation to protect IP.

Design Engineering

1 Discuss how the engineer's responsibility extends beyond meeting the needs of the consumer and manufacturer.

2 Discuss the environmental implications of increased energy use throughout a product's life.

3 Discuss ways in which consumers can help to conserve the environment.

4 Many UK-based companies manufacture their electronic products in other countries. Discuss the moral and/or ethical considerations of the globalisation of product manufacture.

5 Describe two environmental directives relating to reducing the impact of waste electrical products.

6 Discuss the implications of using recycled materials in the manufacture of products.

7 Discuss the role of designing for maintenance and product support in extending the life expectancy of a product.

8 Identify the main technical issues involved in using wind generators or solar panels to power electronic products.

9 Discuss the benefits and drawbacks of the use of legislation to protect IP.

OCR Design & Technology for AS/A Level

4 Design thinking and communication

In this chapter you will find a number of case studies outlining different approaches to design ideas. The case studies are not exhaustive and you will not need to know about these particular examples for the exam, but they may help you in understanding different design ideas. You can undertake your own case studies to support and contextualise your learning.

4.1 How do designers use annotated 2D and 3D sketching and digital tools to graphically communicate ideas?

Introduction

The ability to represent design ideas visually is important for a designer. Sketching enables communication between designers, design teams and different types of stakeholders. A sketch can express more than words and give a greater understanding of an idea and how it is used. Sketches are an easy and fast way to communicate an idea to others.

Ideas are often communicated with quick sketches, and diagrams are also used to help understand the context. Sometimes they are rough and often have limited detail, but they are sufficient to help understand user and stakeholders needs, understand the problem and explore the context where the product will be used. Sketches can be used to explore design concepts. Sometimes quick sketches are used to grasp the overall idea rather than the finer details. With annotation they help explain the design concept, explaining function, structure and form. They communicate to other people and are often produced to get client feedback.

As the design ideas and iterations move closer towards a final design solution, sketches can be used to present more details of a design concept and are often produced using digital methods. Sometimes physical models will also be used at this stage, as they can help put a design into context, give a better 3D representation, explore usability and obtain valuable feedback from users and stakeholders.

Design solutions are normally communicated through a series of visualisations, often incorporating a variety of graphical techniques including digital tools and simulations. The final solution, including constructional details sufficient to inform a third party of the designer's intentions, will normally involve formal CAD (computer-aided design) **working drawings** and diagrams alongside models showing technical details.

How to use annotated sketching and digital tools to graphically communicate ideas

When producing quick sketches in 2D or 3D, the most important thing is to find the method that you are most comfortable with. It really doesn't matter whether it's a pencil, biro, rollerball or fine liner that is used, or a digital graphic drawing tablet and pen. Try some methods you maybe haven't tried before and see which of them suit you best.

2D sketches will look flat on the page, but different line thicknesses, colour and texture can be used to communicate ideas more clearly. A 2D sketch might be used for working drawings such as orthographic projections, sectional drawings or a **lay plan**. They could also be used to show a layout or overall shape of a product or garment. For more details see the section on working drawings.

3D sketches are needed to fully communicate a design idea visually. In order to create realistic 3D sketches, you need to understand **perspective drawing**. Correct perspective can communicate the proportions of an object and show it as it is seen in reality. The most common types of perspective used by designers are two- and three-point perspective. Three-point perspective is the exact translation of the real-life situation. Two-point perspective is mostly used by product designers to create believable sketches and is a faster technique. Three-point perspective is commonly used in architecture. A good perspective drawing will look 'natural' to the eye, so it's worth practising the technique. Designers often present a range of sketches, including 2D and 3D, so that they can communicate an idea clearly.

Software applications are also used widely for creating and exploring ideas. Simple 2D and 3D sketching programs and apps can be used to quickly visualise an idea and variations in the form, shape and colour of a design can be applied.

Isometric drawing is another way of presenting designs/drawings in 3D. Side edges are drawn at an angle of 30 degrees to the horizontal. Designs drawn in isometric projection can be drawn accurately using drawing equipment. However, many designers find 'freehand' sketching in isometric projection allows thoughts to be put down on paper quickly, drawing freehand or using an isometric grid as an underlay.

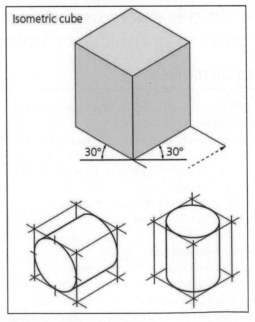

Figure 4.1 Isometric drawing

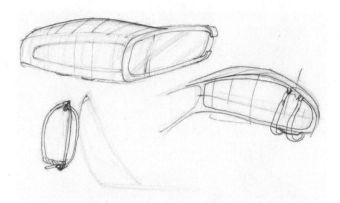

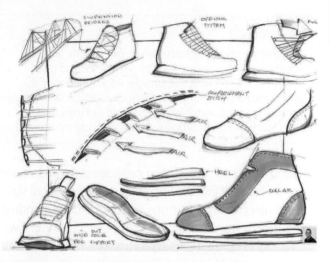

Figure 4.2 Freehand sketching is used by designers to communicate ideas to clients. These are sketches by Mark Hester of the Imagination Factory and accessories and footwear designer Cesar Idrobo

Freehand sketching is like using a pencil and a notepad as an extension of your own brain and thought process. In 1959, Alec Issigonis designed the original 'mini' car. His freehand sketches were said to have been made on a tablecloth in his local restaurant.

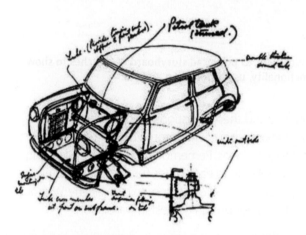

Figure 4.3 Original sketch of the Mini by Alec Issigonis

Fashion designers initially sketch a figure then hand sketch garments onto them to give the idea of form and shape when worn. The figures are sometimes adapted from photographs.

Design students and design professionals sometimes use notebooks to record their iterative design process – recording sketches, **modelling** and the results of testing.

Figure 4.4 Sketches can be adapted from photographs

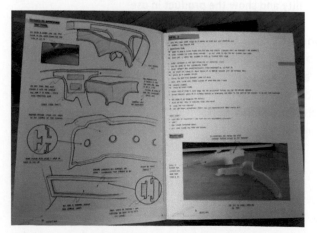

Figure 4.5 Examples of a student sketchbook recording their iterative design process

Sketching a phone or calculator in perspective and isometric can be a good way to get started and build confidence. Start with the straight edges then start to add curves and fillets. The rectangular and circular shapes provide practice sketching ellipses and radiuses. You can then start on more complex shapes and add more detail.

Adding details to 3D sketches provides stakeholders and clients with vital information, indicating the overall size of an object and any buttons, features and mechanisms. Producing a storyboard or scenario sketches can show a product in context and explore how a user will interact with it. Designers attempt to understand the design solution in more detail and any possible issues associated with it, such as user needs, the context and scenario of use, functional and material choices and possibilities for manufacture, while at the same time optimising design solutions.

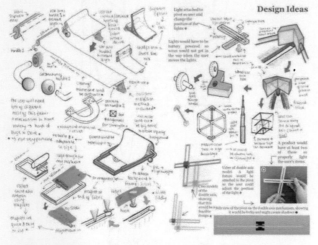

Figure 4.6 Student design work, using 3D sketches to explore iterations of an idea for a portable photo studio. The sketches explore functionality, construction, movement and stability and show the product in use

Annotation

Whenever you sketch or draw an idea it is important to add annotation. Sometimes this will be in detail and sometimes it will just be quick notes to aid communication. Some things to think about when annotating ideas and design iterations are:

- How does the design function?
- How does the design achieve structural integrity?
- Is it designed with the user in mind – is it ergonomic? Is the design safe?
- Will it suit the stakeholders that it is designed for?
- What materials will be used in its manufacture and how much will it cost to produce?

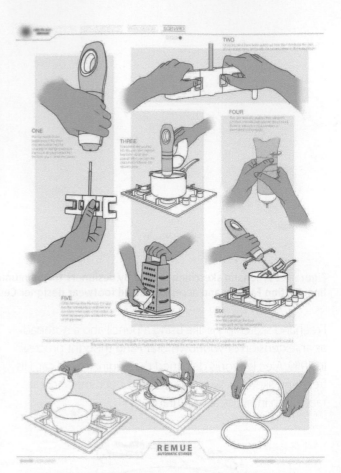

Figure 4.7 CAD produced storyboard of sketches to show functionality, user interaction and use

You may find these questions useful as they will get you thinking about what should be written as notes next to your ideas. Remember that sketches showing parts in relation to each other, functionality and how the product is used can communicate clearly. Without sketches, a design can get lost in translation.

Modelling

Modelling is also commonly used to communicate ideas. At this point, sketch modelling is often enough to show what the idea or concept is, with little detail, so that the general idea can be accepted or rejected by stakeholders and suggestions made for the next design iteration.

Sketch models are not quick sketches but quick models, often of just parts of a design, made from easy-to-work and low-cost materials such as cardboard, **calico** or foam. They help to convey product scale and explore user interaction and are created with the purpose of sharing an idea. Breaking a design down into parts and using sketch models helps to show how parts interrelate and function in order to optimise the design. Sketch models can be tested with users and other stakeholders to get feedback before accurate models are made that focus on details.

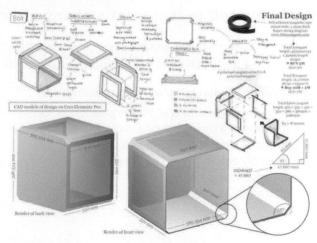

Figure 4.9 **Details of the final design for portable photographic studio and specific material and component details**

Figure 4.8 **Example of vacuum formed components within a circuit model.**

Mathematical modelling is also widely used to optimise designs. It can be used to predict the performance of a design in different situations and to solve potential problems. Charts and graphs can be used to represent data relating to possible design improvements.

Generative design tools that produce optimum forms are used by designers to make aircrafts lighter, cars more aerodynamic, bridges more engineering efficient and footwear more comfortable. Programs mimic the way nature and organisms evolve in the natural world, creating fluid forms – this is often classed as biomimicry or **generative design**. Examples of this are the cellular midsoles customised for each runner using 3D printing by Adidas in their future soles and New Balance and Nervous System's plans to use the technology to produce shoes that are minutely customised to different runners' feet.

Autodesk is a company that creates generative design software. You can find out more about generative design software at their website, www.autodesk.com.

The manufacturing processes that will be used, specific materials and details of any standardised components are often shown in final ideas to be presented to clients.

Rendered life-like drawings are often generated by CAD to give an idea of what the finished product will look like, in terms of materials, colour and textures.

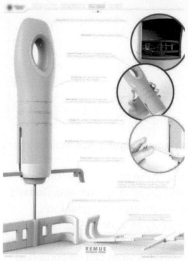

Figure 4.10 **Rendered CAD drawing that communicates a product in use, in this case an automatic stirrer for food**

Computerised simulation software can be used to test mechanical and electrical components and parts without the need to physically build circuits and physical working prototypes.

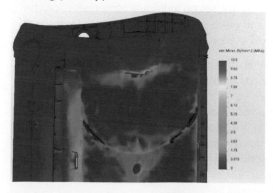

Figure 4.11 **Mathematical modelling to analyse the effects of stress on parts. This example shows a Jurni suitcase. The modelling shows designers where material might need to be strengthened**

113

Methods used to represent systems and components to inform third parties

Constructional diagrams/working drawings

Working drawings contain all the information for the construction of a design solution, including:

- details of individual components
- assembly details
- dimensions and size
- materials.

Working drawings mainly take the form of orthographic projections, with a plan, front and side view and sometimes a sectional view. For some products a 3D drawing may be appropriate, or an **exploded view**, but it is the orthographic drawings of the component parts of a product, with orthographic views of the assembled product, which constitute the primary technical specification for construction or manufacture.

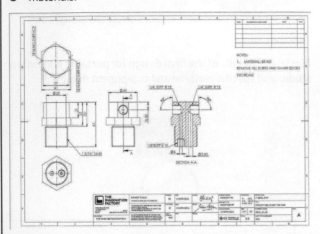
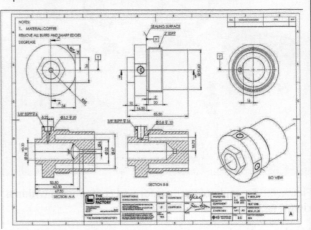

Figure 4.12 **Working drawings of a pump fire extinguisher cap and a pump heat sink**

Traditionally these drawings were produced on drawing boards using T-squares and drawing instruments. Today's designers, architects and engineers use digital tools, as drawings can be more easily amended. A range of software tools enable the creation of complex shapes and designs. The geometries of the shapes can be converted into machine code and used in rapid prototyping/3D printing to manufacture components.

Working drawings are usually accompanied by parts lists which can be useful for planning production. CAD or assembly drawings can show how parts would be assembled in 3D, saving money and speeding up the production process. In the design of structures and moving parts, stresses on individual components can be simulated and predicted and parts can be strengthened before physical prototypes are built.

The standards and conventions for engineering drawings are controlled by British Standards in BS 8888. This UK standard covers all of the requirements for the technical specification of products and their component parts, bringing together international standards. It is the latest version of the standard that was written to supersede BS 308, the world's first engineering drawings standard, first published in 1927.

BS 8888-1 is a guide for schools and colleges to enable teachers and students to familiarise themselves with the British Standard for Technical Product Specifications

(TPS) by introducing students to universally accepted ISO technical drawing practices.

These documents give guidance, for example, on the correct use of line types and the correct use of text. Only pencil or black ink should be used to produce formal engineering drawings and only two thicknesses of lines should be used – thick lines for outlines and visible edges, for example, and thin lines for dimensions, centre lines, hatching, projection and leader lines. Capital letters are used for text. The standard also gives details of standard graphical symbols and standard methods of dimensioning. The standard establishes a 'common language' for technical drawings and specifications to enable clear and unambiguous communications between designers and manufacturers.

Most products, whatever the size or material, will be made up of a number of components, some complex products from many thousands of components. Each component has its own specific material and method of manufacture. It is therefore important that each component is drawn separately and defined fully to specify all the technical details. Dimensioning a drawing may also identify the tolerance (or accuracy) required for each dimension. **Tolerance** is the amount a particular dimension is allowed to vary and achieve functionality.

Exploded views are widely used to show the detail of how the parts of a product are assembled – how they fit together. Accurate exploded views can take a long time to construct but are useful in understanding and communicating the layout of components and parts in relation to each other.

CAD programs and digital tools can be used to create accurate exploded views. Creating an exploded view with isometric or perspective grid paper is relatively straightforward. Figure 4.13 shows the component parts of part of an eye tracker in an exploded view, produced using CAD. Note the numbered parts and separate parts list.

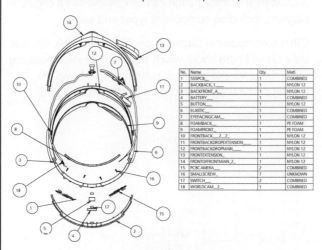

No.	Name	Qty.	Matl.
1	555PCB__	1	COMBINED
2	BACKBACK_1__	1	NYLON 12
3	BACKFRONT_A__	1	NYLON 12
4	BATTERY__	1	COMBINED
5	BUTTON__	1	NYLON 12
6	ELASTIC__	1	COMBINED
7	EYEFACINGCAM__	1	COMBINED
8	FOAMBACK__	1	PE FOAM
9	FOAMFRONT_	1	PE FOAM
10	FRONTBACK__2_2_	1	NYLON 12
11	FRONTBACKDROPEXTENSION___	1	NYLON 12
12	FRONTBACKDROPMAIN___	1	NYLON 12
13	FRONTEXTENSION__	1	NYLON 12
14	FRONTOFFRONTMAIN_2_	1	NYLON 12
15	PCBCAMERA__	2	COMBINED
16	SMALLSCREW__	7	UNKNOWN
17	SWITCH__	2	COMBINED
18	WORLDCAM__2__	1	COMBINED

Figure 4.13 Exploded view example eye tracker project showing positioning of all parts in relation to each other and giving details of materials

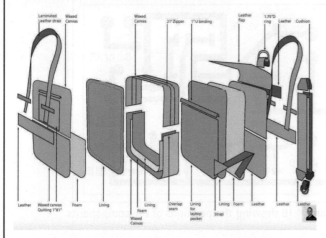

Figure 4.14 Exploded view of bag by Cesar Idrobo, rendered showing detailing of all parts and materials

Flowcharts with associated symbols

Flowcharts are used to represent a process and can be helpful when communicating the sequence of processes involved in the construction or manufacturing of a design solution. In electronic systems, a flowchart is frequently used to show a microcontroller program (see Chapter 6).

Using shaped boxes or symbols, with lines and arrows showing the sequence, flowcharts show different types of actions or steps in a process. Points where checks or decisions are needed are identified as well as the stages needed to correct any problems or get the process back on track.

Many different boxes or symbols are used, but the most common are:

● terminal – the start or end of a process, usually shown as an oval shape
● process – this explains an activity or step in a process and is shown by a rectangular box
● input or output – often shown by a parallelogram
● decision – usually shown by a diamond.

It is important to be familiar with a range of graphical methods that are available to designers and engineers and to select the method that is best suited for the application. For example, simple freehand sketches, which are invaluable in communicating initial design ideas, would be of little use as an engineering workshop drawing. Similarly, such formal drawings would normally be inappropriate for presentation and marketing applications.

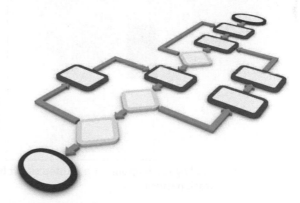

Figure 4.15 Example flowchart

Circuit and system diagrams

Schematic diagrams are used to show the arrangement of components in electrical, electronic and pneumatic circuits and systems. Circuit diagrams and **system diagrams** are examples of schematic diagrams. They are used to indicate the relative points of interconnection of the components only. They do not show the physical positions of the components within the system as a whole.

The illustrations show examples of the different types of diagrams used to communicate information relevant to electronic systems. A system diagram (Figure 4.16) shows the various input sensors and output devices and how they are interconnected to a microcontroller. The lines show signals and the arrows show the direction in which the

signals travel. System diagrams are used to communicate the overall structure of a system and how its subsystems are interconnected. See Chapter 6 for more information.

The circuit diagram of the same system (Figure 4.19) shows how the individual components are connected. Standard symbols are used to represent the components. A great deal of information can be conveyed in a circuit diagram, including component types and values.

Other constructional diagrams specific to electronic systems are a circuit board layout which shows the physical positions of components and the track pattern which shows the routing of the copper tracks on a printed circuit board.

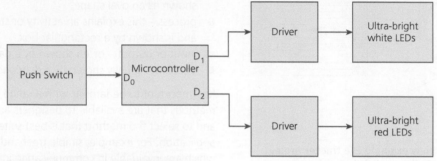

Figure 4.16 Electronic system diagram

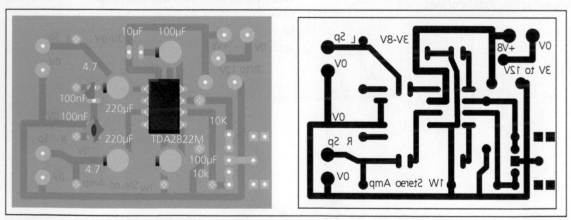

Figures 4.17 and 4.18 Layout diagram for a printed circuit board, 4.18 showing the positions of the components, 4.19 showing the track pattern

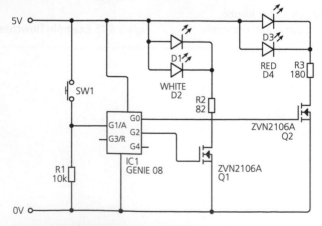

Figure 4.19 Electronic circuit diagram

FASHION AND TEXTILES

Lay plans are used in fashion and textiles products to show how the individual pieces of a garment or other textiles product are to be cut from a piece of fabric. Pattern pieces are laid out on the fabric to determine the most economical way of cutting them out. This must take into account the directional properties of the fabric such as pattern direction, nap or pile, or to match stripes, checks or designs. A lay plan will also show sizes and details such as seam allowances, gatherings and darts.

In a fully automated system, a computer program will calculate and determine the most suitable lay plan and then transmit this to the cutting system

Figure 4.20 shows how a student has created an economical lay plan for the final product of her NEA project. The lay plan shows a range of fabrics required to construct the product and allow for fabric print direction, fabric width and fabric stretch direction for the green T-shirt.

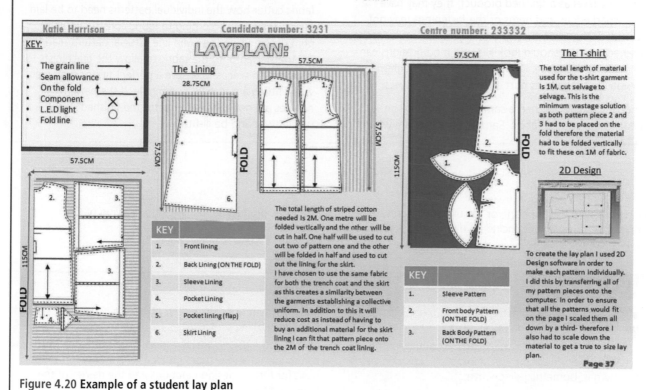

Figure 4.20 Example of a student lay plan

Prototypes and models

Freehand sketching is probably the best way in which to communicate ideas. Initially you may just use the sketches to explore ideas but, as the design develops, there will be a need to use them to communicate with others. Modelling ideas can also do this, whether by a sketch model or at the later stages a more detailed scale model which can be used to demonstrate functionality, any moving parts and a visual idea of the product. It can be used to test the product's usability and features with stakeholders.

Models can be made full scale in materials such as foam, board and card. Styrofoam, wood-based materials and 3D printing can then be used to test concepts.

Figure 4.21 Styrofoam model of a drone

Prototypes and Toiles for Fashion and Textile products

In the fashion industry, toiles are used at the modelling stage. A toile is a test version that can be used before production. It is usually made from calico fabric as it is cost effective to use. However, the fabric used should perform in a similar way to the intended fabric to give a realistic result. Toiles are not usually completed to the same level as a finished product; they may have unfinished edges and hems or the fastenings may not be added. However, they give an impression of what the finished product should look like and modifications can be made at this stage. Designers may go through many versions of the toile as part of the iterative process.

Once the toile is finished a sample garment will be made in the correct fabric. Sample garments are what designers use for their catwalk shows and are essentially the prototype at their final stage to be viewed by potential buyers.

Pattern drafting with relevant cutting and construction symbols

Patterns are a crucial stage of the product development process, as they ultimately dictate the finished fit and look of a garment or textiles product. Patterns inform the fabric cutter how the individual patterns need to be laid out on the fabric according to fabric construction and print design using the grain line. Crucial pattern markings are transferred onto the fabric to allow the machinists to align certain pieces to ensure a perfect fit, for example when fitting a sleeve in an armhole. If the pattern notches are not aligned correctly, the finished sleeve would twist and be ill-fitting.

ACTIVITY

Using the sketch of a phone or calculator you drew earlier in the chapter, produce an engineering drawing or CAD drawing of this product, including dimensions and any additional detail.

KEY TERMS

Calico – a plain-woven textile made from unbleached and often not fully processed cotton.

Exploded view – shows the layout of components and parts in relation to each other. Almost all exploded views are drawn in isometric perspective.

Flowcharts – used in designing to document simple processes. They help with understanding a process and usually use process, input or output symbols.

Generative design – design that mimics nature's approach to design. Generative design software produces many alternative iterations for designers to consider.

Generative design tools – produce optimum forms to make products more engineering efficient. Programs mimic the way nature and organisms evolve in the natural world, creating fluid forms. Biomimicry is an approach that seeks solutions by emulating nature's time-tested forms, structures, patterns and strategies.

Isometric drawing – another way of presenting designs/ drawings in 3D. A 30-degree angle is applied to its sides.

Lay plans – used in textiles to show the pieces to be cut and their sizes and any details such as seam allowances, gatherings and darts.

Mathematical modelling – used to optimise designs and predict performance. Computerised simulation software can be used to test mechanical and electrical components and parts without the need to physically build circuits and physical working prototypes.

Modelling – the preliminary work or construction used to communicate and test design ideas and which forms the basis for further design iterations and the design of the final product.

Perspective drawing – this can communicate a pictorial view of an object. Two-point perspective is mostly used by product designers to create realistic sketches.

Schematic diagrams – used to show the arrangement of components in electrical, electronic, hydraulic and pneumatic circuits, architectural plans and systems.

Sketch models – quick models, made from easy-to-work and low-cost materials such as cardboard, calico or foam.

System diagram – a diagram that shows the interconnections and signal flow between the microcontroller and peripheral devices.

Tolerance – the permissible range of variation in a dimension of an object. Tolerances may also refer to other characteristics such as weight, capacity, quantity or hardness. Sometimes known as allowance.

Working drawings – these mainly take the form of an orthographic projection, with a plan, front and side view and sometimes a sectional view.

KEY POINTS

- Sketches can be used to explore design concepts. Sometimes, quick sketches are used to grasp the overall idea rather than the finer details. With annotation they help explain the design concept, explaining function, structure and form.
- 3D sketches are needed to fully communicate a design idea visually. Perspective drawing and isometric projection can be used to create realistic 3D drawings.
- Whenever you sketch or draw an idea it is important to add annotation. Sometimes this will be in detail; sometimes it will just be quick notes to aid communication.
- Modelling is also commonly used to communicate ideas. Sketch modelling is often enough to show what the idea or concept is. Sketch models can be tested with users and other stakeholders to get feedback before accurate models are made.
- Mathematical modelling is also widely used to optimise designs. It can be used to predict the performance of a design in different situations and to solve potential problems.
- Working drawings mainly take the form of orthographic projections, with a plan, front and side view and sometimes a sectional view. They are usually accompanied by parts lists which can be useful for planning production.
- Exploded views are widely used to show the detail of how the parts of a product are assembled.
- Flowcharts are used to represent a process and can be helpful when communicating the sequence of processes involved in the construction or manufacturing of a design solution.
- Schematic diagrams are used to show the arrangement of components in electrical, electronic and pneumatic circuits and systems.
- Lay plans are used in fashion and textiles products to show how the individual pieces of a garment or other textiles product are to be cut from a piece of fabric.

MATHEMATICAL SKILLS

In examinations and NEA work you will be required to demonstrate mathematical skills to:
- use datum points and geometry when setting out design drawings or patterns
- present accurate 2D and 3D graphics to communicate design solutions
- interpret plans and elevations of simple 3D solids
- construct plans and elevations of simple 3D solids and representations (for example, using isometric paper) of solids from plans and elevations
- construct and interpret scale drawings and scale models.

4.2 How do industry professionals use digital design tools to support and communicate the exploration, innovation and development of design ideas?

LEARNING OUTCOMES

By the end of this section you should have developed a knowledge and understanding of:
- how digital design software is used during design development, such as:
  - visual presentation, rendering and photo-quality imaging
  - product simulation
  - scientific analysis of real-world physical factors to determine whether a product will break or work the way it was intended.

If you are studying at A Level you should also have developed a knowledge and understanding of:
- how designers develop products using digital tools and online collaboration, such as:
  - discussing and exchanging ideas with specialists
  - developing designs concurrently with other designers
  - explaining and communicating their design decisions to stakeholders.

Visual presentation, rendering and photo-quality imaging

Digital tools enable designers to visualise, develop ideas and communicate with their clients more easily, sharing ideas and information using cloud platforms, online forums or email. This in turn is making design a more collaborative process as it enables designers, engineers, manufacturers and stakeholders to have an easy input into the design process. There are now tools that allow CAD files to be edited in real time by more than one person and faster digital prototyping methods are enabling models to be made quickly and more efficiently, allowing a faster time to market and quicker response to trends. Digital tablets are commonly used by designers in all fields to produce quick sketches as apps become more intuitive and allow devices to replicate the freedom that sketching on paper provides. This section draws on real-life examples of products and the design process, collaborative design and methods of communicating ideas to others.

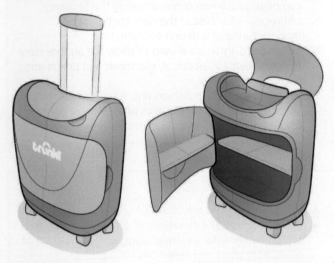

Figure 4.22 Examples of concept sketches produced digitally for the Jurni suitcase

3D CAD design is commonly used by designers and engineers to communicate ideas as it can provide rendered life-like interpretations of ideas. These can be used to share ideas with stakeholders and the software allows working models to be produced that people could see in use or experience hands-on. Most 3D CAD software has rendering functions that allow designs to be rastered to create a photo-like image, in different materials with different backgrounds and lighting. There are also specific rendering software packages such as Keyshot which can provide faster, more realistic results and is easy to use. Mental Ray and VRay are more complex packages that offer a huge variety of options. For your own NEA work you will probably find the rendering tools within the 3D CAD software used at your school or college sufficient.

Product simulation and scientific analysis

Many types of 3D CAD software now used in schools feature tools to perform scientific and engineering analysis on the modelled parts and on the entire assembly. Some of these tools are intended for professional engineers and may be beyond the demands of an A Level course. However, some functions are useful, such as the ability to calculate the weight of parts or the centre of gravity,

which can help predict the stability or balance of a product. The term applied for the use of these particular functions is computer-aided engineering (CAE).

Another CAE function of 3D CAD software is finite element analysis (FEA), which is a method for modelling products and components and simulating conditions to predict how products/components will react to force, vibration, heat and other physical, real-world scenarios. FEA software is used in a wide range of industries, particularly aeronautical, biomechanical and automotive.

'FEA works by breaking down components into thousands to hundreds of thousands of finite elements, such as little cubes. Mathematical equations help predict the behaviour of each element. A computer then adds up all the individual behaviours to predict the behaviour of the actual object.'

www.autodesk.co.uk/solutions/finite-element-analysis

Dynamic response to loads or motions can be used to work out critical loads at which a structure or component becomes unstable or breaks. FEA can also optimise a design and reduce material usage.

For your own NEA work you will probably find the tools within the 3D CAD software used at your school or college sufficient.

Circuit simulation software is another type of CAE. Packages such as Circuit Wizard or Yenka are commonly used in schools to quickly construct and test electronic circuits to see if they work as intended. In addition, they can help students design, test and manufacture printed circuit board layouts and physical component layouts. They provide the ability to write and debug microcontroller software.

Case study: The Imagination Factory

The Imagination Factory is a British product-design engineering agency co-founded by Mark Hester. They are a collaborative team that works with other designers to realise designs. In-house, the team has been working on SwimAR, an augmented-reality display for swimmers that can be attached to a regular pair of goggles. Using a human-centred approach and working with swimmers to develop the product, the team made sketches at the initial concept stage to generate and explore quick ideas. Concepts were then sketched in more detail and designs generated using CAD to develop the ideas through modelling, prototyping and testing.
See more about the project at the company website www.imaginationfactory.co.uk.

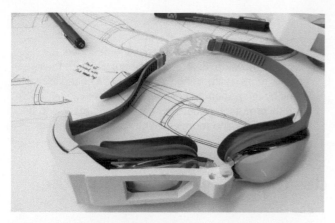

Figure 4.23 SwimAR design images from initial hand-drawn sketches through to use of digital methods and modelling and testing to iterate and optimise the design

Case study: Cesar Idrobo

Cesar Idrobo is a footwear and accessory developer based in the USA. He uses a range of sketches and modelling techniques to communicate his designs for footwear and accessories.

Cesar explores initial concepts with quick sketching and then more detailed sketches where usability is considered. Digital tools and CAD are used to produce life-like renders and modelling occurs as the product is optimised, from paper to fabric and leather. Cesar, in a recent post where he shares trade secrets, explains that conceptualising, as it becomes more technical and complex, makes greater use of digital and generative design to explore new shapes using algorithms. This allows the use of computing power to generate ideas by itself, running through options for the designer to choose the perfect geometry, allowing the designer to evaluate 3D shapes rather than 2D sketches using Grasshopper (a plug in) and Rhino, a 3D parametric CAD package. You can read about these techniques in Chapter 7.

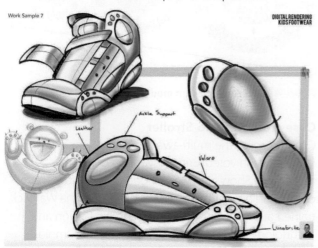

Figure 4.24 Examples of digital sketches produced by Cesar Idrobo for kid's footwear

Case study: Jurni

The makers of Trunki identified that many teenagers and adults really liked the popular children's sit-on suitcase, which generated £8.1 million in revenue sales, despite being rejected by *Dragons' Den*. Jurni was **crowdfunded** on the Indiegogo platform and launched ten years after Trunki. The Jurni features a pop-out pod, a seat, in-line wheels and flexible storage system.

During the process, the team at Trunki used quick sketches to develop concepts, initially as a group, then produced more detailed sketches using digital methods, showing Jurni in context to communicate the idea and how it is used. They used renderings as well as test materials with mathematical modelling techniques to work out tooling, optimise material use and achieve the necessary strength for allowing the design to be a success. In order to create a strong but light structure they developed the i-Beam, which can take up to 220 lb (100 kg) of sitting load. Hours were spent in the design studio where various ideas were tested.

Figure 4.25 Jurni photo post-its were used at the initial concept stage to generate and explore quick ideas

ACTIVITY

Explore the range of digital tools/software available in your area of design. Produce a presentation for your group, of examples of interesting and specific tools/software used by designers.

KEY TERM

Crowdfunding – an online method of raising finance by asking a large number of people for a small amount of money.

4.3 Using different approaches to design thinking to support the development of design ideas

LEARNING OUTCOMES

By the end of this section you should have developed a knowledge and understanding of:
- different strategies, techniques and approaches to explore, create and evaluate design ideas, including:
  - iterative designing
  - user-centred design
  - circular economy
  - systems thinking.
- the importance of collaboration to gain specialist knowledge from across subject areas when delivering solutions in the design and manufacturing industries

- how design teams use different approaches to project management when faced with large projects, such as critical path analysis, Scrum and Six Sigma.

Design Engineering

By the end of this section you should have developed a knowledge and understanding of:
- how design engineers use system design processes to define and develop systems that satisfy specified requirements of users using the three sub-tasks of:
  - user-interface design
  - data design
  - process design.

Different strategies, techniques and approaches to explore, create and evaluate design ideas

Iterative designing

The iterative design process encourages you to be more creative and take design risks. Not all design iterations will be successful, but they are necessary in order to discover whether ideas will work or not.

Many great designers and inventors developed design iterations before perfecting their inventions. Thomas Edison made thousands of unsuccessful attempts before perfecting his design for the incandescent light bulb in 1879. The Wright brothers had hundreds of failed flight attempts before achieving the first flight in 1903. James Dyson spent 15 years creating 5,126 versions of his bagless vacuum cleaner before a final successful design was found.

In both your exam and NEA you will need to demonstrate knowledge, understanding and skills of interrelated iterative processes that explore needs, create solutions and evaluate how well the needs have been met. A simple explanation of iterative design is that the design process is cyclical and repeats over and over to

improve the design again and again (design iterations) until the needs of users and stakeholders are met.

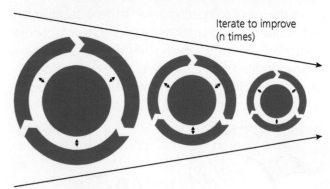

Iterate to improve (n times)

Figure 4.26 Iterative design model

Case study: Omnio Stroller

Omnio Stroller, the multi-award winning, innovative stroller design voted 'British Invention of the Year 2015' at the Gadget Show Live, was funded on the Crowdcube crowdfunding platform. Its designer, Samantha Warwick, trained at Brunel University to become a Design and Technology teacher before establishing the Omnio brand. The idea for Omnio was inspired by Samantha's and her husband (and business co-founder) Markus' own experiences of parenthood. They felt that the marketplace simply didn't offer a stroller for busy families that combined portability and flexibility with unique design. They explored and tested a number of concepts before optimising the design. The stroller is carried like a backpack when your toddler is happy to walk, so it is always with you as a backup. Omnio went through multiple design iterations and tests before the final product was launched on the market.

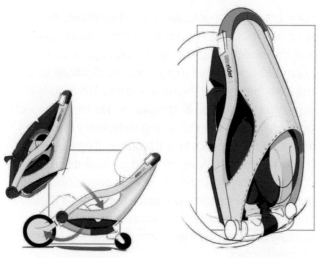

Figure 4.27 Initial concept for the Omnio Stroller (initially called 'Little Rider')

User-centred design

User-centred design (UCD) is based on the understanding of users, the tasks that they do and the environments in which they live and work. In UCD, users are involved throughout the design of products and systems, providing evaluation and feedback at every stage and on all aspects of the design. The process is iterative. The International Usability Standard, ISO 13407, specifies the principles and activities of user-centred design.

> 'The central premise of user-centred design is that the best-designed products and services result from understanding the needs of the people who will use them.'

The Design Council

Some methods commonly used in UCD are:

- focus groups
- usability testing
- **participatory design** (actively involving all stakeholders and users in the design process)
- interviews
- questionnaires
- observation.

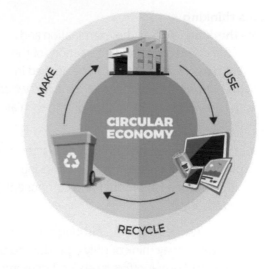

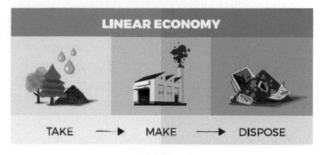

Figure 4.28 Circular vs linear economy

Circular economy

A circular economy is an alternative to a traditional linear economy (make, use, dispose) in which we keep resources in use for as long as possible, extract the maximum value from them while they are in use, then recover and regenerate products and materials at the end of their life. It is a cradle-to-cradle model.

Products that can be easily disassembled at the end of their life are being developed. Many companies have realised the importance of moving towards a circular economy, where all the waste can be put back into the system and reused or recycled, and they are designing products to allow this approach.

ACTIVITY

The Ellen MacArthur Foundation aims to raise awareness and accelerate the transition to the circular economy. Visit the foundation's website at www.ellenmacarthurfoundation.org/ and look for the case studies. Find a product from your area of interest and produce a presentation you can share with your peers.

Systems thinking

Systems thinking involves the examination and analysis of the connections and interactions of the component parts of a system. It can be applied in many fields of business and management, and when designing. This will be important for designers in all of the specifications.

A product or component is often part of a larger system of other products and systems. Systems thinking considers the role of each product and component in the systems around it.

Every product provides a service. Although designers and manufacturers may think of only a product, to the consumer and end user it offers a service. For example, music players provide a service – the enjoyment of listening – and smartphones provide us with tools to enable communication and interaction.

Using a new product is all about the whole experience, from opening its package, using and updating the product, the maintenance and eventual disposal or upgrade. If you think of all products as services that offer great experiences to their owners, you will be able to understand systems thinking when designing.

Systems thinking, user-centred design and consideration of the circular economy are all part of the overall design thinking – the need to consider the whole system from the start, not just the parts or individual products. The stakeholders within the system must be involved with the designing from start to finish and the way we use our resources is of increasing importance.

The importance of collaboration to gain specialist knowledge from across subject areas when delivering solutions in the design and manufacturing industries

Many successful products come into existence because of effective **collaboration** between specialists who are proficient in a particular area of design or manufacture. The key role of a project coordinator is to know how to draw on expertise from specialists and how to communicate effectively with them and set up successful information pathways so that each specialist can work efficiently and without hindrance.

Engineered products often contain electronic or mechanical systems which demand technical expertise from electronic or mechanical engineers, blended with an effective user interface designed with the assistance of product designers and ergonomic experts.

A standard remote control is an example of collaborative design, where electronic engineers and programmers would have focused on the function of the circuit, human factors specialists (perhaps with degrees in psychology) may have been consulted regarding the user interface (layout of buttons, menus, graphics and logos, etc.) and anthropometry specialists may have advised regarding button size and spacing, and the size and shape of the case.

For successful manufacturing, injection moulding specialists (with experience in effective **mould** design and the challenges presented by using particular polymers) would have suggested modifications to the initial designs for the remote control to ensure successful production of the case and of the soft-touch buttons, which have the added complexity of colour and different graphics on each button. The electronics manufacturer may have imposed constraints on size or component layout to ensure reliable operation. Several iterations would have taken place between all these specialists, coordinated by a lead designer, before a suitable compromise was reached to the satisfaction of all stakeholders.

Effective collaboration is critically important in the development of products. Involving users and stakeholders in the design process can mean the difference between success and failure, but collaboration adds complexity to a design project. In order to get products to market, collaboration is key in this modern globalised world.

The Omnio Stroller mentioned earlier in this section is an example of a collaborative company that operates an outsourced business, where some functions of the company's work are carried out by people or other companies from outside the business. With just three full-time members of staff in the business, they work with D2M Innovation, a company which helps people and small companies to develop new product ideas. Their manufacturing partners are in China. Although Samantha, one of the directors, originally wanted to manufacture in the UK, she struggled to get companies to quote for the production of tooling for manufacture. The factory in China produces very high-quality tools and mouldings and all the textiles work is done in-house. Samantha liaises with them directly over product development, taking advice on the use of materials and specialised tooling and mouldings which has cut has costs and timescales to get the product to market. D2M Innovation were pivotal in helping her to get from the initial design stage through to prototyping, tooling and design for manufacture, helping with the full design and commercialisation of her product.

Developing products using digital tools and online collaboration

Online collaboration is used extensively by design professionals, using web/cloud-based software and networking. Designs can be developed concurrently with other designers and specialists around the globe, with ongoing communications with stakeholders throughout the iterative design process. The interactive, real-time collaboration tools are also available as smartphone and tablet apps and cover all aspects of designing from ideas, concepts, brainstorming, prototyping, file sharing and annotating, communications and **project management**.

Onshape is an example of a digital collaboration tool – a fully cloud-based CAD solution that allows two people to look at a CAD model online and as one person spins it around the other sees it moving in real time. They can pass control between them and even edit it in real time while seeing the changes. It also works on any platform and device because it is cloud based.

There are great benefits in discussing and exchanging ideas with specialists in manufacturing, as in the case of Ominio or developing designs concurrently with other designers, as in the case of Puzzle Phone. Using digital tools enables designers to explain and communicate their design decisions to stakeholders and make alterations to these concepts as necessary.

ACTIVITY

Explore online collaborative tools/software and their features and benefits. In what ways might the actual design of products be improved by the use of such tools?

KEY POINTS

- User-centred design (UCD) is based on the understanding of users, tasks they do and environments in which they live and work. In UCD, users are involved throughout the design of products, providing evaluation and feedback at every stage.
- In a circular economy we keep resources in use for as long as possible, extract the maximum value from them while they are in use, then recover and regenerate products and materials at the end of their life. It is a cradle-to-cradle model.
- Systems thinking means considering a product or component as part of a larger system from the start, not as just a component part or individual product in isolation.
- Collaboration is working with others.

DESIGN ENGINEERING

Using system design processes to define and develop systems that satisfy specified requirements of users

When a design engineer decides to incorporate an electronic or software system into a new product, appliance or machine, a thorough understanding of users' needs is essential to ensure that the system will provide the expected function.

System design is covered in detail in Chapter 6. In this section, we will examine the wider aspects of electronic system design that should be considered before getting into technical circuit or programming details, to ensure that a successful outcome is achieved.

User-interface design

A good user interface (UI) will maximise the usability of a product and provide a good user experience, making the product smart and helpful. A user's interaction with a product should be as simple and intuitive as possible. A good UI will be adaptable and will change according to a user's needs.

Figure 4.29 Amazon's Echo has an intelligent human-like VUI (voice user interface) called Alexa

Professionally, effective UI design may be accomplished with the help of graphic designers and human factors specialists. User task analysis may be carried out to find out how users normally carry out tasks linked with the intended new design. Users would be studied to find out their preferences in relation to the look, sound or style of the UI and their familiarity with other technical products.

Iterative development of a UI is achieved through usability testing, where a user's feedback is sought so that the UI can be improved. There is plenty of opportunity for user task analysis if you need to develop a UI for your NEA project.

Inputs to a UI can be through touch (using buttons, keyboard, computer mouse, game controller, touchscreen etc.), speech (using a microphone), gestures (using a camera) or body movements (using accelerometers either worn on the body or handheld). Outputs can be through a graphical user interface (using a display screen or simple LED indicators), sound output or haptics (vibration or 'tapping' on the body).

Data design

The flow of information or data through an electronic system must be considered at the system design stage. In Chapter 6, it is explained how different sensors produce different types of signals which, in the simplest sense, can be digital or **analogue signals**. Each type of signal is processed in a different way by a microcontroller, so it is important to understand which signals are being dealt with. Analogue signals are converted by a microcontroller to numeric data which lies between a minimum and a maximum value.

Microcontrollers process numbers using a type of number storage system called variables. Different microcontrollers have their own characteristics and peculiarities when dealing with variables so it is essential to understand the limitations of a microcontroller when variables are in use.

Failing to get to grips with the data limitations of the microcontroller can result in programming bugs and unexpected operations which will catch out an unsuspecting programmer. This can be avoided by making sure the most appropriate microcontroller is chosen for the application and by reading extensively about the microcontroller mathematics, to build up an understanding of its limitations and potential pitfalls.

Consequently, in system design, it is important to analyse the kind of data to be processed before deciding on the microcontroller and to draft an initial program flowchart to understand how the data will be generated and processed. Some data questions that may need to be asked include:
- Are sensors generating analogue or **digital signals**?
- What number range is generated by the analogue-to-digital conversion inputs?
- Do I need to count up (or down) and up to what number?
- Do I need to deal with time (hours/minutes/seconds)?
- Are negative numbers likely to be encountered?
- Are fractions (for example, 4.7) likely to be used, or generated because of a calculation (for example, $8/5 = 1.6$)?
- Are output devices going to be controlled digitally (for example, switching a buzzer on/off) or in an analogue way (for example, varying the speed of a motor)?

Process design

Chapter 6 describes how a program flowchart is drawn to allow a programmer to 'think like a microcontroller'. Such flowcharts are essential as they illustrate how a complex functioning system can be achieved through a logical step-by-step set of input/output, process and decision instructions.

To assist the programmer, alongside the design of the user interface it is useful to consider the various options and menus which will be presented to the user. This can be done by drawing an option tree flowchart (not to be confused with the program flowchart). Constructing such a diagram allows the designer to monitor the complexity of a system and perhaps simplify the range of options to the user to prevent the system (and the subsequent program) from becoming over-complex.

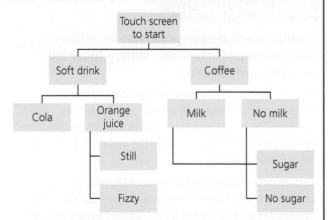

Figure 4.30 An option tree flowchart for a simple drink vending machine

How design teams use different approaches to project management when faced with large projects

Project management

The increasing complexity of products and the advancing technologies in almost every field of design and manufacture have led to the development and growth of project management tools.

Project management ensures the optimisation of all aspects of the design and production process and the satisfying of stakeholder requirements. The purpose of project management is to manage four primary factors of a project: cost, time, scope and quality.

Project management tools are the techniques, procedures and activities that are used to set out and track the individual tasks in a project and to manage the overall project successfully. Most of these use computer software, and charts and various types of diagrams similar to flowcharts. Design teams use management tools such as **Gantt charts** and **critical path analysis (CPA)** through the project as monitoring and management tools. This allows them to react quickly if there is slippage in time at a particular stage and achieve overall delivery schedules.

Some stages of production can be carried out concurrently, for example, preparing different sub-assemblies; others must be in a set order. Before detailed production planning can take place, it is essential to consider timings and sequences of manufacturing activities in detail. Graphical methods and computer software aid complex production analysis. Gantt charts are commonly used in building and construction projects. CPA is used to identify key stages and critical points to aid project management and ensure projects keep to schedule.

Gantt charts

A Gantt chart is a horizontal bar chart developed for production control in 1917 by Henry L. Gantt. It is often used in project management, providing a clear illustration of a schedule that helps to plan and track specific tasks against a time frame. In a Gantt chart, each project task and activity is represented by a bar. The left edge of the bar displays the predicted start date of the activity and the right edge of the bar indicates the planned end date.

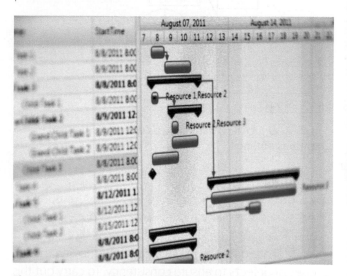

Figure 4.32 **A Gantt chart can be a simple, clear method and you might find it a useful way of planning your NEA work**

Critical path analysis (CPA)

Critical path analysis is a technique used in industry, when faced with completing large projects and orders, to assess and analyse each stage of production and the amount of time needed to complete the range of tasks in the process. CPA determines which activities or points in production are critical. A CPA diagram is produced to calculate the longest duration needed to complete tasks. This should in turn give the quickest time possible to complete the overall project.

FLOW PROCESS CHARTS		FLOW PROCESS CHARTS	
☑ Present method ☐ Proposed method		☐ Present method ☑ Proposed method	
Subject: _Part flow and inspection_		Subject: _Part flow and inspection_	
Chart begins: _Last machining operation in dept._		Chart begins: _Last machining operation in dept._	
Chart ends: _Tote box after final inspection_		Chart ends: _Tote box after final inspection_	
Symbols	Description	Symbols	Description
○⇨□D▽	Injection moulding operation	○⇨□D▽	Injection moulding
○⇨□D▽	Stack in wire baskets	○⇨□D▽	Operator stack in wire baskets
○⇨□D▽	Conveyor to wash	○⇨□D▽	Wait for truck
○⇨□D▽	Wash	○⇨□D▽	Truck to GRWD flash removal
○⇨□D▽	Conveyor to inspection	○⇨□D▽	Wait to synchronize
○⇨□D▽	Inspector	○⇨□D▽	remove flashing
○⇨□D▽	Load in tote box	○⇨□D▽	Conveyor to wash
○⇨□D▽	Wait for truck	○⇨□D▽	Wash
○⇨□D▽	Truck to flash removal	○⇨□D▽	Conveyor to inspect
○⇨□D▽	Wait to synchronize	○⇨□D▽	Inspection
○⇨□D▽	Remove flashing	○⇨□D▽	Stack in tote box
○⇨□D▽	Conveyor to wash	○⇨□D▽	
○⇨□D▽	Wash	○⇨□D▽	
○⇨□D▽	Conveyor to inspection	○⇨□D▽	
○⇨□D▽	Inspection	○⇨□D▽	
○⇨□D▽	Stack in tote box	○⇨□D▽	

Figure 4.31 **The chart shows operations carried out, their type and timings. By examining idle time, the sequence of operations may be improved. Network analysis is another graphical tool for determining the optimum sequence of stages**

The basic method is to show:

- a list of all activities required to complete the project
- the time that each activity will take to complete
- the stages that need completing between the key activities
- logical end points within and at the end of the process.

Each activity within the above schedule is presented as follows:

- start – the earliest time that an activity can start
- duration – the estimated time to undertake the activity
- early finish – the earliest time that an activity can finish
- late start – the latest time that an activity can start without affecting the overall project duration
- float – the time by which an activity may be delayed without affecting the overall project duration
- late finish – the latest time that an activity can finish without affecting the overall project duration.

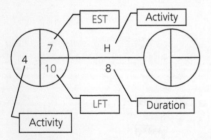

Figure 4.33 A critical path analysis with activities and durations outlined

Six Sigma

Six Sigma is a set of techniques and tools to ensure process improvement and consistent output within manufacturing. It seeks to improve the quality of the output of a process by identifying and removing the causes of defects to ensure consistency. To carry out this process, a number of experts are employed to ensure quality and efficiency at each manufacturing stage within an organisation, using statistical data to analyse results and get as close to 'zero defects' as possible. Each Six Sigma project carried out within an organisation follows a defined sequence of steps and has specific targets, for example:

- reduce process **cycle time**
- reduce pollution
- reduce costs
- increase customer satisfaction
- increase profits.

Companies can be awarded a Six Sigma certificate which is beneficial within the business and manufacturing world as it projects an image of quality and efficiency.

Many companies successfully use this approach. Some of the best known are Amazon, Boeing, Caterpillar and Motorola. Six Sigma recognises that people are at the centre of a business or company and gives rewards to recognise training and experience. Key people are given extensive training in advanced statistics and project management to be designated 'Black Belts'.

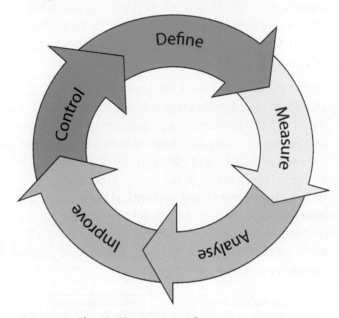

Figure 4.34 The Six Sigma approach

Define: Identify the project goals and all customer deliverables.

Measure: Understand current performance.

Analyse: Determine the root causes of any defects.

Improve: Establish ways to eliminate defects and correct the process.

Control: Manage future process performance.

Scrum

Scrum is a project management tool used mainly in software development but these days also in consumer product design and development. The name comes from the scrum (or scrummage) in the game of rugby, which is used to restart the game.

Scrum uses the iterative design process and involves collaboration, use of software, teams organising themselves and flexibility to adapt to changed stakeholder requirements. Scrum meetings are held daily to monitor and manage progress. The design of each feature of a product or system is tested, refined and completed before starting the next, ensuring that stakeholder feedback is incorporated.

ACTIVITY

Consider how you might use or adopt some of the project management tools above in the planning of your own NEA.

KEY TERMS

Analogue signal – a voltage which can be any value.

Collaboration – the act of working with others to produce a product or design solution.

Critical path analysis (CPA) – a project management tool that lists and charts all tasks that must be completed as part of a project. This then gives identification to the quickest route to completion.

Cycle time – the time taken during the manufacturing of a product for the process to go through a range of steps and return to the same point.

Digital signal – a voltage which can only have two values.

Gantt chart – a horizontal bar chart often used in project management, providing a clear illustration of the schedule that helps to plan and track specific tasks against a time frame.

Mould – a hollow shape used to pour materials into, like a jelly mould.

Participatory design – design which actively involves all stakeholders and users in the design process.

Project management – the act of defining, initiating, planning, executing, monitoring and completing the work of a team to meet a specific target. It is the planning and control of everything involved in delivering an end result to meet the stakeholder requirements; getting the job done.

Scrum – a project management tool that uses the iterative design process. It involves collaboration, use of software, teams organising themselves and flexibility to adapt to changed stakeholder requirements.

Six Sigma – a set of tools and techniques to ensure consistent output in manufacturing where little variation is crucial.

Systems thinking – the understanding of a product or component as part of a larger system of other products and systems.

KEY POINTS

- A good user interface (UI) will maximise the usability of a product and provide a good user experience, making the product smart and helpful.
- Project management tools are the techniques, procedures and activities that are used to set out and track the individual tasks in a project and to manage the overall project. Design teams use management tools, such as Gantt charts and critical path analysis (CPA), throughout the project as monitoring and management tools.

- Six Sigma is a set of techniques and tools to ensure process improvement and consistent output within manufacturing. It seeks to improve the quality of the output of a process by identifying and removing the causes of defects to ensure consistency.
- Scrum is a project management tool used mainly in software development but these days also in consumer product design and development.

Further reading on design thinking and communication

- Koenig, P. (2011) *Design Graphics: Drawing Techniques for Design Professionals*, 3rd ed., Pearson, ISBN-10: 013713696X
- Eissen, K. and Steur, R. (2013), *Sketching: Drawing Techniques for Product Designers*, Bis Publishers, ISBN-10: 9063691718
- Eissen, K. and Steur, R. (2013), *Sketching: The Basics*, Bis Publishers, ISBN-10: 9063692536
- Henry, K. (2012), *Drawing for Product Designers (Portfolio Skills)*, Laurence King, ISBN-10: 1856697436
- Hongkiat publishes tips and tutorials for designers at www.hongkiat.com. Search for the guidelines on basic sketching
- Infographic on user-centred design: http://paznow.s3.amazonaws.com/User-Centred-Design.pdf
- Doctor Disruption is a blog looking at principles of design, design methods and user-centred design: www.doctordisruption.com
- Industrial Designers Society of America: www.idsa.org. Search for their guidance on communicating design
- Scrum reference card: http://scrumreferencecard.com/
- Critical Path Analysis: www.tutor2u.net/business/reference/critical-path-analysis
- Lidwell, W. Holden K. and Butler, J. (2003, 2011) *Universal Principles of Design (125 Ways to Enhance Usability, Influence Perception, Increase Appeal, Make Better Design Decisions, and Teach through Design)*, Rockport Publishers, Inc. ISBN: 978-1-59253-587-3

PRACTICE QUESTIONS: design thinking and communication

Design Engineering

1 Produce a system diagram of an engineered product of your choice, then produce a sketch and rendering that could be presented to stakeholders or clients to communicate your design.

2 Discuss the importance of collaboration in a design engineering team.

3 Suggest an engineered product where you feel systems thinking has been part of the design process. How does this approach affect the design of a product?

4 Design a new user interface for a keyless door entry system and draw an option tree flowchart for the system.

Fashion and Textiles

1 Complete a quick sketch of a fashion item or accessory of your choice, then produce a rendering of this product.

2 Discuss the importance of collaboration in a fashion design team.

3 Suggest a textiles-based product where you feel systems thinking has been part of the design process. How does this approach affect the design of a product?

4 Choose a textiles or fashion designer and try to find examples of the communication techniques they use to communicate their ideas to stakeholders and clients at various stages of the design process.

Product Design

1 Complete a quick 3D sketch of a consumer product of your choice, then produce a rendering of the product to show more detail. Produce a board to include user and in use sketches to communicate how and in what context the product is used.

2 Discuss the importance of collaboration in a product design team.

3 Suggest a consumer product where you feel systems thinking has been part of the design process. How does this approach affect the design of a product?'

4 Choose a product designer and try to find examples of the communication techniques they use to communicate design ideas to stakeholders and clients at various stages of the design process.

Material and component considerations

PRIOR KNOWLEDGE

Previously you could have learnt:
- to apply in depth knowledge of the working and physical properties and characteristics of a selection of materials when making design and manufacturing decisions
- how incremental to the design process it is for designers to make informed decisions when selecting materials and components
- how to select and apply the correct terminology when describing a wide range of materials
- how to select and work with materials and how well they fulfil the required functions of products in different contexts.

TEST YOURSELF ON PRIOR KNOWLEDGE:

1 Select a product of your choice and explain the functional performance characteristics of the materials and components in that product.
2 Choose two products made from different materials. Name the materials used in each product and discuss the factors that will have influenced their selection.

3 Identify a product you have in your home and explain how the selection of materials and components in that product may impact on the environment.
4 Explain the meaning of the terms 'hardness' and 'toughness' when used to describe the properties of materials.

5.1 What factors influence the selection of materials that are used in products?

LEARNING OUTCOMES

By the end of this section you should have developed a knowledge and understanding of:
- how the selection of materials and components is influenced by a range of factors, including:
  - functional performance
  - aesthetics
  - cost and availability
  - properties and characteristics
  - environmental considerations
  - social, cultural and ethical factors.

One of the first tasks in the design process is to decide which materials and components will perform and meet certain criteria to best effect. This selection process is a crucial step in the designing of any physical object and can dictate how successfully a product functions and performs. A well-designed product may fail to be a profitable product if the most appropriate material and component combinations are not used. This may not necessarily be the most expensive material and in fact, it is usually a combination of factors that influence the decisions. Within the selection process, testing would be carried out to establish fitness for purpose and **material characteristics**. All materials have different properties that make them suitable for a range of uses. Some of those properties will ultimately be essential, while others could be desirable. Most **material properties** and characteristics can be measured by carrying out **standardised tests**. For example, it could be essential that the material has a high tensile strength and to establish this, measurable tests would have been carried out.

Figure 5.1 shows an exploded diagram of a Russell Hobbs jug kettle. Table 5.1 on page 134 explains which materials have been selected for the kettle's individual parts and their functional performance.

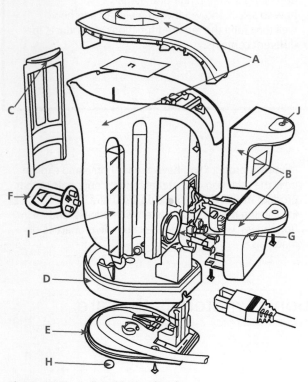

Figure 5.1 Russell Hobbs jug kettle

It is very important that you understand the correct terminology required to explain the properties and characteristics of materials, whether this is in a mechanical, physical or aesthetic context. To help you with this in your NEA project and for revision, refer to Section 5.3.

Functional performance

The **functional performance** of any material or component refers to how it actually works and performs rather than the aesthetic aspect. The functional properties relate to the practical use of the product, such as strength, weight, durability, elasticity, etc. The choice of materials and components will very much depend on the product's end use and how it has to perform. The performance and success of any product is largely dependent on the choices made when selecting the materials and components. Every product will have, during the development stage, a list of crucial criteria that the material and component selection will depend on. For example, for a component that must support a specific load, the minimum stress that is required for the component's material will be determined. This will be one of the material selection criteria. It is important to understand and recognise the difference between the functional and aesthetic performance characteristics of materials and components.

Aesthetics

Aesthetics play a vital role in the development of a product. This not only relates to how the finished product looks in terms of shape, form and colour, but also how the material and component aesthetic properties and characteristics contribute to the success of the end product. Aesthetics links into the customer's reaction and response when selecting a product to purchase and as such is an integral part of its success.

The aesthetic properties also relate to touch, feel and style; and while it may appear that the material selection may be appropriate for a product from an aesthetic point of view, the function and performance of the material or component is crucial. It is a combination of both that leads to a product's longevity, through the pleasure and satisfaction it gives.

DESIGN ENGINEERING

In design engineering, functionality tends to be more important than aesthetics as electrical and mechanical systems tend to be concealed. However, where there is user interaction with system interfaces, aesthetics can be important.

FASHION AND TEXTILES

In garments, it could be that a fabric is required with excellent drape, soft handle and on-trend colours and prints.

Table 5.1 **Functional performance of materials used for a Russell Hobbs jug kettle**

Component	Label	Material	Function
Main body kettle	A	Brushed and lacquered stainless steel	• The high chromium content of stainless steels provides the unique stainless and corrosion resisting properties and helps to resist scaling inside the kettle. • Durability and hardness to make it resistant to knocks and dents • Hot strength to maintain its strength and shape when boiling water inside • Low maintenance as the stainless steel is smooth and easy to clean • Stainless steel is not a good conductor of electricity and is therefore beneficial to the electric kettle; if a fault did occur with the appliance then it is less likely to become live and cause serious injury.
Handle	B	Low density polyethylene (LDPE)	• Heat resistant • Colour options • Can have a smooth or textured finish • Thermal and electrical insulation properties • Can easily be moulded to an ergonomic shape • Cost effective
Water filter	C	Polypropylene (PP)	• Heat resistant • Easy to shape by injection moulding • Can be coloured • Durable
External base	D	Polypropylene (PP) Decorative brushed steel band around the perimeter	• Heat resistant • Easy to shape by injection moulding • Can be coloured • Durable
Internal base	E	Anodised stainless steel comprising of two sections that are stamped out and punch pressed to contain the heating element	• The stamping that houses the electrical components is anodised to ensure that mineral deposits will be easier to remove. This protects the steel parts from corrosion.
Heating element	F	Nichrome, 80% nickel and 20% chromium, wire, ribbon or strip Outer copper housing that is chrome plated	• A protective coating of chromium oxide forms when heated for the first time which prevents the wire inside from getting damaged. • Conducts heat very effectively • Very high melting point • Vastly resistant to heat and oxidation at high temperature
Thermostat	G	Bimetallic strip made from brass and iron	• Iron expands less than brass as it gets hotter, so the bimetal strip curves inward as the temperature rises, which eventually breaks the circuit to switch the kettle off.
Rubber feet	H		• Protects the work surfaces from heat damage • Good electrical insulator • Water repellent
Water level clear window	I	Low-density polypropylene (PP)	• Flexible • Translucent • Durable • Excellent chemical resistance
On/off switch	J	LED light in polycarbonate	• Naturally transparent amorphous thermoplastic • High impact resistance • Very good heat resistance • Easily formed by injection moulding

Solid hardwood timber such as oak, used in finely finished furniture, suggests craftsmanship and skill. Wood is a natural material with grain and a surface texture, pattern, colour and feel that other materials do not have.

Metals are precise, clean and cold in appearance and feel. There is a solidity and strength in how they look and feel and a sense of reliability and trust.

Figure 5.2 **An example of how the aesthetic properties of different woods are important to the design of a product**

Polymers are easily moulded, coloured and printed; they can be transparent, opaque, textured, rigid or flexible, providing product designers with infinite possibilities in form and colour, with the opportunity to imitate other materials, such as natural timber.

Figure 5.3 **Metals can be used in a wide range of products to give different aesthetic effects. This image illustrates how stainless steel splashback tiles give the impression of a clean, contemporary, hygienic and industrial look**

The component selection is of equal importance as the selection of the dominant material when designing and manufacturing a product. The designer has to consider the function and the aesthetics of the component. It could be that the manufacturer requires the component to blend seamlessly within the product so it is barely visible and does not hinder the overall aesthetics of the product; or alternatively, it could be that the component needs to complement the design and be a feature in itself. There are many companies that supply and manufacture components to satisfy the designer's requirements and that also offer bespoke services such as colour matching.

Figure 5.5 shows a plan of a park bench and illustrates the components used for construction. While the seating area is made from hardwood, the fixings and components are stainless steel to ensure durability and

Figure 5.4 **Colour matching zips**

a clean appearance for many years. The timber will be more susceptible to wear, damage and weathering but can easily be replaced when required.

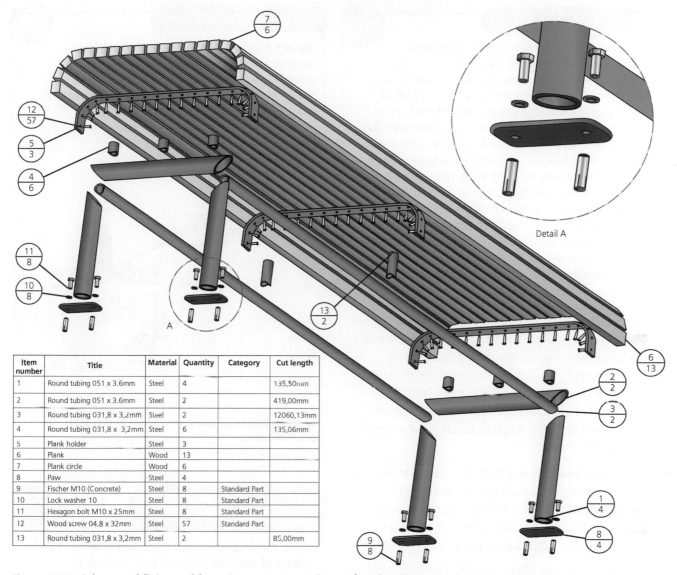

Item number	Title	Material	Quantity	Category	Cut length
1	Round tubing 051 x 3.6mm	Steel	4		135,50mm
2	Round tubing 051 x 3.6mm	Steel	2		419,00mm
3	Round tubing 031,8 x 3,2mm	Steel	2		12060,13mm
4	Round tubing 031,8 x 3,2mm	Steel	6		135,06mm
5	Plank holder	Steel	3		
6	Plank	Wood	13		
7	Plank circle	Wood	6		
8	Paw	Steel	4		
9	Fischer M10 (Concrete)	Steel	8	Standard Part	
10	Lock washer 10	Steel	8	Standard Part	
11	Hexagon bolt M10 x 25mm	Steel	8	Standard Part	
12	Wood screw 04,8 x 32mm	Steel	57	Standard Part	
13	Round tubing 031,8 x 3,2mm	Steel	2		85,00mm

Detail A

Figure 5.5 **Stainless steel fixing and fastening components for outdoor furniture**

Cost and availability

Cost and availability are critical factors to consider when selecting materials and components for a certain product. When estimating costs, all the associated cost factors must be considered to get a more realistic price. This may involve the transportation and processing of materials and components. Cost and availability will determine the choice of materials and components in the development stage. Materials and components must be sourced to appropriate quality standards, in the right quantity, for the right price and at the right time, to ensure effective production.

Many materials and components can be purchased pre-prepared, which can lead to savings overall as it could cost more to manufacture them in-house rather than contracting out.

DESIGN ENGINEERING AND PRODUCT DESIGN

Production of complex items such as electrical goods and cars are not entirely carried out in one location. Several industries may contribute to the production of a variety of parts and then these items are brought together as the final item is assembled. For example, most automobile manufacturers depend on numerous suppliers for pre-manufactured materials. This could be from the car chassis material such as aluminium or steel to the polypropylene dashboard. This requires outside parts vendors to subject their component parts to rigorous testing and inspection audits similar to those used by the assembly plant.

The majority of textile products are constructed in one place but the fabrics and components will have been outsourced from elsewhere to get the most cost effective price. China accounts for over 60% of world chemical and fibre production and India is the world leader for cotton production, so historically it has been cheaper to source fabrics from there. Labour costs have also been lower in China and India but low energy costs elsewhere have meant that it is now not necessarily cheaper to source fabrics in South Asia and the Far East. Recently, some European clothing companies have sourced fabric from Italy as Italian fabric producers are closer to market, so not only is transport cheaper, but timelines between orders and sales are shortened, and in the fast-paced world of high fashion, this matters.

Using pre-prepared components can lead to savings in storage, handling, processing and other costs. Whole processing steps may be eliminated by selecting specific materials, such as pre-coated sheet materials, or by processing materials prior to manufacture. It is important that manufacturers consider all potential savings when deciding whether to carry out material preparation in-house.

Standard components and parts are used extensively. These are usually manufactured to comply with international standards for size and specification to ensure interchangeability and to enable repair and maintenance of products as required during their life.

The costs can also be affected as the choice of materials may require specialist machinery to process or manufacture. If the manufacturer has certain constraints to adhere to due to the machinery available, this will dictate the selection of material because the materials must be compatible with the processes used to make the product.

DESIGN ENGINEERING AND PRODUCT DESIGN

Components to be joined using a specific welding, brazing or soldering process must be made of materials that enable good joints to be formed using the specific joining process.

In the development of new materials or products that are significantly different or ground breaking, using specific manufacturing processes may not be fully compatible with the process and therefore specialist machinery may be required.

Figure 5.6 shows an industrial sewing machine required to join neoprene fabric. Regular industrial machines will not join neoprene correctly due to its thickness and the rubber back which can prevent the fabric from feeding through the machine.

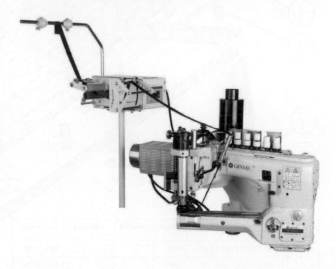

Figure 5.6 Specialist sewing machine for stitching neoprene fabric

DESIGN ENGINEERING AND PRODUCT DESIGN

This is a remote laser welding machine that has a range of uses, but has been used for both the automotive and aerospace industries as they continue to move towards lighter, more fuel-efficient vehicles. Components could be intentionally designed to take advantage of the laser joining process. Producing a laser weld with the identical strength of a resistance spot weld requires only a fraction of the time.

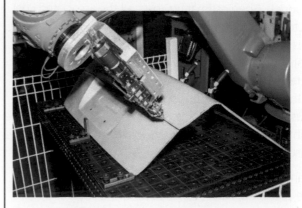

Figure 5.7 Remote laser welding machine

DESIGN ENGINEERING AND PRODUCT DESIGN

If we look at the steel industry it is easy to see how important it is for manufacturers to be up to date with material availability and cost. Much of the current state of the steel industry can be linked to the rate of Chinese production. Because of China's massive infrastructure needs, it has significantly increased production over the past decade. The increased production and resulting low prices have had an impact on the steel industry in other parts of the world, including the United States, the UK and Japan. However, Chinese officials have stated that they will decrease steel production by more than 165 million tons by 2020. That's a 20 per cent decrease in production for the world's leading steel producer. As China is dropping its steel output so dramatically, global steel supply is expected to be less than demand, resulting in higher steel prices. See Figure 5.8 to see the predictions up to 2020.

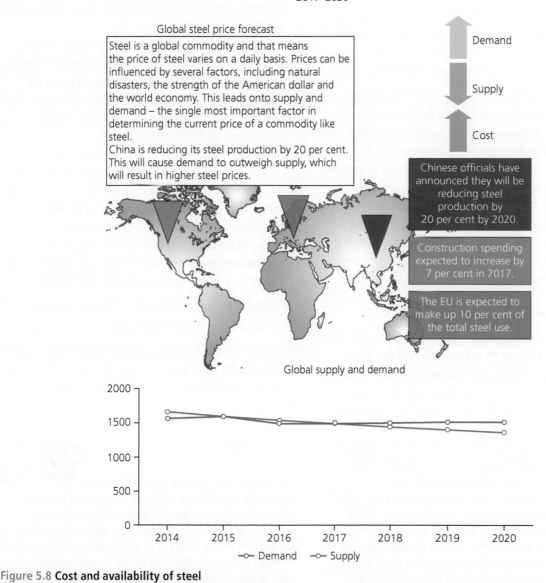

2017–2020

Global steel price forecast

Steel is a global commodity and that means the price of steel varies on a daily basis. Prices can be influenced by several factors, including natural disasters, the strength of the American dollar and the world economy. This leads onto supply and demand – the single most important factor in determining the current price of a commodity like steel.

China is reducing its steel production by 20 per cent. This will cause demand to outweigh supply, which will result in higher steel prices.

Demand

Supply

Cost

Chinese officials have announced they will be reducing steel production by 20 per cent by 2020.

Construction spending expected to increase by 7 per cent in 2017.

The EU is expected to make up 10 per cent of the total steel use.

Global supply and demand

-o- Demand -o- Supply

Figure 5.8 Cost and availability of steel

The cost of materials and components also dictate the final price, along with other crucial factors, so it is vital that the selection of both, while meeting the **performance criteria** required, also satisfies **cost constraints**.

Selection of the right materials invariably means that a product will perform better and for longer.

The thorough and rigorous testing of materials and components as part of the iterative design process will enable the correct selection.

In environments such as a hospital where critical life-saving and life-giving equipment is often in constant use by many different people, this is particularly important. The correct materials and components with properties and characteristics to fully match requirements often involve higher costs initially but reduce breakdowns and maintenance costs with the associated equipment downtime. Hospitals therefore gain operational efficiency through availability of equipment and minimised administrative resources to manage repairs or replacements.

The use by designers of standard 'bought-in' components can reduce costs.

Whether the product is something as simple as a torch or as complex as a car, standard components are available for such things as simple fixings (screws, bolts, etc.) to bulbs, switches, seat belts, car seat covers, electric motors and pumps and even whole power units.

Designers must consider the availability of materials and components in the quantities required. It could be disastrous for a company if the consistency of availability of materials is unsure or unreliable, as it could halt production. The product manufacturer needs to ensure there are no constraints in ordering the quantities required and that they can rely on the quality. The material or component manufacturer may specify a minimum quantity for orders, which could result in the product manufacturer needing large storage facilities, which can affect costs.

Properties and characteristics

A knowledge and understanding of materials is essential for all designers. In order to make informed decisions when selecting the most appropriate material for the product, designers need to consider a range of factors including properties and characteristics. The main aim of material selection is to meet product performance criteria and minimise cost.

When selecting a material for a product, the material's properties must satisfy the function and the operating conditions of the component or structure being designed. The properties, which directly influence the choice of material, can be summarised under the following categories:

- Mechanical properties – how the material reacts to forces acting upon it; for example, strength, ductility, hardness, toughness, elasticity, flexibility, fatigue limit, coefficient of friction.
- Physical properties – the handling characteristics of a material, including its feel; for example, density, conductivity (thermal, electrical), melting point, flammability, thermal expansion, optical properties and acoustic properties (for example, sound absorption and reflection).
- Chemical properties – how the material reacts and changes when in contact with other substances; for example, corrosion resistance, reactivity to other materials, hygroscopy (absorption of water).
- Manufacturing properties – the processing properties of the material for the conversion processes needed to change the material into the required shape; for example, formability, machinability, fusibility.

A material's properties determine the methods of manufacture that may be suitable.

Applying a certain finish to a material can make it more appropriate for its purpose. By applying a surface finish, a material's functional and aesthetic properties can be improved.

DESIGN ENGINEERING AND PRODUCT DESIGN

A component such as an alloy wheel for a car, manufactured by casting, requires a material which will flow readily in the molten state (flowability), will fill all areas and details of the mould and on solidification will not produce undesirable cracks or surface imperfections. A secondary stage of manufacture involves machining of the alloy wheel, so machinability of the material is also important.

The alloy has to support the weight of the loaded car plus the torque from the power unit, plus added stress from impact from potholes in the road. Safety must be maintained across a wide temperature range from below freezing to a high temperature reached through the disc brakes system which is immediately next to the wheel. Light weight is important to achieve economy for the car.

FASHION AND TEXTILES

If permanent pleats are required, some fabrics must have a resin finish applied to them to heat set the pleats into place.

PRODUCT DESIGN

A steel bathroom tap can be coated with chrome to give it a more durable and shiny finish. Galvanised coatings can be applied to iron and steel to protect the base metal from corroding. This is often used in products that are to be used outdoors that need to withstand varying elements.

Environmental considerations

There has been a major increase in public interest in environmental issues and many people take this into account when purchasing a product. This should therefore be a crucial factor when manufacturing products in industry. When selecting the materials and components it is important to ensure that they can be recycled, or at least there will be a consideration of pollution levels that could be affected. Any responsible designer should seek to conserve the environment wherever possible.

Many governments run schemes that encourage informed choices about materials, components and the range of manufacturing processes that are available.

FASHION AND TEXTILES

The Action Plan for the Sustainable Clothing Roadmap has been run by Defra (the Department for Environment, Food and Rural Affairs) since 2007. It sets out agreed stakeholder actions in the following five key areas to improve the sustainability performance of clothing:
- improving environmental performance across the supply chain
- consumption trends and behaviour
- awareness, media, education and networks
- creating market drivers for sustainable clothing
- traceability along the supply chain (ethics, trade and environment).

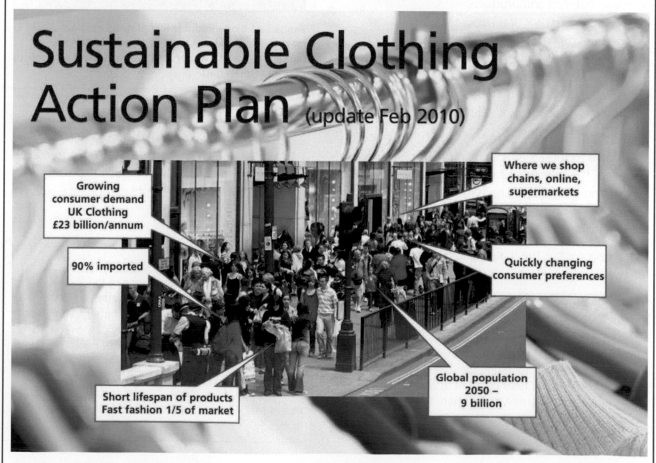

Figure 5.9 **The Sustainable Clothing Action Plan run by the UK government**

The choice of materials and components should limit the use of finite resources whenever possible. However, this can sometimes be challenging for companies when they are faced with decisions relating to cost and material choice. For example, using recycled plastic may be more favourable on a finite resource like oil, but when oil prices are low, companies may find it more economical to use virgin plastic.

As future designers, we need to consider the environmental, social and economic implications of our decisions and choices.

Many products are produced where workers are subject to poor working conditions and low wages, another social issue. Recycling has environmental benefits but also has economic costs and uses energy and can cause pollution, which can result in social issues.

We can also try to:
- rethink what materials and energy we use and the way we use them
- reduce the materials we use and try to create products that are multifunctional
- reuse materials and products
- repair existing products rather than buying new ones; design products that can be repaired easily
- recycle materials where possible; perhaps use recycled materials in new products
- refuse to use certain materials or to buy certain products if they are not needed.

Within the product development process, the material selection is a crucial stage. To be more responsible designers, this selection process should include considering wider implications in relation to the environment and sustainability. (For more information on sustainability see Chapter 3.)

Wherever possible, the materials should be from renewable or replaceable sources.

FASHION AND TEXTILES

Most natural textile fibres are renewable, for example, cotton. However, due to the impact on the environment of processing the fibres, alternatives or organic versions that have not used as many chemicals in their production may now be selected. Hemp, bamboo and coconut are a few examples.

PRODUCT DESIGN

Figure 5.10 is an example of how one company has managed to offer an alternative solution to plastic plates and containers. The company is called Leaf and they have developed a system to make plates from leaves. The bottom and top layer are made from leaves, stitched together with fibres from palm leaves. In between is a layer of paper made of leaves. There are no plastics, additives, oil, glue or chemicals. The plates, once disposed of, are completely biodegradable within 28 days. This compares to the usual 730,000 days it takes for a plastic plate to degrade.

Figure 5.10 **Biodegradable leaf plates. Leaf have managed to develop a process of manufacturing packaging that is 100 per cent biodegradable and utilises renewable materials**

- Recycled materials – the designer and manufacturer should consider whether recycled materials could be used to minimise the use of depleting resources. Most modern materials can include high levels of recycled content, for example, cardboard boxes, metals and most plastics.

PRODUCT DESIGN

The Innocent Drinks bottle was one of the first to be made from 100 per cent recycled PET. However, this had to be reduced to 50 per cent after trialling the bottles, as they found that the quality of the recycled plastic had declined to an unacceptable level. The company's philosophy involves using as little material as possible, not using new materials and to making use of as much recycled or renewable material as possible, as well as making sure the materials and packaging are easy to recycle.

- Materials that are in plentiful supply – it is not hard to imagine that if current and projected material consumption trends continue, we will face scarcity and depletion of certain resources. In the past 50 years, humans have consumed more resources than in all previous history. The issue of depleting resources is one that should be addressed when selecting materials and one solution is to select materials that are in plentiful supply. The world shares a finite set of materials and other natural resources and we should not presume they will remain available indefinitely. The rate of deforestation in the tropics is approximately one

acre per second. Most metal ores are in abundance within the Earth's crust but there is a large impact on the environment during extraction and processing. Lifecycle assessment (LCA) is a technique that is widely used to assess and evaluate the impact of a product from its conception and realisation through to its disposal at the end of its usable life – see Chapter 3.

Moral responsibilities of the designer

All designers should share a moral responsibility to ensure that the products they design contribute towards our environment in a positive way.

FASHION AND TEXTILES

Flyknit® by Nike

Fashion is a complicated business involving long and varied supply chains of production, raw material, textile manufacture, clothing construction, shipping, retail, use and ultimately disposal of the garment. This industry is the second largest polluter in the world, second only to oil. The damage is caused by the use of pesticides in cotton farming, the toxic dyes used in manufacturing and the huge amount of waste discarded clothing creates; there is also the extravagant amount of natural resources used in extraction, farming, harvesting, processing, manufacturing and shipping. Nike has tried to develop its products to address some of these issues by creating Flyknit®. This technology prevents millions of pounds of waste from ever reaching the landfill, delivering the lightest, strongest and most adaptive footwear ever created by Nike. The environmentally sustainable benefit to Nike Flyknit® is

that it reduces waste because the one-piece upper does not use the multiple materials and material cuts used in traditional sports footwear manufacture.

Figure 5.11 **Nike Flyknit® technology reduces waste**

DESIGN ENGINEERING AND PRODUCT DESIGN

The Model U SUV is a concept car, introduced by Ford. The list below illustrates how the designers have approached environmental issues:
- recyclable polyester on seats, dashboard, steering wheel, headrests, door trim and armrests
- a corn-based biopolymer for the retractable canvas roof and carpet mats
- corn-based fillers in the rubber tyres as a partial substitute for rubber black to improve rolling resistance and thus fuel economy
- soy-based composite resins to form the rear tailgate and side panels
- soy-based composite foam for seating
- interchangeable armrests for ease of maintenance and remanufacturing
- lightweight, recyclable aluminium body.
- versatile body shape and design, converts from a sedan to a utility truck.

Figure 5.12 **Cradle-to-cradle Ford Model U vconcept car**

Some designers and manufacturers have looked at how a product can cause minimal impact on the environment at the end of its life by making it easier to recycle.

FASHION AND TEXTILES

Natural and synthetic fibres are typically not woven together in sustainable fabric design as it can be difficult to pull them apart at the end-of-use phase. One company that has managed to overcome this is Designtex. They created Vox which is an upholstery fabric that employs a weave structure that brings together both natural and synthetic materials in a manner that allows for easy disassembly. The natural layer (composed of wool and Lenzing rayon) can easily be separated from the synthetic Cradura® layer using a simple industrial process that has been established worldwide. Once separated, the synthetic fibres can be recycled while the natural fibres are able to biodegrade.

Social, cultural and ethical factors

Designers and manufacturers have to make **ethical choices** in the development of a new product. This is particularly important when selecting the materials and components and, as such, they need to be aware of both the positive and negative **social impacts** of the manufacture of their product, and design appropriately. When selecting materials and components for the product, designers and manufacturers have a responsibility to ensure that other people's quality of life and human rights are not compromised. Socially responsible and ethical designers should consider whether traditional skills may be encouraged or lost through the material extraction or manufacture. Will the making of the materials and components have a positive or negative impact on the quality of life of the people concerned or those living in that country or area? Will the making of the components or products infringe any basic human rights, for example, fair pay and decent working conditions?

DESIGN ENGINEERING

One example where designers are having to consider whether they are being ethically responsible is with the mining of columbite tantalite (coltan) in the Democratic Republic of Congo. Coltan is used primarily for the production of tantalum capacitors, used in many electronic devices. It is important in the production of mobile phones, but tantalum capacitors are used in almost every kind of electronic device. For the Congolese, mining of Coltan is an easy source of income because the work is consistent. However, the mining of it has resulted in political instability and conflict and it is difficult to distinguish between legitimate and illegitimate mining operations.

Figure 5.13 Coltan mining

Social factors

There are a multitude of choices that the designer and manufacturer have to make to ensure their selection of materials will cause minimum impact. Selecting the right materials can have a positive impact on people's lives and the environment as it can encourage and maintain traditional knowledge and skills.

FASHION AND TEXTILES

Becky Cocker is an example of how designers can maintain traditional skills using traditional materials. She is the founder of CARV and the designer and creator of CARV bags. Every CARV bag is made by hand from start to finish using centuries-old leatherworking techniques, natural vegetable tanned leather and products sourced from local suppliers to support a diverse UK skills base and manufacturing economy. Becky is passionate about the environment, sustainable fashion and supporting UK-based skills, manufacturing and businesses. For this reason, CARV is based on the principles of slow fashion, with each bag being made in small numbers, allowing the focus to be on lasting quality and individuality.

Figure 5.14 CARV bags are made by hand from start to finish using centuries-old leatherworking techniques and materials and components from local suppliers

Social issues can arise when a new product has unforeseen side effects on a group of people, which can be both positive and negative. For example, the rise in the use of mobile phones to send text messages has increased the demand for such phones. This provides employment, enables us to keep in touch easily but means there is less need for social face-to-face interaction.

Inclusivity should also be considered when selecting materials and components to ensure the product meets the needs of as many people as possible. It is impossible to create something that will work for all people, but keeping a diverse group of people in mind during the design and material-selection process will improve the end result. The designer should consider age, culture and physical ability/disability. These aspects will affect the user's reaction to a product and therefore affect the detailed design of a product and the selection of components and materials.

PRODUCT DESIGN

One company that has based the design of its product on the ethos of universal design is OXO. Sam Farber, the founder, chose the name OXO because whether it is horizontal, vertical, upside down or backwards, it always reads OXO. Noticing that his wife, Betsy, who suffered from mild arthritis in her hands, was having difficulty gripping ordinary kitchen tools, he saw an opportunity to create more comfortable cooking tools that would benefit users. Their philosophy is dedicated to providing innovative consumer products that make everyday living easier. They study all genres of the population, left-handed and right-handed, male and female, young and old, and encourage them to interact with products to identify opportunities for meaningful improvement. In Figure 5.15 we can see an example of the OXO Easy Grip range developed to be more inclusive through the use of materials and by design. The company selected the thermoplastic elastomer Santoprene as it is extremely durable in both extreme hot and cold environments and is resistant to harsh chemicals, grease and oil, making it an excellent choice for cooking utensils.

Figure 5.15 **OXO Easy Grip kitchen scissors**

Many products do not take into account the problems of the disabled, old or very young.

DESIGN ENGINEERING

Components such as displays and controls are often too small or too difficult for some people to operate. Modern digital and technology components can be threatening to older people, making them feel insecure.

Good choices by designers will take into account as many users as possible.

Cultural factors

Care needs to be taken to ensure that products do not cause offence to a particular race or culture. Good design incorporates appropriate cultural features with materials and components that are acceptable and preferred. Materials all have associations in terms of where they are sourced or manufactured, which may not be acceptable in all cultures.

The importance of selecting the right materials and components

Designers and manufacturers have a responsibility to analyse the choices, constraints and conflicts when selecting appropriate materials and components. The decisions they make are crucial and, for a successful and commercially viable product, must be based on a thorough investigation and consideration of all of the factors we have covered so far in this chapter. However, these factors will have varying degrees of importance depending on the product. For example, the strength of a particular material or component may take precedence over the aesthetics. It may be that the designer has to make compromises throughout product development. It is interesting to analyse products to recognise all the materials and components that have been selected and, more crucially, why.

A car is a helpful example to illustrate all of the factors a designer needs to consider when selecting materials and components.

There are occasions when a particular material for a product or component is selected for aesthetics or cost over performance or durability, or simply the incorrect material choice is made. It is very important that the correct materials and components are selected, as an incorrect choice can compromise the function and performance of the product. The wrong decision can have disastrous effects for a company or the user. The product can be unsafe for the user, which could result

in the manufacturer being liable for personal injury or they could have a case for product recall which again, has massive cost and reputation implications for the company.

ACTIVITIES

1   Examine the development of a product of your choice over the last 20 years. Explain how the selection of materials and components for the product has changed from when the product was first introduced to the present day, with reference to:
    –   functional performance
    –   aesthetics
    –   cost and availability
    –   properties and characteristics
    –   environmental considerations
    –   social, cultural and ethical factors.
2   Describe how the change in materials and components through the product's life has impacted on the product and consumer.

KEY TERMS

Cost constraints – the cost limits that the product has to adhere to, to ensure commercial viability.

Ethical choices – decisions that have to be made that have no clear right or wrong answer but will affect people's lives or the environment.

Functional performance – how well a product performs; its function and use.

Material characteristic – how the material behaves (can be variable).

Material property – how well the materials performs (often definite). For example, the density or strength of a material.

Performance criteria – the list of essential or desirable standards that a material needs to meet. The standards against which the performance can be measured.

Social impact – how the use of a material or manufacturing method could impact on people's lives and the community.

Standardised test – a test that can be carried out in a consistent manner and will give a quantitative result.

KEY POINTS

- Designers and manufacturers have to consider a long list of criteria when selecting materials and components to ensure the success of a product.
- The cost of materials is a crucial factor in material and component selection. It is not always the most expensive material that is suitable for the end use of the product.
- Materials can be combined and finished to make them more suitable for the product's function.

- It is important that designers consider the environmental impact of their selection of material.
- The selection of the correct material and component can have both negative and positive wider implications to society at any stage of the product lifecycle.
- The incorrect selection of material or component can have a detrimental effect on the manufacturers and can lead to costly repercussions.

5.2 What materials and components should be selected when designing and manufacturing products and prototypes?

LEARNING OUTCOMES

By the end of this section you should have developed a knowledge and understanding that products consist of multiple materials and that product designers are required to discriminate between them appropriately for their use.

Design Engineering
These materials include:
- ferrous, non-ferrous and alloy metals, such as mild steel, aluminium and brass
- thermosoftening and thermosetting polymers, such as HIPS (high impact polystyrene), ABS (acrylonitrile-butadiene styrene) and polyester resin, epoxy resin and polyamides
- timbers and manufactured boards, such as oak, plywood and MDF
- textiles used for reinforcement and coverings, such as **geotextiles** used in civil engineering and construction
- composite materials, such as fibre-reinforced plastics, glass-reinforced plastics (GRP) and carbon fibre reinforced polymer (CFRP)
- smart materials, such as shape memory alloy, self-healing materials, thermochromic, photochromic and electrochromic materials
- modern materials, such as sandwich panels, e-textiles, rare earth magnets, high performance alloys and super-alloys, graphene and carbon nanotubes.

Fashion and Textiles
These materials include:
- natural and synthetic textiles
- polymers used in component parts, blended textiles
- metals used for jewellery, component parts and conductive threads
- wood used for component parts
- rubber used for performance and functionality.

Demonstrate an understanding of the classification and source of textile fibres and materials, including:
- natural animal textiles, such as wool, silk, leather, cashmere
- natural plant textiles, such as cotton, flax, modal and acetate
- natural mineral textiles, such as glass fibre and metallic threads
- synthetic textiles, such as nylon, polyester and acrylic.

Demonstrate an understanding of the classification of different yarns, including:
- single fibre spun yarns
- mixed blended fibre spun yarns, such as cotton/polyester and wool/acrylic
- filament yarns
- fancy yarns, such as boucle, chenille and Lurex
- bulked and textured yarns.

Demonstrate an understanding of the classification of different structures of fabrics, including:
- knitted fabrics, including warp knits and knitted pile fabrics, hand and machine knitting
- structured fabrics, such as knotted and braided fabrics/structures, 3D novel structures
- woven fabrics, such as brocades, jacquards, plaid, tartans and crepe
- non-woven fabrics, such as felt and heated, mechanical and adhesive bonded fabrics
- microfibres.

Product Design
These materials include:
- hardwoods and softwoods, such as oak, teak, beech, spruce and fir
- manufactured boards, such as plywood, MDF and blockboard
- ferrous and non-ferrous metals, such as cast iron, mild steel and stainless steel, aluminium, copper and silver
- metal alloys, such as brass and bronze
- thermoplastics and thermosetting plastics, such as PET, HDPE, PVC, LDPE, polypropylene, polystyrene and ABS, urea formaldehyde, epoxy resin and polyester resin
- natural and synthetic fibres, such as cotton, wool and silk, polyester and nylon
- textile fabrics, such as woven, non-woven, knitted and blended textiles
- composite materials, such as fibre-reinforced plastics, glass-reinforced plastics (GRP) and carbon fibre reinforced polymer (CFRP)
- modern materials, such as e-textiles, super-alloys, graphene, bioplastics and nanomaterials
- smart materials, such as thermochromic, photochromic and electrochromic materials, shape memory alloy and shape memory polymers, conductive paints and e-textiles.

The selection of materials is a complex matter and quite often you will need to model different developments before selecting a material suitable for a given application. You will need to consult technical data to be able to make informed decisions.

Analysing the suitability of the materials or components used in existing products is a great way of finding out how materials perform in different applications. The use of a material shows the properties and characteristics and its potential suitability for similar or different applications.

To make informed decisions when selecting materials for a given application it may also be necessary to consider some of the following issues:

● How the product's functional, aesthetic and economical requirements need to be understood by the designer when selecting materials and components.
● The variety of combinations of materials that the designer has to make choices from to avoid time-consuming iterations.
● That material choice can become limited if there are restrictions on cost and availability.
● That material choice can radically affect the design possibilities or design outcome.
● That the designer must be able to identify a product's critical requirements to make optimal material selections.

The crucial choices that need to be made by the designer and manufacturer may involve numerous material tests depending on where the material is going to be used on the product and how it is expected to perform. Many products have to pass certain tests and meet criteria before they can be sold.

Most products are manufactured using multiple materials, components and manufacturing processes. As product designers may have many commissions or design tasks that involve considering numerous materials and components, it is beneficial for you to refer to sections of this textbook outside of your chosen qualification to ensure you are able to recognise why multiple materials may have been selected. The sections covering different material categories below include some examples of how these materials may be used in Design Engineering, Fashion and Textiles and Product Design contexts. As a fashion designer, for example, while you may not need to have a detailed understanding of each different thermopolymer and theromosetting polymer, you should know how polymers are used in component parts when using blended fibres for fashion and textile products.

Similarly, if you are a design engineer you will not need to know how yarns and fabrics are classified or constructed in detail, but you should have an understanding of the use of geotextiles in engineering. Make sure you refer to the learning outcomes at the start of this section, which provide an outline of what knowledge and understanding you will need about each of the materials below, depending upon the qualification you are studying.

Materials selection charts

A number of materials selection charts such as those shown in Figure 5.16 are referred to throughout this chapter. These charts provide comparative data on materials and their properties in an easy-to-use format.

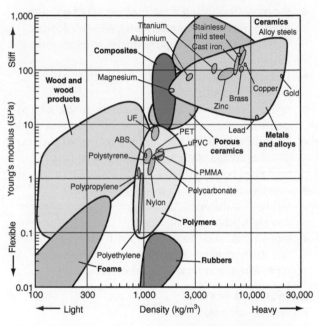

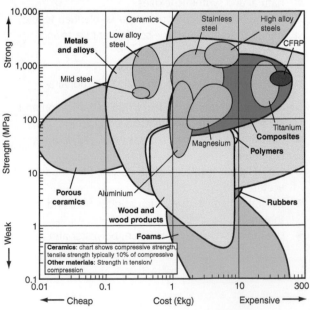

Figure 5.16 **Stiffness vs density chart and Strength vs cost chart**

You will not be asked to draw any of the charts in an examination, but it is possible you could be asked to draw conclusions from a chart, compare possible materials or suggest an alternative material for a given application. You might also need to cross reference different charts where combinations of different properties are required. The main aim of this chapter is to enable you to identify the materials used in products and give you the knowledge and tools to suggest suitable materials for specific applications, including in your NEA iterative design project.

You need to note that the axes on the charts are logarithmic, so a very small difference in position can represent a very large change in properties. Similar groups of materials are shown in 'bubbles' with individual materials identified within.

When describing material properties and characteristics you will need to refer to Section 5.3 for the correct terminology and detailed explanations.

Timbers

Hardwoods and softwoods

Natural timbers are classified into two groups: hardwoods and softwoods. The group a particular timber belongs to has nothing to do with whether it is hard or soft to the touch. It is a biological classification that is related to the way in which the tree produces its seed. Although not strictly accurate, for the purposes of a design technologist, it is reasonable to say that hardwoods come from broad-leaved (deciduous) trees and softwoods from coniferous (cone bearing or evergreen) trees.

DESIGN ENGINEERING

Wood is frequently used in the construction of buildings, especially houses. It can be used to quickly fabricate structures for supporting walls, roofs, and it is sometimes used to construct the support framework of a house itself. Wood is not often used in the construction of larger building projects because there is a limit on the sizes available, although large beams can be made by laminating thin strips of wood with adhesive to produce longer lengths.

FASHION AND TEXTILES

Wood used for component parts

The most common use of wood within fashion and textiles is through the use of fastenings. Buttons and toggles are still used today and are as much a functional part of a textiles product as they are aesthetic. Most woods used for fastenings are hardwoods, such as rosewood, walnut, ash and maple, as they need to be very durable. The wood is usually finished with varnish to make it waterproof.

Hardwoods such as beech and birch are used for the frames for canvas storage units found in the home, as well as for the carrying handles of bags.

Figure 5.17 **Wooden toggles**

PRODUCT DESIGN

There are hundreds of different types of hardwoods and softwoods with a wide range of appearances, **physical properties** and different applications. As a product designer you will be expected to know the main characteristics and typical applications of the timbers set out in Tables 5.2 and 5.3.

Table 5.2 Common hardwoods

Timber	Origin	Description	Physical characteristics	Typical uses
Oak	Europe and North America	European oak is typically yellow brown in colour. American oak is either darker in colour (red oak) or lighter (white oak). All usually have attractive grain patterns.	Strong and tough. Resistance to outdoor conditions varies greatly with species – European oak is generally very durable, whereas American white oak is not. Density lies between 720 and 790 kg/m³.	Heavy structural use, cladding, exterior joinery, interior joinery, furniture, flooring
Beech	Europe, especially central Europe and the UK	The wood is typically straight grained with a fine, even texture. Pink/pale red, reddish brown (after steaming), white/cream.	Beech is stronger than oak in bending strength, stiffness and shear by some 20 per cent and considerably stronger in resistance to impact loads. Density is around 720 kg/m³.	Interior joinery, furniture, workbenches, mallets, kitchen ware, chopping boards
Teak	Burma, Indian peninsula	Yellow/brown in colour. Straight grained with a rough, oily texture. Very expensive; other varieties (for example Iroko) are used as alternatives.	A little less resistant to impact than oak but slightly stronger and stiffer. Very durable – it resists outdoor conditions well because of the high content of resin. Density about 660 kg/m³.	Furniture and joinery (both interior and exterior), favoured for outdoor chairs and benches

Table 5.3 Common softwoods

Timber	Origin	Description	Physical characteristics	Typical uses
Pine (several varieties available, also known as redwood and/or deal)	UK, northern Europe, Russia	White/creamy coloured timber. Straight grained but sometimes marked by knots.	Reasonably strong and stiff but the quality varies because of the wide and varied distribution. Density is around 510 kg/m³.	Exterior joinery, interior joinery. Heavy structural use (for example roof trusses and timber-framed buildings). Can be treated for exterior applications such as fences and garden sheds.
Spruce	The spruce is native throughout Europe and Canada. Spruce grows in lowlands and mountainous regions.	Spruce is straight-grained with thin and regular texture. Resin canals are rather common. It is not very susceptible to shrinkage and when dried remains stable.	The wood is soft, low in weight and has medium density. The strength and elasticity properties are good. Density is around 450 kg/m³.	The typical end uses for spruce wood are structural, indoors and outdoors, thus it is the most important building and construction timber in Europe.
Fir	The fir tree is native to central and southern Europe and is most common in the mountainous forests of the Alps and mountainous ranges.	The colour is white with a little tendency to grey. The texture is fine to medium according to growing speed. There is no resin in the wood.	Soft, low in weight and has medium density. Strength properties are good. Sawing, machining and assembling are easy. It is elastic and flexible. It is less susceptible to shrinkage when dried and is strong and stable. Density is 530 kg/m³.	Fir and spruce wood are often mixed for structural end uses, indoors and outdoors: general carpentry, interior construction, windows and doors.

Making the right choice

There are many factors to consider when choosing a species of timber for a particular application. The factors that need to be considered vary enormously from one product to another. For example, in some cases appearance may be the most important consideration, but in others a physical property, such as strength or stiffness, may be most significant. Typically designers will consider:

- short-term physical properties: strength (in bending, in tension, in compression, in shear), stiffness, density, hardness
- long-term physical properties: stability (how much the timber expands and contracts, warps or twists as moisture levels in the atmosphere change), suitability for outdoor conditions, resistance to chemicals and so on
- working properties: ease of machining, ability to be glued, ease of finishing
- aesthetic properties: colour, grain pattern, lustre and surface blemishes
- commercial factors: cost, availability of sizes, consistency, wastage.

The physical properties of natural timber

In common with all other resistant materials, the physical properties of timber can be measured in terms of strength (in bending, in tension, in compression, in shear), stiffness and elasticity, density, impact resistance and resistance to chemicals.

However, because timber is a natural material (rather than being man-made) many of its physical properties are influenced by natural features in the timber such as localised variations in grain pattern, natural growth defects, drying defects, moisture content or biological decay. Also, as it is natural, it inevitably produces variable quality.

Manufactured boards

Wood-based boards are categorised into two main groups:

- laminated boards (including plywoods)
- compressed boards (including blockboard and MDF).

Man-made boards offer a number of useful advantages to designers and manufacturers compared to natural timbers:

- They are available in large sheets – 1220 × 2440 mm is a standard size, although considerably larger panels can be obtained.
- They are relatively dimensionally stable – they do not shrink or expand, warp or twist as much as natural timbers.
- Their properties are consistent across the whole board.
- Their properties are consistent from one board to another.
- Many are available pre-finished using laminated plastic sheet (melamine), foil (PVC) or natural veneer surfaces.
- They may be treated with flame-retardant chemicals.

For many products these advantages are so overwhelming that man-made boards have effectively eliminated all alternatives. In particular, because of their consistency and stability, they are so suitable for mass-produced furniture such as kitchen cupboards, that an extensive range of assembly fittings (usually referred to as knock-down (KD) fittings) have been developed.

Man-made boards also offer some environmental benefits because they can be made from lower grade timbers and so avoid much of the wastage associated with manufacturing with natural timbers.

Laminated boards

Laminated boards are produced by gluing layers together in a well-organised structure. The adhesive used is usually a synthetic resin that sets rapidly during manufacture, using heat and pressure.

Plywood

Probably the most well-known and widely used laminated board. Thin layers (or veneers) of wood are laid one on top of another with the grain direction at right angles in successive layers. There are always an odd number of layers so that the grain on the two outer faces of the board is parallel. Alternating the direction of the grain through the thickness means that the board has similar structural properties in all directions, unlike the natural timber from which it is made. The number of layers depends upon the thickness of the board, which for most commercial applications lies between 4 mm (three plies) and 27 mm (19 plies).

Blockboard and laminboard

These are similar to plywood in that there are layers of timber through the thickness of the board, but in each case the centre of the board is made from parallel strips of timber running perpendicular to the surfaces of the board.

In blockboard the core is made from sawn timber battens approximately 25 × 30 mm wide; in laminboard the core is made from vertically placed veneers, usually made from the same veneer as the facing surfaces.

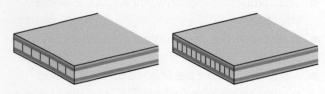

Figure 5.18 Sections showing blockboard (left) and laminboard (right)

This construction means that the boards are stiffer and stronger parallel to the core strips than perpendicular to them. For this reason, blockboard and laminboard are usually used for load-bearing shelves and similar applications where the direction of loading is constant.

Compressed boards

Compressed boards are made from chips, strands or particles of timber that are glued together, as the name implies, under a compressive force to produce large, flat sheets.

Chipboard

Chipboard is a cost-effective panel product made from small (2 × 3 mm) chips of timber often from the waste by-product of another production process. Compared to plywood it is rather weak and flexible and has very poor durability. Nevertheless, it is commonly used for flat-packed furniture where it is pre-finished with a surface

veneer or coating that disguises its humble origins and gives the panel some durability in situations where moisture may be present, such as kitchen cupboards or worktops.

Figure 5.19 Cut sections of chipboard

Medium- and high-density fibreboards (MDF and HDF)

In some respects MDF/HDF is similar to chipboard but it is manufactured from much finer fibres of wood rather than chips. As the name implies, HDF is more dense than MDF (approximately 800 kg/m³ compared to 600 kg/m³), giving a harder surface, greater stability and an ability to be shaped more precisely. The finer fibres mean that the face surfaces of these boards are considerably smoother than that of chipboard and this allows a high-quality painted finish to be obtained with little preparation.

Recent developments in timbers

Although timber is one of the most traditional of all materials used to manufacture artefacts and products, modern developments in manufacturing methods allow designers to use wood creatively.

Increasingly 'engineered wood products' allow innovation in commercial-scale projects. One example of this is flexible sheet materials used to produce curved shapes – especially in furniture but also in interior design.

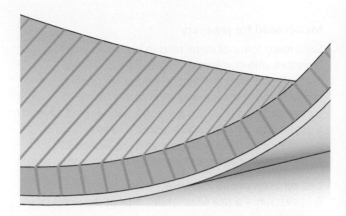

Figure 5.20 Neatflex™, a flexible board manufactured by Neat Concepts

Another development in the manufacture of sheet material is cement-bonded particleboard for use in the building industry. This overcomes problems of poor durability associated with most compressed boards, so cement-bonded boards can be used for exterior as well as interior applications.

A major area of innovation is the use of lamination techniques where strips of solid timber are glued together with their grains running parallel to produce structural members. The lamination process allows timber to be used for longer spans, heavier loads and complex shapes. It is two-thirds the weight of steel. The combination of great strength in comparison to weight gives the opportunity to produce curved forms with the aesthetic properties of solid timber. This technique, known commercially as 'glulam' (glued lamination), has been used to produce some spectacular shapes.

Figure 5.21 Glulam arches in Sheffield Winter Gardens

New types of engineered timber that are considerably stronger and more stable than natural solid timber are allowing architects to build bigger and higher, with timber skyscrapers now a real prospect.

Metals

Ferrous metals

The term 'ferrous' comes from the name ferrite (iron). Ferrous metals are therefore those that contain iron, and non-ferrous metals are those that do not.

Iron is relatively soft. A range of steels, which are harder, are produced by adding carbon.

Steels, alloys of iron and carbon, are classified by their carbon content:
- mild steel: up to 0.25 per cent
- medium carbon: 0.25–0.5 per cent
- high carbon: 0.5–1.5 per cent

FASHION AND TEXTILES

A wide range of metals can be incorporated into fashion and textile products, as a fibre within the material, as part of the aesthetics, as a component or as a conductive thread.

The most well-known use of metals in fashion and textiles is for the production of lame and brocade fabrics. Originally gold and silver was used as a fibre but now aluminium, copper and steel are commonly used to create the same effect.

Metals used for component parts

Metal is used for structural components in textiles such as:

- underwires in construction garments – tends to be mild steel or nickel titanium. (shape memory alloy)
- boning used to create shape in corsetry – made from steel or cheaper alternatives made from polymers
- wires for hoops in crinolines.

Steel is the favoured metal as it has better shape-retention properties than polymers. It can also be tempered to have spring-like behaviour which allows it to better resist stress. Steel has some of the highest strength values for its size, and it can endure heat/cool wash cycles.

Metal is also used for functional components such as:

- metal buttons – aluminium, brass, steel and sterling silver
- zips – brass, nickel and aluminium
- magnetic fastenings – the majority are made from steel
- buckles and D rings – titanium, brass and zinc
- eyelets and poppers – brass, iron and stainless steel.

Decorative components also make use of metals, such as using precious metal threads to create embroidery and motifs. These threads tend to be silver or gold and are historically used in heraldic work, and military and ceremonial embroidery.

Figure 5.22 Shoes featuring metal components

Metals used for jewellery

Since many forms of metal tend to rust, most jewellery is created with non-ferrous metals. Base metals are relatively abundant and tend to oxidise or corrode relatively easily; some examples include iron, nickel, copper and titanium. Precious metals such as gold, silver and platinum are particularly desirable for jewellery because they are less reactive than most elements, possess a higher lustre and are easier to work with.

Less common metals in jewellery making include:

- palladium – a rare silver white metal of the platinum family
- rhodium – a rare silver white metal of the platinum family. It is particularly hard and is the most expensive precious metal
- titanium – a natural element which has a silver white colour. Titanium is the hardest natural metal in the world yet is very lightweight. Pure titanium is also 100 per cent hypoallergenic, which means that it is safe for anyone to wear.
- tungsten – stronger than gold and scratch resistant.

Metals used for conductive threads

Conductive thread is ideal for introducing electronics into fashion and textiles projects. The thread looks and behaves like conventional sewing thread with the added bonus of being conductive, allowing the thread to be used in the place of wires, with conventional electronics. Most threads are metallised with an alloy of various metals, which can include silver, copper, tin, nickel and stainless steel. The core is normally cotton or polyester. Conductive threads are uninsulated and sewing them tightly to metal usually makes for a good connection. (For more information on conductive threads, see e-textiles on page 184.)

Figure 5.23 shows the properties for steels with different carbon contents. On the left side, steels are known as 'mild steels', in the mid-range, 'medium-carbon steels' and on the right side, 'high-carbon steels', also known as 'tool steel'.

Steel is an alloy of iron and other elements, primarily carbon, which is widely used because of its high tensile strength and low cost. Mild steel is also known as plain-carbon steel and is now the most common form of steel because its price is relatively low. Low-carbon steel contains approximately 0.05–0.25 per cent carbon, making it malleable and ductile.

Mild steel is often used when large quantities of steel are needed. The density of mild steel is approximately 7.85 g/cm³. Low-carbon steels contain less carbon than other steels and are easier to cold-form, making them easier to handle.

A wide range of alloy steels are available for specialist purposes, containing a wide range of metals and other elements in addition to carbon, such as manganese, nickel, chromium, molybdenum, vanadium, silicon, lead and tungsten.

PRODUCT DESIGN

Product designers use metals for a wide range of applications, including those outlined in the more detailed sections on **ferrous** metals, non-ferrous metals and alloys below. As a product designer, you will need to know about each of these categories.

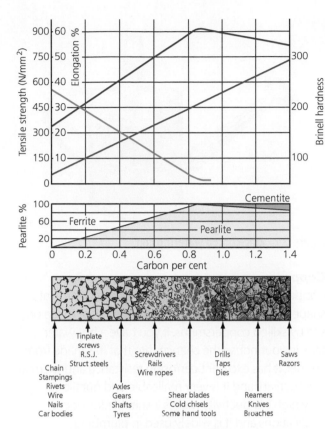

Figure 5.23 **Properties of steels with different carbon contents**

DESIGN ENGINEERING

Metals are widely used in design engineering. Their resilience, stiffness, and machineability, along with their electrical and thermal conductivity, make them ideal for specific design applications. You will need to be aware of a range of metals and their properties which make them useful in engineered products and systems.

Cast iron

Cast iron was used by the Victorians for lampposts and railings, but we still use it today for machine beds, engine castings and engineering vice bodies.

- Cast iron is cost effective but very heavy.
- It is very rigid and strong in compression.
- It casts well and can be easily machined.

Non-ferrous metals

Non-ferrous metals are those that do not contain iron.

Aluminium

Aluminium is the most widely used non-ferrous metal. It is well known for its low density and its ability to resist corrosion. Aluminium alloys are widely used in the

Table 5.4 **Different types of steel**

Metal	Description	Properties	Uses
Mild-steel	Carbon is the primary alloying element. Mild steel has no more than 1.65% manganese, 0.6% silicon or 0.6% copper. Mild steel is available with varying levels of formability.	Tough, ductile, malleable, good tensile strength, easily welded Poor resistance to corrosion Cannot be hardened except by case hardening	General purpose engineering and structural material
High-carbon steels	High-carbon steels contain from 0.60 to 1.00% carbon, with manganese contents ranging from 0.30 to 0.90%. The composition is very fine which makes the steel very hard. Unfortunately this also makes the steel quite brittle and much less ductile than mild steel.	High resistance to wear Brittle Poor resistance to corrosion	Hand and machine cutting tools and blades that retain their sharp edge
Stainless steel	Stainless steel is an alloy of iron with a minimum of 10.5% chromium. Chromium produces a thin layer of oxide on the surface of the steel which prevents corrosion. The range of stainless steels contain other alloying elements such as carbon, silicon, manganese, nickel and molybdenum in varying amounts to suit applications.	Corrosion resistant Work hardening can make it magnetic Ductile	Cutlery, kitchen utensils, appliances, sinks, surgical instruments, to name a few

aerospace industry, building construction (windows, doors, door handles) and increasingly in transportation and car construction. Aluminium has about one-third the density and stiffness of steel and is a good thermal and electrical conductor. It is easily machined, cast, drawn and extruded.

As a raw material, aluminium is relatively soft with low strength and high ductility. Strength is increased by alloying with silicon, manganese, magnesium, tin, copper and zinc. Aluminium is quite reactive, but protects itself very effectively with a thin oxide layer. The surface can be anodised to resist corrosion and to give decorative effects.

The production of aluminium uses a great deal of energy, the electrical energy consumption going into the extraction of aluminium is more than 3 per cent of the world's entire electrical supply. From a positive environmental perspective, aluminium is easily and widely recycled (for example, drinks cans) using 1 per cent of the energy needed to produce the metal. The use of aluminium in cars gives large weight reductions, leading to higher fuel economy and lower emissions.

Figure 5.25 Zinc-coated watering can

Copper

Copper in its pure form is soft and can be beaten into shape. In the past, many copper vessels were made on site by skilled coppersmiths. If you have a hot water tank in your house the chances are that it is made from copper. It has excellent corrosion resistance and thermal conductivity and it is very malleable and ductile. Its very high conductivity makes it suitable for electrical applications and it is widely used in plumbing and heating applications.

Tin

This metal is rarely used just by itself. Look at the cans sold in supermarkets. Though we commonly say they are, are they really made from tin? What do you think the solder used to join the parts of the tin cans is made from?

Tin cans are made from mild steel, but they have a very thin coat of tin – the material is called tin plate. This protects the inside from food contamination and the outside from corrosion. Like **galvanising**, mild steel is dipped in a bath of molten tin; squeeze rollers then remove any excess. Solders used are approximately 99 per cent tin and 1 per cent copper.

Figure 5.24 Jaguar F-Type with cut-away body showing the aluminium alloy chassis

Zinc

Zinc is the 'altruistic' metal – it sacrifices itself for the good of others! The process of coating steel with zinc is called **galvanising.** Without its zinc coat, the watering can in Figure 5.25 would have disappeared long ago. This process is only effective if it is not scratched.

Tungsten

Tungsten is a chemical element with the symbol W and atomic number 74. It is one of the toughest materials found in nature. It is super dense (about 1.7 times that of lead, 19.3 times that of water, comparable to the density of uranium and gold), extremely durable, resistant to corrosion and, with a melting point of 3422°C, almost impossible to melt.

Tungsten has the highest tensile strength of any element. Its strength comes when it is made into compounds and alloys, because pure tungsten is very soft. The hardest alloys are shaped using diamonds, being the only compounds harder than some tungsten alloys.

One of the most common, and hardest, tungsten compounds is tungsten carbide. Because of its strength when made into compounds, tungsten is used to make high-speed cutting tools and tips for circular saw blades and masonry drill bits.

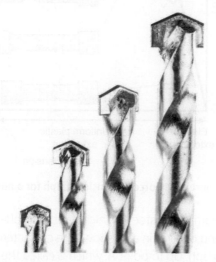

Figure 5.26 **Drill bits with tungsten carbide tips used for drilling into concrete, bricks and stone**

Non-ferrous metal alloys
Alloys are metals mixed with other metals and other substances to improve certain properties; this might be to improve strength, hardness, strength-to-weight ratio or aesthetic.

Brass
Brass is an alloy of copper and zinc. The proportions of copper and zinc are varied to create a range of brasses with varying properties. Its bright gold-like appearance makes it suitable for decorative metalwork. It is suitable for applications where low friction is required such as in mechanisms like locks, gears, door handles, plumbing taps, valves, electrical fittings and zip fasteners. It is used extensively in brass musical instruments where a combination of high workability and durability is desired. The relatively low melting point of brass (900 to 940 °C depending on composition) and its flow characteristics make it a relatively easy material to cast and extrude. The density of brass is 8.4 to 8.73 g/cm³

Bronze
Bronze is an alloy with a dull gold/brown colour that consists primarily of copper with the addition of other

ingredients. In most cases the ingredient added is tin, but arsenic, phosphorus, aluminium, manganese and silicon are also used to produce different properties in the material. All of these ingredients produce an alloy much harder than copper alone.

One example of a bronze alloy is phosphor bronze (tin bronze) which typically has a tin content ranging from 0.5 per cent to 1.0 per cent, and a phosphorous range of 0.01 per cent to 0.35 per cent. These alloys are noted for their toughness, low coefficient of friction, high fatigue resistance and fine grain. The tin content increases the corrosion resistance and tensile strength, while the phosphorous content increases the wear resistance and stiffness. Some typical end uses for this alloy are electrical products, bearings and components in mechanisms and machines.

Figure 5.27 **A large bronze worm gear wheel in an engineering works**

Although manufacturers may refer to bicycle and window frames as being made from 100 per cent aluminium, in practice most products will contain an aluminium alloy. Tubes, panels and 'hollow ware' are typically made from 99 per cent aluminium with 0.1 per cent manganese and 0.5 per cent silicon. This can give a strength-to-weight ratio five times that of pure aluminium. Aluminium also comes in other forms: for example, the brake lever on a bike could be made from die cast aluminium containing 10 per cent silicon. There is a reason for this, as a 10 per cent silicon alloy will change from a liquid to a solid very quickly and will not shrink.

The mixing of metals and other substances to improve properties has been explored since the Bronze Age and even today scientists and metallurgists create new alloys. In 2016 a super-hard metal was made in the laboratory by melting together titanium and gold. Beta-Ti3Au, as it is known, is too hard to be ground in a diamond-coated mortar and pestle, and this hardness may be vital in extending the life of dental implants and replacement joints. South Korean researchers have recently developed a lightweight aluminium-steel alloy that is flexible and ultra-strong, and are now exploring how to protect the material from corrosion out in the real world.

Physical properties of ferrous and non-ferrous metals

Strength, hardness, toughness, elasticity, **plasticity**, brittleness, ductility and malleability are **mechanical properties** used as measurements of how metals and other materials behave under load. The properties are described in terms of the types of force or stress that the metals withstand and how they are resisted.

Table 5.5 illustrates which metals and metal alloys have the best mechanical properties in comparison. They are ranked in descending order according to the property in each column heading. The table does not include mild steel. Where do you think it might fit in each of the columns?

Table 5.5 Mechanical properties of metals

Toughness	Brittleness	Ductility	Malleability
Copper	Cast iron	Gold	Gold
Nickel	Hardened steel	Silver	Silver
Iron	Bronze	Platinum	Aluminium
Zinc	Aluminium	Iron	Copper
Aluminium	Brass	Nickel	Tin
Lead	Zinc	Copper	Lead
Tin	Tin	Aluminium	Zinc
	Copper	Tungsten	Iron
	Iron	Zinc	
		Tin	
		Lead	

Figures 5.28 and 5.29 show force–extension graphs for carbon steel and a non-ferrous alloy. You can see by the shape of the graphs the properties associated with that material.

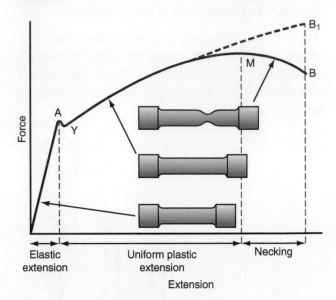

Figure 5.28 Force–extension graph for a non-ferrous alloy

Normalised steel – which has been softened by heating and cooling in air – shows the characteristics of the graph. Up to point A, which is called the elastic limit, the steel will spring back to its original shape if the force is removed. After point Y – the **yield point** – plastic deformation takes place and the material stretches until it breaks at B.

Hardened tool steel does not elongate much before it breaks. Hardened and tempered, it becomes much less brittle. A non-ferrous alloy annealed (softened by heat and left to cool slowly) behaves in a plastic fashion and elongates considerably before it breaks. Cold working – by hammering or rolling – will cause an alloy to be much harder.

You can compare the ferrous and non-ferrous graphs on the same axis.

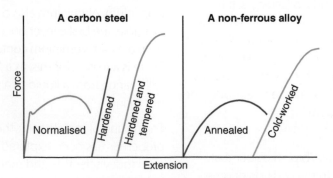

Figure 5.29 Force–extension graph for carbon steel

Figure 5.30 The Crooked Spire in Chesterfield

The image of the Crooked Spire in Chesterfield seen in Figure 5.30 is an example where the incorrect choice of material could have potentially had a disastrous effect on the building. The spire was added to the fourteenth-century tower in about 1362. It is both twisted and leaning, twisting 45 degrees and leaning 2.90 m from its true centre. It is now believed that the twisting of the spire was caused by the lead that covers the spire. The lead causes this twisting phenomenon because, when the sun shines during the day, the south side of the tower heats up, causing the lead there to expand at a greater rate than that of the north side of the tower, resulting in unequal expansion and contraction. This was compounded by the weight of the lead (approximately 33 tons) which the spire's bracing was not originally designed to support.

Polymers

Polyethylene, polyvinyl chloride, polystyrene…why do they all start with 'poly'? Polymers … what on earth is a 'mer'? 'Poly' simply means many and 'mer' is the name of the simple single units from which thermopolymers are made. Polymer is therefore a name for many of these units joined together.

The units are usually joined together in long chains. These chains can become entangled during a chemical reaction and a solid is formed that is both strong and rigid. When the solid is heated, the forces of attraction between the molecules decreases, making the material less rigid. This makes it easy to mould and is the source of the name 'thermo' polymer.

Long-chain molecules behave in an entirely different way to the cross-linked structure found in thermosetting polymers. The process of forming these long-chain molecules is called polymerisation.

Figure 5.31 The gas ethylene is a monomer with a double bond between two carbon atoms

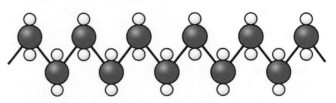

Figure 5.32 The gas ethylene is 'polymerised' to form polyethylene with a single bond between each carbon atom in the chain

The monomer (a single 'mer') ethylene is shown in Figure 5.31. The gas ethylene is 'polymerised' to form polyethylene (Figure 5.32). Individual carbon atoms are shown in red, with two white hydrogen atoms attached.

DESIGN ENGINEERING

Polymers are very useful to a design engineer. They are excellent electrical and thermal insulators, and they can be moulded to create bespoke shapes. They also have a good strength-to-weight ratio and useful mechanical properties such as flexibility and good impact resistance. The casings of most electrical products are moulded from polymers for all these reasons. The specific choice of using a polymer for an engineering design is often based on the properties of the polymer, i.e. its stiffness, ability to withstand high (or low) temperatures and resistance to chemicals, but the choice may also be influenced by aesthetic and manufacturing considerations.

FASHION AND TEXTILES

Polymers used in component parts

Textile materials are made from natural or synthetic fibre-forming polymers. Natural polymers exist as short fibres which need to be twisted to make them into useable lengths. Most synthetic polymers are manufactured from petrochemicals, using the process of polymerisation. To convert the polymers into workable fibres, the solution is fed through spinnerets.

Nylon is the most commonly used polymer in the fashion and textiles industry and is actually not one single substance but the name given to a whole family of very similar materials called polyamides. Nylon can be moulded into components or drawn into fibres for making fabric. Due to the fact it is reasonably durable, resists sunlight and is waterproof, it is very suitable to be used for a wide range of components. Buttons and zippers with polymer resin parts, for example, are widely used in the production processes of companies involved in the textile industry. (For more information on synthetic fibres see page 165.)

Blended textiles

The main idea behind blended fabrics is to combine fibres with certain qualities, with fibres of other qualities that complement each other and ultimately improve the end fabric for the purpose required. Many fabrics today are combined with a polymer to improve overall performance. Cotton and polyester is one example of a blended fabric, but there are many others. (See page 177 for more information on blended textiles).

PRODUCT DESIGN

In Product Design you will need to know about each of the thermopolymers and thermosetting polymers covered in the sections below.

Thermopolymers

This group of polymers deform whenever heated. They lose their rigidity and can be remoulded many times. This makes them suitable for vacuum forming, injection/blow moulding and recycling. Thermopolymers generally have a lower safe working temperature than thermosetting polymers.

Table 5.6 **Thermopolymers**

Thermopolymer	Abbreviation	Safe working temperature	Characteristics	Common uses
Polystyrene	PS	80°C	Stiff and hard, brittle	Disposable cups, construction kits
Polyethylene (low density)	LDPE	75°C	Flexible, soft and waxy	Detergent squeeze bottles, bin liners
Polyethylene (high density)	HDPE		Fairly stiff, hard	Domestic bleach bottles, milk crates
Polymethyl methacrylate	PMMA (acrylics)	95°C	Crystal clear, glossy	Car rear light units, illuminated signs
Polypropylene	PP	100°C	Stiff, hard, flexible (thin)	Child car seats, integral hinges, storage crates, car interior panels, textiles, stationery, carpets, reusable containers
Polyvinyl chloride	PVC	95°C	Available in rigid and flexible forms	Cable insulation, hosepipes, rainwater and underground pipes
Acrylonitrile-butadiene-styrene	ABS	80°C	Opaque, stiff, hard High impact resistance	Casings for cameras, electrical appliances
Polyethylene terephthalate	PET		Translucent, hard	Water and fizzy drinks bottles

ABS

ABS (acrylonitrile-butadiene-styrene) is used when products are subject to impact and weight isn't a critical factor. ABS combines good qualities of strength and toughness and would be the natural choice for a pre-fitted, injection-moulded 13 A plug for a vacuum cleaner, where the plug is often dropped or dragged along behind the cleaner.

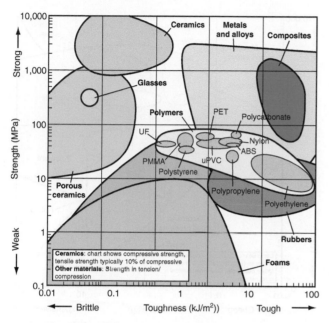

Figure 5.33 Material selection chart showing strength vs toughness

Polypropylene (PP)

Polypropylene is used in a wide variety of applications including packaging, textiles, stationery, packaging, carpets, reusable containers and car parts, to name a few. Polymer banknotes have recently been introduced in the United Kingdom.

Figure 5.34 Polymer banknotes

Polyethylene (PE)

This is a very widely used plastic, particularly in various forms of packaging. PE has excellent chemical resistance. It is available in two forms:

- low-density polyethylene (LDPE) – fairly flexible and has a soft, waxy feel; used for fertiliser bags, bin liners and the insulation for co-axial television aerial leads
- high-density polyethylene (HDPE) – fairly stiff and hard; used for domestic bleach bottles.

Polyvinyl chloride (PVC)

PVC is available in two forms:

- rigid/unplasticised – used for hosepipes and wire/cable insulation (excluding aerial cables) and to produce thin, coloured adhesive-backed films referred to as vinyl, which are cut to produce labels and signs including vehicle markings
- plasticised – widely used for domestic window frames, cladding and rainwater guttering and pipes. Some of the PVC is foamed (or expanded) which provides extra thickness without added weight and cost.

PVC has good chemical resistance. Additives are used to enhance performance and durability, for example, heat and UV (ultra-violet) stabilisers to prevent degradation with prolonged exposure.

Figure 5.35 Section of a PVC window frame

Polystyrene (PS, GPPS, HIPS and EPS)

This is a very common and low cost plastic. It is available in three forms:

- polystyrene (PS) or general purpose polystyrene (GPPS)
- high impact polystyrene (HIPS)
- expanded polystyrene (EPS).

PS is hard, rigid and has low shrinkage. It is easy to mould and process, with fine detail possible. It is used for toys and novelties, rigid packaging and refrigerator trays and boxes.

HIPS has similar properties to PS and is easy to machine and fabricate. Having up to seven times the impact strength of GPPS, it is often specified for low strength structural applications when impact resistance, machinability and low cost are required. It is frequently used to machine pre-production prototypes and modelling.

EPS foam is widely used to protect or package delicate items. It is made of 98 per cent air, making it one of the lightest packaging materials, so transport costs and fuel emissions are kept to a minimum. You have probably used some of this Styrofoam as a modelling material.

Polymethyl methacrylate (acrylic, PMMA)

Acrylics, sometimes known by their trade name Perspex, are a very important group of plastics. PMMA, the abbreviation of the full, correct name is made by the polymerisation of methyl methacrylate.

Acrylic is much tougher than glass and has the advantage that it can be easily moulded. It was used for machine guards and was only replaced when tougher polycarbonate became available. Acrylics can transmit more than 90 per cent of daylight and are used for roofing panels (it can eventually degrade in sunlight after many years). Due to excellent optical properties it is used extensively for motor car rear light units and illuminated advertisement signs.

Acrylics have a fairly good chemical resistance and are used for baths, sinks and dentures.

Polyethylene terephthalate (PET)

Figure 5.36 **PET bottles**

PET is a relatively recent development and its unique qualities have led it to replace other polymers for many functions. PET has taken over from polyethylene and PVC for the following applications:

- fizzy drinks bottles
- clear washing up detergent bottles
- milk containers
- chocolate box liners.

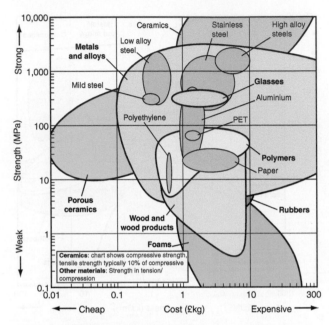

Figure 5.37 **Material selection chart**

PET is one of the strongest plastics. It is more expensive than PE but its much greater strength means that less of it is needed to withstand the gas pressure in fizzy drinks bottles. PET bottles are thin and could be even thinner if the gas didn't seep through very thin sections.

Polyamide (PA)

Polyamides occur both naturally and artificially. Examples of naturally occurring polyamides are wool and silk. Artificially manufactured (synthetic) polyamides are nylons, aramids and sodium poly(aspartate).

The best-known synthetic polyamide is nylon which as a thermoplastic is widely used in the automotive and electrical fields for plastic fasteners such as screws and washers, cable ties, gears and machine components.

With a relatively high melting temperature of 256°C, its heat resistance makes it suitable for cooking utensils and components next to engines.

Thermosetting polymers

Once these polymers have been shaped and hardened, they cannot be reshaped or softened by heating. There are five main thermosetting materials in common use for products that you need to know about – they all have high safe working temperatures.

Urea formaldehyde

This is a light-coloured thermosetting plastic which is hard and brittle and has a high heat-distortion temperature. In synthetic resin form (powder mixed with water) it is a widely used wood glue used in plywood, chipboard and laminated structures. It is inexpensive, sets hard and rigid and is water and heat resistant. It combines good insulation properties with relatively low cost. It can also be easily formed into complex shapes by compression moulding.

Table 5.7 **Thermosetting polymers**

Thermosetting plastic	Abbreviation	Safe working temperature	Characteristics	Uses
Urea formaldehyde	UF	80°C	Opaque, light in colour	Electrical fittings – 13 A plug cases, sockets, switches
Phenol formaldehyde	PF	120°C	Opaque, dark in colour	Domestic iron and saucepan handles
Melamine formaldehyde	MF	130°C	Opaque, multi-coloured	Cups, plates, buttons, kitchen worktop surfaces
Polyester resin		95°C	Clear liquid resin	With GRP and CFRP; boats and vehicles
Epoxy resin – high melting point		200°C	Two resins in tubes	'Potting' of electronic circuits
Epoxy resin – low melting point		80°C	One tube has activator	Adhesives

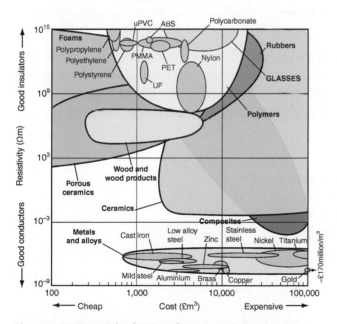

Figure 5.38 **Material selection chart: resistivity vs cost**

Phenol formaldehyde

Phenol formaldehyde has a higher safe working temperature than urea formaldehyde and is used in applications exposed to heat. PF resins are used as the basis of 1 mm thick decorative plastic laminate sheets used to give a hardwearing and heat and water-resistant work surface to a kitchen worktop.

When urea reacts with formaldehyde in the first instance a 'syrup' is formed; this is mixed with filler and allowed to dry out. The powder formed is the raw material used in the compression moulding process. The syrup can also be used as a coating on metals on washing machines or fridges. In all cases, the application of heat sets the material permanently.

Figure 5.39 **Thermoplastic pan handle**

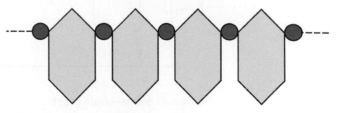

Figure 5.40 **Formaldehyde molecules**

Epoxy resin

Epoxy resin is a thermosetting polymer with high mechanical properties, temperature and chemical resistance. A cross-linking agent (hardener) is mixed with the epoxy resin. It is used in epoxy powder-coating, typically applied electrostatically and then cured under heat to allow it to flow and form an even coating. Epoxy resin gives a very hardwearing and durable finish for metal which is used for furniture frames.

Figure 5.41 Epoxy powder coating of metal surfaces (note the resin and hardener feeds to the spray nozzle)

Epoxy resin is also available as a two-part high-strength adhesive which is particularly useful in structural applications and for joining metal where products are subjected to heat. A two-part resin is also used in **composite materials** reinforced by glass, Kevlar and carbon fibres.

Being an excellent insulator, epoxy resin is used in many electronic and electrical components, notably the encapsulation of transistors and integrated circuits.

Polyester resin

Without fillers or reinforcements, polyester resin is a very brittle resin. It is primarily used in glass-reinforced plastic (GRP) and CFRP composites to produce car bodies, boat hulls, canoes and surfboards and building products such as cladding and roof sheeting.

Polyester resin offers ease of handling, low cost, dimensional stability, as well as good mechanical, chemical-resistance and electrical properties. Polyester resins are the most economical way to incorporate resin, filler and reinforcement to mould both large and small components. The resins do have a limited storage life.

Physical properties of polymers

The material selection charts will give you most of the information you need to know about the comparative mechanical properties of the plastics you need to study.

In addition to the selection charts, you need to know how plastics perform under tension and understand about elasticity and plasticity. Figure 5.42 shows comparative information on different groups of plastics on a stress–strain graph.

These are nearly, but not quite, the same as the force–extension diagrams used when metals are tested. In general terms, the more the plastic elongates before breaking, the softer the plastic. Very hard plastics have steep graphs

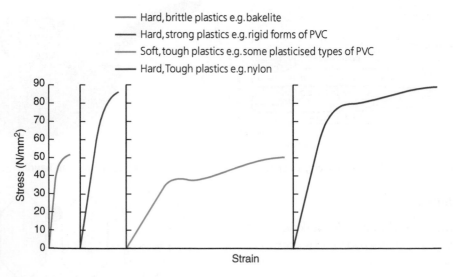

Figure 5.42 Stress–strain graph

then suddenly break without elongating. Most plastics, as their name suggests, behave in a plastic fashion – they elongate for a long time before breaking. This property is called plasticity. This property also makes thermopolymers suitable for a variety of moulding processes.

Elasticity is the ability to return to an original state when a load is removed. This condition only usually occurs during the straight line part of the graph. After this point, plastic deformation occurs, which is non-reversible.

Polymers are mainly known for their ability to withstand corrosion and impact. Most thermosetting polymers are brittle in tension but stronger in compression. Care is needed to reinforce designs where this might be a problem (the body of a 13 A plug is a good example of this). Most thermopolymers cut very easily and are therefore susceptible to shear (slicing) forces.

Textiles

It is very important that you can differentiate between a range of natural and synthetic textiles to understand which materials will be most suitable for your projects. By examining a wide range of products you will be able to develop your knowledge on material properties and characteristics that make them suitable for given situations.

New materials and processes are continually being developed to meet a wide range of needs and situations. Natural fibres have many natural properties and characteristics, but many products benefit from combining both natural and synthetic fibres and fabrics. This could be due to cost or the performance of the product.

PRODUCT DESIGN AND DESIGN ENGINEERING

Textiles offer interesting and useful properties to all designers and engineers.

Textiles can provide very high tensile strength in a very thin, lightweight material. They can be used on their own to provide tensile support and to act as a barrier against weather, such as the glass fibre O2 Arena roof shown in Fig 5.56 on page 172. Boat sails and parachutes made from ripstop nylon are examples where a textile really is the only solution, as they provide high strength, the

ability to be manufactured to take up a specific shape during use, low weight and they can be packed down and stored in a small space.

Textiles can be combined with other materials such as resins to produce GRP or CFRP, or they can be laminated with other materials to increase abrasion resistance (for example, Kevlar), fire resistance (for example, Nomex), or to reduce flexing on surfaces such as an aeroplane wing.

Fibres are classified according to their origin: some are from plants and animals, for example, cotton, flax, wool and silk; some are from plant material, for example, viscose and lyocell; and some are made entirely by chemical reaction, for example, polyester and polyamide.

A fibre is a fine and flexible raw material which could be short or very long depending on where it comes from and how it is manufactured. There are three main fibre types:

- Staple fibres – staple fibres are short, ranging from a few millimetres, such as in cotton, to around a metre, as in flax.
- Continuous filaments – continuous filaments are an indefinite length. All synthetic fibres start off as continuous but they may be cut up to form staple yarns. Silk is the only natural continuous filament.
- Microfibres – microfibres are very fine fibres, around 60 to 100 times finer than a human hair!

Table 5.8 **Classification of fibres**

Natural or manufactured	Origin	Examples
Natural	Vegetable	Cotton, linen (flax), kapok, jute, hemp, sisal
Natural	Animal	Wool, silk, camel, cashmere, mohair, angora
Manufactured	Regenerated	Viscose, lyocell
Manufactured	Synthetic	Polyester, polyamide (nylon), acrylic, elastomeric, flourofibres, chlorofibres
Manufactured	Inorganic	Glass, metal

Natural animal textiles

Natural animal fibres are formed from proteins.

Wool

The most common animal fibre is wool, usually from sheep but also from goats, llamas and angora rabbits.

Wool fibres are very similar to human hair; they are made of protein molecules (keratin) formed into bundles and can vary in length. The best woollen yarns are made from **staple fibres** that are 50 to 120 mm in length, highly crimped and very fine. These are used to produce worsted cloth, which is used for fine suiting.

Coarser, longer fibres with less crimp are used to produce woollen cloth for heavier, more robust clothing. Carpets and upholstery are produced from the most coarse, long fibres.

Wool fibres have a very distinctive structure; when wool fibres absorb moisture and the temperature rises, the fibres swell and the protein bundles move apart. The bonds between the fibres are broken down. These bonds reform as the bundles dry and cool. This makes the fibres very elastic and gives wool its recovery and reshaping properties.

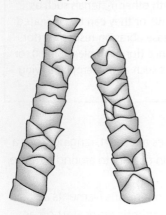

Figure 5.43 Wool fibres

Performance characteristics of wool:
- Three factors make wool a good insulator: the scales, the crimp and the length of the fibres.
- Wool is highly absorbent (**hydrophilic**): it can readily absorb 33 per cent of its weight in water without feeling wet. This also makes it anti-static.
- The fibres are **hydrophobic** because they are covered with a very thin skin – the epicuticle – which causes liquid to roll into droplets while allowing the passage of water vapour.
- The softness of wool gives it a comfortable feel next to the skin.

- Textiles made from wool are not particularly durable.
- Elasticity and springiness are excellent in wool, so creases easily drop out.
- Wool can be formed into durable shapes because the molecular structure can be adjusted under heat.
- Laundering can be a problem as wool will stretch when it is wet due to the high amount of absorbed water and felting can occur under the influence of mechanical action, heat and water.
- Wool is biodegradable and recyclable.
- Wool has high natural fire resistance and built-in UV protection.

Typical wool fabrics include flannel, herringbone, tartan, tweed and Viyella.

Figure 5.44 Wool jumper

Silk

Silk fibre is the only naturally produced continuous filament. It is produced by the caterpillar of the silk moth when it pupates. The long protein fibre comes from a spinneret below the mouth of the caterpillar which then forms the cocoon. Silk fibre is made from two long protein filaments glued together. The physical, chemical and comfort properties of silk fabric depend on the way the long chain protein molecules lie inside each filament of the fibre. The more closely they are layered together, the stronger and more resilient the fibre. The direction of the protein chains affects the ability to absorb moisture and the lustre of the fabric.

Cultivated filament silk is made into fine fabrics. Any broken fibres or short lengths are spun into spun silk, which is not as strong or lustrous as filament silk. Wild silk cocoons are also made into spun silk.

Figure 5.45 Different colours of woven silk fabrics

Performance characteristics of silk:

- Silk is both cool and warm. Filament silk lies smoothly on the skin and gives a cooling effect, but it is also a good insulator because the layer of warm air between it and the skin cannot escape through the compact silk fabric.
- It can absorb and hold about 33 per cent of its weight in water vapour without feeling wet.
- The next-to-skin feel is excellent because of its fineness, softness and elegant drape.
- It is very strong, durable and light. Silk fibres can be up to 1 km long.
- Creases fall out because it has excellent resilience.
- It does not build up any electrostatic charge because the moisture contained in the fibres conducts any charge away.

Typical cultivated silk fabrics include chiffon, damask, organza, satin and taffeta. Wild silk fabrics include dupion and shantung.

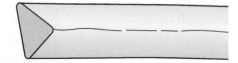

Silk fibre cross-section

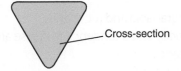

Cross-section

Figure 5.46 Silk fibres

Leather

Leather is a durable, flexible and versatile material that usually originates from cattle. There are many different types and qualities and it can be used for a huge variety of products from the mass-produced to luxury market.

The process used to turn the raw hide of the animal skin into useable leather is called tanning and many different methods are used to achieve this. Leather is generally sold in one of the following four forms:

- full grain
- top grain
- corrected grain
- split.

Leather can originate from many animals; some of the less common ones include:

- ostrich (one of the strongest commercial leathers and one of the most expensive)
- fish (a unique appearance and texture, very strong)
- kangaroo (can be cut very thin, but still retains strength)
- snake (an exotic product with a scaly appearance).

Figure 5.47 A leather skirt

Performance characteristics of leather:

- Leather has a very high tensile strength and therefore is resistant to tearing under strain. This makes it suitable for fashion and textile products that have to withstand a lot of wear; for example, upholstery for cars seats and sofas and footwear.

- Leather has good heat insulation so leather garments will keep you warm.
- Leather fibres are permeable to water vapour and can hold large quantities of water vapour. This property enables leather to absorb perspiration, which is later dissipated. This is a significant factor in comfort.
- Leather has thermostatic properties; it is warm in winter and cool in summer.
- Leather has mouldability; it can be moulded and will retain its new shape.

Cashmere

Cashmere was considered a luxury fibre and fabric as it is time consuming to produce. Over recent years, the cost of cashmere has lowered as it is being mass-produced and sold in many high-street stores. Prices have been lowered as many manufacturers are using the coarser hair or mixing it with yak hair. Cashmere is also often blended with wool, silk or polyester. This still gives a luxurious quality but at a lower price. It is most commonly used as a luxury yarn to be woven into expensive fabric for suits and coats or, more commonly, knitted to produce luxury knitwear. It can also be used for interiors for soft furnishing products and for the textiles used in cars, planes and yachts.

Performance characteristics of cashmere:
- Cashmere is known for its extreme softness.
- The fibre has similar properties to wool and will therefore keep you warm.
- It has a lustrous, silky quality as a result of its extremely fine fibres. The fineness of the fibre is typically between 7 and 19 microns. In comparison, ordinary sheep's wool has a diameter of 36 microns.
- It is very light weight.
- It is not very durable due to its soft downy texture.

Natural plant textiles
Cotton

Cotton is a natural vegetable fibre from the seed pod of the cotton plant. It is a staple fibre and is the most widely produced natural fibre. The fibres are formed from plant cellulose.

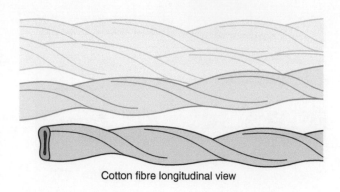

Cotton fibre longitudinal view

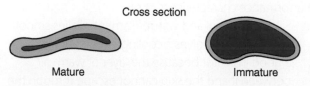

Cross section

Mature   Immature

Figure 5.48 Cotton fibres

Cotton production has considerable environmental implications as well as being a crop of major global importance. Cotton production accounts for 25 per cent of total global pesticide use and there are concerns about health risks for farm workers and soil contamination.

Performance characteristics of cotton:
- Because of the cavities in the fibres it can absorb up to 65 per cent of its own weight in moisture.
- It always contains some moisture so it is non-static.
- The fibres are strong when wet because they become swollen with water and this distributes stresses more evenly along the length of the fibre.
- Its soft handle, good drape and moisture absorption make it a comfortable choice for clothing.
- Durability and abrasion resistance combined with strength make it a good choice for a range of products.
- It is biodegradable and recyclable.
- Poor elasticity and a tendency to crease are disadvantages.
- Cotton dries slowly.

Typical cotton fabrics include calico, chintz, corduroy, denim, drill, poplin, terry, velvet and lawn.

Flax (linen)

The flax plant is the source of fibres for linen yarns and fabrics; it is an annual crop grown from seeds in many parts of Europe. The fibres are obtained from the inner fibrous parts of the stem. They are long fibres, which make it suitable for producing crisp, attractive, cool fabrics. It is produced as a staple fibre.

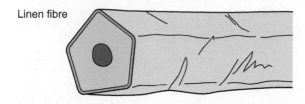

Linen fibre

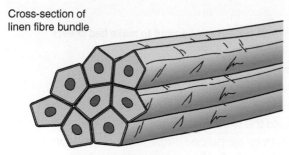

Cross-section of linen fibre bundle

Linen fibre (flax) cross-section

Figure 5.49 Flax fibres

The flax fibres, like cotton fibres, are long-chain cellulose molecules. They are about 25 to 40 cm long and are cemented together by a mixture of ligins, pectins and hermicelluses. The properties of flax are influenced by this composition. Flax is stiffer than cotton because of the cement holding the fibres together. It also has a smoother surface and a darker colour than cotton.

Performance characteristics of flax (linen):

- Flax is strong, durable and long lasting. It is strong when wet, cool to wear and is fast drying because it is highly absorbent but releases water quickly.
- It is anti-static as it always contains some moisture.
- It has a crisp, stiff handle due to the cements holding the fibres together.
- It is easy to wash and dry and also shrink proof as it takes up water rapidly and releases it quickly.
- It has low elasticity, so it creases badly.
- Poor insulation as the smooth fibres do not trap any air, making it cool to wear.
- Linen fabric has a low lustre and does not soil easily.

- The coarse fibres give linen a firm handle and affect its next-to-skin feel.
- It resists micro-organisms.
- It is biodegradable and recyclable.

Regenerated vegetable fibres

Regenerated fibres are made by chemically changing natural materials that come from plants.

Viscose

Viscose fibres are produced from eucalyptus, pine or beech wood that is dissolved in a solvent and extruded through a spinneret with fine holes. Viscose and acrylics are extruded into a bath containing chemicals that solidify the filaments. This is wet spinning. The holes in the spinneret can be circular or any other shape, which affects the lustre and handle of the fabric. The long filaments are smooth, fine and soft. Staple viscose fibres are produced by cutting short lengths as the fibres are extruded. It is cheap to produce.

The chemical structure of viscose is similar to cotton but the molecules are shorter, which is the main reason for the lower strength of viscose.

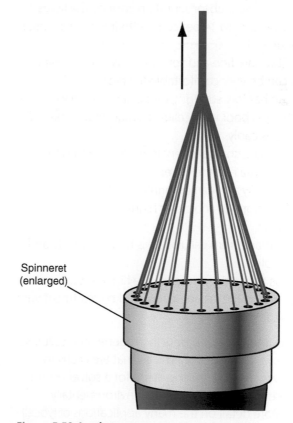

Spinneret (enlarged)

Figure 5.50 A spinneret

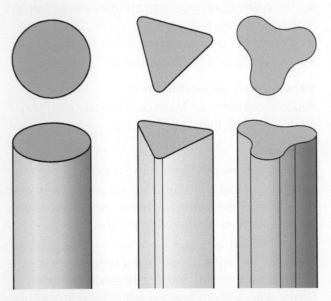

Figure 5.51 **Spinneret shapes and fibre cross sections**

Performance characteristics of viscose:

- Viscose filament yarns are made into smooth fabrics and therefore have poor thermal insulation. When cut into staple fibres the resulting fabric can trap air and improve insulation.
- Viscose is more absorbent than cotton. The fibres can absorb 11 to 14 per cent of their weight in water vapour.
- The fibres are fine and soft, so it has a good drape and can be very comfortable to wear.
- Viscose has low strength, particularly when wet.
- Elasticity is poor in all cellulose, man-made fibres, so it creases easily.
- Electrostatic charge is very low because the fibres always contain some water.
- It dyes easily and can be printed with bright colours, so it is a fabric of choice for high-street fashion designers.
- As it is made from cellulose it is biodegradable and recyclable.
- It tends to shrink in water as the fibres absorb water and swell, but this can be reduced by treatment with synthetic resins.

Traditional methods of producing regenerated cellulose fibres involve the use of chemicals that are costly to recycle. Lyocell is the generic name for a solvent-spun regenerated cellulose fibre that is environmentally friendly. More detail on the many applications of lyocell and modal can be found on the following website: www.lenzing-fibers.com/en/tencel/.

Figure 5.52 **Lyocell can be used to make bed sheets**

Modal

Modal is a semi-synthetic cellulose fibre made by spinning reconstituted cellulose from beech trees. To produce the modal fabric, the wood fibres taken from beech trees are pulped into liquid form and forced through tiny holes (in a spinneret) to make the fibre, which is then woven. It's now widely used throughout the fashion world and the fibre can be found in many of the collections of well-known brand manufacturers. Due to its properties, modal is a favourite fibre in lingerie and loungewear, as well as for children's clothes and baby wear.

Performance characteristics of modal fibre:

- Modal fibres are skin-friendly.
- They are extra soft and easy to care for.
- They are dimensionally stable and do not shrink or get pulled out of shape when wet like many cellulose fibres.
- They are also wear resistant and strong while maintaining a soft, silky feel.
- Modal fibres have a wide variety of uses in clothing, outerwear and household furnishings.
- They are often blended with cotton, wool or synthetic fibres.

Acetate

Acetate derives from cellulose from wood pulp. After it is formed, cellulose acetate is dissolved in acetone for extrusion. As the filaments emerge from the spinneret, the solvent is evaporated in warm air (dry spinning), producing fine filaments of cellulose acetate. Acetate is versatile but can be difficult to dye and therefore special dyes have to be used to give the fabric colour. The main uses for acetate are:

- blouses, dresses
- linings
- curtains
- upholstery.

Performance characteristics of acetate fibre:

- It has a luxurious natural feel and appearance, and is soft and gentle on the skin.
- It drapes lightly and flexibly.
- It dries relatively quickly.
- Compared to many other fibres, acetate fibre has greater dimensional stability, as well as strong pleating durability, because it does not swell much in water.
- It does not stain easily and stains are easy to remove.
- Acetate has poor fibre strength, but by combining it with other fibres that have sufficient strength, such as polyester, it has become possible to use acetate

fibres for practical applications. Polyester fibre has a mottled appearance and elasticity, so it can add to the texture of acetate fibre.

Figure 5.53 **Modal production from tree to fibre**

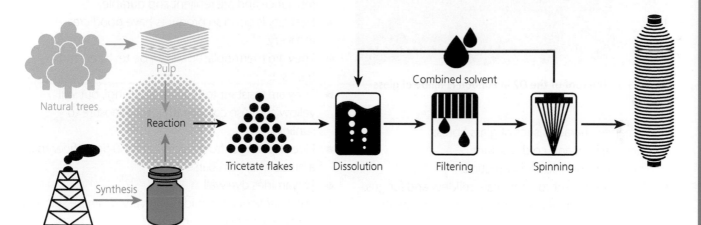

Figure 5.54 **Schematic of acetate production**

Rubber

Natural (latex) and synthetic rubber is used in many fashion and textile applications and products, either alone or in combination with other materials. In most of its useful forms, rubber has a large stretch ratio and high resilience and is extremely waterproof. This makes it particularly useful for flooring, waterproof coverings and wellingtons. Because of its electrical resistance, soft rubber can be used as insulation and for protective gloves, shoes and blankets. A large

percentage of rubber is used into the automotive industry.

Natural rubber is:

- warm, pliable and soft
- antistatic, antibacterial and anti-slip
- not breathable
- joined by stitching (but this can puncture the rubber) or by adhesive
- recyclable.

Natural mineral textiles

Glass fibre

Glass is an example of an inorganic fibre that can be manufactured into filament yarns and staple fibres to produce long-lasting, durable, woven-mesh glass fabrics. Although hard and rigid by nature, glass can be made into fine, shiny and translucent fibres which more or less look and feel like silk fibres. These glass fibres are commonly known as fibreglass. Glass textiles are used by the aerospace and military industries to produce flame and heat barriers, as well as light and ultraviolet filters. A glass textile was used for the roof of the O2 (Arena) in London.

Figure 5.55 **The roof of the O2 in London is made of glass textiles**

Performance characteristics of glass fibres:
- They are heat and cold resistant.
- They are non-toxic and non-stick.
- They are resistant to chemicals, mildew and fungus.
- They can bring reflective properties to garments.

Uses of glass textiles include architectural roof coverings, sterile wall coverings for hospitals and protective garments.

Synthetic textiles

Synthetic fibres are formed entirely by chemical synthesis from oil or coal sources. They were developed during the 1940s and 1950s and are now used to produce fibres that can be used in products as diverse as a soldier's protective headwear to an elegant prom dress. Man-made, synthetic fibres can be designed and engineered to meet whatever functions and properties are required; they are the most innovative of all the fibres.

They are made from chemical units called polymers that are formed from single-molecule units called monomers, linked together like a bead necklace.

All synthetic fibres are produced from a non-renewable source and are not biodegradable.

Manufactured fibres

Polyamide (nylon)

Polyamides are produced from chemical chips by melting and extruding into fibres. As they pass through the spinneret into a stream of cold air, the **filament fibres** are stretched to give a fine, strong yarn. Staple fibres can be produced by chopping up the filament fibres. The shape of the spinneret can be varied to give textured fibres.

Performance characteristics of polyamides:
- Insulation depends on the type of fibre. Flat filament fibres do not trap air and have low insulation. Texturing increases the volume and gives better insulation. Staple yarns may be either fine and smooth or more bulky.
- Polyamides absorb very little water, which means that they are susceptible to electrostatic charge. It also means that they are easy to wash and dry.
- They are very strong, have excellent abrasion resistance and are resilient and durable.
- Elasticity is good so products have good crease recovery.
- They are thermoplastic so can be textured and heat set.
- They are resistant to moulds and fungi, but will yellow and lose colour with long exposure to sunlight.
- Fineness ranges from microfibres (60 times finer than a human hair) to coarse fibres.
- Polyamides dye well.
- The handle of polyamides can vary. It can be soft and lightweight or coarse and firm.

Other uses of nylon fibres and fabrics include parachute canopies, tents, sleeping bags, sails, rope and fishing lines, due to its elasticity and strength (tear resistance), availability and price. It is also found in backpacks, work wear, camera cases and luggage.

Aramids

Aramids are a group of fibres in the polyamide category. They are highly flame retardant and able to withstand very high temperatures. Nomex and Kevlar are expensive fibres to produce and are used for firefighters and racing drivers' clothing. Kevlar also has a high strength-to-weight ratio – four times that of steel wire – which makes it suitable for bullet-proof vests, high-performance tyres and protective helmets.

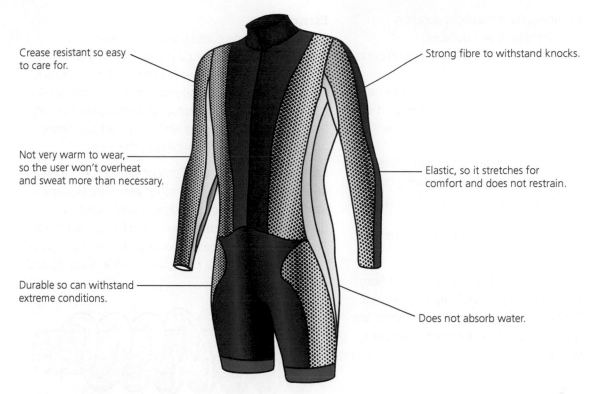

Crease resistant so easy to care for.

Strong fibre to withstand knocks.

Not very warm to wear, so the user won't overheat and sweat more than necessary.

Elastic, so it stretches for comfort and does not restrain.

Durable so can withstand extreme conditions.

Does not absorb water.

Figure 5.56 **A nylon cycling suit**

Figure 5.57 **Kevlar fabric and use**

Polyester

Polyester is the most important synthetic fibre because it is very versatile and has a wide range of uses. The fibres, derived from petrochemicals, are macromolecules produced by a melt spinning process. The molten polyester is extruded into an air stream, which cools the melt and solidifies the filaments. It is inexpensive to manufacture and about 60 per cent of it is produced as staple fibres. It can be engineered to produce a wide variety of properties and characteristics. Now, half of all the world's clothing is made from polyester.

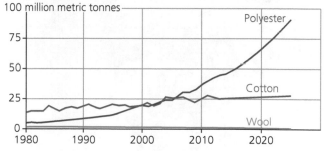

World fibre production

Figure 5.58 **Graph showing the huge increase in polyester production**

Although polyester is made from a non-renewable source, the fibres can be produced from plastic bottles (it takes 25 bottles to make one Polartec® jumper).

Performance characteristics of polyester:

- Insulation depends on the type of fibre. Flat filament fibres do not trap air and have low insulation. Texturing increases the volume and gives better insulation. Staple yarns may be either fine and smooth or more bulky.
- Polyesters scarcely absorb water, which means that they are susceptible to electrostatic charge. It also means that they are easy to wash and dry.

171

- They are very strong, have excellent abrasion resistance and are resilient and durable.
- Elasticity is very high so products have a good crease resistance.
- Polyester fibres are thermoplastic so can be textured and heat set. It is used for permanently pleated garments.
- They are resistant to most acids, alkalis and solvents, as well as moulds and fungi.
- Fineness ranges from microfibres to coarse fibres. They can be fine and soft or firm depending on the fibre fineness, fabric construction and finishing.
- Polyesters dye well.
- The handle of polyester can vary. It can be soft and lightweight or firm and stiff.

Staple fibres are used mainly as blends with other fibres, especially wool, cotton and viscose. They are used in a wide range of clothing and bedding. Filament yarns are usually textured and are used for dresses, blouses, ties, rainwear and linings.

Acrylics

Acrylics are made from the petrochemicals propylene and ammonia and can be either wet or dry spun into staple fibres. They have a wool-like handle, low density and a good resistance to light and chemicals. Acrylic can be spun as microfibres. Acrylic fibres are made from a non-renewable source and are cheap to produce.

Figure 5.59 **100 per cent acrylic fancy dress costume**

Performance characteristics of acrylics:
- They are voluminous, very soft and warm and have a wool-like handle with good drape.
- Acrylics are wrinkle resistant, machine washable and dry quickly as they do not absorb moisture.
- Like all synthetic fibres they are thermoplastic and can be easily deformed by heat; they shrink easily and must be ironed quickly.
- Acrylics are used for blankets, furnishing fabrics, knitting yarns, fake fur and fleece.

Elastomeric

Elastane is manufactured from segmented polyurethane. Its outstanding property is its elastic recovery. An elastomeric yarn can stretch by up to 500 per cent and recover its original length after the tension is released. Its molecular structure is composed of soft, flexible segments bonded with hard rigid segments, which allow for the extensibility of the yarn.

Elastane is produced as a filament yarn; it is very fine and is usually covered by another yarn. Bare elastane filaments are used in sheer hosiery and foundation garments. Covered yarns are used in fashion garments. They are unseen in the garments and never come into contact with the body, providing invisible elasticity and wrinkle resistance.

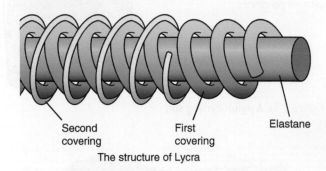

Second covering — First covering — Elastane

The structure of Lycra

Figure 5.60 **The construction of an elastomeric yarn**

Performance characteristics of elastomeric fibres:
- It has the capacity to stretch up to five times its original length, making the fabric drape well.
- It adds comfort, softness and crease resistance to garments.
- Body shape is improved by wearing garments with elastane in them.
- It dyes well and is easy to care for.
- It can be manufactured to enhance the performance of other fibres, for example chlorine resistance and comfort in swimwear.

Flourofibres

Flourofibres (PTFE) are produced from a milky-white synthetic polymer that can be produced in films, staple fibres or filament yarns. It is produced in films as a microporous membrane about 0.02 mm thick (the thickness of domestic cling film) containing microscopic pores (holes). It is laminated onto textiles or interleaved between two fabrics as in the production of Gore-Tex, a breathable, water-repellent outdoor fabric.

Performance characteristics of flourofibres:
- They are chemically resistant to water so are water repellent.

- Because they do not absorb any moisture, they are practically undyeable.
- They are windproof and stain resistant.
- They are flexible and durable.
- They are breathable when used a as a microporous film.
- They are a non-renewable resource but the polymer is water based so it doesn't harm the environment.

Flourofibres are used in the production of garments, upholstery, sportswear, work wear and shoes.

Chlorofibres

Polyvinylchloride (PVC) is manufactured as filament yarns and staple fibres but has only a limited use in clothing. Knitted thermal underwear is sometimes made from chlorofibres as they can be very warm to wear. They are also used in protective clothing as they have a high chemical resistance.

Performance characteristics of chlorofibres:
- They are strong, flexible and durable.
- They provide good insulation.
- They are thermoplastic, so care must be taken when laundering.

Foamed rubber

Foamed rubber is a more recent development and is a synthetic rubber that is produced by forming gas bubbles in a polymer mixture. Neoprene is a well-known example of this that has been used traditionally for wetsuits but has now been used to make high-fashion garments. Neoprene was originally intended to serve as an oil-resistant substitute for natural rubber, but its other properties have enabled its use as a rubber alternative in a wide range of applications.

The main characteristics of Neoprene:
- It is water resistant, which makes it an ideal outdoor material and an excellent choice for surf suits, wet (diving) suits and dry suits.
- It is weather resistant as it resists degradation from all weather conditions.
- It provides thermal insulation which again makes it suitable for wetsuits.
- It is stretchable so it conforms to forms of varying sizes and shapes.
- It provides cushioning and protection and is available in a wide range of thicknesses which makes it ideal for protective covers not only for many pieces of equipment such as cameras and cellular phones, but also for the human body, such as knee and elbow pads.

- It is chemical and oil resistant which makes it suitable for protective gear and clothing, such as gloves and aprons.

Figure 5.61 Wetsuit made from Neoprene

Yarns

A yarn is a continuous length of fibres, with or without a twist. The fibres have to be combined or twisted to give them the strength to be manufactured into fabrics. It is the process or method that is used to create the yarn that gives them their characteristics. The thickness and twist tightness of the yarn ultimately affects the weight, appearance, handle and texture which then can dictate its end use.

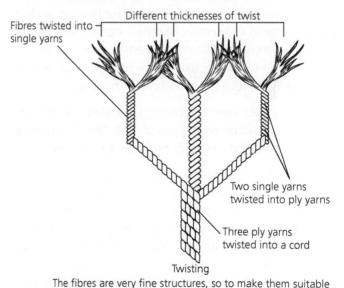

Fibres twisted into single yarns

Different thicknesses of twist

Two single yarns twisted into ply yarns

Three ply yarns twisted into a cord

Twisting

The fibres are very fine structures, so to make them suitable to use for fabrics they have to be spun into yarns.

Figure 5.62 Process of converting fibres to yarns

Single fibre spun yarns

A single yarn is a continuous single thread of twisted or continuous filament yarns. When more than one yarn is twisted together to make it stronger or fit for purpose, the yarn is called 2 ply, (2 yarns twisted together), 3 ply (3 yarns twisted together) and so on.

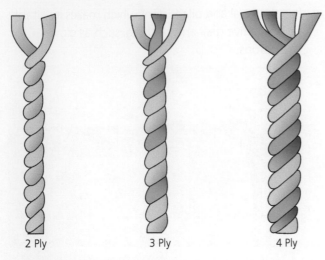

2 Ply   3 Ply   4 Ply

Figure 5.63 Single and plied yarns

Spun yarns are made by twisting together (spinning) staple fibres. Spinning is used for staple (short) fibres from cotton, wool and flax; any short fibres from silk; and any broken or cut synthetic fibres.

Mixed and blended fibre spun yarns

It is very important that the end fabric functions as required for its purpose. For this reason it is usual to mix or blend fibres so that the optimal performance characteristics of the chosen fibres make the end product more suitable for purpose. The main idea behind blended fabrics is to combine fibres with certain qualities with fibres of other qualities that complement each other and ultimately improve the end fabric for the purpose required.

Blending is achieved by spinning yarns from a blend or mixture of two or more types of fibre. This can improve the strength and weight or produce special effects.

The reasons for blending, mixing and laminating are:
● To combine the most desirable properties of different fibres into one fabric, for example, polyester cotton.
● Effects of texture and handle can be enhanced, for example, the addition of wool increases warmth and fullness.
● Novelty effects can be achieved by dyeing fabrics from blended yarns. One fibre component may remain undyed or dyed a different colour: this is known as cross dyeing.
● The cost of producing fabric is always important and blending can be used to control the price, usually to reduce the cost.

Some of the more common blends are cotton/polyester and wool/acrylic. However, the combinations are only limited by the fibres available.

Table 5.9 Blended textiles

Fibre	Typical blends	Typical end use
Cotton	Polyester	Shirts, bed linen, trousers, underwear
Cotton	Polyamide	T-shirts, knitwear
Cotton	Viscose	Upholstery, clothing
Cotton	Modal	Bedding, upholstery, towels and outerwear
Cotton	Elastane	Jeans, T-shirts, sportswear, swimwear, medical products, e.g. bandages
Linen	Cotton	Bed linen, tablecloths and serviettes, clothing, upholstery
Linen	Polyester	Upholstery, table linen, clothing
Linen	Acrylic	Knitwear
Wool	Polyester	Blankets, suits, industrial felts, soft toys
Wool	Acrylic	Carpets, upholstery, jumpers, faux fur, knitting yarns
Wool	Elastane	Clothing
Silk	Polyester	Clothing, linings, tablecloths, curtains, bed linen
Silk	Elastane	Clothing, underwear

Core spun yarns

Core spun yarns are multi-component yarns but the core stays at the centre of the yarn. Other fibres are then spun around it to cover it. A well-known example of this, and one that is used in many fashion and textiles products today, is a core of elastane with man-made or natural yarns spun around it.

Core spun yarns enhance fabrics, making them more comfortable to wear and are used in a wide variety of woven and knitted fabrics.

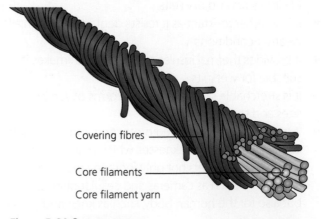

Covering fibres

Core filaments

Core filament yarn

Figure 5.64 Core spun yarn

Table 5.10 Yarn construction

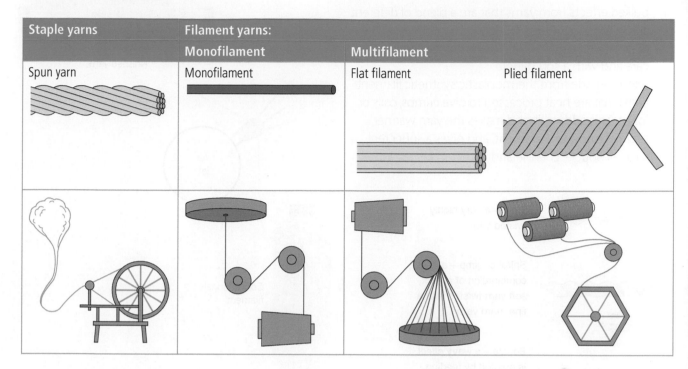

Staple yarns	Filament yarns:	
	Monofilament	Multifilament
Spun yarn	Monofilament	Flat filament / Plied filament

Filament yarns

Filament yarns can be made from one or more filaments; they can be made from silk or synthetic fibres.

- Monofilament yarn consists of a single continuous filament.
- Multifilament yarn is a yarn made from several filaments with or without a twist.

Twist

Twist is put into yarns during spinning to make them stronger so that they are suitable for weaving or knitting. Yarns made from twisted fibres vary according to the direction of the twist and the twist level. Yarns can be spun clockwise (Z twist) or anticlockwise (S twist). Light is reflected in opposite directions from Z and S yarns, so striped effects can be achieved in fabrics.

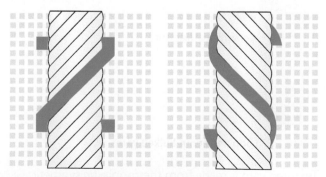

Figure 5.65 Z twist (left) and S twist (right)

The twist level determines how the yarn can be used. Low-twist yarns are softer, weaker and more bulky, making them suitable for weft yarns and for knitting. High-twist yarns are strong and hard, making them suitable for warp yarns in weaving as they can stand the tension in the loom.

Core-spun yarns are multi-component yarns with one filament in the centre and staple fibre yarn spun around it. A good example of this is stretch yarns with elastane filament as the core covered by non-elastic natural or man-made yarn such as cotton, polyamide or wool. Core-spun yarns enhance fabrics, making them more comfortable to wear. Think about your pairs of jeans and how comfortable they are when they have five per cent elastane added to the fabric.

Sewing threads are often made from a polyester core covered with a cotton yarn.

In the design of textile products, yarns are first selected on the basis of the performance characteristics of the fibres they are made from. However, yarns may also be selected for their appearance.

Fancy yarns

Yarns are usually chosen for their functional characteristics but also for their appearance. The yarns can be spun to be fancy, plain, thick or thin.

Fancy yarns are produced by special spinning processes to give:

- texture in the construction, for example slub, loop, chenille and bouclé
- colour effects by mixing different coloured yarns

- lustre effects, for example, mixing with Lurex
- bulked effects from yarns that are a blend of different staple fibres, for example, acrylic and cotton to give an inexpensive yarn that is warm, lightweight, easy care and with a soft handle
- texture, made from thermoplastic synthetic filament yarns that are heat processed to give crimps, coils or loops. This adds bulk and makes the yarn warmer, more elastic and absorbent, and gives a softer feel, such as boucle, chenille and Lurex.

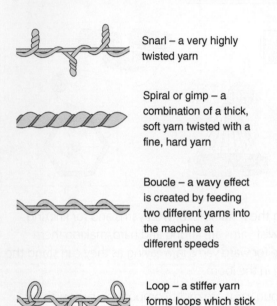

Snarl – a very highly twisted yarn

Spiral or gimp – a combination of a thick, soft yarn twisted with a fine, hard yarn

Boucle – a wavy effect is created by feeding two different yarns into the machine at different speeds

Loop – a stiffer yarn forms loops which stick out from a core yarn

Figure 5.66 Fancy yarns

Bulked and textured yarns

Yarn can be given extra properties to make it more suitable by bulking or texturing.

Bulked yarns are produced by taking staple fibres and making them thicker, bulkier and softer by applying water or heat. Figure 5.67 illustrates two methods that are used to create the bulk. The figure on top illustrates how the jet air process creates loops to bulk the yarn. The figure on the bottom illustrates the stuffing box process that is used to crimp the yarns. This is achieved by compressing regular filament yarns in a stuffing box, causing individual filaments to take on a crimp. The yarn is then heat-set.

Acrylic/cotton fibres, for example, can be bulked up using a heat process to shrink the acrylic fibre and fluff up the cotton. The result is a yarn that is warm and with a soft handle that makes it particularly suitable for knitwear.

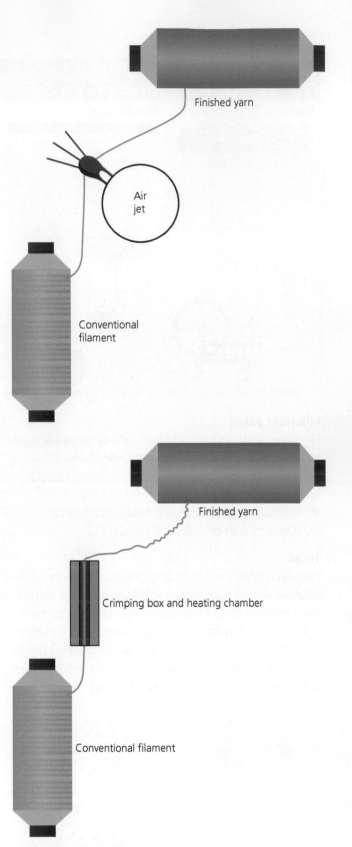

Finished yarn

Air jet

Conventional filament

Finished yarn

Crimping box and heating chamber

Conventional filament

Figure 5.67 Loop and crimp bulked yarns

Textured yarn is made from thermoplastic synthetic filament yarn such as polyester or nylon. The texturing is achieved by a heat process which makes yarns more opaque, improves appearance and texture and increases warmth and absorbency. Texturing is the forming of crimp, loops, coils or crinkles in filaments; see Figure 5.68 for examples of textured yarns. Such changes in the physical form of a fibre affect the behaviour and handle of fabrics made from them. Textured yarns are used for manufacturing a wide variety of textile products: hosiery, knitted underwear and outerwear and shape-retaining knitted fabrics for

men's and women's suits and overcoats. They are also used in the production of artificial fur, carpets, blankets and drapery and upholstery fabrics.

In the production of abraded yarns, the surfaces are roughened or cut at various intervals and given added twist, producing a hairy effect. Bulk is frequently introduced by crimping to give waviness and curling to produce curls or loops at intervals.

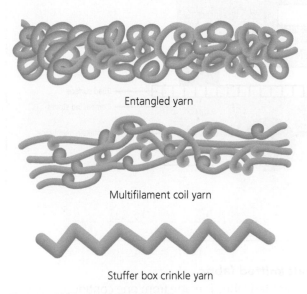

Entangled yarn

Multifilament coil yarn

Stuffer box crinkle yarn

Knit-deknit crinkle yarn

Monofilament coil yarn

Core-bulked yarn

Figure 5.68 **A range of textured yarns**

Fabric construction

Geotextiles

Geotextiles are fabrics that are used in civil engineering, construction or landscaping applications (such as roads, airfields, reservoirs and dams) to provide control of the ground. They are permeable which means that they allow water and gases to pass through them while blocking the larger particles of soil, silt or stones.

Geotextiles can be used to:
- reinforce and stabilise the ground
- keep layers of ground materials separate from each other
- promote drainage and allow filtration
- provide protection from invasive, unwanted plants such as weeds and tree roots.

In many applications, a geotextile will be performing several of these functions at the same time.

Geotextiles are usually manufactured from polypropylene or polyester and they can be either woven or non-woven.

Permeability is achieved in non-woven geotextiles by punching needle holes through the material. The size of the needle holes is selected to control the permeability of the textile for specific applications.

Geotextile membrane can be used in the construction of roads, runways, railways, car-parks and patios as shown in Figure 5.69.

In section 2 of this diagram, the geotextile is reinforcing the road by keeping the compacted stone layer separate from the soft base soil. Without the geotextile in place (section 1), the stones will eventually disperse into the soil and the unsupported road surface will sag under the weight of vehicles. The permeability of the membrane allows drainage of surface water into the base soil and the membrane protects the road surface by supressing the growth of unwanted weeds which could damage the road as they grow through it.

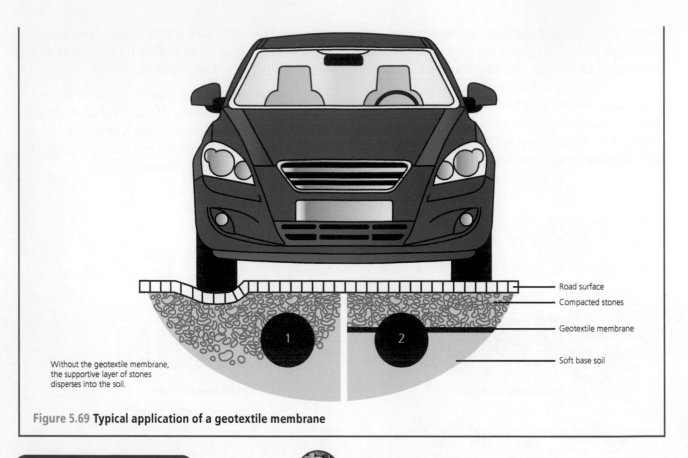

Without the geotextile membrane, the supportive layer of stones disperses into the soil.

Road surface
Compacted stones
Geotextile membrane
Soft base soil

Figure 5.69 **Typical application of a geotextile membrane**

FASHION AND TEXTILES

If you are completing the fashion and textiles qualification, you will need to know about each of the different fabric construction methods detailed below.

PRODUCT DESIGN

Product designers often use fabrics and textile material in the design of consumer products. It might be to provide comfort to the primary user, as in a case of headphones where leather has been used. The product and the context it is used in affects the choice of material, be it a fabric, a metal or polymer. The material is chosen for its specific properties, cost and availability.

Consider consumer products that use fabrics and identify why the material is suitable, for example the seat of a camping chair or the casing of a portable speaker. As a product designer you will not be directly tested in depth on textile material but you may find you use them for part of your NEA work and, as with any material, reasons for choice and relevant constructional techniques and fittings would need to be justified.

Knitted fabrics

Knitted fabrics are made from interlocking loops, formed from either a single yarn or many. They are classified into weft knitted and warp knitted.

Weft knitted fabric

Weft knitted fabric is made from one continuous length of yarn which is fed across the width of the fabric. The disadvantage of weft knitted fabric is that it can unravel and ladder which means it cannot be cut like woven fabric. The knitted fabric is soft and comfortable and can have variable elasticity depending on the structure.

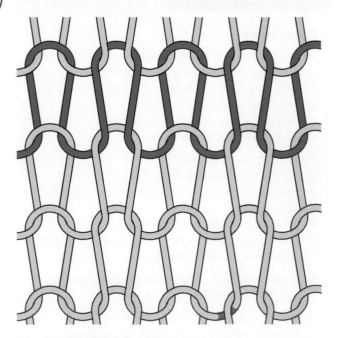

Figure 5.70 **Weft knitting diagram**

Hand weft knitting can be used to make one-off products and can be very time consuming and expensive to produce.

Machine weft knitting is generally produced by one of three main types of machine:

- Straight bar or fully fashioned machines produce high-quality knitwear using natural fibres; usually wool.
- Single or double flatbed machines are used to knit rectangular panels.
- Circular knitting machines are used mainly for producing socks, hosiery and T-shirts.

Warp knitted fabric

Warp knitted fabrics are made on straight or circular knitting machines and use more fabric than warp yarn. A separate warp yarn is fed to each needle, producing a loop that interlocks with the loop below along the entire length of the fabric.

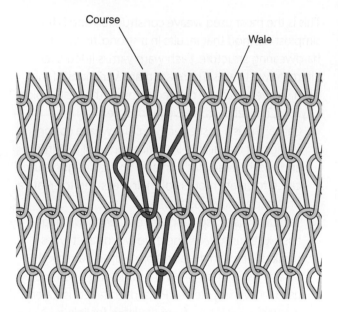

Figure 5.71 Warp knitting

Warp knits cannot be unravelled and do not usually ladder, but their use is limited mainly to swimwear linings, laces, ribbons and trimmings. Geotextiles used in the construction industry are often warp knitted.

Examples of warp knitted fabrics are:

- tricot, which uses mainly synthetic yarns to produce fabric used for gloves, lingerie and lightweight furnishings
- locknit, a combination of tricot and plain knit stitches, which produces a lightweight, smooth, non-laddering warp knit used for lingerie and linings

- velour, a knitted pile fabric made from continuous filament fibres, which has a raised fleecy surface formed from cut loops of yarn that stand up from the fabric, and is used a lot for leisure wear.

Structured fabrics
Knotted

Macramé is a traditional technique that uses the looping and knotting of yarns rather than weaving or knitting. The decorative qualities are created by the selection and use of ornamental knots.

Figure 5.72 Macramé

Netting is an open-mesh form of fabric construction that is held together by knots or fused thermoplastic yarns at each point where the yarns cross one another. Netting may be made of any kind of fibre and may be given a soft or stiff sizing.

Braiding

Braiding is a simple form of narrow fabric construction. A braid is a rope-like structure, which is made by interweaving three or more strands, strips or lengths, in a diagonally crisscrossing overlapping pattern. There are primarily two types of braiding: round or flat. 3D braiding is a less common form but is an interesting concept of creating a two-dimensional array of interconnected 2D circular braids. 3D braids are formed on two basic types of machines.

The fibre architecture of three-dimensional braided fabrics provides high strength, stiffness and structural integrity, making them suitable for a wide array of applications. 3D braided fabrics have found applications in areas including medicine, aerospace and automobiles.

3D novel structures

3D woven fabrics can be manufactured both with 2D and 3D weaving. Ordinary fabrics also have length, width and breadth, but in the three-dimensional fabrics, the thickness is much more than that of ordinary fabric. The thickness is achieved by forming 'multiplayer' using multi series of warp and multi series of weft, which intersect at regular 90 degree angles as in usual cloth-weaving principles.

In the textile industry, 3D printing is being continually developed and allows the designer to quickly create new structures using innovative new materials. You can read more about the use of 3D printing in Chapter 7.

Figure 5.73 Example of a 3D printed product

Woven fabrics

Woven fabrics are made by interleaving two sets of yarns, warp and weft, at right angles on a loom.

The warp yarns are those that lie in the length direction of a fabric while it is being woven. These are the stronger yarns. The weft yarns are those that are introduced between the warp yarns across the width direction of the fabric. Woven fabrics tend only to stretch diagonally (on the bias) unless the yarns used are elastic. Woven fabrics are classified according to the weave or structure in which the warp and weft yarns cross each other. The three primary weaves, of which others are variations, are plain, twill and satin.

Plain weave

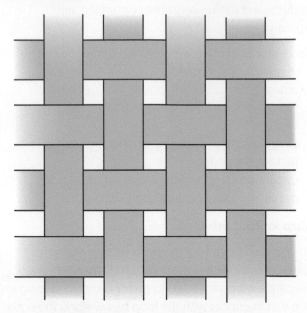

Figure 5.74 Plain weave

This is the most used weave construction and is the simplest method that results in a strong, firm and hardwearing structure. Each warp yarn is lifted over alternate weft yarns. There are many variations of the plain weave which are created through the vast choice of yarns. Typical plain weave fabrics are:

- Calico: this is a cost effective, plain weave cotton fabric that is available in a wide range of weights and widths. Many fashion and textile designers use calico to create models of their ideas and to create a prototype, known as a **toile**.
- Taffeta: this is a crisp, plain woven fabric and it frequently has a lustrous surface. There are two distinct types of silk taffeta: yarn dyed and piece dyed. Yarn dyed taffeta has a stiff handle and a rustle, and is commonly used for evening dresses. Piece dyed taffeta is softer and washable, making it a favourite fabric for linings.
- Poplin: this is a durable, tightly woven cotton fabric, primarily intended for making clothes. The weight varies but it is generally light to medium weight. Poplin is commonly used for shirts, dresses and skirts.
- Muslin: this is a fine cotton fabric of plain weave. It is made in a wide range of weights from delicate sheers to coarse sheeting.
- Rip stop fabric: this is a high-performance, plain weave fabric in which reinforcement yarns are doubled up at intervals. This fabric has a favourable strength-to-weight ratio and small tears cannot easily spread. Fibres used to make rip stop include cotton, silk, polyester and polypropylene, with nylon. Rip stop fabrics are commonly used in yacht sails,

kites and parachutes. It is also used for high-quality camping equipment such as lightweight tents, sleeping bags, camping hammocks and rucksacks.

Twill weave

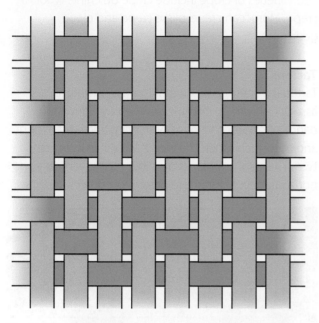

Figure 5.75 Twill weave

Twill weave makes a pattern of diagonal lines. Each warp yarn lifts over or remains under more than one weft. The twill order of interlacing causes diagonal lines to appear on the fabric. Weaving twills in opposite directions produces weave variations such as herringbone.

The most common twill weave fabric is denim, woven from strong, hardwearing cotton. Usually the warp is dyed blue and the weft white.

Satin weave

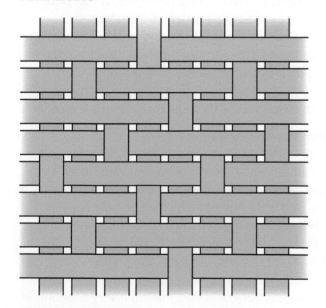

Figure 5.76 Satin weave

Satin weave has a lustrous appearance and good drape due to the fact that the warp floats over four or more wefts and remains under only one. Satin fabric has a smooth, shiny finish; the weft only shows on the back, which is dull. The pattern of the weave is random so that there is no twill line generated. Examples of satin weave fabrics include satin, sateen and duchesse.

Loop and cut pile

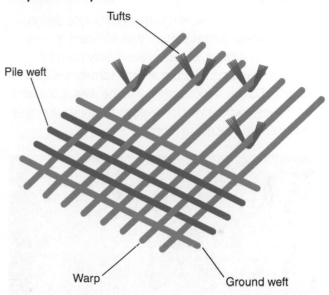

Figure 5.77 Velveteen

Loop and cut pile fabrics are made by creating a loop of yarn from an additional warp or weft. An additional warp is used in velvet or a weft yarn is used in velveteen to form a cut pile on the face of the fabric, which is called a nap. The third yarn forms the cut pile. For example, in velveteens, the pile weft floats on the surface of the fabric; after cutting, the tufts are held by the weft yarns. If the pile yarns are between the same warp yarns a corded effect is achieved.

When making products, pile fabrics all have to be cut in the same direction, which can make the textile products more expensive to produce as it is difficult to tessellate the pattern pieces due to them all having to face in one direction.

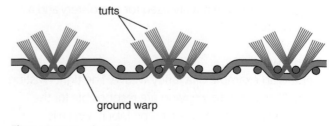

Figure 5.78 Corduroy

Woven terry towelling is a plain-woven cotton fabric made from two warps. The basic warp (ground warp) is highly tensioned and the pile warp is loose. The loose pile warps form loops on either one or both sides of the fabric as it is being woven.

The loop increases the absorbency of the cotton towelling by increasing the surface area.

Brocades

Brocade fabrics are woven with an elaborate design, having a raised overall pattern with different ground and pattern weaves. Brocades are mainly used for upholstery and curtains but are still sometimes used for clothing. Due to the complexity of the weave, the fabric can be very expensive. However, today they are mainly produced on a jacquard loom.

Figure 5.79 **Example of brocade fabric used for a cushion**

Jacquards

Jacquard is a fabric woven on a special loom called the jacquard loom. This loom allows individual control on interlacing of up to several hundred warp threads that can create complex and unique patterns. It uses all types of fibres and blends of fibres and is capable of creating complex patterns on fabrics. The jacquard loom can be programmed to raise each warp thread independently of the others. Some of the examples of jacquard fabric include satin fabrics, brocade fabric and damask fabric. Jacquard fabrics are mainly used for upholstery and as drapery fabrics.

Crepe weave

Crepe is an example of a fancy weave. The tight twisting of the fibres prior to weaving are responsible for the characteristic surface of the crepe fabric, which is puckered or pebbly. Fibres such as silk, wool, polyester or cotton can be used to produce crepe fabrics. Crepe fabric that uses polyester fibres is one of the most extensively used fabrics in the garment industry, as the resultant garments are wrinkle free and comfortable. The varieties of crepe include crepe de chine, wool crepe, Moroccan crepe, plisse crepe and crepe georgette.

Tartans

Tartan is a checked fabric that is most commonly associated with Scotland. It is woven by alternating bands of coloured yarns as both warp and weft at right angles to each other. The weft is woven in a simple twill. The resulting blocks of colour repeat vertically and horizontally in a distinctive pattern of squares and lines.

Figure 5.80 **Highland tartan fabric on a traditional weaving loom at Lochcarron Weavers in the Highlands of Scotland**

Plaid

Basically all tartans are plaid, but not all plaids are tartan! Plaid refers to a fabric woven of differently coloured yarns in a cross-barred pattern. With most tartans, the pattern on the stripes running vertically is exactly duplicated on the horizontal axis. Basically, this matching pattern in both directions will create a grid. When looking at a simple plaid, the stripes are not the same in both directions and therefore the pattern is less uniform.

Non-woven fabrics

Non-woven fabrics are made directly from fibres that are initially entangled to produce a web. There are two types of non-woven: felts and bonded. In felts, the web is strengthened by the entanglement of fibres, while in bonded, the web is strengthened by adhesives. Felt has no strength, drape or elasticity, but it does not fray and it is warm and resilient. It is cheap to produce and, as it has no grain, pattern pieces can be placed in any direction, making it very economical to use.

Felted wool

Wool and other animal fibres are entangled to produce a fibre web using a combination of alkaline chemicals, water, heat, pressure and repeated mechanical action. It is used for hats, slippers and toys.

Needle felt

Needle felt can be made from any type of fibre but mostly synthetic fibres. The bulky fibre web is repeatedly punched with hot barbed needles that drag the fibres to the lower side of the web. This is used for waddings, upholstery, mattress covers and filters.

Bonded fabric

Bonded fabrics are made from a web of fibres bonded with adhesives, solvents or by heat. This kind of fabric is used mainly for interlinings, disposable cleaning cloths and disposable hospital clothing.

Adhesive bonded webs are made from a web of fibres, bonded together with adhesive. This can be applied by spraying, dipping or foam pressing.

Thermally bonded webs are made from a web of fibres that are bonded together under pressure and heat, which softens and fuses the fibres together.

Microfibres

Figure 5.81 Microfibres can be used for sportswear

Microfibre technology combines a high number of very fine fibres into one extremely fine yarn. A microfibre is around 60 to 100 times finer than a human hair.

- Microfibres can be specially engineered to have specific qualities and functions, and can be:
  - manufactured from polyester, polyamide, acrylic, modal, lyocell and viscose
  - blended with other synthetic or natural fibres
  - used in smart and technical fabrics for active wear, all-weather wear and industrial uses.
- Microfibres are well suited to blending with other man-made and natural fibres.
- Microfibres are soft, durable and drapeable, possess high absorbency and are used for a wide range of purposes within the textile industry. They are used a lot for high-performance garments and accessories, particularly for sportswear.

Composite materials

Composite materials are a mixture of two or more materials. Particles, fibres or layers of one material are used to strengthen another. The resultant material has a better combination of properties than either of the separate materials.

DESIGN ENGINEERING AND PRODUCT DESIGN

You will need to know about each of the composites in the following section if you are completing a Design Engineering or Product Design qualification.

FASHION AND TEXTILES

You are less likely to need the information in this section if you are studying a Fashion and Textiles qualification.

Carbon fibre reinforced plastic (CFRP) and glass reinforced plastic (GRP)

The main bulk material in both of these materials is polyester resin, although epoxy is also widely used. These resins are thermosetting polymers in liquid form which set when an activator is added. The resins are quite strong in compression but weaker in tension. The reinforcement gives the CFRP its strength and rigidity.

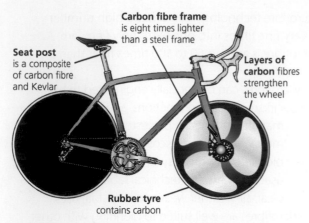

Seat post is a composite of carbon fibre and Kevlar

Carbon fibre frame is eight times lighter than a steel frame

Layers of carbon fibres strengthen the wheel

Rubber tyre contains carbon

Figure 5.82 **Carbon fibre composite materials used in a racing bicycle**

CFRP is much stronger than many metals, but takes longer to form. Figure 5.83 shows CFRP to be much stronger than mild steel and aluminium. It is very expensive and is only used where high performance is essential and cost is not an issue.

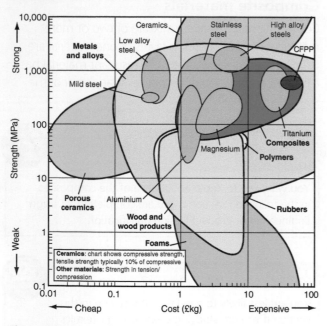

Figure 5.83 **Strength vs cost chart**

Carbon fibres have an incredibly high tensile strength – 2900 N/mm² – but are of little structural use, resembling silk strands. Mixed with polyester they become strong and rigid. They provide a light, hard-wearing surface with a thin section, which is resistant to corrosion and stronger than glass fibres and is also lighter in weight. They are used for protective helmets, performance bicycles, sports cars and sporting equipment. They are also stronger and lighter than today's steel and aluminium parts, but the higher cost of carbon fibres and longer production times limit this use.

Glass fibre is lower in cost and also commonly used for canoes and boat hulls. It is available as a woven mat or loose strands. As the lay-up process of glass fibre is time consuming it is usually used for one-off production making only one or a small number of products. The lay-up technique for glass reinforced plastics (GRPs) involves a simple profile mould made of metal, wood or plaster. Liquid polyester resin is then mixed with a catalyst (or hardener) and is applied to the mould to form a pre-gelled coat. Glass fibres in mat or woven fabric form are then laid on the first gelcoat and covered with a liquid polyester resin, or a catalyst mix is sprayed, rollered or painted on until the fibre layer is saturated. When the resin mix has hardened, the moulding is removed from the mould. This curing (setting) can take place in the cold or can be speeded up by heating. Two other techniques used with GRP are the rubber bag and matched-die moulding methods, where various compositions of polyester resin, catalyst or glass fibre are used to produce mouldings in pressurised processes.

Chopped strand glass matting (a ceramic material) can be mixed with polyester resin to form a strong and durable composite used for car and boat bodies (GRP).

Modern materials

Modern materials are developed to perform particular functions and have specific properties; they are intentionally developed, rather than being naturally occurring changes.

DESIGN ENGINEERING AND PRODUCT DESIGN

You will need to know about each of the composites in the following section if you are completing a Design Engineering or Product Design qualification.

FASHION AND TEXTILES

E-textiles and nanomaterials are the main modern materials influencing fashion designers. If you are studying a Fashion and Textiles qualification you are less likely to need to know about the other materials in this section.

E-textiles

E-textiles or electronic textiles are often referred to as smart garments, smart clothing, smart textiles or smart fabrics. They are innovative textile materials (fabrics, yarns and threads) that incorporate conductive fibres or elements directly into the textile itself. These materials eliminate wires and hard electronics, so all that is seen is the textile itself.

E-textiles can be used to create sensors, **thermochromic** (smart colour) displays, communications and data transfer systems, antenna and heating elements, and to release medication. Athletics garments can monitor the wearer's heart rate; fabric keypads can control smartphones and other devices, such as heating products. Hats and gloves can generate power through movement, producing energy to keep the wearer warm. Fibre optics can be woven into garments to act as radios or mp3 players and lights can be incorporated into clothing for safety purposes. E-textiles are commonly found in wearable products, but also in interior design and soft furnishings in homes and vehicles.

Super alloys

These are alloys developed for use in extreme applications, having significantly enhanced characteristics such as mechanical strength, performance at high temperatures, resistance to creep (deformation under continuous stress), resistance to corrosion or **oxidisation** and wear resistance.

Figure 5.84 Turbine blades in a Jet engine

Turbine engines are the main context for super alloys, where they are used for turbine blades and components in jet/rocket engines and marine (ships and submarines) applications. They are also used in chemical processing plants and power stations, for example, nuclear reactors. These super alloys are widely used in the oil and gas industries.

Nickel is one of the primary metals, with other alloying elements such as chromium, tungsten, titanium, cobalt, tantalum, molybdenum, aluminium and vanadium each serving a specific purpose in each alloy.

Some super alloys are formed as single crystal materials, so there are no grain boundaries in the material. This further enhances the stability and performance of the material at high temperatures. The method of manufacture is important to maximise the performance of these super alloys, with investment casting and sintering being common methods. Sintering is a form of powder metallurgy where metal powders or particles are formed by being layered into components.

High-performance alloys

High-performance products and demanding applications need high-performance materials. Manufacturers engage the services of specialist companies, metallurgists and material scientists to develop unique alloys tailored to meet specific performance requirements.

Figure 5.85 High-performance alloys for pharmaceutical mixing tanks on a production line

Increasingly stringent regulations in healthcare and the pharmaceutical industry which relate to the preparation of drugs and medicines are placing higher demands on the materials that come into contact with them, where any kind of contamination can lead to risks to consumers, as well as considerable financial loss. Stainless steel has been used for tools, machines and work areas for many years due to its resistance to corrosion. Higher standards of cleanliness require surfaces of materials that come into contact with drug products to withstand higher temperatures and chemical erosion. This has led to the use of high-performance nickel and molybdenum alloys to increase the resistance to corrosion and pitting of the stainless steels.

Graphene

Graphene is probably the most exciting material development of recent years, with products and components using this amazing material now starting to be available. It is early days with methods to process and manufacture Graphene in quantity still being devised, but we will be seeing and hearing a lot more about the world's thinnest and strongest material.

Graphene is an extraordinary and revolutionary new material that was discovered by researchers at Manchester University in 2004. It consists of carbon atoms arranged in a perfect hexagonal lattice. It is a 'super material' because of the combination of its unique properties. It is a one atom thin layer of graphite, the material used in pencils.

It is the world's strongest material with amazing characteristics:

- the world's thinnest material - the first 2D crystal which is one atom thin
- harder than diamond
- extremely light
- 300 times stronger than steel
- conducts electricity and heat better than almost any other material
- transparent to the extent it is almost invisible to the naked eye (97% transparent)
- more flexible than rubber and, like an elastic band, will extend to 120% of its original length
- when combined with other elements provides superior material properties to meet specific requirements.

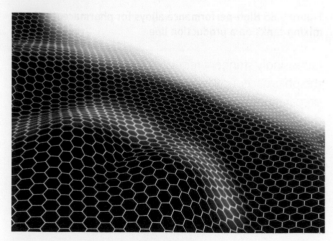

Figure 5.86 Graphene – hexagonal lattice of carbon atoms

Graphene is starting to appear in products. It is already used in:

- ink-based products in smart security tags in retail stores
- tennis rackets, where it makes the shaft stronger and lighter, giving greater control

- skis, where it gives lightness and increased durability (and adds 20 per cent to the price)
- cycle helmets with increased safety and impact absorption
- li-Ion batteries that operate at a lower temperature and have a longer (double) operation time.

This material has almost limitless potential, and applications currently being developed include:

- membranes to purify water, for use in third world countries
- protective coatings for use on steel and masonry
- smaller batteries for energy storage, e.g. for use in electric vehicles
- delivery of drugs in the human body – graphene is attracted to cancer cells
- attaching strands of DNA to graphene to form sensors for early disease diagnosis
- flexible and foldable phones and tablets
- displays within glass and other transparent materials in car windscreens and domestic applications.
- smart contact lenses and night vision
- an invisible electricity-conducting layer on materials and components
- using graphene sheets as electrodes in ultra-capacitors which will have as much storage capacity as batteries but will be able to recharge in minutes
- high strength composite materials

Scientists are working on methods suitable for manufacturing high-quality graphene in quantity. It is very difficult to work with, particularly because of the incredibly small size, but also because any imperfections or contamination in the process significantly reduce the quality and compromise its conductivity.

DESIGN ENGINEERING

Carbon nanotubes

If graphene is rolled into a tube, it forms a carbon nanotube which is a cylindrical structure just a few nanometres (10^{-9} m) wide. Carbon nanotubes exhibit some extraordinary properties:

- They are over 100 times stronger than steel but only 1/6 of the weight.
- They are excellent **conductors** of heat and electricity – better than copper.
- The tubes can be millions of times longer than their diameter.

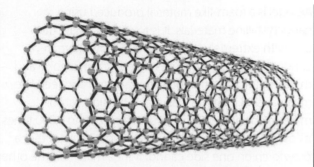

Figure 5.87 The structure of a carbon nanotube

Various structures of nanotubes are possible, differing in length, diameter, number of walls and the way they are rolled. The exact configuration gives the nanotube specific properties; for example, some structures behave like metallic conductors, others like semiconductors and it is possible that these will form the basis of the next generation of computer chips.

Bioplastics

Bioplastic is a form of plastic derived from organic and renewable biomass sources, such as vegetable oil or corn starch. Starch-based bioplastics are the most widely used, with other types including cellulose or protein based.

Bioplastics are made from plant-based sources, not from fossil fuels and petroleum, so they have a much lower carbon footprint. Their manufacture also emits less carbon dioxide and requires less energy, making them more environmentally acceptable than traditional plastics.

Bioplastics are mostly used for disposable items, such as packaging, crockery, cutlery, pots, bowls and straws. They are also often used for bags, trays, fruit and vegetable containers, egg cartons, meat packaging and bottling for soft drinks and dairy products. The use of bioplastics in non-disposable applications includes mobile phone casings, carpet fibres, insulation, car interiors and plastic piping. New electroactive bioplastics are being developed that can be used to carry electric current.

The constituents of bioplastics are referred to as ingredients. Many ingredients are readily available and amateur production of bioplastic and their application to products for the home are common. Ingredients include agar (the polymer substance found in seaweed and algae), vegetable glycerine, purified water, food colouring, and flavouring. The ingredients of the bioplastic are all natural, non-toxic, and edible. They are also fully biodegradable and compostable.

Glycerol, also called glycerine, acts as a plasticiser in bioplastics – it is a material that makes the polymer chain molecules bend and slide past each other more easily, which adds to the flexibility of the plastic. It is made by fermentation of sugar, or from vegetable and animal fats as a by-product in the manufacture of soaps and fatty acids. An alternative to glycerol in bioplastics is sorbitol.

Those promoting bioplastics make many claims for the material, and there are debates about the production, use and disposal of bioplastics:

- Biodegradation is a process where organic materials are broken down by natural processes and micro-organisms. All bio-plastics and conventional plastics are technically biodegradable. Some bioplastics can break down in a matter of weeks. The corn starch molecules they contain slowly absorb water and swell up, causing them to break apart into small fragments that bacteria can digest more readily. However, some break down very slowly, many bioplastics are only biodegradable under certain circumstances, and some will not biodegrade at all, not even in landfill, one of the problems which the development of bioplastics aims to address.

- There are problems relating to the large-scale mono-cropping (the agricultural practice of growing a single crop year after year on the same land). The methods by which the base materials are grown, and the processing involved, both impact the footprint of the materials and products. Many bioplastics also release carbon dioxide or monoxide when biodegrading. Nevertheless, their overall environmental impact is considered lower than that of conventional plastics.

- Most bioplastic manufacturers use oil products to fuel the production process. Polylactic acid (PLA) is a transparent plastic produced from corn or dextrose. Its characteristics are similar to conventional petrochemical-based plastics (like PET, PS or PE), and it can be processed using standard plastic production equipment. PLA is generally supplied in the form of granules, and these are used in the plastic processing industry for the production of films, fibres, plastic containers, cups and bottles. Making PLA saves nearly 70% of the energy needed to make conventional plastics, and produces almost 70% less greenhouse gases when it degrades in landfills.

Nanomaterials

A material's properties can be altered by combining different materials together. Nanotechnology involves altering the material's property by changing the individual atoms that form the material. It involves manipulating matter at the nanoscale (down to 1/100,000 the width of a human hair) to create new and unique materials and products.

The incredibly small structure of nanomaterials means they have a greater relative surface area than other materials and this can alter properties such as strength/elasticity, thermal or electrical conductivity, absorbency and optical performance.

Figure 5.88 **NanoCrystalCoat on a camera lens. The technology is effective in reducing ghost and flare on ultra-wide-angle lenses**

Aerogel is a foam-like material produced using nanocrystalline materials. It is a solid material but one with extremely low density and low thermal conductivity. It is sometimes called 'frozen smoke' and this gives an indication of its appearance and feel. The material is 98.2 per cent air and is the lightest solid material on Earth, yet is able to support up to 4,000 times its own weight. Aerogel is such a good insulator that a blowtorch on one side cannot light a match on the other side. Aerogel lines the walls of many buildings in the form of insulation. Clothing companies use it to create super light-weight and warm ski jackets. It's also found inside some tennis rackets. Polymer-based aerogels make excellent **insulators** for refrigerators and clothing.

A nanocomposite is a material to which nanoparticles have been added to improve a particular property of the material, for example, structural components with a high strength-to-weight ratio.

Figure 5.89 **The insulating properties of Aerogels**

Nanocomposites are used to make flexible batteries. A nanocomposite of cellulous materials and nanotubes is used to make a conductive paper. When this conductive paper is soaked in an electrolyte, a flexible battery is formed.

The Airbus 380 uses about 20 per cent composite materials containing nanomaterials, saving fuel and minimising environmental impact.

Sandwich panels

A sandwich panel is a composite material made from three layers: a lightweight core material sandwiched between two thin skin layers. This arrangement achieves good rigidity with minimum weight. The skin layers provide the tensile strength for the panel and the core holds the skins a constant distance apart. These components interact to create a panel which is very effective at resisting bending.

Aluminium composite sandwich panels are used extensively as cladding for the facades of modern buildings, for roofing and for signage and advertising boards. They are used for walls in ship building. As the core material is often a polymer foam, these panels offer a good level of thermal insulation which can be useful in architectural applications.

Sandwich panels are increasingly being used in aeronautical applications as a very lightweight, rigid and tough outer skin for aeroplanes, and also for floor and wall panels in larger airliners. In such crucial applications, the core material is often an aluminium honeycomb which, while expensive to manufacture, gives the panel exceptionally high rigidity.

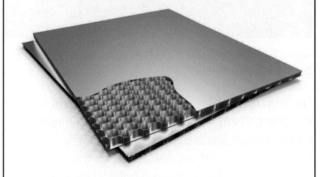

Figure 5.90 **An aluminium honeycomb panel**

DESIGN ENGINEERING

Rare Earth Magnets

Rare earth magnets are extremely strong permanent magnets made from an alloy containing the element neodymium, or from an alloy of samarium and cobalt. They produce much stronger magnetic fields than magnets made from conventional iron materials. Rare earth magnets will attract any ferro-magnetic material (for example, iron or steel) and they are capable of lifting thousands of times their own weight. Rare earth magnets are very brittle and prone to corrosion, so they are usually plated with nickel or tin to protect them and to stop them from crumbling.

Rare earth magnets are used in electric motors to achieve high torques in very compact packages. Examples include electric car drive motors and the brushless motors used in computer disk drives and in flying drones.

In Design and Technology project work, rare earth magnets are useful for creating magnetic catches and to enable parts of a project to be attached and detached by the user as required.

Due to the large forces that these magnets can exert, care must be taken when they are handled and used. It is very easy to pinch body parts between two magnets and there have been serious injuries caused when children have swallowed more than one magnet which have then stuck together inside the body, pinching and destroying body tissue.

Figure 5.91 **Neodymium magnets**

Smart materials

Smart materials respond or react to changes in temperature, light, electrical current or some form of stress, and adapt in some way. Some smart materials have a 'memory' and can revert back to their original state. An example of this type would be medical threads, which knot themselves. Temperature and/or light is used to activate the 'memory material' to revert to its original shape.

Smart or modern materials may be high-performance materials, such as genetically engineered dragline spider silk, which is used to produce super-strong, super-light military uniforms. The materials that will have the greatest impact are those that sense conditions in their environment and respond to those conditions. These materials function both as sensors and actuators.

Thermochromic materials

Thermochromic materials, including thermochromic pigments, change colour reversibly with changes in temperature. They are usually in the form of semi-conductor compounds, liquid crystals or metal compounds. These are used on contact thermometers made from plastic strips, on test strips on the side of batteries (where the heat comes from a resistor under the thermochromic film) and in kettles. They are also used as food packaging materials that show when the product they contain is cooked to the right temperature and in textiles and fashion.

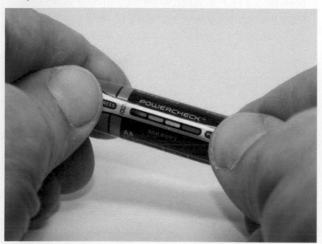

Figure 5.92 **Thermochromic material for a test strip**

Phosphorescent materials

Phosphorescent smart materials are sometimes called 'afterglow' materials because they glow in the dark after being 'charged up' in the day.

Phosphorescent pigments absorb natural or artificial visible light energy and store it in their molecules. This energy is slowly released over a period of time, which may be just a few seconds or several hours. It is therefore a cheap and convenient rechargeable light source for particular applications. For example, phosphorescent pigments have been used in emergency signs, novelty decorations, lighting for the garden, toys and glow-in-the-dark watches.

Photochromic materials

Photochromic materials change colour reversibly with a change in light intensity. Usually they are colourless in a subdued light but when sunlight or ultraviolet radiation is applied, the molecular structure of the photochromic material changes and produces a different colour. The colour disappears when removed.

Applications include light sensitive sunglasses that darken in response to increased intensity/brightness of sunlight to reduce glare, for example, when driving a car or when skiing at high altitude when the snow reflects extra light into your eyes. Textiles and clothing can incorporate photochromic inks with patterns that change in different light conditions. They are used for security markers that can only be seen in ultraviolet light.

Electrochromic materials

Electrochromic materials change their colour or opacity when a voltage is applied. This effect is used in liquid crystal displays in digital technology and in smart goggles and motorcycle helmets, controllable by the touch of a button. In some airliners, instead of window blinds, the cabin windows use electrochromic smart glass, with different tint settings.

Figure 5.93 **An electronic dimming system on Air Canada's Boeing 787 Dreamliner**

The term electroluminescence is also used to describe this effect. Electroluminescent materials can produce brilliant colours if stimulated by an AC current. Electrochromic materials can be used in the illumination of buildings and for displays on public transport.

Piezochromic materials

Piezochromic materials change colour when pressure is applied. Powders or paints made from piezochromic pigment change their colour due to pressure or other mechanical effects such as bending or scratching. The colour change can be either reversible or irreversible. These powders and paints are used for security coatings on buildings and structures where attempts to damage need to be evident.

Solvatochromic materials

Solvatochromic materials are usually in dye form. The 'smart' dyes change colour when they are dissolved with a liquid. The colour is dependent on the specific solvent, meaning there is potential for using these dyes to detect the presence of fluids with certain characteristics in the fields of chemistry and biology.

Shape memory alloys (SMAs)

The most common shape memory alloys are made up of nickel and titanium. These are known as nitinol alloys. Nitinol is used by the military and in many specialist medical products. When heated, they retained their original shape. Gold-cadmium and some alloys of brass all have a memory. Available in wire or sheet form, shape memory alloys are used in robotics (copying muscle function) and are being developed to operate the wing flaps on aeroplanes. They are used in modern buildings in temperature control through thermostats and automatic air vents. Nitinol in smart-wire form changes length when heat is applied. It is used for dental braces where body heat shortens the wire and pulls the teeth back into shape.

Nitinol shape memory alloys are also used to make stents – fine cylindrical mesh tubes used to treat constricted blood cells restricting the flow of blood. They are inserted into the blood vessels and as they warm up they expand to allow the flow of blood.

Shape memory polymers

Shape memory polymers (SMPs) are polymer materials in which deformation can be created and removed by a change in temperature or stress. Recent developments have created SMPs able to store up to three different shapes in memory, each triggered by a different temperature.

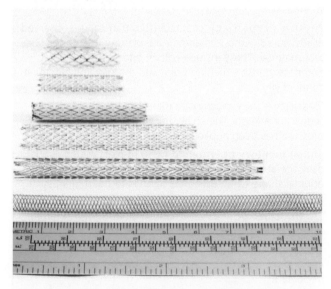

Figure 5.94 Stents of various sizes for endovascular surgery

An SMP shows a significant change from a rigid polymer to a very elastic state, then back to a rigid state again. In its elastic state, it will return to its memory shape if left unrestrained. But while soft it can be stretched, folded or otherwise conformed to other shapes, and will tolerate up to 200 per cent elongation without damage to the material. The memory feature comes from the stored mechanical energy gained during the shaping and cooling of the material. Applications for this smart material include reusable moulds, furniture or toys that can be re-shaped, specialised adjustable containers and packaging and sensors.

Polymorph

Polymorph (polycapralactone) is a thermoplastic polymer, usually supplied in the form of plastic pellets. It is heated by immersing the pellets in hot water or by using a hairdryer. It becomes easily mouldable at 62°C and when cooled takes on a solid form, very similar in performance to nylon. It can be cut, drilled and machined. Polymorph can be coloured by adding pigment while soft and mouldable, or painted using acrylic paints when moulded. It is often used as a modelling material and is reusable and 100 per cent biodegradable.

Figure 5.95 Polymorph

Coolmorph™ is a new development, a very similar material to Polymorph, but with a lower melting point of 42°C, making it much easier and safer to use.

Conductive paints

Not strictly smart materials, these are paints that conduct electricity. You can paint 'wires' directly onto almost any material, including paper, wood, plastic, glass, plaster, brick and textiles. They are mostly water based and pure silver is used in the paint formulation to give excellent conductive properties. The paint is flexible and tough and adheres to most materials and surfaces, drying in less than 30 minutes. Standard acrylic or water-based paints can be used alongside conductive paint to act as insulation or to create multiple layers of circuitry.

The paint is widely used to repair broken track on printed circuit boards or car heated rear windscreens. It is also used to provide radio frequency interference shielding and reduce noise and hum by connecting to electrical ground in audio systems.

Conductive ink delivers a printed item that conducts electricity. It is created by adding graphite or silver material to ink. This presents an alternative to etching copper to form printed circuit boards, with little or no waste being produced. Inks can be applied by screen printing, pad-printing, flexography and rotogravure methods. Membrane switches in computer keyboards and similar products use conductive inks.

Smart and performance textiles

Fabrics are available that not only look and feel good, but which also respond to the environment, providing added properties such as being waterproof, windproof and breathable, meeting the needs of the user.

Simpatex® fabric, made from polyester and polyether, allows water molecules of perspiration to pass through but does not allow even the smallest drop of water to penetrate.

Other applications include first-aid plasters and dressings with antiseptic incorporated, or that change colour to match the skin tone of the wearer (called Chameleon Bandages).

Fastskin®, developed for use by Olympic swimmers, combines fibres, knit and garment design to replicate the skin of a shark, enabling movement through water with less resistance.

Stomax® fabric takes its inspiration from the way that leaves breathe and regulate temperature – it keeps the wearer dry and comfortable during exercise or exertion.

ACTIVITY

When choosing materials for a product the designer has to consider:
- the function of the product
- the quantity to be manufactured
- budget limitations.

Choose one product within your material specialism and in a small group discuss how these three factors will affect the choice of materials.

KEY TERMS

Composite materials – a material made from two or more constituent materials with significantly different physical or chemical properties. The composite material has characteristics different from the individual materials, usually improved properties.

Conductors – materials through which an electric current can pass. In general, metals are good conductors. Copper or aluminium is normally used to conduct electricity in commercial and household systems.

Electrochromic – able to change colour when charged with electricity.

Ferrous – metals that contain iron. They may have small amounts of other metals or other elements added, to give the required properties. All ferrous metals are magnetic and give little resistance to corrosion.

Galvanising – a process in which steel is coated in zinc.

Geotextiles – any permeable textile materials used to increase soil stability, provide erosion control or aid in drainage.

Hydrophilic – substances that have an affinity for water, often because of the formation of hydrogen bonds.

Hydrophobic – able to repel or unable to mix with water.

Insulators – materials that do not allow energy such as electricity or heat to transfer through them. Wood, plastic, rubber and glass are good insulators.

Mechanical properties – characteristics that indicate the behaviour of a material under pressure (force), such as, brittleness, flexibility, ductility, toughness and tensile strength. Mechanical properties determine the range of usefulness of a material.

Oxidisation – when a substance is oxidised or when it oxidises, it changes chemically because of the effect of oxygen on it, affecting the quality of the surface of the material. Rust is an iron oxide, formed by the reaction of iron and oxygen in the presence of water or air moisture.

Phosphorescent – a glow or soft light which is produced in the dark without heat.

Photochromic – able to change colour due to light stimulus.

Physical properties – characteristics that can be observed or measured without changing the identity or composition of a material. These include colour, odour, texture, hardness, density, melting point and thermal and electrical conductivity.

Plasticity – the capacity of a material to be moulded, shaped or altered. The ability to retain a shape once pressure has been applied.

Shape memory polymers (SMPs) – smart materials that have the ability to return from a temporary shape to their original shape when activated by an external trigger, such as temperature change.

Thermochromic – able to change colour due to external temperature stimulus.

Toile – an early version of a finished garment made from inexpensive material so that the design can be tested and perfected.

Smart materials – these respond or react to changes in temperature, light, electrical current or some form of stress.

Staple fibres – fibres can be short in length or long, dependent on their origin or how they have been manufactured. Staple fibres are relatively short in length.

Yield point – or yield strength, the stress at which a material begins to deform. Before the material reaches its yield point it will return to its original size.

KEY POINTS

- There are many factors to consider when choosing materials and components for a particular application. The factors that need to be considered vary enormously from one product to another.
- By analysing products and identifying the materials used we can understand the choices that the designer has had to make and constraints they may have had to adhere to when selecting materials and components.
- Performance characteristics relates to the features or characteristics of products, materials or systems and how they perform under certain conditions or as a result of tests.
- Material selection is one of the foremost functions of effective product design as it determines the suitability of the design in terms of industrial aspects and commercial viability.

- A great design may fail to be a profitable product if it is not possible to find the most appropriate material combinations.
- Alongside the selection of materials and components for a product, the ability to manufacture must be considered.
- Cost is a critical factor to consider when selecting materials for a design. Most products face competition and manufacturing constraints. The cost factor can be neglected when performance is given top priority.
- On some occasions, particular properties of a material may become the dominant factor over other properties. For example, electrical conductivity is vital for an electrical application so it must be given priority. In product design, designing for light weight is important for certain body parts of vehicles, where aluminium alloy is used instead of steel.

5.3 Why is it important to consider the properties/characteristics of materials when designing and manufacturing products?

LEARNING OUTCOMES

By the end of this section you should have developed a knowledge and understanding of:
- how the function of a product influences material selection
- the terminology relating to material performance characteristics
- how to establish the required property characteristics when designing and developing a product and use this information to make informed decisions when selecting materials

- how to select materials and components that ensure optimal product performance and recognise the implications associated with incorrect choice
- how to use and differentiate between the technical terminology when referring to material and component choices.
- how the available forms, costs and properties of materials contribute to the decisions about suitability of materials when developing and manufacturing your own products.

The following sections on the consideration of the properties and characteristics of materials relate to each specific material area and therefore you should be able to differentiate between each.

APPLICATION AND EXAMPLES FOR DESIGN ENGINEERING

Understand the characteristics and properties of materials that are significant in Design Engineering, such as density, tensile strength, strength-to-weight ratio, hardness, durability, thermal and electrical conductivity, corrosion resistance, stiffness, elasticity, plasticity, impact resistance, malleability and ductility and machinability.

Table 5.11 **Properties/characteristics of materials used in Design Engineering**

Property/ characteristic/ term	Meaning	Contextual application
Density	The mass per unit volume of a material. It is the compactness or closeness of the structure of the material. $$\text{Density (kg m}^{-3}) = \frac{\text{mass (kg)}}{\text{volume (m}^3)}$$	Density is important in relation to product weight and size (for example, for portability).

Tensile strength	The maximum tensile stress (stretching force divided by cross-sectional area) a material can withstand before it breaks $$\text{Tensile strength (MN}^{-2}) = \frac{\text{breaking force (N)}}{\text{cross-sectional area (m}^2)}$$	Tensile strength is important in selecting materials for ropes and cables, for example, for an elevator.
Strength-to-weight ratio	The strength-to-weight (S/W) ratio of a material is defined as: $$\text{S/W ratio (Nm kg}^{-1}) = \frac{\text{strength of material (Nm}^{-2})}{\text{density of material (kg m}^{-3})}$$ S/W ratio is also known as the 'specific strength' of the material. For similar components made from different materials, S/W ratio is a way of comparing the maximum force the component can withstand to the weight of the component itself.	Materials with the highest strengths are typically fibres such as carbon fibre, glass fibre and various polymers, and these are frequently used to make composites (e.g. carbon fibre-epoxy). These materials are widely used in aerospace and other applications where weight savings are worth the higher material cost.
Hardness	The ability of a material to resist permanent indentation. Because there are several methods of measuring hardness, the hardness of a material is always specified in terms of the particular test that was used to measure this property.	Hardness is important where resistance to penetration or scratching is required.
Durability	The ability to withstand wear or damage. Also refers to the ability to resist weathering and to last.	Durability is important where abrasion and cutting may take place and for materials used in harsh outdoor environments.
Thermal conductivity	The ability of a material to conduct/transfer heat.	Aluminium and copper have high thermal conductivity and are widely used to dissipate heat from components and processors in computers. Materials with low thermal conductivity, such as expanded polystyrene, are used as thermal insulation.
Electrical conductivity	A measure of a material's ability to conduct electricity – how easily an electrical current can flow through it.	Consideration of this property is particularly important when selecting materials as conductors or insulators.
Corrosion resistance	The definition of corrosion resistance refers to how well a substance can withstand damage caused by oxidation or other chemical reactions.	Oxidation-resistant materials include steel alloys with chromium (stainless steel) and alloys of titanium, molybdenum and tantalum. An example of corrosion control is a steel watering can treated by galvanising to prevent rust.
Stiffness	The resistance of an elastic body to deflection or bending by an applied force. The rigidity or firmness of a body, the opposite of flexibility or pliability.	Stiffness is important when maintaining shape is crucial to performance, for example, an aircraft wing or a shelf or plank between two supports.

Elasticity	When a material has a load applied to it, the load causes the material to deform. Elasticity is the ability of a material to return to its original shape after the load is removed. The elastic limit is the point beyond which permanent deformation takes place. **'Young's modulus'** measures the resistance of a material to elastic (recoverable) deformation under load. $$\text{Young's modulus (Nm}^{-2}) = \frac{\text{stress (Nm}^{-2})}{\text{strain}}$$ $$\text{Stress (Nm}^{-2}) = \frac{\text{force (N)}}{\text{cross-sectional area (m}^2)}$$ $$\text{Strain} = \frac{\text{extension (m)}}{\text{original length (m)}}$$	A stiff material has a high Young's modulus and changes its shape only slightly under load (e.g. diamond). A flexible material has a low Young's modulus and changes its shape considerably (e.g. rubbers).
Plasticity	Plasticity is the ability of a material to deform permanently without breaking or rupturing.	Plastic deformation is a property of ductile and malleable materials. Brittle materials, such as cast iron, cannot be plastically deformed. Most engineering materials show more plasticity when hot than when cold.
Impact resistance	This is a measure of the ability of a material to absorb high impact forces (which occur over a very short period of time) without fracturing.	Certain polymers such as ABS have a very good impact resistance in that they can absorb the energy of an impact without the plastic cracking.
Brittleness	Brittleness is the opposite of the property of plasticity. A brittle metal is one that breaks or shatters before it deforms. White cast iron and glass are good examples of brittle materials.	Tough materials such as mild steel are not brittle.
Malleability	Malleability is the property that enables a material to deform by compressive forces without developing defects. A malleable material is one that can be stamped, hammered, forged, pressed or rolled into thin sheets.	Aluminium is shaped into cans without breaking.
Ductility	The ability of a material to be drawn into a length of material with a certain cross section. It often refers to the ability to be stretched into a wire. Similar to malleability. Brittle materials are not malleable or ductile.	Copper is an example of a ductile metal, making it highly suitable for electrical wires and pipes. The ductile properties of gold, silver and platinum make them suitable for jewellery. Most metals become more ductile with increasing temperature.
Machinability	The capability of a material to be cut or shaped by machine tools. Materials with greater performance characteristics and enhanced properties are most likely to have poorer machinability.	Materials with good machinability cut easily and quickly, achieving a good finish, and tooling is not quickly dulled. Such materials are said to be 'free machining'. The carbon content of steel affects its machinability. Carbon in steels abrades the cutting tool. Additives such as tin and lead increase machinability.

When you examine a design engineered product you should be able to identify the material content, the structure and the properties and characteristics of the materials chosen for each component part. You should be able to recognise which terminology is required to best describe the characteristic of the materials selected. Recognising the choices that have been made helps to clarify how the choice of materials is critical to the product's use.

Table 5.12 **Properties/characteristics of fashion and textile materials**

Property/characteristic/term	Definition
Tensile strength	The strength shown by a fibre, yarn or fabric to resist breaking when a force is applied to it. It is determined by the actual amount a fabric will give before the material is broken on the testing machine.
Softness	How comfortable the fibre or fabric is against the skin.
Texture	The visual and especially **tactile** quality of a material or fibre on its surface. This is determined by the fibre structure are how the material is woven or knitted.
Durability	The ability a material has to resist abrasion and how tough and hardwearing it is. This usually depends on the durability of the fibres used and how the fabric is constructed.
Drape	The supple and flexible character of a fabric; how it falls and behaves when pleated, gathered or folded.
Handle	The way the fabric feels, which depends on the way the fabric has been constructed.
Resilience	The ability a material has to enable it to resume its original shape or position after being bent, stretched or compressed.
Weight	A measure of the fabric's quality, determined by yarn count and fabric construction.
Stiffness	A material's tendency to keep standing without any support. This can be tested to determine a fabric's drape.
Elasticity	How a material can be stretched but then return to its original shape.
Flammability	How a fabric or fibre burns once ignited. The degree of difficulty required to cause the combustion of a substance is quantified through fire testing. Materials and fibres can be finished to make them more resistant to burning.
Absorbency	The ability of the fibre to take up moisture. Different fibres have varying levels of absorbency.
Washability	The ability to wash the fibre or fabric without it deteriorating.
Breathability	How the material or fibre allows perspiration to evaporate. Breathable fibres 'wick' moisture away from the body.
Thermal conductivity and insulation	The ability of a fibre or fabric to effectively maintain normal body temperature under different conditions.
Electrical conductivity	The ability of a fibre or fabric to conduct electricity.
Resistance to decay	The ability of a fibre or fabric to resist microorganisms. Natural fibres are more prone to decay than man-made/synthetic ones.
Biodegradability	The ability of a fibre or fabric to break down and decompose by the action of microorganisms. Natural fibres and fabrics tend to be biodegradable.

When you examine a textile product you should be able to identify the fibre content, the fabric structure and the properties and characteristics of the fabrics chosen. You should be able to recognise which terminology is required to best describe the characteristics of the fibre or fabric selected. Recognising the choices that have been made helps to clarify how the choice of materials is critical to the product's use.

For example, when you look at an everyday item like a towel in Fig 5.97, ask yourself: what fibre has been used? What is the fabric's structure and more importantly why has it been selected?

Figure 5.96 **Example of terrycloth used for towels**

The majority of towels are made from 100 per cent cotton fibre. This is because the fibre must be absorbent as this is the main function of a towel. The towel will be used frequently and therefore also washed frequently. It needs to be easy to wash. Therefore the fibre needs to be durable and easy to care for. Cotton fibres meet all of these criteria. However, this only covers the fibres. You then need to consider the fabric. The fabric of towels can be structured in a variety of ways but the most efficient at performing the job required is terrycloth. This is a loop pile fabric with the loops visible on both sides. It is selected as it is incredibly absorbent and soft to touch.

APPLICATION AND EXAMPLES FOR PRODUCT DESIGN

Table 5.13 **Properties/characteristics of materials used in Product Design**

Property/ characteristic/ term	Meaning	Contextual application
Density	The mass per unit volume of a material. It is the compactness or closeness of the structure of the material.	Density is important in relation to product weight and size (for example, for portability).
Strength	The ability of a material to stand up to force, stress or pressure being applied without it bending, breaking, shattering or deforming in any way.	This is a generic term that tends to be used to summarise a number of material properties. The type of 'strength' usually needs to be stated.
Compressive strength	The capacity of a material or structure to withstand pressure or loads tending to reduce size.	In buildings and structures, concrete provides compressive strength.
Tensile strength	The amount of longitudinal stretching or drawing out a material can take before failure or breaking.	Tensile strength is important in selecting materials for ropes and cables, for example, for an elevator.
Hardness	The ability of a material to resist scratching, wear or indentation by other materials.	Hardness is important where resistance to penetration or scratching is required. Ceramic floor tiles are extremely hard and resistant to scratching.
Durability	The ability to withstand wear or damage. Also refers to the ability to resist weathering and to last.	Durability is important where abrasion and cutting may take place and for materials used in harsh outdoor environments.
Strength-to-weight ratio	This compares the weight of the structure itself to the amount of weight (or force) it can support without collapsing. It is also known as the 'specific strength', a material's strength divided by its density.	Materials with the highest strengths are typically fibres such as carbon fibre, glass fibre and various polymers, and these are frequently used to make composites (e.g. carbon fibre-epoxy). These materials are widely used in aerospace and other applications where weight savings are worth the higher material cost.
Stiffness	The resistance of an elastic body to deflection or bending by an applied force. The rigidity or firmness of a material, the opposite of flexibility or pliability.	Stiffness is important when maintaining shape is crucial to performance, for example, an aircraft wing, or a shelf or plank between two supports.
Elasticity	The ability of a material to return to its original shape and size after a force is applied. The elastic limit is the point beyond which permanent deformation takes place. 'Young's modulus' measures the resistance of a material to elastic (recoverable) deformation under load.	A stiff material has a high Young's modulus and changes its shape only slightly under load (e.g. diamond). A flexible material has a low Young's modulus and changes its shape considerably (e.g. rubbers).

Impact resistance (or toughness)	The capability of a material not to break or shatter when receiving a blow or sudden shock. Toughness is the opposite of brittleness.	The use of ABS (a tough polymer) in children's toys to give impact resistance.
Plasticity	The ability of a material to be moulded or changed in shape permanently.	Plastic deformation is a property of ductile and malleable materials. Brittle materials, such as cast iron, cannot be plastically deformed. Many materials show more plasticity when hot than when cold.
Malleability	The ability of a material to be formed into a shape by bending, pressing, rolling or hammering, without breaking or fracturing.	Aluminium is shaped into cans without breaking.
Ductility	The ability of a material to be drawn into a length of material with a certain cross section. It often refers to the ability to be stretched into a wire. Similar to malleability. Brittle materials are not malleable or ductile.	Copper is an example of a ductile metal, making it highly suitable for electrical wires and pipes. The ductile properties of gold, silver and platinum make them suitable for jewellery. Most metals become more ductile with increasing temperature.
Corrosive resistance	How well a substance can withstand damage caused by oxidisation or other chemical reactions.	Oxidation-resistant materials include steel alloys with chromium (stainless steel) and alloys of titanium, molybdenum and tantalum. An example of corrosion control is a steel watering can treated by galvanising to prevent rust.
Flammability	How a material burns once ignited. The degree of difficulty required to cause the combustion of a substance is quantified through fire testing. Materials can be treated or coated to make them more resistant to burning.	Many pieces of furniture have to adhere to flammability regulations. For example, the fabrics and wooden frame in a sofa.
Absorbency	The ability a material has to absorb moisture, heat, sound or light. Different materials have varying levels of absorbency.	Woods in general are porous and are prone to movement (warping) when exposed to moisture. However, this characteristic is used in steam bending. Acoustic tiles and panels are designed to absorb unwanted sounds such as echos in recording studios.
Washability	The ability to wash the material without it deteriorating, shrinking or fading.	This property is important for hospital and hotel bedding and clothing items which are washed repeatedly.
Thermal conductivity	The ability of a material to conduct/transfer heat.	Aluminium and copper have high thermal conductivity and are widely used to dissipate heat from components and processors in computers. Materials with low thermal conductivity such as expanded polystyrene are used as thermal insulation.
Electrical conductivity	A material's ability to conduct electricity – how easily an electrical current can flow through a metal.	Consideration of this property is particularly important when selecting materials as **conductors** or **insulators**.

Resistance to decay	To decay means to rot or decompose, deteriorate or disintegrate. Decay is usually caused by microorganisms such as bacteria and fungi, but can also be caused by chemical reaction.	Natural materials are more prone to decay than man-made/synthetic ones. Pressure-treated timber (tantalised) will be the most resistant to decay and will provide the longest service. This is often used for outdoor decking.
Biodegradability	The ability of a material to break down naturally and decompose by the action of microorganisms. Natural materials tend to be biodegradable.	Items like plastic bags, tin cans, computer hardware and Styrofoam are not biodegradable. Most of these products take decades and more to degrade, if they degrade at all. Biodegradable materials include cotton, silk, wool, wood, paper and bioplastics.

The use of the correct terminology is important when describing the properties of materials and explaining their suitability.

When you examine a product you should be able to identify the materials used and the properties and characteristics of the materials chosen for each component part. You should be able to recognise which terminology is required to best describe the characteristics of the materials selected. Recognising the choices that have been made helps to clarify how the choice of materials is critical to the product's use.

For example, when you look at a children's playground such as the one in Figure 5.97, ask yourself: what materials have been used? What are the properties of the materials, and more importantly, why have they been selected?

Figure 5.97 **Materials in a children's playground**

The available forms, costs and properties of materials

Design, regardless of material area, is about having a well-developed understanding of what the product is trying to solve, who the user of the product is and therefore which materials, components and manufacturing methods are best suited to meet the requirements identified.

There are numerous factors that you will need to consider for your NEA project when selecting materials and components, but some of the relevant ones arising from this chapter are:

- Stakeholder and user requirements – the designed product must satisfy stakeholders and users. Quality, performance, reliability and durability are essential when selecting materials and components.
- Raw materials to be used – you need to consider the raw materials to be used, their availability and the forms and quantities in which they are supplied (for

example, sizes, profiles, liquids or resins); similarly for components such as fasteners and fittings. All of the information in this chapter should help you to make informed material and component decisions to assist with your selection. You must have proper knowledge about latest materials. Reference to materials and component stockists' catalogues should be useful.

- Production facilities – you must ensure that you have the correct tools, equipment and all necessary facilities to produce your product. When selecting materials and components you must consider all the processes required to convert them into your design proposal.
- Cost and commercial viability – cost is one of the main factors which will influence the design of a product. This is particularly crucial when you are selecting the materials and components to be used. The majority of designers have to consider cost implications and will be given strict cost limitations.

It may be that you will have to consider a range of materials and methods required to manufacture the product and then conclude which will be the most cost effective choice without sacrificing performance. The use of standard sizes of materials and standard components where possible will help to reduce costs.

MATHEMATICAL SKILLS

When selecting materials and components you will need to apply your mathematical skills in a variety of ways. Some of these could be:

- calculating the amount of materials and components required to complete a prototype and for volume production and working to a budget
- recognising how to maximise the use of materials to ensure economical production, including tessellating
- using maximum stress and deflection calculations in terms of material selection with relation to yield strength

- using mathematical ideas of ratio and proportion, including similarity and scale and trigonometry
- enlarging or reducing designs, introducing ideas of multiplication, scale and ratio
- understanding graphs and calculations of measurements such as percentages, ratios, areas, volumes, geometry, statistics and probability
- using percentages in relation to economies of scale.

ACTIVITIES

1 Explain why mass-produced products are cost effective to manufacture.
2 Examine two similar products, one for the mass market and one for the luxury market. Compare the two products for the following:

– the selection of materials and components, including technical terminology
– the assembly processes used in relation to the materials and cost
– the designer's and manufacturer's considerations with reference to the selling price.

KEY TERMS

Compressive strength – the capacity of a material or structure to withstand loads under compression (or pressure), as opposed to tensile strength, which withstands loads tending to stretch or elongate.

Conductors – materials through which an electric current can pass. In general, metals are good conductors. Copper or aluminium is normally used to conduct electricity in commercial and household systems.

Insulators – materials that do not allow energy such as electricity or heat to transfer through them. Wood, plastic, rubber and glass are good insulators.

Tactile – having a surface that is pleasant or attractive to touch.

Young's modulus – the resistance of a material to elastic deformation under load.

KEY POINTS

- Designers have to consider both the mechanical and physical properties of materials.
- Design is about having a well-developed understanding of what the product is trying to solve, who the user of the product is and therefore what materials, components and manufacturing methods are best.

- The performance criteria of materials and components need to be considered in relation to properties and finish, quality, cost, life span and reliability.
- Materials can have a combination of properties to describe them, dependent on their composition.

Further reading on material and component considerations

- www.gov.uk/browse/business/waste-environment
- Lefteri, C. (2014), *Materials for Design*, Laurence King, ISBN13: 9781780673448
- Dent, A. (Foreword), Sherr, L. Chochinov, A. (Introduction), Caniato, M. (Preface) (2014), *Material Innovation: Product Design*, Thames and Hudson Ltd, ISBN13: 9780500291290
- Pfeifer, M. (2009), Materials Enabled Designs, 1st Edition Butterworth-Heinemann, ISBN: 9780323164924
- Plastipedia resource. Comprehensive information on plastics materials and manufacturing. www.bpf.co.uk
- Fletcher, K. (2014), *Sustainable Fashion and Textiles: Design Journeys*, Routledge, ISBN: 9781844074815
- Miodownik, M. (2014), *Stuff Matters: The Strange Stories of the Marvellous Materials that Shape Our Man-made World*, Penguin, ISBN: 9780241955185

Design Engineering

AS Level

1 Mild steel and brass are two metals commonly used by design engineers. For each metal:

 a) describe its composition

 b) list two properties of the metal which make it useful in engineering applications

 c) identify two drawbacks of using the metal

 d) give two specific application examples and justify why the metal is suitable.

2 List four reasons why aluminium is popular in engineering applications.

3 Explain the key properties of copper which lead to its use in engineering applications.

4 Explain the benefits offered by using manufactured wood boards rather than timber in design engineering applications.

5 Explain the difference between thermosoftening and thermosetting polymers. For each category:

 a) identify two polymer materials

 b) give one specific application example for each polymer in a) and justify why the polymer is suitable.

A Level

1 a) Describe the properties of a woven geotextile which make it suitable for use in ground-stabilising applications.

 b) Explain how a geotextile is used to stabilise the ground beneath a footpath.

2 Describe the process of laying-up a fibre-reinforced plastic part.

3 Describe the nature of thermochromic, photochromic and electrochromic materials and describe an application for each one.

4 Describe the key properties of an aluminium sandwich panel and explain how the structure of the panel achieves these properties.

5 Describe, with examples, how high-performance alloys and super-alloys have influenced the design and manufacture of modern engineered parts.

6 Describe the current and future applications for graphene and carbon nanotubes.

7 Discuss factors that designers need to consider to reduce environmental impact when selecting materials and components.

8 Give one example of a ductile material and describe a situation where its ductile behaviour is desirable.

9 Calculate the mass of a rectangular brass block, measuring 120 x 50 x 15mm. The density of brass is 8.5 g cm^{-3}.

10 Calculate the minimum diameter of steel wire required to withstand a tensile force of 5 kN. The tensile strength of steel is 841 MPa (1 Pa = 1Nm^{-2}). Hint: calculate the minimum cross-sectional area of the wire, then calculate the diameter.

11 A mass of 10 kg is hanging vertically from a length of 1.2 mm diameter copper wire.

 a) Calculate the tensile force in the wire. (g = 9.8 Nkg^{-1}).

 b) Calculate the tensile stress in the wire.

 c) Calculate the strain in the wire. Young's modulus for copper = 117 GPa (1 Pa = 1 Nm^{-2}).

 d) The original length of the wire is 2 m. Calculate the extension of the wire when the mass is hung on it.

Fashion and Textiles

AS Level

1 With reference to fibre blending and properties, suggest two blends that would be suitable for each of the following products:

 a) swimsuit

 b) car upholstery

 c) ski salopettes.

2 Explain the properties and give examples of their use for:

 a) Kevlar

 b) PTFE

 c) modal

 d) polyester.

3 Discuss the LCA of cotton production and how the environmental impact can be reduced.

4 The properties of materials can be improved by applying finishes. Give two examples where a finish would make the fabric more suitable for a specific purpose.

A Level

1   Explain the difference between technical and performance requirements in fashion and textile products. Give three examples to support your answer.

2   Compare the properties of polyamide and polyester. In your comparison explain why you think polyester is one of the most widely used fibres. Give examples of products to support your answer.

3   Explain the properties that carbon, glass and ceramic fibres can bring to materials and components. Give examples of how they can be used to enhance protective clothing.

4   When selecting materials in the product development stage, what factors need to be considered?

5   Discuss factors that designers need to consider to reduce the environmental impact when selecting materials and components.

Product Design

AS Level

1   As a designer sourcing materials and components, what social, moral and environmental factors do you need to consider and be aware of?

2   Advances in modern materials have dramatically expanded the range of options available to designers. One of these is graphene. Explain why graphene is particularly suitable for the manufacture of cycle helmets.

3   Give two examples of recent developments in the timber industry and how they have enhanced the manufacture of innovative products.

4   Give two examples of ferrous metals and stage two products that are manufactured from the metals you have given, with reference to their mechanical and physical properties.

A Level

1   Developments in plastics have increased their use in furniture design, in place of traditional materials for:
    –   construction
    –   surfaces
    –   fixtures and fittings.

    For each of these, show how designers have taken advantage of the material properties offered by plastics in comparison to traditional materials.

2   When selecting materials in the product development stage, what factors need to be considered?

3   Discuss factors that designers need to consider to reduce the environmental impact when selecting materials and components.

6 Technical understanding

PRIOR KNOWLEDGE

Previously you could have learnt:
- what gives a product structural integrity
- how materials and products can be finished for different purposes
- how controlled movement is introduced to products and systems
- how electronic systems provide functionality to products and processes
- about developments in modern and smart materials, composites and technical textiles
- how and why specific fabrics have to be reinforced
- some of the processes that can be used to ensure structural integrity in fabrics, including:
  - applying **interfacing** to stiffen fabrics to improve shape

- applying interlining to add structure and insulation
- applying linings and underlinings to add extra body and hide construction methods
- how to deform and reform fabric using a range of techniques, including some methods to dispose fullness
- how to select a range of standardised components to use with fabrics
- some of the processes used for finishing and adding surface treatments to materials for specific purposes
- how knowledge of mathematics and science can be applied to Design and Technology thinking.

TEST YOURSELF ON PRIOR KNOWLEDGE:

1 Describe two different methods by which a 1.6 mm sheet of steel can be made more rigid.
2 Describe the options open to a designer for increasing the rigidity of a moulded plastic part.
3 Describe how a hole in textile material could be reinforced to prevent tearing.
4 Draw a diagram to show how triangulation is used to increase rigidity in a rectangular framework.
5 a) Explain the ways in which wood can degrade when used outdoors.
  b) Describe two ways that wood can be protected for use outdoors.
6 Identify the four different types of motion.
7 Describe the principle of a lever.
8 Describe ways that rotary motion can be transferred with and without a change in speed.

9 Identify two analogue and two digital sensors for electronic systems.
10 Draw the four different symbols used when drawing flowchart programs.
11 Identify specific output devices to produce sound, light and movement.
12 Describe which method of reinforcement would be used when manufacturing a shirt and where on the shirt it would be used.
13 What are the main reasons for reinforcing fabrics?
14 Explain the method for disposing fullness using gathers.
15 Which structural components can be used to add structure to a fashion or textiles product?
16 Give one example each of a mechanical, chemical and biological fabric finish and explain how each is achieved and what products they are used on.

6.1 What considerations need to be made about the structural integrity of a design solution?

LEARNING OUTCOMES

By the end of this section you should have developed a knowledge and understanding of:

● how and why some materials and/or system components need to be reinforced or stiffened to withstand forces and stresses to fulfil the structural integrity of products.

Product Design and Design Engineering

● processes that can be used to ensure structural integrity of a product, such as
  – triangulation
  – reinforcing.

Fashion and Textiles

● how constructional solutions can be used to make fabrics suitable for purpose, including:
  – the difference between whole garment knitting and fully fashioned panels

  – shaping through the addition of boning for structural integrity
  – reduction of fullness according to the design; darts, gathers, elastic, pleats
  – quilting to add thermal insulation.

● how a variety of components fulfil functional requirements through their application in the manufacture of a textiles product, including:
  – fastenings, such as button and buttonholes, zips, poppers, Velcro, hooks and eyes, parachute clips, eyelets and ties and toggles
  – decorative components, such as appliqué motifs, ribbon, lace, braid, beads, sequins and piping
  – constructional components, such as shoulder pads, cuffing and interfacing.

Reinforcing or stiffening to fulfil structural integrity

A well-designed product will make efficient use of materials to achieve resilience, **structural integrity** and product functionality, and look aesthetically good. A poor approach to design sometimes occurs when a product is over-engineered to withstand **stresses** without much thought about the nature of these stresses. For example, structural parts might simply be made thicker to cope with higher forces, but a more careful analysis of the forces might achieve the same structural strength with less material; a wasteful use of materials results in high cost and excessive weight. Good design aims to provide the structural integrity required using the least amount of material.

When designing a product, care should be taken to consider the extreme range of stresses that might reasonably occur during use, including those which perhaps were not part of the original design specification. Some examples of the stresses that products might reasonably encounter are listed below and, while any one of these can be dealt with by careful design, failure to do so may result in the product breaking early in its life:

● A table leg might be hit sideways by a vacuum cleaner when a house is being cleaned, which

could over-stress the attachment between leg and frame.
● A user might lean back on two legs of a chair, which will cause bending stresses in the rear legs.
● A user might insert batteries the wrong way around, causing damage to the electronic circuit in a product.
● A piece of outdoor furniture will be exposed to large amounts of sunlight and, over time, this can cause some polymers to degrade and become brittle.
● The pockets in a garment might be used to carry coins, which can abrade the fabric, leading to holes in the pocket.

Careful analysis of a product will sometimes reveal that stresses are concentrated on certain parts of the product and, therefore, it is sometimes only necessary to strengthen those particular parts. Strengthening can be achieved by:

● **reinforcing** or protecting the material
● reinforcing the assembly of parts within the product.

In order to protect a product from complete failure if it is over-stressed, some products make use of **sacrificial parts** which are intentionally designed to be the weakest and to fail first to protect other more significant parts. Sacrificial parts are usually easy to replace. Examples include fuses in electrical systems, shear pins in mechanical systems or elbow pads in garments.

FASHION AND TEXTILES

One of the most common methods used to reinforce and strengthen fabrics is by applying interfacing. Interfacing is used to provide shaping, reinforcement, firming and support. Collars, cuffs, facings and plackets are the most commonly interfaced areas but are far from being the only places where interfacing is used.

Without interfacing, buttonholes could tear; facings, collars and cuffs would have no structure and be unable to keep their form.

Interfacing is available in a wide range of weights, composition and method of application. The choice depends on the fabric it is being applied to and the product function. For example, chiffon would require a very lightweight interfacing whereas a structured blazer would require a heavy weight type.

The majority of interfacing used today is fusible which is applied to the wrong side of the fabric using heat. There is also non-fusible which needs to be stitched in. The interfacing can be:

● Non-woven – this is a bonded structure and therefore has no grainline
● Woven – needs to be cut in the same manner as the fabric it is being applied to
● Knitted interfacing – used for fabrics that will stretch.

DESIGN ENGINEERING AND PRODUCT DESIGN

Process to ensure structural integrity

Material reinforcing may be applied to parts of a product which undergo high levels of wear. For example:
● corner protectors are used on speaker cabinets
● a plain bearing is used to support a shaft.

Figure 6.1 **Corner protector screwed on to an amplifier**

Material **stiffening** can be achieved by adding folds, ribs or webs. Indents and corrugations can be pressed or formed into sheet metal (or sheet plastic) during the manufacturing process of a product. These methods allow thin sheets to be made more rigid with no increase in material quantity.

Figure 6.2 **Ribs moulded in to the base of a monitor to increase rigidity**

Material protection can be achieved by:
● applying a suitable finish to metals such as paint or lacquer, or by galvanising or **anodising** to prevent corrosion
● adding a stabiliser to a polymer, such as Ultra Violet (UV) stabiliser to uPVC, widely used in window frames, to prevent degradation by sunlight
● using a preservative on wood to prevent biological decay.

Consideration should be given to the methods of assembling the parts within a product so that structural rigidity can be achieved. For example, a rectangular framework is inherently unstable and can collapse when subject to sideways forces.

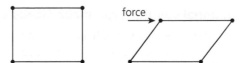

Figure 6.3 **A rectangular frame is naturally unstable**

An effective way to keep a rectangular framework rigid is to use strong joining methods at the corners, such as large screws, knock-down fittings and/or high-grade adhesive. In some designs this is the only method available, but a better approach is to use **triangulation**, which is the principle of adding cross-members to a rectangular frame to give it intrinsic rigidity. A triangular frame cannot change its shape and, therefore, does not rely on the stiffness of the joints to keep the structure rigid. If the design prevents a full cross-member being used, then corner-bracing or gusset plates also help provide rigidity.

Adding a cross-member Corner-braces Gusset plates

Figure 6.4 Methods of triangulating a framework to increase its rigidity

ACTIVITIES

1 Draw diagrams to show how each of the frameworks below could be made more rigid by triangulation.

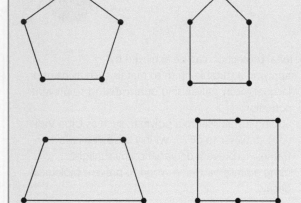

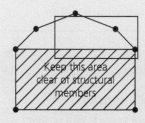

Keep this area clear of structural members

2 Photograph and annotate examples of products or frameworks where cross-members have been added to achieve triangulation.
3 Photograph and annotate examples of gusset plates or other corner reinforcement methods used in products.

FASHION AND TEXTILES

How constructional solutions can make fabrics suitable for purpose

Constructional methods can be used to help a fashion or textile product perform as intended. They may improve the aesthetics of the product but ultimately are selected to serve a purpose.

Whole garment knitting

Producing a garment through whole garment knitting (knitting the whole piece as one construction) means there is virtually no requirement for seams, which can be a weak point or hinder the aesthetics of a garment. Using this method, parts of a knitted jumper such as the front, back and sleeves can be knitted in one piece.

There is growing interest in whole garment knitting for applications other than fashion. Of particular interest are opportunities in technical textiles provided by whole garment technology for knitting seam-free tubes and performing 3D shaping.

Smart textile applications using conductive yarns is an ever increasing market within a wide range of industrial sectors. With no seams to break the conductivity of yarns throughout the garment, whole garment knitting can maximise the potential for such wearable technology.

Manufacturer benefits include:
- elimination of labour-intensive cutting and sewing
- little waste due to no fabric cutting
- a reduction in manufacturing lead time as garments are quicker to produce.

Consumer benefits include:
- reduced need for seams, meaning less bulk
- stronger garments due to fewer seams
- garments with improved draping characteristics due to less bulk created by the seams.

Whole garment knitting methods can be applied to a wide range of industries.

OCR Design & Technology for AS/A Level

Table 6.1 Industries using whole garment knitting

Industry	Application
Sports	● Seamless wetsuits for better comfort, warmth and mobility ● Seamless running footwear ● High-performance garments with no seams for maximum comfort and mobility
Automotive	● Thermal covers for hoses made with yarn that can resist temperatures up to 500°C ● Muffler covers
Medical	● Supportive braces and circulation socks, easy-to-dry fabrics that draw sweat away from the body ● Garments for people with disabilities that are easy to wear, comfortable, thermal and help circulation ● Wheelchair covers
Space exploration	● Parts for shuttles
Industrial	● Knitting with Kevlar ● Glass fibre covers for pipes that can resist temperatures up to 500°C
Furniture	● Upholstery for couches and chairs

Fully fashioned knitting

Fully fashioned knitting means that each piece required to construct a fashion or textiles product can be produced individually and then joined together. This allows for complex shapes to be created for a perfect fit. Creating each piece individually before joining the pieces together reduces the amount of waste as no cutting out is required. It also gives the designer the ability to create complex styles without the usual visible seam lines.

Main factors to know about fully fashioned knitting include:
● Each piece is knitted to the shape required.
● It is very distinctive and easily identifiable by the fashioning marks which normally run parallel to the garment seams. The fashioning marks are created as the shaping takes place when knitting the panel.
● The panels are assembled using 'linking'. This involves sliding individual knitted loops onto metal points from each of the two panels that are to be joined. The advantage of this method of construction is there is no visible seam inside the garment and the linking join lies perfectly flat.
● Because the pieces are knitted to each **template** shape, joining together still needs to take place.

Fully fashioned knitting has many uses today and some of these include:
● Undergarments – the seamless technique lends itself to underwear that does not produce unsightly seamlines under garments.

● Sportswear – the fewer the seamlines or at least flat seams, the more suitable for close fitting sportswear where friction could be uncomfortable.
● Upholstery – seamless technique is also used for the upholstery purpose. This is generally used in office chairs. It is also used in the automotive industry for seat covers.
● Medical textiles – orthopaedic supports, bandages and medical stockings.

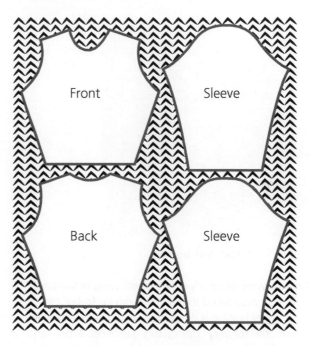

Cut and sew knitwear

Fully fashioned knitwear

Figure 6.5 Cut and sew versus fully fashioned knitting techniques

Shaping through the addition of boning for structural integrity

The use of boning to add structural integrity was originally used in undergarments to 'hold' the body in and give the silhouette that was in fashion at the time. The boning supports the desired shape and prevents fabric from wrinkling, giving a very smooth image. It also gives a strapless garment the support to stay up.

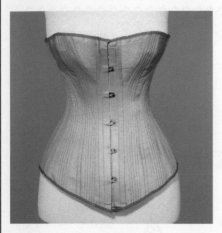

Figure 6.6 Corset with boning

Corsets were originally constructed using whale bone and they were laced to pull the torso in. Today, the majority of boning is made from nylon, although steel is favoured for very high-quality garments. More cost effective polymer boning is used for mass produced garments, but this does not have the strength of steel and has a tendency to warp and bend. Modern steel bones come in two varieties:

- flat boning – bends in only one direction
- spiral boning – bends in two directions and therefore can be used on curved channels.

Traditionally, boning is added to garments once the main bodice pieces are stitched. Channels are created to 'thread' the boning through and this is usually parallel with the garment seams. Today, with the need for fast and cost effective manufacture, boning can be purchased that is already attached to a fabric ribbon for quick application. This avoids having to create the channels and is stitched directly to the garment. The boning can be hidden within the structure of a garment; for example, by attaching it to the lining. However, fashion over the years has resulted in the structural components being a main feature of the garment and therefore key to the aesthetics. Jean Paul Gaultier famously introduced traditional corsetry into his collections by redesigning the unfashionable garment and recreating it as outerwear. Boning is now not restricted to undergarments and can be seen in jackets, shirts and dresses.

Reduction of fullness according to the design

The desired shape and definition can be achieved in fashion and textile products not only through the cutting but also by techniques that reduce the fullness in the fabric.

Darts

Darts are folds in the fabric that end in a tapered point to give a smooth shape. They are used to shape fabric to give a desired form. They are primarily used in garments but can be seen in upholstery to allow fabric to fit around given shapes. There are three main types of darts:

- single pointed – commonly used at the waistline of skirts and trousers; also used to give shape over the bust
- double pointed – generally used on dresses where there is no waist seam
- dart tuck – an opened dart used as a feature on blouse shoulders or waists.

Sometimes the excess dart fabric is trimmed away when the fabric is bulky. This can be seen in soft-toy production and upholstery.

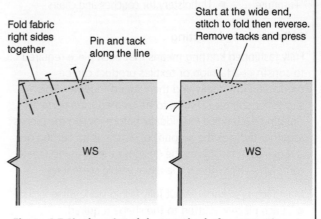

Figure 6.7 Single pointed dart method of construction

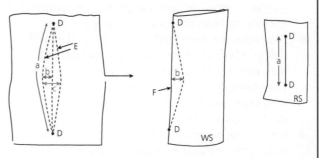

Figure 6.8 Double pointed darts

Gathers

Gathering is widely seen in dresses, skirts, blouses and sleeves. It can be used as a decorative technique to create frills for soft furnishings and is used to reduce fullness at the top of curtains. The gathers are created by hand or machine by stitching two parallel lines along the edge. No reverse stitching should be done and the machine should be set to the longest straight stitch possible. The two threads are then pulled up to evenly draw in the fullness to the required amount. The same principle is used when making curtains by stitching curtain tape, which is available to create a range of effects, at the top of the curtains.

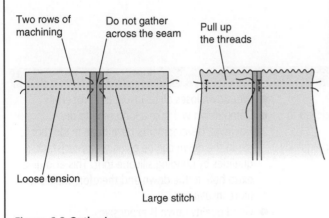

Figure 6.9 Gathering

Elastic

Elastic is used as an alternative to gathering where increased flexibility is needed. It is often used in the waist of children's garments where no other fastening is required. A casing is made to a depth dependent on the depth of the elastic. Elastic is then threaded through the casing. It may be necessary to add elastic to a section of a garment, for example, the back of a jacket. This is achieved by stitching the elastic directly onto the fabric rather than using a casing. The elastic has to be stretched during machining.

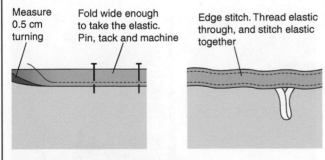

Figure 6.10 An elasticated casing

Shirring elastic can be used to give a softer all-over effect. The sewing machine bobbin is threaded with shirring elastic which gently gathers the fabric.

Figure 6.11 A product made using the shirring technique

Pleats

Pleats are used to give shape with fullness by forming folds in the fabric. Variations of pleats can be achieved by the folding methods used. Types of pleats include:

- knife pleats
- box pleats
- inverted pleats
- kick pleats.

Figure 6.12 Pleated material

To create a pecermoplastic fabrics are used and heat is applied along the fold where required. When creating pleats, the measurements have to be carefully considered and the **pattern markings** accurately transferred onto the fabric. The pattern will indicate the direction of the fold to create the pleat.

(a)

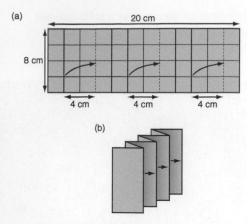

(b)

Figure 6.13 **Knife pleating**

Quilting to add thermal insulation

Quilting involves the sewing of two fabrics with a layer of wadding in between. The method is explained in further detail in Chapter 7.2. Quilting is essentially functional, to add insulation to the product. The wadding (otherwise known as batting) adds the insulation but the stitching traps air within the quilted sections. Insulation works by creating numerous pockets of air close to the skin. The more air trapped; the warmer it will be. The type of insulation selected depends on the end function of the product, and the manufacturer will also have to consider the cost. Examples of insulation can be natural, synthetic or a mix.

Table 6.2 **Pros and cons of types of insulation**

Insulation	Pros	Cons
Down	• Unparalleled warmth-to-weight ratio • Excellent compressibility allowing it to be packed to take up little space • High resilience so returns to its original shape quickly after compression • Drapes well and feels luxurious	• Wet down loses all its insulation properties • Damp down will take a long time to dry. However, developments have been made that mean the down can be given water-resistant qualities by bonding silicone to it. This also gives extra bulk to the down and therefore makes it more insulating • Good quality down is expensive
Synthetic	• Good level of warmth even when wet • Synthetic dries out quicker than down • Offers value in comparison to down products • Easier to maintain than down	• Less warm than down • Won't compress as well as down • Extra bulk makes garments less comfortable to wear and carry

How components fulfil functional requirements in a textiles product

Fashion and textile products include a range of components to enhance their style and performance. The majority of manufacturers use pre-manufactured standardised components to ensure quality. These are purchased for specific market segments.

Fastenings

Fastenings are mainly functional, to close an opening, but they can also be used as a decoration. They are a temporary method for joining fabrics. Designers will have to consider how the fastening has to function. For example, it would be pointless to use buttons as a closure for a tent as this would not provide protection from the elements.

Button and buttonholes

Buttons can be functional, decorative and fashionable. They are available in a multitude of options that includes size, shape, material and application methods. They are one of the most commonly used functional methods

of opening and closing fashion and textile products. Button selection is dependent on the type and style of the product. This selection must be finalised prior to making the product as it will dictate the size of the corresponding buttonhole or loop.

Figure 6.14 **Buttonback chair**

Zips

Zips are a neat and secure way of giving an edge-to-edge fastening. For lightweight and fine fabrics, a plastic zipper should be used. For clothing that is fitted and where the zip needs to be concealed, an invisible zip is used. Metal zippers for casual and sportswear are heavier, broad and firm.

Poppers

Poppers function in the same way as press studs but have more strength to hold fabrics together and tend to be decorative as well as functional due to them being visible from both sides of the fabric. They are not suitable for thicker fabrics as the two parts will not lock together. Poppers can be made from metal or polymers and are attached using a special tool and hammer. In industry, specific machinery is used to attach the poppers.

Velcro

Velcro is a 'hook and loop' fastener made from two layers of polyamide. One layer has a furry surface and the other has lots of tiny hooks. Lightly pressing the two together forms a strong attachment, which can be peeled apart. It makes an easy fastening for children's garments and shoes. Velcro is available in different widths and colours and can be sew on or non-sew on. It is also available in different peel strengths and an 'anti-snag' type to meet different functional requirements and applications.

Hooks and eyes

These tend to be used where there is likely to be strain; for example at the top of a zip. However, they can be used as a fastener independently. Hooks and eyes are usually stitched on by hand individually. However, hook-and-eye tape can be used and machine stitched on where a longer strip of hooks and eyes is required.

Parachute clips

Parachute clips are commonly used on bag straps to attach around the waist, for example. They are usually made from polymer and most commonly available in a range of sizes. They consist of two parts and function as a very secure fastening by pushing one side into the other where they clip together. To open them, the inner section has to be squeezed together and then pulled apart.

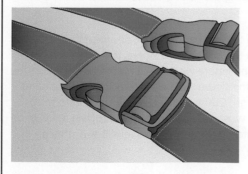

Figure 6.15 Parachute clips

Eyelets and ties

Eyelets are used to thread a tie through; for example, cords or ribbons. They are typically made from stainless steel. Eyelets are attached to the fabric by pressure. Once fixed onto the fabric they should be permanent and take the strain of the tie passed through them without damaging the fabric. Eyelets also form the closure on belts. Eyelets can be attached to the fabric using a special attachment that the eyelet is placed in; pressure is then applied to 'seal' the eyelet to the fabric. In industry, semi-automatic and fully automatic punch machines are used.

Figure 6.16 A method of using eyelets

Toggles

Traditionally toggles are seen as a method for fastening duffle coats and were made from oak. However, today they can be made from various materials and ones made from polymers are commonly used for sports and outdoor pursuits clothing and equipment. These are usually attached to the end of a drawstring and usually have a locking mechanism which allows the drawstring to be pulled up and secured in place.

Decorative components

The following are components that are used to make a product more visually appealing but are not required to be functional.

Appliqué motifs

Appliqué motifs are commercially produced decorative motifs. Some of the more decorative appliqués can be seen in couture garments and wedding dresses and can be very expensive. The lower cost ones are mass produced. There are various methods for attaching the appliqué (see Chapter 7).

Ribbon

Ribbon is usually used for decorative tying or binding. Ribbon can be patterned, printed, woven, braided, decorated with embroidery, pearls or sequins, shaped like ric-rac and edged with metal so it can be moulded and shaped. It can be made of natural fibres such as silk, cotton and jute, and of synthetic materials, such as polyester, nylon and polypropylene. Ribbons tend to be designed in much the same way as fabrics. Materials are selected based on use, wear ability, cleaning requirements and fabric trends that the ribbons must match.

Lace

The term 'lace' can be applied to a number of textile fabrics and components. It is a delicate, open structure that can be produced by hand or machine. There are many different types, which tend to be classified by how they are made. For example:

- bobbin lace – traditionally the lace is made by weaving threads held on bobbins and pinning them on top of a pattern pinned to a pillow
- needle lace – this is similar to embroidery and consists of thousands of buttonhole stitches
- embroidered lace – the embroidery is done on a tulle base.

Lace is available in a variety of widths and bought by the metre. It can also be bought as a trim to finish the edge of a fashion or textiles product.

Figure 6.17 Bobbin lace

Braid

Braid is traditionally a woven decorative trim that is finished on both edges. It can be made from almost any fibre (similar to ribbon) and comes in a variety of weights and widths. Flat braid may be used to produce a decorative border effect on garments or home furnishings or may be folded over raw fabric edges as a finishing method.

Beads and sequins

Beads are available in a variety of shapes and sizes. In modern manufacturing, the most common bead materials are wood, plastic, glass, metal and stone. Usually a small hole in the centre allows the bead to

be stitched on or threaded. Bead embroidery is a type of beadwork that uses a needle and thread to stitch beads to a surface of fabric.

Sequins are disc-shaped beads used for decorative purposes. Originally they were made from metals but are now usually made from plastic. They are available in a wide variety of colours and geometrical shapes. Sequins are used as embellishment on clothing, jewellery, bags, shoes and soft furnishings. They can be stitched flat to the fabric so that they do not move or they may be stitched at only one point so they move easily to catch more light. Sequin fabric can also be bought by the metre.

Piping

Piping is a type of trim consisting of a strip of folded fabric that is covering a cord to form a 'pipe'. This is then inserted into a seam by 'sandwiching' the covered piping between the fabric to define the edges of a garment or other textile object. Usually the fabric strip is cut on the bias. Piping is usually seen as a trim on sportswear, bags and soft furnishings. It can be used as decoration and constructed in a coordinating colour to the main fabric or can add strength to a product and give the seam a firmer structure. When stitching the piping, a **zipper foot** attachment must be used on the machine to allow the needle to get close to the cord. This gives a neat finish.

Constructional components

These components perform a function within the product. They may give shape and structure, and improve the aesthetics but primarily they serve a specific purpose.

Shoulder pads

Shoulder pads are commonly used in tailoring for both men's and women's wear and are available in different shapes and sizes according to the style of sleeve on the garment. They can be used to disguise figure faults such as sloping shoulders. They are usually stitched at the shoulder seam and sandwiched between the outer fabric and lining to conceal them. Their inclusion in women's wear is dependent on the fashion taste of the day. They were popular additions to clothing (particularly business clothing) during the 1940s, 1980s and late 2000s/2010s.

Cuffing

Cuffing is a tube of knitted fabric designed to be attached to the bottom of a sleeve. It is commonly seen on sports and leisure wear. It is a method to finish off the bottom of the sleeve which stretches to allow the hand to go through. Cuffing will fit snugly and comfortably around the wrist. When attaching the cuffing it has to be machine stitched using a stitch that will also stretch.

Interfacing

Interfacing is used in fashion and textile products to strengthen areas and is attached to the wrong side of the fabric so it remains hidden. Its most common use is in shirt collars, cuffs and plackets and on waistbands. The interfacing helps to maintain shape and give structure. On the placket it is required to give strength where the buttons are attached and also to enable the stitching of a neater buttonhole. See Chapter 7 for how to apply interfacing.

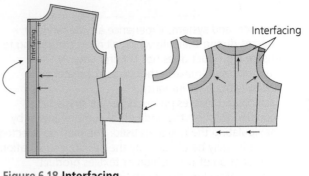

Figure 6.18 Interfacing

ACTIVITIES

1 Examine a range of fashion and textiles products and identify the components that have been used.
   a) List the components and identify if they are functional or aesthetical.
   b) What materials have been used to make the components?
   c) Why have the components been selected?
   d) Create a flowchart, with diagrams, to explain how the component has been added to the main product.
2 Whole garment knitting is a relatively recent development.
   a) List the advantages to both the manufacturer and consumer of whole garment knitting.
   b) Examine and name products from different industrial sectors that are manufactured by the whole garment knitting process. Why are these products superior to previous ones?
3 Examine a baby carrier (you can look at an actual product or find one online) and identify the stress points on the product. How has the carrier been designed and manufactured to ensure product longevity and address the stresses?
4 Carry out research on fashion and textile products that have used some form of reinforcement or stiffening. For each one, explain the technique used and why it has been selected.

ACTIVITY

Investigate a wide range of products to identify the diverse ways in which the product achieves structural integrity. This could include looking at rigidity, ability to hold loads, resisting wear and tear and protection from the environment. The techniques used will vary depending on the materials and on the product's design, so look at a variety of products made from woods, metals, polymers and textiles.

KEY TERMS

Anodising – coating (a metal, especially aluminium) with a protective oxidised layer by electrolysis.

Feasibility study – this investigates the implications of a project before getting involved and investing resources. In the iterative design process, the feasibility of proposed designs can be assessed through experiments, trials, mock-ups, testing and modelling.

Interfacing (textiles) – a bonded web fabric that can have an adhesive side that is added to fabric to add strength and structure.

Pattern markings – the markings on all pattern templates to indicate the position of crucial areas on the product.

Reinforcing – adding to, or altering the form of, a material or component to strengthen it or reduce its wear.

Sacrificial parts – weak parts of a product designed to fail first in order to protect more significant parts.

Stiffening – reducing flexibility in materials or structures.

Stresses – the broad range of factors that may cause a material, product or system to fail.

Structural integrity – the ability of a product or system to hold together when stressed or under load.

Template – can be drawn around to mark a shape onto material.

Triangulation – adding cross-members to increase the rigidity of a framework.

Zipper foot – a foot attached to the sewing machine to allow the machinist to stitch close to the teeth of a zip.

- Products and systems experience a wide range of stresses during their life which all need considering to ensure a product does not fail prematurely.
- Careful design makes efficient use of materials to achieve structural integrity.
- Fashion and textiles products can be shaped and moulded through the addition of components or by manipulating the materials used. The method selected will ultimately be dictated by the design, end situation and function of the fashion or textiles product.
- Constructional methods can be used to help a fashion or textile product perform as intended. They may improve the aesthetics of the product but are selected to serve a purpose.

- Whole garment knitting involves knitting the whole piece as one construction. Fully fashioned panels are produced by knitting the individual shapes to construct the garment which are then stitched together.
- Boning supports the desired shape within a garment and prevents fabric from wrinkling, giving a very smooth image.
- Shape and definition can be given to a fashion and textiles product not only through the cutting but also by techniques that reduce the fullness in the fabric.
- The main function of components when included in fashion and textile products is to enhance their style and performance.

6.2 How can products be designed to function effectively within their surroundings? (Fashion and Textiles and Product Design only)

LEARNING OUTCOMES

By the end of this section you should have developed a knowledge and understanding of:

Fashion and Textiles

- surface finishes, decorative techniques and surface pattern technology that can be used to enhance the aesthetic qualities of products, including:
  - printing and dyeing techniques, such as screen, block, roller and discharge printing and methods of resist and vat dyeing
  - biological techniques, such as the use of natural enzymes to create stone wash effects on jeans
  - embroidery and appliqué techniques
  - mechanical processes, such as embossing and heat setting used on thermopolymer fabrics to shape or create pleats
  - digital technologies used to print, emboss and cut designs, such as: **dye sublimation** printing and use of a laser cutter.
- how materials and products can be finished in different ways to prevent corrosion or decay, or enhance their performance for their intended purpose, including:

  - methods of laminating to strengthen fabrics
  - chemical finishes used to improve a fabric's performance, such as water repellence, stain resistance, flame resistance, antistatic, mothproofing, anti-pilling, rot proofing, anti-felting, hygienic (sanitised)
  - breathable coatings for high-performance wear
  - transparent coatings on fine fabrics.

Product Design

- how surface finishes and coatings can be used to enhance the appearance of products and the methods of preparing different surfaces to accept finishes to deliver a decorative, colourful and quality outcome.
- how materials and products can be finished in different ways to prevent corrosion or decay in the environment they are intended for, such as:
  - paints, varnishes, sealants, preservatives, **anodising**, plating, coating, galvanisation, cathodic protection and electroplating.

FASHION AND TEXTILES

Surface finishes, decorative techniques and surface pattern technology

Printing and dyeing techniques

Producing colours and designs on a fabric is achieved by dyeing or printing. Dyeing gives a solid colour to the fabric whereas printing is the application of dye on specified areas to create designs. Today we see fabric in all tints and shades of colours, small and big prints, woven in colourful designs. All these are possible because of dyeing and printing.

There is a wide range of printing and dyeing techniques available to create different effects. Further information on the processes used to create these techniques can be found in Chapter 7 and it is important that you familiarise yourself with each method. This section will focus on how these methods enhance the aesthetics of the fabrics and products.

Some print styles may be classic and change little over time but many are influenced by any number of factors that dictate the patterns used. It is usually the job of a **colour/print forecaster** to imagine and predict the colours and prints that will be in fashion. The colour and print trends that designers, manufacturers and buyers will be working with in their respective industries each season are influenced by the forecasters. It is down to them to reflect the mood of our changing culture at any point in time.

Discharge printing

Also known as extract printing, discharge printing produces a white or coloured effect on a previously dyed base. This is achieved by an oxidising and reducing the agent capable of destroying colour. It is similar to bleaching, except it does not damage the fibres. The agent used to remove the colour can be chlorine or hydrosulphite. In colour discharge printing, a dye that is resistant to the bleaching agent is combined with it, producing a coloured design instead of white on the dyed ground and resulting in an extremely soft print. You can discharge print on knit or woven fabrics. The solution used to create the discharge will work on natural fibres but not on synthetic ones. For example, with a 50 per cent cotton, 50 per cent polyester blend fabric, the cotton threads will discharge while the polyester threads will not. This can create interesting effects.

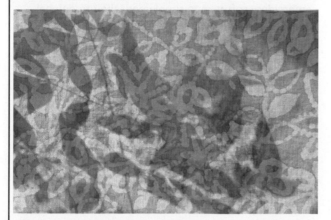

Figure 6.19 **A discharge dyed pattern with most of the original colour removed**

Resist dyeing

Resist dyeing is a traditional method of applying colours or patterns to fabric. It basically involves applying a substance that is resistant to the dye, which prevents those areas of the fabric being dyed, while other parts are free to take up the dye colour. Interesting and unique results can be achieved using this technique.

Tie dyeing

There are a wide range of complex patterns that can be achieved by using different methods for tying the fabric and by repeating the process using numerous colours. The resist is created by tying the fabric tightly using knots, or by applying string or elastic bands to prevent the dye from being absorbed into the fabric.

Ikat

Ikat is a resist dyeing technique that involves tying the yarns and dyeing prior to weaving. The resist is formed by binding individual yarns with a tight wrapping applied in the desired pattern. The yarns are then dyed. The bindings can be altered to create a new pattern and the yarns dyed again with another colour. This process may be repeated multiple times to produce elaborate, multi-coloured patterns. When the dyeing is finished all the bindings are removed and the yarns are woven into cloth. The benefit of this process is the pattern is visible on both sides once woven. Ikat fabrics are used widely for interior furnishings as well as for fashion.

Batik

Batik uses wax to create a barrier. The wax is melted and drawn onto the fabric using a tjanting tool. The fabric is then dyed and the process can be repeated numerous times to create complex multi-coloured designs. The parts covered in wax resist the dye and remain the original colour. Batik is unique as the wax cracks when handled, creating an interesting crackled effect which differentiates this method from others. This resist technique offers immense possibilities for artistic freedom as patterns are applied by actual drawing rather than by weaving with thread. Although it is common to see the mass production of batik with machines, it is still made by hand in many parts of the world.

Silk painting

Silk painting involves using a gel called gutta to outline designs and create the resist. This is applied to white silk that has been pre-washed, dried and stretched onto a frame. Once the gutta has dried it acts as a barrier against the dye, keeping the colour within the outlined areas. After the dye has dried the gutta is removed and a defining line remains the colour of the gutta that has been used. Silk painting can be seen today used on scarves and soft furnishings products like cushions. Due to the laborious technique and use of silk it is a costly process and is therefore used on smaller items.

Vat dyeing

Vat dyeing is a process that refers to dyeing that takes place in a bucket or vat. Vat dyes are mostly used for colouring cellulosic fibres and, in particular, cotton. The technique is widely used for dyeing cotton fibres to manufacture jeans. Indigo is an example of a vat dye: it changes from yellow, in the dye bath, to green and then blue as the air hits it.

Digital printing

The majority of designers are inclined to print digitally. This allows for small orders and an endless number of colours. With digital printing technology, it is possible to go from the design stage to finished fabric in a matter of weeks or days. This gives the designer the capability to stay on trend and even change prints or colours mid-season.

Digital print technology gives a **photorealistic** image and vibrant colours. However, there will still be the need for the more traditional screen and hand-print and patterning processes as there will always be a demand for individuality and uniqueness. Digital tends to sit flat on the surface. Other techniques penetrate through the cloth and give texture and individual patterning.

Biological techniques

Biological finishes use natural enzymes to change a fabric's appearance. As the textiles industry has always been one of the most polluting, an attempt to develop a suitable textile processing method has been made which has resulted in bio-finishing. It includes:

- **bio-stoning** – traditionally, to get the look of stonewashed jeans, pumice stones were used which could damage the garment. However, thanks to the production of cellulose enzymes, the jeans industry can reduce and even eliminate the use of stones
- **bio-polishing** – fabrics containing natural cotton fibres often have tiny loose yarns on the surface, which usually become tangled with repeated wearing and washing, making the garments look worn. The bio-polishing process targets the removal of these fibres and reduces the hairiness of the fabrics.

Embroidery and appliqué techniques

Embroidery

Many embroidery and appliqué techniques have been around for years and the decoration and techniques used may be specific to a particular culture. For example, **Shisha embroidery** originated in India in the eighteenth century, but Figure 6.20 illustrates how it has been used to reflect a particular modern trend. The same effects can be achieved using computerised sewing machines. For example, shisha motifs can be purchased ready to apply and will save the manufacturer many hours in production.

Figure 6.20 An example of shisha embroidery

Hand-stitched embroidery gives an **artisan** appearance to products and is still used for luxury products. The time-intensive, specialised nature of embroidery requires highly talented designers and technicians to achieve the results required by the couture industry. However, there are suppliers of fashion and textile products that still support traditional embroidery techniques from other countries. For example, Christina Lynch has developed a range of fashion and textile products that are embellished using traditional Mexican embroidery techniques. Many designers now use embroidery software to supplement their traditional designs and give them a modern twist without the need for time-consuming, costly hand embroidery.

Appliqué

The method for achieving appliqué is covered in Chapter 7. When the trend for wearing garments adorned with motifs and embroidery hits the catwalks, it is emulated by the high-street stores by using appliqué motifs to create the same effect. The use of appliqué creates a hand-crafted, customised look that lends itself towards folklore, bohemian and hippie styles. However, appliqué motifs can also be extremely delicate and are favoured by wedding dress designers as a method of adding detail and decoration.

Figure 6.21 A catwalk trend for appliqué on denim reflects a bohemian influence

Mechanical processes

Embossing

Embossing is a technique in which images and patterns are created on the surface of the fabric through the application of heat and pressure. During the process of embossing, the surface of the embossed material is raised, adding a new dimension to the object. Engraved calendar rollers are used to emboss relief patterns on the fabric surface. Virtually any kind of fabric can be embossed as long as it can withstand the heat and pressure. One of the most recognisable embossed fabrics is moiré. This gives the fabric a rippled, wavy effect and is used for both fashion and soft furnishings. Many different effects can be created, dependent on the engraving.

Figure 6.22 Embossed moiré fabric

Heat setting used on thermopolymer fabrics

The heat-setting process is mainly used for thermoplastic fibres to improve their aesthetic requirement for a design. With the application of heat and pressure, the fabric can be moulded and remoulded. This technique enables a flat fabric to be transformed into structural and sculptural forms. This is particularly advantageous to create permanent pleats. The fabrics can be given a 3D form regardless of construction methods and interesting effects can be created.

Figure 6.23 The heat setting of fabrics can be extremely versatile

Digital technologies

Printing

Digital fabric printing is basically the method of transferring digital files onto fabric. In addition to clothing fabrics, the home furnishing industries can now produce custom made and limited edition lines. There are a number of different types of digital fabric printing and each works on specific fabric types and offers various qualities of print. The method used also dictates the price of the fabric.

Digital print methods include:

- Dye sublimation/disperse dye – this process is easy to work with and only requires heat and a little pressure once printed. It produces bright colours, but the limitation is that it only works with polyesters. Polyester mixes can be used, but the lower the amount of polyester in the mix, the duller and more prone to fading in the wash the image will be. The colours are quite durable, wash-resistant and **colour fast**. Dynamic graphics can be created for branding and smooth colour graduations, dense colour and fine details lend it to creating excellent fabric prints for soft furnishings and fashion. The process can also be used for customisation as small runs can be completed cost effectively.
- Transfer methods – when you print onto paper you get a sharp image. The transfer method also lets you print to stretchable polyester.
- Acid dye and reactive dye – cotton and silk require reactive dyes and nylon requires acid dyes. These

inks are printed directly onto textiles in inkjet printers using piezo print heads. After printing with either acid or reactive inks, the dye has to be fixed at a high temperature, usually by washing in a combination of hot and cold.

Figure 6.24 Fashion and textile products use digitally printed fabric to reflect the latest trends

Embossing

Digital embossing creates the same effects as traditional embossing without the need for tooling; in other words, the engraved rollers are not required and therefore the whole process is quicker and smaller orders can be met. It uses digital inkjet printing to add a thick, clear UV curing liquid polymer to create a 3D pattern to selected areas on a design, which adds a tactile feel that can normally only be achieved by mechanical embossing. The technique can be used on a range of fabrics and the designs can be extremely complex.

Laser cutting

Laser cutting can replicate the same cut of design in any given number of fabrics. By having the facility to create very complex designs on different kinds of materials, laser cut technology has enabled fashion and textile manufacturers to produce products at a premium level. The advantages include:

- complexity of shapes – lasers are able to cut detailed patterns
- sealing of the edges – laser cutting seals the edges of most textiles, virtually eliminating the problem of fraying
- precision cutting – cutting precisely with scissors or blades can be very challenging, especially if high volumes are needed
- repeat process – lasers offer precise and clean cuts whether cutting 10 shapes, 1,000 shapes or 10,000 shapes. The cut is the same every time. This means high volumes of the same design are possible
- minimum waste.

Figure 6.25 Example of laser cut fabric used on a dress

Laser engraving

Laser engraving is very similar to laser cutting and is being used increasingly in the textile world. The laser removes layers from the fabric to engrave a design created in bitmap format. This works particularly well on leather.

Finishing products to prevent corrosion or decay or enhance performance

Methods of laminating to strengthen fabrics

Laminated fabrics combine two or more layers of materials which are usually bonded together using a pre-prepared polymer film or membrane by using adhesives, heat and pressure. This process enables a material to have properties that it would not normally have when functioning independently. It can be carried out on virtually every textile form including yarns, fibres and fabrics. These can be woven, knitted or bonded.

Chemical finishes used to improve a fabric's performance

Chemical finishes involve the application of chemical solutions or resins to either the face or the back of the fabric. This improves the appearance, handle or performance of a fabric.

Table 6.3 **Uses for laminated textiles**

Industry	Application
Automotive and aerospace	Vehicle interiors – textiles often laminated onto interior components such as door panels
Medical and hygiene	Waterproof breathable **hydrophilic** membranes
Construction and engineering	Tarpaulins – laminating waterproof HDPE fabric to woven laminated PE
Interiors	Furniture/upholstery – laminating foam to furnishing fabrics for added strength and comfort
Underwear industry	Bras – laminating the outer fabric to foam for added strength and support
Technical apparel	Waterproof breathable membranes; for example Gore-Tex
Apparel	Use of interfacing to strengthen collars, waistbands, plackets and buttonholes
Orthomedical supports, sports supports, back supports, equestrian underwear, surface water sports, wetsuits, survival suits, surfing wetsuits, warm-up suits, dive suits, liners and footwear	Stomatex 'breathable neoprene' is made from closed-cell foam neoprene. These fabrics can be applied as laminates. The product is suitable for use wherever thermal insulation or body protection is required and comfort would normally be compromised by sweating

Table 6.4 **Chemical finishes used to improve a fabric's performance**

Finish	Brand	Process	Application
Water repellence	Teflon® Scotchgard®	Water-repellent chemicals are applied, for example fluorochemicals and silicones, to the face or back of the fabric. The finish can be permanent or temporary	All fibres used for all-weather wear and use: leisurewear, bags, shoes, tents
Stain resistance	Teflon Scotchgard Huntsman®	Silicones are applied to resist water-based stains. Synthetic resins are applied to fabrics to resist oil-based stains	All fibres Used in garments, ties, upholstery, curtains, carpets
Flame resistance	Proban® Pyrovatex®	Chemicals are applied to make the fabrics difficult to ignite	All fibres Especially children's nightwear Cotton/viscose furnishings must, by law, be given a flame-resistant finish
Antistatic		Chemicals are applied to the fabric to stop it clinging and attracting dust. The disadvantage is that it eventually washes out	Carpets Lingerie Clothing
Mothproofing	Mitin®	Chemicals are applied and fixed that make the fibres inedible to moth grubs	Wool products, especially those stored for long periods
Anti-pilling		Chemical treatments are applied to the surface to reduce the forming of pills on fabrics and knitted products. This aims to suppress the ability of fibres to become loose and form a ball or 'pill'	All fabrics but particularly knitted products that are susceptible to pilling
Rot proofing		Chemicals are applied to the fabric to preserve and protect materials used in industrial manufacturing or production to prevent biodegradation and chemical decomposition	Particularly cellulosic fibres which are particularly vulnerable to fungi and microorganisms
Anti-felting	Teflon®	A finish is applied that enables wool items to be washed in the washing machine, limiting the felting and shrinkage while maintaining the original properties of the surface of the fabric	Wool products

Hygienic (sanitised)		Anti-bacterial chemicals that hinder the growth of bacteria are applied to the fabric. It also prevents stains and odours, retaining the hygiene and freshness of the product The same effect is now created through silver ion technology	Hospital fabrics and wall coverings, sports shoes and garments, medical products

Breathable coatings for high-performance wear

The term 'coating' can apply to the adherence of a textile membrane to the fabric surface, or to a coating of micro or nanoparticles that adhere to the fibre surface, forming a layer. The particles are so small that they are not visible on the fabric and do not make a lot of difference to the fabric's handle. To achieve a breathable coating, a polymeric layer is applied in liquid form in a solvent or water base, which evaporates off to leave the polymer behind, applied to one or both surfaces of the fabric. The thickness of the coating, or amount of product applied, is controlled. Where a thicker coating is required, this may be built up by applying successive layers. Bonding of the polymer to the fabric is achieved by a drying process or through a curing process.

The breathable coatings make the fabric ideal for a broad range of uses including sportswear, outdoor pursuits and high-performance protective work wear. With the development of nanotechnology, breathable polyurethane coatings have advanced thanks to microporous structures.

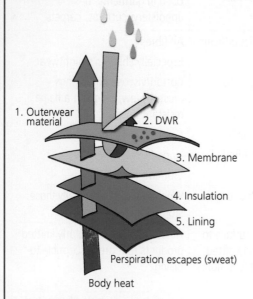

1. Outerwear material
2. DWR
3. Membrane
4. Insulation
5. Lining
Perspiration escapes (sweat)
Body heat

Figure 6.26 Breathable to allow moisture vapour and heat to escape

Transparent coatings on fine fabrics

Silicone coatings provide unique functional benefits in a variety of apparel items. Their anti-slip properties make them suitable for strapless bras and stockings. This coating provides an invisible layer that enables the fabric to adhere to the skin so there is no slippage. It also gives the fabric increased strength to reduce tearing.

Coatings can be applied to fine wool fabrics used for tailoring to incorporate an advanced moisture management system which creates a 'hydrophyllic' surface to improve the wicking qualities of the fabric. Moisture is conducted away from the body and dissipated through the fabric. The quick-drying properties of the finish keep the cloth drier and more comfortable, even at summer temperatures.

Nanotechnology can be used on delicate and luxury fabrics to create a unique finish that repels liquids and resists stains and yet retains the luxurious handle and drape expected of the finest fabrics. Liquids bead up and roll off and even ground-in stains wash out easily, keeping fabrics looking new for longer. The nanopolymer coatings can be sprayed onto a surface to create a completely water- and oil-repellent material. This prevents the accumulation of water on the material, which prevents the growth of bacteria.

ACTIVITIES

1 Examine a printed fabric that is currently on trend for either fashion or soft furnishing purposes. Explain what has influenced the fabric designer in the creation of the print. What method of printing has been used and why?

2 Research the benefits of using biological techniques to create fabric finishes.

3 What are the benefits to the manufacturer and consumer of using digital printing?

4 Explain the difference between mechanical and chemical finishing with examples of how and why they are used.

5 Explain the difference between coating and lamination of fabrics with examples of how and why they are used.

PRODUCT DESIGN

Surface finishes and coatings

Materials are often finished or coated in order to protect a product or material against wear, dirt, damage, corrosion and decay, and enable easy maintenance by dusting, washing or polishing. They also can be used to change and enhance the appearance of the end product by applying a surface coating or painted finish.

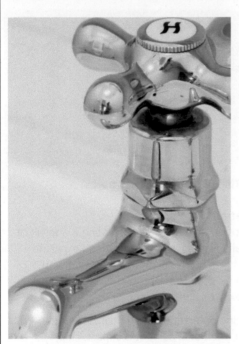

Figure 6.27 Bathroom fittings are often chrome plated to improve appearance and durability; the surface finish also protects against corrosion in the hot steamy environment

Any product designer or design engineer will need to consider if a surface finish is required to protect a material or appropriate to enhance appearance. The first stage of any finishing process is preparation.

Most woods decay or rot if they aren't protected in some way. Surfaces should be made smooth using a smoothing plane and/or glass paper; dents and scratches will look worse if painted or varnished over. Metals containing iron are prone to rust, and those that don't can oxidise, so they need protecting, cleaning or polishing if the original finish is important. Surfaces need to be degreased and cleaned with white spirit before painting or applying a plastic coating. Polymers are generally self-finishing but textures are often used in plastic moulded products to achieve different surface finishes.

Using finishes to prevent corrosion and decay

Varnishes and painting

Polyurethane varnish and lacquers give an attractive and hardwearing finish to woods. Available as clear or colour, with a matt, satin or high-gloss finish, they provide a tough surface finish that can be both heat- and water-proof and will stand up to hard knocks. They can be applied with a brush or spray. Metals are sometimes lacquered to protect the shiny surface finish.

Finishing timber

Timbers are sanded before a finish is applied. Glass paper, garnet paper and wire wool are used to prepare the surface for painting, oiling or varnishing. Sanding sealer is a solvent based product that seals the surface and raises the wood fibres so they can be sanded. It is often used before applying a varnish or wax.

Paint and varnish

Paint and varnish provide colour and protection and are used on both wood and metal. Acrylic paints are water-based and are generally not waterproof. Oil-based paints are more expensive but are waterproof and tougher. Polyurethane paint is particularly hardwearing. Softwoods tend to be painted whereas hardwoods cover less well and look more interesting when left naturally. Several different types of paint are available from flat non-shine (matt) through satin to very high gloss.

Non-toxic paints and varnishes are available for children's toys and furniture. A painted or varnished finish sits on the surface of a material and requires maintenance as the coating can peel or flake off.

Wax/polish and sealants

When a natural appearance on wood is required, one method is to apply oil to the wood surface. This highlights the wood's own colour and grain, making it water-resistant with a non-shine finish. There are many types of oils but commonly used are:

- olive oil – used as a finish when the wood is going to come into contact with food (for example salad servers)
- Danish or linseed oil – can be used on most woods; it needs two to four coats for the best protection
- teak oil – as the name suggests, this is ideal for such woods as teak and iroko; it is based on linseed oil with additives such as silicone to give a harder wearing surface.

Beeswax polish is often used in interior wooden furniture to achieve a natural-looking finish. Wax finishing will not stand up to heat and many liquids may stain or mark the final surface. Specialist hard wax oils can be used on floors; the benefit of these over varnish is that the wax can be reapplied without the need for re-sanding and preparing the wood.

Figure 6.28 Polished wooden floor

Preservatives

The effectiveness of any preservative depends on the penetration. Brushing or spraying is not recommended because it only gives superficial surface protection to the timber. These methods are, however, useful where timber needs to be treated in situ and can extend the life of a product by several years if applied regularly. The internal structure of the wood can be impregnated with a preservative, by means of pressure treatment.

Methods of preserving timber include:

- creosote (tar oil) – this is highly water repellent which gives it excellent weathering characteristics. Creosote-treated wood usually has an odour, but creosote is an excellent preservative for applications such as bridges, telephone poles, railway sleepers and marine piles
- immersion treatments – timber is fully immersed in a bath of preservative. Hot and cold bath treatment in open tanks is a more controlled method of immersion treatment; timber is immersed in the bath and the temperature is raised to about 85°C then allowed to cool, or the timber is transferred to a separate cold preservative bath
- pressure impregnation – wood is placed in a cylinder which is then sealed. A vacuum is applied and the cylinder flooded with preservative. The pressure is raised until the timber refuses to absorb further preservative
- Tanalising – this involves impregnation with Tanalith E, an environmentally friendly wood preservative that is applied in a vacuum pressure timber impregnation plant. The chemicals become chemically fixed into the timber and cannot be removed.

Figure 6.29 Timber used for railway tracks

Powder coatings

The rapid growth of wood powder coating is a result of technological advances and manufacturing expertise. Finishes can have the quality of a grand piano with a smooth high gloss, or simulate solid surface granite in a smooth or textured appearance.

Powder coating on manufactured boards is more durable than traditional paint. It is impact, chip, temperature and stain resistant, so it is well suited to hot, wet conditions like gym changing rooms and kitchen and bathroom cabinets because it won't fade in sunlight or warp in humidity. It involves the use of an electrostatic charge, both positive and negative, which offers optimum adhesion in the bond and significant strength.

Finishes for metals

Metals are also finished by painting or lacquering. The metal needs to be prepared and cleaned before a paint finish is applied. Paints can be water, solvent or oil based. Enamelling involves a drying or curing process when, after spraying, the component is baked in an oven at a temperature of 150–200°C. This process creates heat resistant parts for metal products such as stoves or radiators. Enamelling is also used for decorative jewellery where powdered glass is used to create decorative coatings (vitreous enamel).

Some non-ferrous metals have an attractive surface finish but still require polishing or buffing. Polishing removes deep scratches with a rough abrasive then continues using finer abrasives. Buffing can be done by hand or with a polishing wheel; different compounds are used with different mops to achieve a desired finish. Polymers are also polished in the same way.

Figure 6.30 Powder coating wood requires substantial investment in state-of-the-art technology and expertise

Table 6.5 A polishing compound colour chart

	Polymers			Silver, gold and thin polymers			Nickel and chrome plate			Copper, brass, aluminium, pot metal and other soft metals			Steel and iron			Stainless steel		
Buff type	A	B	C	A	B	C	A	B	C	A	B	C	A	B	C	A	B	C
Sisal										X			X			X		
Spiral sewn								X			X			X			X	
Loose												X			X			X
Canton flannel				X			X											
String	X	X	X															
Compound	A	B	C	A	B	C	A	B	C	A	B	C	A	B	C	A	B	C
Black										X			X			X		
Brown											X							
White							X					X	X					
Blue	X	X	X			X			X						X			
Green																	X	X
Red						X			X						X			

Dip coating/plastic coating

Plastic coatings are generally applied to metals using either plastisol (a liquid vinyl) or powder (fluidised bed). Metal parts are cleaned and degreased then preheated, dipped and then heated. During the dip, heat in the parts gel the surrounding plastic material. The hotter the metal parts and the longer the dip, the thicker the gelled coating. During the 'cure', the plastic coating fuses. A fluidised bed is a common way of coating a product in plastic and many schools and colleges have plastic dip coaters that use a fluidised bed. Plastic powder is fluidised by gently blowing air into the bottom of the tank. Before the powder is fluidised it is about as penetrable as fine beach sand. Once the air is switched on, the bed rises about 30 per cent and the plastic powder looks like it is boiling at the surface.

Metal is preheated to a temperature to suit the thickness of the material and dipped into the fluidised bath. The plastic melts onto the hot metal and coats the surface. After the part is withdrawn, the residual heat in the metal fuses the coating and creates a smooth surface. For this to happen, the metal must be heated to a sufficiently high temperature to be able to retain enough heat after fluidising. Otherwise the part is returned to the oven for a very short time.

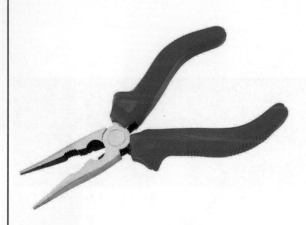

Figure 6.31 These pliers have plastic-coated handles

Electroplating

Electroplating involves the coating of an object with a thin layer of metal by use of electricity. The metals commonly used are gold, silver, chromium, tin, nickel and zinc. The object to be plated is usually a different metal, but can be the same metal or a non-metal, such as a plastic part for a car.

Electroplating usually takes place in a tank of solution containing the metal to be deposited on the work. This metal is in a dissolved form called ions. An ion is an atom that has lost or gained one or more electrons and

is thus electrically charged. You cannot see ions, but the solution may show a certain colour; a nickel solution, for example, is typically emerald green. The object to be plated is negatively charged and attracts the positive metal ions, which then coat the object to be plated and regain their lost electrons to become metal once again.

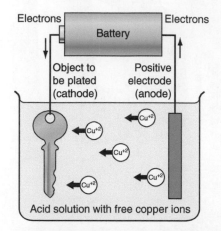

Figure 6.32 Positively charged copper ions are free in the solution, but are being attracted by the negatively charged key

Table 6.6 **Plating is done using many different metals**

Metal plate	Metal plated over	Uses and applications
Gold plating	Copper Silver Nickel	Jewellery, electrical connectors, printed circuit boards
Silver plating	Tin Antimony Copper	Jewellery, cutlery, candlesticks, musical instruments
Chrome plating	Steel Nickel Copper	Car parts, bath taps, kitchen gas burners, wheel rims
Zinc plating	Steel	Screws, nails, bolts, roofing sheets, outdoor building hardware
Tin plating	Copper Nickel Iron	Food cans

Galvanising

Galvanising is a process in which steel is immersed in a molten bath of zinc to provide the steel with a metallurgical bond between the zinc and the steel, creating a barrier of protection from the elements.

Figure 6.33 Hot dip galvanising

The process consists of three steps: surface preparation, galvanising and final inspection. Molten zinc will not react or bond properly to the steel if it is not perfectly clean. Caustic is a hot alkali solution that removes surface dirt, paints, grease and oil. It is followed by a hot sulphuric acid bath, more commonly referred to as the pickle, which removes mill scale and rust. The final step in the cleaning process is the flux. The flux tank contains aqueous zinc ammonium chloride. The flux prevents further oxides from forming on the steel surface prior to galvanising. Between each step in the cleaning process, the material is rinsed in water to prevent cross-contamination.

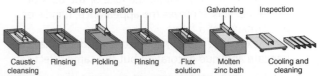

Figure 6.34 A typical galvanising line

ACTIVITIES

1 In groups, discuss the factors that a manufacturer must consider before deciding on the most appropriate surface finish for metal components.
2 Look at some domestic kitchen appliances. Consider the reasons why a surface finish is necessary on metal panels used in domestic kitchen appliances.
3 Find examples of wooden and metal products used outdoors and identify the material and what, if any, surface finish has been used.

KEY TERMS

Anodising – coating (a metal, especially aluminium) with a protective oxidised layer by electrolysis.

Artisan – a person skilled in an applied art; a craftsperson. A person or company that makes a high-quality or distinctive product in small quantities, usually by hand or using traditional methods.

Bio-stoning and bio-polishing – biological finishes that use natural enzymes to change a fabric's appearance.

Colour fast – refers to dyed fabric that will not fade or lose colour when washed.

Colour/print forecaster – an individual who attempts to accurately forecast the colours, fabrics and styles of fashionable garments and accessories that consumers will purchase in the near future, approximately two years ahead.

Dye sublimation and transfer – methods of digital printing.

Electroplating – involves the coating of an object with a thin layer of metal by use of electricity.

Galvanising – a process in which steel is coated with zinc.

Hydrophilic – substances that have an affinity for water, often because of the formation of hydrogen bonds.

Oxidation process – the chemical combination or reaction of a substance with oxygen.

Photorealistic – resembling a photo in a very realistic way.

Resist dyeing – a traditional method for dyeing fabrics; something is applied to the fabric to prevent the dye from entering the fabric in order to create a pattern.

Shisha embroidery – a type of embroidery which attaches small pieces of mirror to fabric.

Vat dyeing – dyeing that takes place in a bucket or vat.

6.3 What opportunities are there through using smart and modern technologies within products?

LEARNING OUTCOMES

By the end of this section you should have developed a knowledge and understanding of:

Design Engineering

- how smart materials change the functionality of engineered products, such as:
  - colour changes, shape-shifting, motion control, self-cleaning and self-healing.
- how programmable devices are used to add functionality to products relating to coding of and specific applications of programmable components, such as:
  - how they incorporate enhanced features that can improve the user experience and solve problems in system design
  - how they use basic techniques for measuring, controlling and storing data and displaying information in practical situations
  - electronic prototyping platforms and integrated development environments (IDE) for simulation in virtual environments
  - the use of programmable components and microcontrollers found in products and systems such as robotics arms or cars
  - creating flowcharts to describe processes and decisions within a process to control input and output components.

Fashion and Textiles

- how smart materials change the functionality of products, such as:
  - colour changes using thermochromic, photochromic and electrochromic fibres
  - shape-shifting such as shape memory alloy
  - breathable membranes like Gore-Tex.
- how e-textiles are innovative wearable textiles that incorporate conductive fibres or elements directly into the textile itself, integrating functional performance into products. Consider developments, such as:
  - a range of conductive threads and pigments
  - fibretronics
  - a range of programmable controllers using a range of sensors
- how technical textiles are developed for a range of industry sectors including:
  - geotextiles used in civil engineering, coastal engineering and the construction industry
  - fabrics for hi-tech clothing, such as superfabric.

Product Design

- how smart materials change the functionality of products, such as:
  - colour changes, shape-shifting, motion control, self-cleaning and self-healing
  - smart materials used in medical procedures to act in a way that conventional materials and processes would not previously have permitted.
- how modern technologies can support the function of products, such as:
  - programmable components that can be built into a product and coded to respond to inputs that command an action.

Smart materials

A smart material is a material that exhibits a useful property, reacting in a controlled way to a specific **stimulus**. Smart materials can be used to add functionality to a product, sometimes avoiding the need for complex or possibly expensive techniques such as electronics. For example, a product could be designed to be worn on a user's wrist, which uses **photochromic** pigments to change colour when exposed to the ultra-violet light present in sunlight; therefore, the product could indicate to the user when sun cream should be applied to protect their skin.

Colour changes

Thermochromic materials change colour in response to a change in temperature. The basic thermochromic pigment can be integrated into a number of materials including paint, polymers and textile threads and fabrics. There are different technologies available to achieve this; the commonly available pigments contain a dye which is strongly coloured at lower temperature but which becomes colourless at higher temperatures.

The temperature boundary is not very accurate but usually occurs above room temperature and can provide a useful indication of 'too hot', 'too cold' or 'temperature OK'.

Figure 6.35 Colour-changing product

The pigment is usually mixed with another pigment (or a colourless carrier material) so that, at low temperature, the colour produced is that of the base pigment mixed with the thermochromic pigment, and at higher temperatures the colour is that of the base pigment alone.

Table 6.7 Colour changes produced by mixing thermochromic pigments with other colours

Thermochromic pigment	Base pigment	Colour when cold	Colour when hot
Blue	Yellow	Green	Yellow
Orange	White	Pale orange	White
Green	Clear	Green	Transparent

ACTIVITY

Using thermochromic pigments, design and make a colour-changing sign to indicate a hot surface. For example, when it is hot the sign might display 'HOT' or display an icon to indicate 'Don't Touch', or the sign might display 'SAFE' when it is cold; there are lots of possibilities. The sign could even be painted directly onto a polystyrene drinks cup.

Photochromic materials

Photochromic materials change colour in response to a change in light level. In a similar way to thermochromic materials, the photochromic material is mixed with a carrier medium which may contain another pigment. A well-known application of photochromics is in spectacle lenses which automatically darken to become sunglasses in bright sunlight.

Photochromic pigments and dyes can be mixed with paints or solvents to produce a wide range of finishes, including colour-changing nail varnish. Polymers are available which are colourless in artificial light but which transform rapidly to a deep colour when exposed to sunlight. Photochromic threads are available for creative applications in colour-changing textiles products.

Electrochromic materials

Electrochromic materials change colour when an electric voltage is applied to them. They can be found in 'smart glass' – a material whose transparency can be controllably changed by an electric signal. Smart glass is increasingly being used in windows, mirrors and privacy screens.

- In the health and medical industry, temperature indicators can be used to monitor changes in body temperature. Colour-change fabric that reacts when exposed to UV radiation, for example, enables completely new products in the field of adaptive sun protection.
- In the military, a chameleonic effect can be achieved to match with the surroundings. Colour changing camouflage nets can be used to provide protection depending upon the appearance of the surrounding environment.
- Novel promotional products like T-shirts can be created. Can be used on garments to print brand name or logo to prevent its duplication.

DESIGN ENGINEERING AND PRODUCT DESIGN

Controllable tinted glass used for windows in buildings can reduce the need for expensive and unreliable mechanical systems to open ventilators on sunny days and eliminates the need for blinds or curtains, providing new design opportunities in architecture and interior design. Future developments include car windows which automatically darken to reduce solar heating in the cabin; additionally, all the glass can become completely opaque when the car is locked to help keep the car cool on a sunny day and to increase security.

Electroluminescent (EL) panels are thin, flexible sheets which, when connected to a small battery-powered 'inverter', give out a pleasant soft glow in various colours. EL panels can be cut to shape, making them attractive to designers for signage or to improve the user-interface by providing illumination to a product.

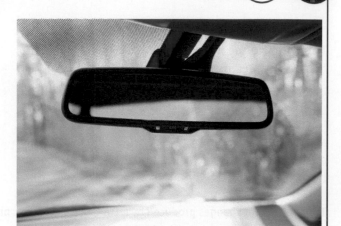

Figure 6.36 **Some cars have rear-view mirrors that automatically darken in the presence of a car's dazzling headlights**

ACTIVITY

Make an illuminated badge by cutting a shape from an EL (electro luminescent) panel and placing this behind a piece of translucent acrylic. Connect the EL panel to an inverter power pack and the acrylic will light up; some inverters can even be set to flash. Take care not to receive an electric shock – the inverter generates a high voltage. Read the manufacturer's safety instructions.

Shape-shifting

Shape-shifting materials change their shape in response to a physical stimulus such as an electric voltage or **current**, a change in temperature or the presence of a magnetic field. Some of these materials will expand, contract or twist and can be used to construct artificial muscles which find applications in robotics. Tiny fluid pumps and valves are being developed to mimic or replace biological systems. Research engineers are conducting experiments on shape-shifting aircraft wings to copy the way birds flex their wings to control flight.

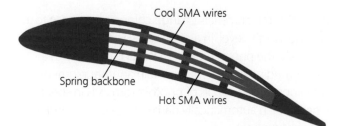

Figure 6.37 **Operating principle of a shape-shifting aeroplane wing**

Shape memory alloys (SMA) are already widely available. 'One-way' SMAs can be 'programmed' to remember a particular shape. They can be bent and twisted to a new shape, but when they are triggered by heating above the transition temperature they will return to their programmed state, exerting a large force as they do so. Two-way SMAs remember two different programmed shapes, one at low temperature and one at high temperature.

Shape memory polymers (SMPs) have similar properties to SMAs but with the added advantages of using polymers, in that they exhibit even greater elasticity, are low cost, are low density, have the ability to be moulded and have high chemical resistance. While the transition stimulus is usually a rise in temperature, there are SMPs which are triggered by light, electricity or magnetism.

DESIGN ENGINEERING

SMAs can be used in actuators to replace conventional motors although, as they are normally heated by passing an electric current through them, they are quite slow to respond as they take time to heat up and cool down.

Nonetheless, their simplicity, light weight, small size and low power have resulted in them being increasingly used in some quite specific engineering applications to replace mechanical actuators.

FASHION AND TEXTILES

- Healthcare – garments have been developed for the management of diabetes, to improve blood circulation, in burn management and for post-surgical compression to aid in recovery after a surgical procedure.
- Athletics – garments have been developed for improvement of oxygen circulation. An SMA component can control shock absorption at a given temperature.
- Emergency medicine – garments have been developed to tourniquet active wounds.

- Space and travel – compressive space suits have been developed to provide the required pressurisation to an astronaut's body. Leg compression bandages have been developed to deter deep vein thrombosis.
- Construction garments – body shape wear has been developed. Memory foam is used for bra cups for comfort.
- Fashion – clothing has been made to be affected by the user's heat. A shirt has been developed with vents on it that react to the high temperature of the body. These vents open, allowing air in so sweat can evaporate quickly.

PRODUCT DESIGN AND DESIGN ENGINEERING

Motion control

The **viscosity** of a liquid is a measure of its 'thickness' or its resistance to an object moving through it. Some liquids are rheopectic, which means that their viscosity increases as the liquid is stressed. So the faster an object moves through the liquid, the more viscous the liquid becomes and this has the effect of controlling the speed of the moving object. Such liquids are used in motion-control gels which can be applied to the bearings of moving objects to provide self-regulating speed control. This provides a very easy way to 'slow down' a spring-opening mechanism which would otherwise pop open uncontrollably, or a closing mechanism which would otherwise slam shut. Examples include soft-close toilet seats.

The viscosity of magnetorheological (MR) fluid can be controlled by applying a magnetic field. MR fluid can be

mixed with oil and used in dampers or shock absorbers, as shown below.

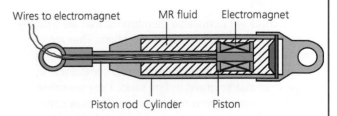

Figure 6.38 **MR fluid damper**

When the electromagnet is switched on, the viscosity of the MR fluid increases, making it 'thicker' which provides a greater resistance to the piston as it moves in the cylinder. Therefore, the damper can be made harder or softer under electronic control. Such systems are increasingly being used in the automotive industry to provide adjustable suspension stiffness, and in earthquake-damping systems for buildings.

Self-cleaning

Much research has been carried out into developing surfaces, or coatings for surfaces, which make it difficult for contamination to stick to the surface. Such surfaces are described as self-cleaning. The nanoparticles used to create **hydrophobic** (water-hating) surfaces are highly water repellent and dirt simply cannot stick to them; so, as water flushes across the surface it takes any dirt with it. Hydrophobic coatings can be applied to many surfaces, including glass or textiles, in the form of a 'stain protector'. This can help keep a textile product free of stains for longer but the coating tends to wash off eventually, especially if detergents are used to launder the product.

Smartphone and tablet screens have an oleophobic (oil-hating) coating which helps stop fingerprints sticking to the glass. Fingerprints are still deposited, but they can be wiped off with a cloth rather than being smeared around which would happen without the coating.

Much progress has been made on self-disinfecting, antimicrobial surfaces which are ideal for use in hospitals, kitchens or areas such as public toilets where there is potential for germ transfer. Embedding small amounts of copper or silver into the surface gives it antimicrobial properties which makes it hygienic and contamination free. Similar antimicrobial products can be impregnated into textiles for use in nurses' uniforms.

Self-healing

Self-healing materials have the ability to repair themselves if they become damaged. A variety of different technologies currently exist. Some polymers have a tendency to re-form molecular bonds between adjacent edges when they are punctured or cut. You will probably have used a so-called 'self-healing' cutting mat in Design and Technology lessons. Over a period of time, the small scores in these mats are able to heal themselves.

Self-healing concrete uses limestone-producing bacteria to repair small cracks as they develop. The bacteria is held dormant inside micro capsules within the concrete which dissolve and release the bacteria when water enters the cracks. A new asphalt containing fine steel fibres is being developed for road manufacture. This material can be heated and softened by electromagnetic induction to heal small cracks before they develop into potholes. The induction device is fitted to a lorry which drives slowly over the surface of the road.

Self-healing technologies are being developed to manufacture bullet-resistant fuel tanks and for use in space vehicles to repair the microscopic puncture holes created as the vehicle encounters space particles.

Figure 6.39 **A Self-healing cutting mat**

Technical textiles developed for a range of industry sectors

Geotextiles

Geotextiles are permeable membranes that are typically made from manufactured synthetic fibres. They are available in a range of thicknesses, widths and lengths. The permeability of geotextile sheets is comparable in range from coarse gravel to fine sand. They are either woven from continuous monofilament fibres or non-woven and made by the thermal or chemical bonding of **continuous fibres** and pressed through rollers into a relatively thin fabric. They are used in contact with soil or rocks and have the ability to separate, drain, filter, reinforce or protect.

Geotextile products can be applied to numerous engineering tasks including airfields, roads, railways, retaining structures, canals, bank protection, reservoirs, coastal engineering and dams.

Figure 6.40 **Geotextile applications**

Breathable fabrics

Some fabrics are said to be hydrophobic, such as acrylic, polyester, nylon and Nomex®. This means they resist water. If a drop of water is put on the fabrics, it will sit there until it evaporates. Therefore, these fabrics are particularly unsuitable for sportswear, for example, as there is nowhere for perspiration to go. The fabrics can be combined with a natural fibre or can be coated with chemicals to make the water pass through the fabric. One example of this is breathable ripstop nylon. This is extremely durable and lightweight due to the fabric's structure but it can then be coated to make it breathable.

To take this one step further, the fabrics can also be developed to make them not only breathable but also waterproof. This results in a fabric that is suitable for a wide range of applications. One of the most well-known versions, and the first to be introduced, is Gore-Tex. The use of expanded polytetrafluoroethylene (ePTFE) as a thin membrane (0.01 mm thick) enabled the company to introduce innovative outerwear fabric. The ePTFE membrane is laminated to an outer shell fabric to create a two-layer textile that can then be cut and stitched. Most of the time, a third layer is added to further improve durability. This makes these fabrics particularly suitable for outerwear, high-performance clothing, footwear, sleeping bags and accessories, and they have even been used for astronaut spacesuits.

E-textiles

Electronic textiles (e-textiles) are innovative textile materials that incorporate conductive fibres in the fabrics, yarn or thread. This enables wearable computing, or electronic devices worked into garment designs. The technology uses materials that eliminate the need for wires and hard electronics, making the product more wearable and useable. There are many kinds of conductive textile materials available, from yarns to woven and coated fabrics. E-textiles can be used in a wide range of applications. They can be used for data storage or provide a physical interface through control elements in the fashion or textile product.

Figure 6.41 Example of an e-textile

Conductive pigments

One of the main issues with incorporating electronics within the structure of textiles is that electronics incorporate materials that are not as flexible as fabric. One solution is being able to print onto the fabric with a **conductive pigment**, as this opens the possibility of producing comfortable wearable electronics. It enables the creation of 'smart' functional fabrics without limitations resulting from motion, folding and wearing. The end result is a thin, form-fitting circuit that can be seamlessly fused with standard fabrics, allowing for comfort and freedom in wearable electronics design.

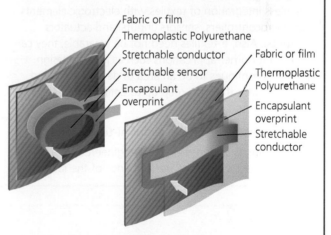

Figure 6.42 Stretchable thin membrane electronics

Printed electronic materials usually include a combination of conductive silver, carbon and silver chloride inks. Conductive ink technologies are compatible with many surfaces, including polyester, and can be applied using screen printing as well as other processes. Varying the viscosity of a conductive ink makes it possible to print on textiles. The ability to print directly on textiles gives the ability to integrate electronic devices, **sensors** and actuators on textile substrates or pre-fabricated garments.

Fibretronics

Fibretronics explores how electronics can be integrated into textile fibres. This requires the construction of electronic capabilities on textile fibres using conducting and semi-conducting materials. The electronic components are inserted into the fibres, making them invisible. There are a number of commercial fibres that include metallic fibres mixed with textile fibres to form conducting fibres that can be woven or sewn.

One of the most important issues of e-textiles is that the fibres should be washable. Electrical components need to be insulated during washing to prevent damage. The fabric should carry the transmission lines and connectors so that clothing can be flexible and washable enough to be wearable.

Use of programmable microcontrollers with a range of sensors

Electronic textiles are different from wearable computing because the emphasis is placed on the seamless integration of textiles with electronic elements like microcontrollers, sensors (inputs) and actuators (outputs). Also, e-textiles need not be wearable; they can be used for other applications such as interior design. E-textiles continues to be an area of development and it is important that you keep up to date with the latest developments.

The development of fabrics for hi-tech clothing

Hi-tech fabrics are ones that are engineered for a wide variety of uses where the performance of the fabric is the major factor. They are commonly used for all active wear, sportswear, summer and winter wear, mountain activities, trekking, work wear, military and protective wear.

Developments include a self-healing fabric. The principle behind the fabric is a special coating that, when damaged, melts at a very low temperature to seal the gap and restore the waterproof properties. During testing, the fabric could stand up to 100 scratches with a razor blade before the waterproofing was seriously compromised, and over 200 wash cycles.

SuperFabric is created with a base fabric such as nylon, polyester, neoprene, crepe, etc. and is overlaid with tiny, hard guard plates in a specific pattern. Spacings between the guard plates allow a degree of flexibility and breathability and are small enough to keep most sharp objects from penetrating. This guard plate technology protects the base fabric and contributes to the durability of the material. The geometry, thickness and size of the guard plates, as well as the base fabric, vary depending on industry requirements.

The sports industry is one area that constantly demands the latest technology in fabrics. Nike is just one manufacturer among many to have developed a number of hi-tech fabrics to fulfil specific functions when worn. Just a few examples of these are Nike Dri-FIT and Flyknit.

Dri-FIT is a high-performance, microfibre, polyester fabric that moves sweat away from the body and to the fabric surface, where it evaporates. As a result, Dri-FIT keeps athletes dry and comfortable.

Flyknit, made entirely of polyester yarn in a precise knit construction process, is a featherweight, form-fitting and virtually seamless shoe upper. The knit varies for more flexibility or breathability where required and is tighter in areas that demand support. Flywire cables are also knit-in for enhanced support.

ACTIVITIES

1 Keep up to date with the latest developments in the fashion and textiles industry. Create a folder and collect articles from newspapers, magazines and the internet. Include further investigations into some of the smart materials and technologies mentioned in this chapter Make sure you understand how the developments function and what the benefits are of particular applications.
2 Explain the benefits and restrictions of e-textiles within the fashion and textiles industries.
3 Study the development of geotextiles and how they have benefitted civil engineering, coastal engineering and the construction industry.

KEY TERMS

Actuators – electrical devices that control the flow of material or power.

Conductive pigment – an ink that results in a printed object which conducts electricity.

Continuous fibres – otherwise known as filament fibres. These are opposite to staple fibres and are long in length.

Current – a measure of the actual electricity flowing.

Electrochromic – able to change colour when charged with electricity.

Fibretronics – explores how electronic and computational functionality can be integrated into textile fibres.

Geotextiles – any permeable textile materials used to increase soil stability, provide erosion control or aid in drainage.

Hydrophobic – able to repel or unable to mix with water.

Photochromic – able to change colour due to light stimulus.

Sensors – electronic components that detect changes in their environment and send the information to other electronics.

Shape memory polymers (SMPs) – smart materials that have the ability to return from a temporary shape to their original shape when activated by an external trigger, such as temperature change.

Stimulus – the physical quantity that causes a smart material to change.

Thermochromic – able to change colour due to external temperature stimulus.

Viscosity – the thickness of a liquid or a measure of how resistant a liquid is to flowing.

KEY POINTS

- Smart materials provide many characteristics enabling more effective use of a product and can be developed for specialised applications.
- Thermochromic, photochromic and electrochromic fabrics can change colour in reaction to external factors.
- Shape memory is the ability of a product to remember its original shape upon application of an external stimulus such as chemicals or temperature.
- To make fabrics breathable or waterproof they can be combined with a natural fibre or coated with chemicals to make the water pass through the fabric.
- To ensure a fashion or textiles product is truly waterproof, the seams must be sealed.
- Electronic textiles (e-textiles) are innovative textile materials that incorporate conductive fibres in the fabrics, yarn or thread.
- The ability to print directly on textiles using conductive pigments gives the ability to integrate electronic devices, sensors and actuators on textile substrates or pre-fabricated garments.
- Geotextiles are used in contact with soil or rocks and have the ability to separate, drain, filter, reinforce or protect.
- Hi-tech fabrics are ones that are engineered for a wide variety of uses where the performance of the fabric is the major factor.

Programmable devices

Programmable **microcontrollers** are electronic devices which can be **embedded** into a product or a system to add functionality. Designers are increasingly using the potential of programmable microcontrollers in products. Robotics, for example, make full use of a microcontroller's ability to interface to sensors and actuators and to process information at very high speed. A modern car may contain up to 100 microcontrollers to control the various electrical and engine systems in the vehicle and these microcontrollers will also be able to communicate and share information with each other.

There is a booming market in wearable technology such as intelligent watches, smart glasses, fitness/activity monitors, health monitors and jewellery. Electronics manufacturers are working with the fashion industry to create ergonomic and fashionable technology which is highly desirable. Such technology has the ability to be connected wirelessly to networks, uses very little power so batteries last a long time and, in the case of clothing, can even be washed.

Programmable microcontrollers offer a number of advantages. They

- can add functionality and control to a product or system, yet they are relatively simple to use
- can interface to a huge range of **peripheral devices** including sensors and output devices, and a host of other units such as Global Positioning Systems (GPS), Wi-Fi, accelerometers (to detect movement), mobile phones, the internet etc.
- can, with the appropriate choice of sensor, measure physical quantities
- can interface to displays so that information and data can be clearly presented
- can be networked with other devices, allowing products to communicate with each other and with the internet
- can store information even when the power is switched off, allowing them to memorise settings or quantities
- are small, relatively low cost and will run off small batteries, enabling compact products to be developed

- require minimal equipment to develop a fully-functioning solution, just a computer and a connecting lead, and the integrated development software (IDE) is usually free to download
- can be reprogrammed which is vital during the iterative designing stages and can be useful once the product is in use for later product upgrades.

A microcontroller needs a **program** (sometimes called the code), which is a unique set of instructions written by the designer, to tell the microcontroller what to do in their product. The program is developed with the aid of a computer but the program itself resides within the microcontroller which is embedded in the product, so, once the program is finalised, the computer is no longer required.

DESIGN ENGINEERING

Designing a microcontroller system

When planning to use a microcontroller, it is essential to draw a **system diagram** which shows the various input sensors and output devices and how they are interconnected to the microcontroller. For example, suppose we wish to design a road safety product which uses white and red light emitting diodes (LEDs) and is controlled by a single push switch:

- 1st press of switch turns on the white LEDs
- 2nd press of switch flashes the red LEDs
- 3rd press of switch turns all LEDs off.

The system diagram might look something like this:

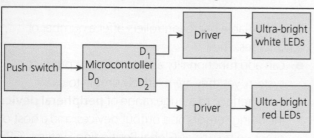

Figure 6.43 System diagram for the road safety device

Notice that the input (push switch) is always on the left, the outputs (LEDs) are always on the right, with the microcontroller always in the centre. The lines show signals and the arrows show the direction in which the signals travel. Signals are electrical voltages, and can be digital or analogue:

- A **digital signal** is either on or off, nothing in between.
- An **analogue signal** represents the value of a **physical quantity**, such as temperature or weight. It can have any voltage between zero volts and the **power supply** voltage.

It is important to identify if the signals are digital or analogue as different signals need connecting to different **input/output (i/o) pins** on the microcontroller and your program will need to deal with each type of signal in a different way. In this system, all the signals are digital, because the push switch is either on or off and the LEDs will either be fully on or completely off, in this application at least.

The **driver** is a device to convert the signal from the microcontroller into an electric current to operate the LEDs. Further technical information about electronic systems can be found in Section 6.6.

The program should initially be drawn as a flowchart.

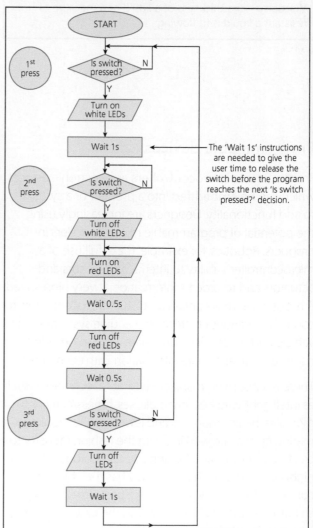

Figure 6.44 Flowchart program for the road safety device

Flowcharts are a simple way of breaking down the microcontroller task. A microcontroller can only perform one step at a time, in sequence. When the program is run, each step is carried out extremely quickly, so the user would not notice any delay between the steps even though they are only executed one at a time. Drawing a flowchart helps the designer think like a microcontroller.

Later, the flowchart is of great help when writing the actual program for the microcontroller in the appropriate programming language.

Notice the range of symbols used for drawing flowcharts:

Symbol	Name
 ⬭	Start/end
 →	Arrows
 ▱	Input/Output
 ▭	Process
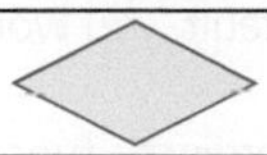 ◇	Decision
○	Connector

Figure 6.45 Flowchart symbols

Designing a microcontroller system

In this next example, suppose we are designing a cooling fan which runs at a speed dependent on the room temperature – the hotter the room, the faster the fan. For this application we require a temperature sensor and an electric motor to drive the fan, so the system diagram would look like this:

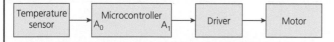

Figure 6.46 System diagram for a cooling fan

The signal from the temperature sensor is analogue (because temperature can have any value) and, since we want to vary the speed of the fan, the output signal is also analogue. So we require a microcontroller with at least one analogue input and one analogue output.

The flowchart for this system might look like this:

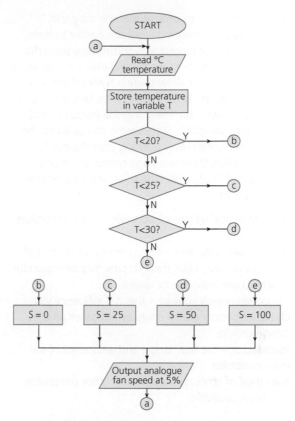

Figure 6.47 Fan flowchart

Choice of programmable device for project work

The range of devices available is constantly developing and each type has its advantages and disadvantages. Many microcontrollers have been specifically designed for use in experimental projects and this makes them ideal in the iterative design process where you will continually be testing and improving. Some are specifically aimed at beginners, others are aimed at textile or wearable technology projects. Some are intended to be connected temporarily with crocodile clip leads, others with conductive thread and some of them will need you to design your own printed circuit board (PCB) and solder them in place.

It is necessary to download and install software onto a computer to help you develop a program for the microcontroller. This software is called an **integrated development environment (IDE)** and, apart from the program editor, it will probably contain other useful tools including, perhaps, a simulator which lets you see the effect of running your program without actually uploading it to the microcontroller, a debugger to help you find errors in your program, a monitor which lets you see the values of parameters inside the microcontroller when it is running, and resources such as a program language library and links to online support.

Each microcontroller has its own programming language, and some are able to use more than one language allowing the designer to choose the one they feel comfortable with. Some languages are harder to learn, but even the simplest language is surprisingly powerful and will be sufficient for many A Level projects. The more complex languages will allow more advanced projects and permit greater product functionality, but only if you are prepared to invest time learning the language and debugging your program. Start simple and build up the complexity as you iteratively develop your design. For all microcontrollers there is a huge online community offering tutorials and project ideas, so you can draw on these for support.

The main technical differences between microcontrollers boil down to:
● the number of i/o pins – if you plan on using a lot of switches or many LEDs then i/o pins may be important
● which i/o are analogue or digital
● the power supply voltage – this will influence your choice of battery which could affect other product design criteria
● form factor – the size, shape and aesthetics of the microcontroller
● the range of sensors, outputs and other peripherals which are available.

Success with microcontrollers

Getting a microcontroller to perform a simple function should be relatively easy – lights, colours, sounds and simple movement should be quickly achievable even by beginners. More advanced functions require some experience and time to learn some more advanced programming skills, and time spent debugging programs when they don't work! So don't try to run before you can walk. Remember, as with many aspects of Design and Technology, you don't need to learn everything at once – master the basics and draw upon deeper technical information as and when you need it.

Tips for success:
● start every project with the most basic function – for example, flashing a single LED
● only make one change at a time to your project, and always test after each change – if it stops working, undo the change and get it working again
● there is plenty of support online and someone has probably already done the thing you are trying to do, so search for answers to your problems – this is all part of the iterative designing process
● you might be tempted to copy a code you find online and paste it into your own program – this is OK but try to understand how the code works, otherwise, if your project develops a software fault, you won't have a clue how to fix it!
● ask independent users to test your project's function – they will almost certainly do things you would not predict so they are ideal for finding program bugs you didn't know existed.

ACTIVITIES

1 Experiment with a microcontroller. If you have never used one before, choose a microcontroller that can be connected with crocodile clips. Install the IDE on a laptop PC if you need to. Start by programming the microcontroller to flash a single LED. Then flash a few LEDs in a repeating pattern. Program it to play a tune.

2 Search for online tutorials to help you develop your programming skills and increase your electronics knowledge.

KEY TERMS

Analogue signal – a voltage which can be any value.

Digital signal – a voltage which can only have two values.

Driver – a current amplifying subsystem which ensures an output device operates correctly.

Embedded – when a microcontroller, acting as a miniature computer, is programmed to solve a particular design problem and then placed permanently into the product or system.

Input/output (i/o) pins – the microcontroller connections to the peripheral devices.

Integrated development environment (IDE) – software installed on a computer to help develop and upload the microcontroller program.

Microcontroller – an electronic device that can be programmed to carry out specific tasks.

Peripheral devices – devices that connect to a microcontroller to complete the whole system.

Physical quantity – a physical property that can be quantified by measurement. Examples include temperature, light, force, density, sound, length and position.

Power supply – the voltage supply which powers the electronic system.

Program (or code) – the set of instructions that tell the microcontroller what to do.

System diagram – a diagram that shows the interconnections and signal flow between the microcontroller and peripheral devices.

6.4 How do mechanisms provide functionality to products and systems? (Design Engineering only)

LEARNING OUTCOMES

By the end of this section you should have developed a knowledge and understanding of:
- the functions that mechanical devices offer to products, providing different types of motion, including:
  - rotary
  - linear
  - reciprocating
  - oscillating

- devices and systems that are used to change the magnitude and direction of forces, including:
  - gears, cams, pulleys and belts, levers, linkages, screw threads, worm drives, chain drives and belt drives
  - epicyclic gear systems
  - bearings and lubrication
  - efficiency in mechanical systems.

The basic machine principle

A mechanism is a system designed to control forces and motion. Controlled motion is essential in a variety of engineered products. An ideal mechanism transfers **power** without adding to or subtracting from it. Since power is the result of multiplying force and speed, this means that a mechanism simply trades off forces against movement, so:

> (input force x input speed) = (output force x output speed)

Therefore, a mechanism can either:
- reduce speed but increase force, or
- increase speed but reduce force.

The ratio of the output force to the input force is called the **mechanical advantage** (MA) of the mechanism:

$$MA = \frac{output\ force}{input\ force}$$

For mechanisms that don't continuously move (for example, a lever) it is more useful to consider the energy transfer, rather than power transfer, in which case:

> (input force x distance moved by input) = (output force x distance moved by output)

Types of motion

Motion can be broadly classified into four different types:
- rotary motion, which follows a circular path, such as revolving bicycle wheels or the output shaft from a motor
- linear motion, which follows a straight line path, such as a moving robotic vehicle or a conveyor
- reciprocating motion, which moves back and forth in a straight line, such as the blade on a jigsaw
- oscillating motion, which moves back and forth in a circular path, such as the head of an electric toothbrush.

Devices and systems used to change the magnitude and direction of forces and torques

Design engineers make use of a huge variety of systems to generate, control and change motion. Rather than trying to memorise every possible combination of mechanism and motion, try to build on the knowledge you gained at GCSE and develop an understanding of the principles and 'rules' of the basic mechanisms so that you are able to apply these to analyse the more advanced mechanisms you may encounter.

Levers

A lever consists of a bar which pivots on a fulcrum. The input force is often called the effort and the output force is called the load.

$$\text{mechanical advantage (MA)} = \frac{load}{effort}$$

By applying the **basic machine principle**:

(effort x distance moved by effort) =
(load x distance moved by load)

you should be able to see that it is also true to say:

$$MA = \frac{distance\ moved\ by\ effort}{distance\ moved\ by\ load}$$

which, by considering the geometry of a lever, is also the same as:

$$MA = \frac{input\ arm\ length}{output\ arm\ length}$$

(Note that the arm length is always the right-angled distance between the force and the fulcrum.)

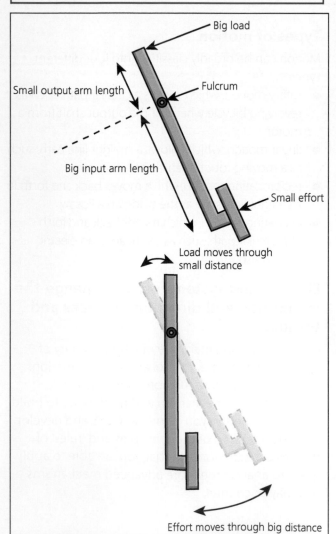

Figure 6.48 **A brake pedal is a simple lever**

$$MA = \frac{input\ arm\ length}{output\ arm\ length}$$

which equates to 4. So an effort of 75 N would be amplified to a load of 300 N.

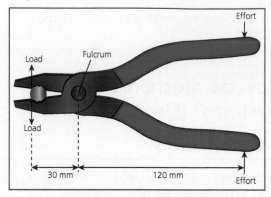

Figure 6.49 **Pliers are a pair of simple levers**

Depending on the relative positions of effort, load and fulcrum, a lever can amplify or reduce forces, and also reverse the direction of motion.

ACTIVITY

Study the lever classes in the diagram and complete a summary table like the one below.

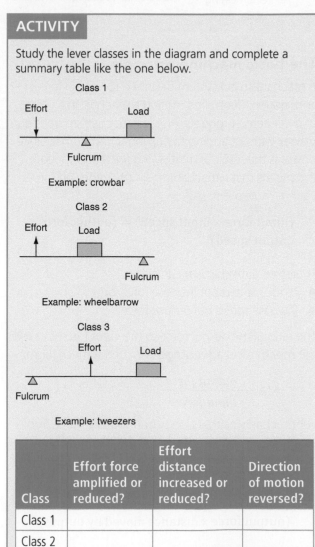

Class	Effort force amplified or reduced?	Effort distance increased or reduced?	Direction of motion reversed?
Class 1			
Class 2			
Class 3			

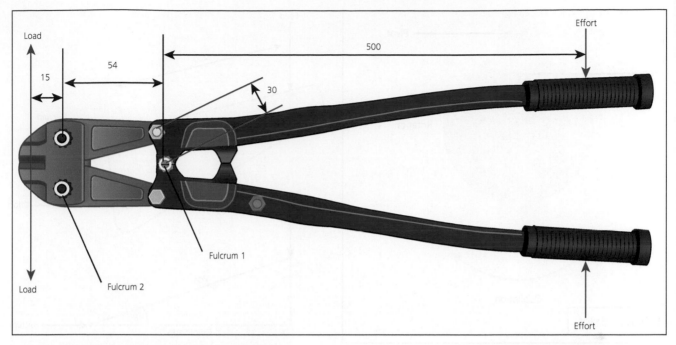

Figure 6.50 A pair of compound levers in a bolt cutter

A compound lever consists of two (or more) simple levers working together, usually so that the input force is amplified by a very large amount. The overall mechanical advantage is:

overall MA = (MA of lever 1) × (MA of lever 2)

An example of this can be found in a bolt cutter, as shown in Figure 6.50.

The overall mechanical advantage is:

$$MA = (\frac{500}{30} \times \frac{54}{15}) = 60$$

Therefore, an effort of 250 N will be amplified to a load of 15 kN.

Linkages

A linkage is an assembly of parts used to direct forces and movement to where it is needed. A linkage will often change the direction of motion, and it can also be used to convert between **types of motion**. In many cases, the individual parts of a linkage behave like a lever and can be analysed as such. Some examples of linkages are shown in Figure 6.51.

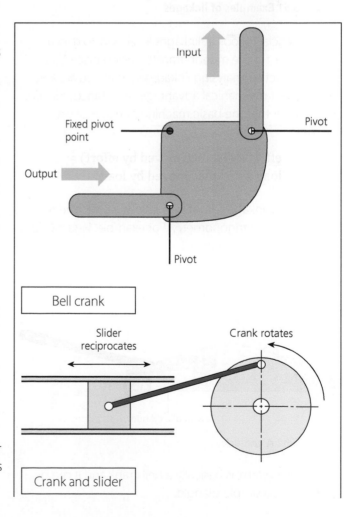

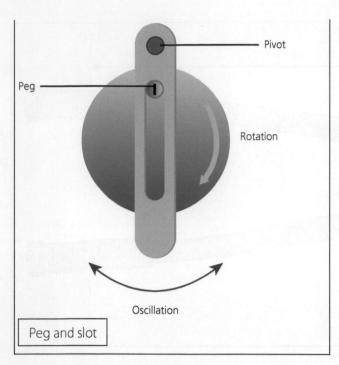

Figure 6.51 Examples of linkages

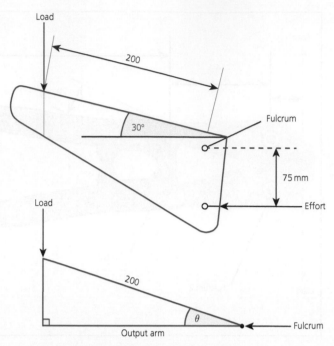

Figure 6.53 The bell crank linkage in the jack

It is beneficial to look at linkages 'in action' to gain a better 'feel' for the motion transfer taking place. You should practise analysing linkages so that you are able to work out mechanical advantage or distances moved. Remember that the basic machine principle always applies:

> (effort x distance moved by effort) =
> (load x distance moved by load)

Analysing a linkage sometimes involves using the rules of geometry or trigonometry. For example, Figure 6.52 shows a linkage in a car jack.

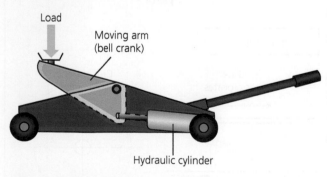

Figure 6.52 A car jack

The 'moving arm' is basically a bell crank and it can be redrawn as a simple triangle.

This can be analysed as a simple lever, but remember that the arm length is measured at right angles to the force so, while the input arm length is clearly 75 mm, the output arm length needs to be calculated using trigonometry:

$$\cos \theta = \frac{output\ arm\ length}{200}$$

$$output\ arm\ length = 200 \times \cos (30)$$

$$= 173\ mm$$

Therefore, the mechanical advantage of the linkage is:

$$MA = \frac{75}{173} = 0.433$$

So, for a load of 3000 N, the effort would be 6928 N

Using the basic machine principle, we can also approximate how much the load will be lifted if the effort moves by 1 mm:

$$distance\ load\ moves = \frac{distance\ effort\ moves}{MA}$$

$$= 2.3\ mm$$

(In this linkage the load is less than the effort, but the trade-off is that the load moves more than the effort. Also, the reason this is only an approximation is that, as the arm moves, the angle θ will change which also causes the MA to change, but it is a good approximation if θ is quite a small angle.)

ACTIVITIES

1 A bicycle braking system is shown below.

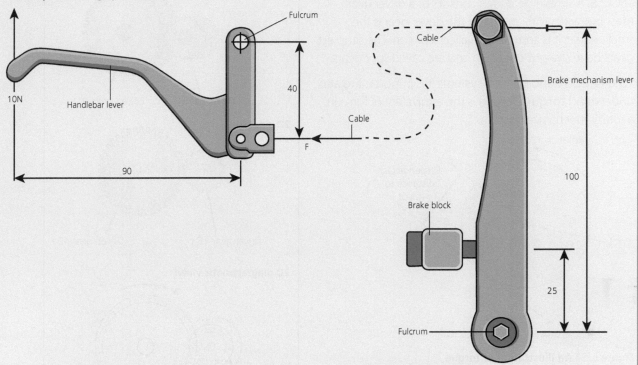

a) Calculate the force F exerted on the brake cable.
b) The cable transfers force F to the end of the brake mechanism lever. Calculate the force applied to the brake block.

2 A lifting arm is supported by a hydraulic ram, as shown below.

a) Calculate the force exerted by the hydraulic ram. (Hint: use trigonometry to work out the input and output arm lengths, then work out the mechanical advantage.)
b) Calculate the approximate distance the load will be lifted if the ram extends 10 mm. (Hint: use the basic machine principle.)

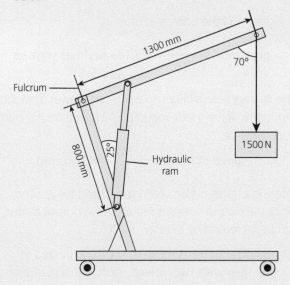

Gears

You will have been introduced to simple gear systems at GCSE. A simple gear train consists of a driver gear meshing with a driven gear. If the driver gear is the smaller gear, it is sometimes called a pinion. The simplest gears have straight-cut teeth and are called spur gears.

When understanding gear systems it is important to also understand **torque**. Torque is the equivalent of force in a rotary mechanical system.

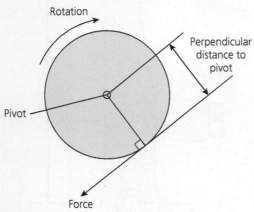

Figure 6.53 An illustration of torque

torque = force x perpendicular distance to pivot

The units of torque are newton-metres (Nm).

You may have noticed that torque is the same quantity as 'moment'. Moment is usually the term used to describe the 'turning effect' in stationary systems and torque is usually the term used in rotating systems, but the words may be used interchangeably.

The basic machine principle applies as always, but, in a rotating system, we deal with torque instead of force:

(input torque × input speed) = (output torque × output speed)

There are two further simple rules to remember:
- The smaller gear will rotate faster than the larger gear.
- The gears will rotate in opposite directions.

The **gear ratio** is defined as:

$$\text{gear ratio} = \frac{\text{number of teeth on driven gear}}{\text{number of teeth on driver gear}}$$

Once this ratio is known, the above rules and the basic machine principle can be used to work out speeds and torques.

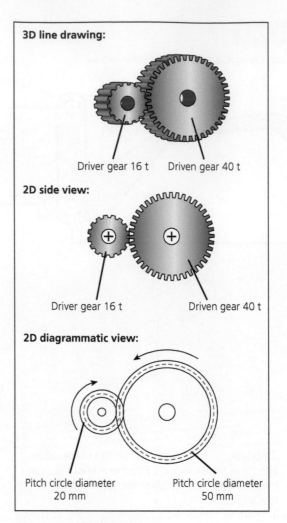

Figure 6.54 Simple gear train

In Figure 6.54 the gear ratio is:

$$\text{gear ratio} = \frac{40}{16} = 2.5 \text{ (sometimes written as 2.5:1)}$$

The output gear is larger so will rotate slower than the input gear. If the input speed is 500 rpm then the output speed will be:

$$\text{speed of output} = \frac{500}{2.5} = 200 \text{ rpm}$$

There is a trade-off between speed and torque and, since the output speed is reduced, the output torque will be increased by a factor of 2.5.

When two spur gears mesh they can be treated as though they were two wheels, each with a diameter called the pitch circle diameter.

The pitch circle diameter is used in order to determine the separation of the drive shafts. In Figure 6.54, the input and output shafts would be separated by a distance:

$$\frac{20}{2} + \frac{50}{2} = 35 \text{ mm}$$

For two gears to mesh properly they must have the same tooth pitch (the same number of teeth per millimetre around their circumference). This is specified by the module of the gears:

$$module = \frac{pitch\ circle\ diameter\ (in\ mm)}{number\ of\ teeth}$$

which, in Figure 6.54 would be:

$$module = \frac{50}{40} = 1.25$$

Gears with a larger module have fewer teeth for a given diameter of gear.

A compound gear train consists of several pairs of gears working together, usually so that there is a very large overall gear ratio.

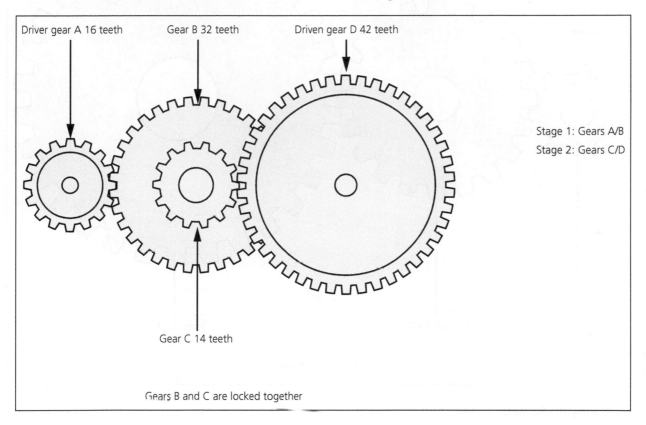

Driver gear A 16 teeth   Gear B 32 teeth   Driven gear D 42 teeth

Stage 1: Gears A/B
Stage 2: Gears C/D

Gear C 14 teeth

Gears B and C are locked together

Figure 6.55 Compound gear train

The overall gear ratio of the compound gear train in Figure 6.55 is:

overall gear ratio = (gear ratio of stage 1) × (gear ratio of stage 2)

$$= (32/16) \times (42/14)$$

$$= 6$$

For an input speed of 3000 rpm the output would turn at:

$$output\ speed = \frac{3000}{6} = 500\ rpm$$

Compound gear trains are used extensively in mechanical systems, such as the gearbox on the DC motor shown in Figure 6.56.

Figure 6.56 This motor/gearbox features a compound gear train

An idler gear is a gear which is inserted between two other spur gears.

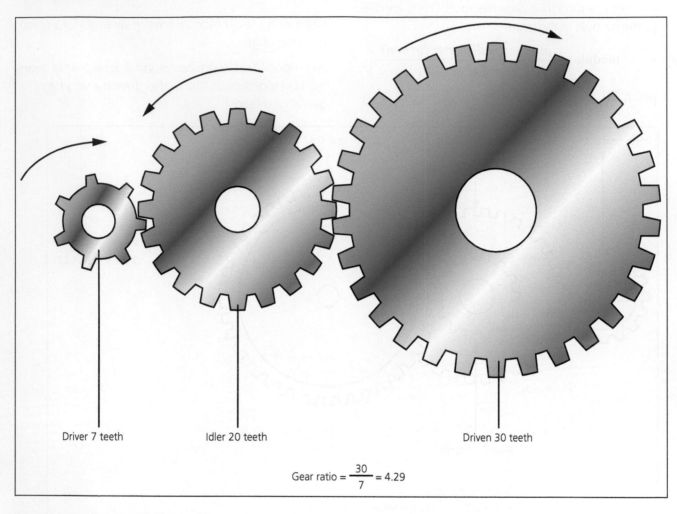

Driver 7 teeth   Idler 20 teeth   Driven 30 teeth

$$\text{Gear ratio} = \frac{30}{7} = 4.29$$

Figure 6.57 Using an idler gear

The idler gear has no effect on the overall gear ratio, but it will reverse the direction of rotation of the output shaft. Idler gears are sometimes also used to increase the spacing between the input and output shafts without affecting the gear ratio.

ACTIVITIES

A compound gear train is used on the output of a DC motor as shown in the diagram.

1   Calculate the rotational speed of the output.
2   State the direction of rotation of the output.
3   Calculate the module of the gears.
4   Calculate the separation of shafts B and C.

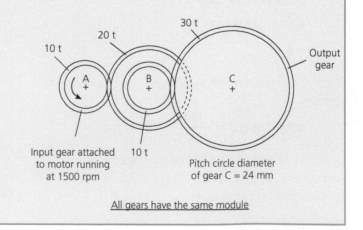

Input gear attached to motor running at 1500 rpm

10 t

Pitch circle diameter of gear C = 24 mm

All gears have the same module

Spur gears are used for many applications, but you should be aware of other types of gear and the reasons for using them.

- Bevel gears can be used to transfer the direction of the drive shaft in a mechanical system, usually by 90°, but other angles are possible.

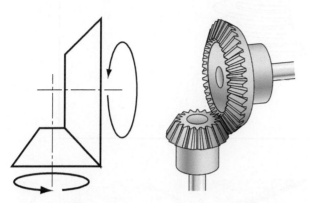

Figure 6.58 Bevel gears

- Mitre gears are a pair of identical bevel gears, providing a 1:1 gear ratio.
- Helical gears have their teeth cut at an angle and this allows the teeth to mesh more gradually as the gears rotate (compared to spur gears, where the teeth meet suddenly as they rotate).
- A rack and pinion is a system to change between rotary and linear motion. In a stairlift, the rack is stationary and the pinion 'climbs' up it. In applications such as automatic sliding gates, the pinion stays in position and the rack moves.

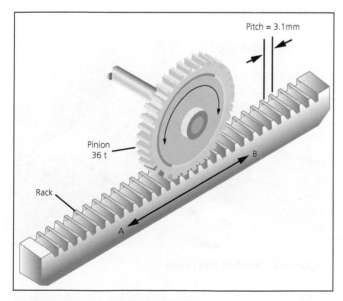

Figure 6.59 Rack and pinion system

- To calculate the distances moved it is necessary to know the number of teeth on the pinion and the **pitch** of the rack. In the diagram shown, for every revolution of the pinion, the rack will move along 36 teeth which is a distance of: 36 x 3.1 = 111.6 mm.

Figure 6.60 Helical gears (with angle-cut teeth) run quieter than spur gears (with straight-cut teeth)

Worm drives

A worm drive is a system in which a worm screw engages with a worm wheel (which resembles a spur gear, although the teeth are cut concave to match the curve of the worm screw) to produce a very high gear ratio in a compact system. The drive direction is also transferred through 90°.

The simplest worm screw is called a 'single start' (it has a single thread running along its length) and for every revolution of the worm screw, the worm wheel will advance by one tooth. Therefore, the reduction ratio is simply the number of teeth on the worm wheel, for example, a worm wheel with 60 teeth can achieve a huge 60:1 speed reduction in a single stage.

Worm drives are used extensively in mechanical systems but they require the input and output shafts to be held precisely in position. Worm drives experience large frictional forces between the teeth and, therefore, tend to wear quickly. An advantage of this high friction, however, is that the worm drive is 'self-locking', meaning that whilst the input shaft can drive the output, the output cannot drive the input, which is useful in some applications.

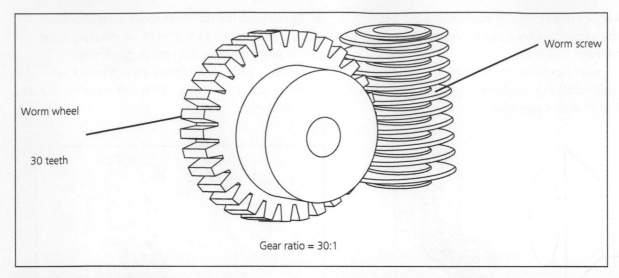

Worm screw

Worm wheel

30 teeth

Gear ratio = 30:1

Figure 6.61 **Worm drive system**

Chain and sprocket drive

A chain and sprocket drive is similar to a spur gear system, with the following differences:

- The input and output shafts can be placed a long distance apart, linked by a chain.
- The input and output shafts rotate in the same direction.

Figure 6.62 **Chain and sprocket drive**

$$\text{gear ratio} = \frac{number\ of\ teeth\ on\ output\ sprocket}{number\ of\ teeth\ on\ input\ sprocket}$$

Belt and pulley drives

A belt and pulley drive works on the same principle as a chain drive, but belts tend to be quieter and are often used in lower load applications. A common type of belt drive uses a V-shaped belt and pulley to provide a large area of contact and, therefore, less likelihood of slippage.

$$\text{gear ratio} = \frac{diameter\ of\ output\ pulley}{diameter\ of\ input\ pulley}$$

Figure 6.63 **V-belt in use**

A toothed belt and pulley is used where slippage must never occur, such as in a laser-cutter drive, as shown in figure 6.64.

Figure 6.64 **Toothed belt in use**

When using chains or belts, provision must be made to keep the tension correct as the chain/belt will stretch with use. If it is too slack there is a risk of a belt slipping or a chain jumping off the sprockets, and if it is too tight there are high friction losses. One of the pulleys can be made movable, or a tensioner can be used to achieve this.

Screw threads

A screw thread is another mechanism which changes rotary motion into linear motion. A linear actuator is a device in which a screw thread is rotated by a motor, and this causes a nut (which is not allowed to rotate) to slide along the screw.

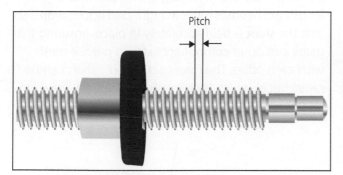

Figure 6.65 The principle of a lead screw

For a single-start thread, every rotation of the screw causes the nut to move along by one pitch distance. For a thread with a 2 mm pitch, the number of rotations needed to move the nut through a distance of 10 mm would be:

$$\frac{10}{2} = 5 \text{ rotations}$$

Figure 6.66 Lead screw in use

A screw thread used in this way is sometimes called a lead screw and these can be found in CNC (computer numerical control) machines and 3D printers where accurate linear positioning is required. In such

applications, the screw would be turned by a stepper motor so that precise and controlled rotation is achieved.

Cams

A cam and follower is another mechanism to convert rotary to linear (usually reciprocating) motion. With a little thought, you should be able to predict the type of linear motion produced by a particular shape of cam, such as those shown Figure 6.67:

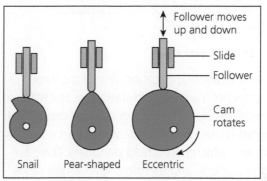

Figure 6.67 Three of many cam types

The dimensions of the cam determine the following parameters:
- stroke – the maximum distance that the follower rises
- rise interval – the number of degrees the cam rotates to produce the full stroke
- dwell – the number of degrees of rotation during which no follower movement occurs.

For example, in Figure 6.68:
- the stroke will be 28 – 15 = 13 mm
- the rise interval will be 90°
- the dwell will be 180°.

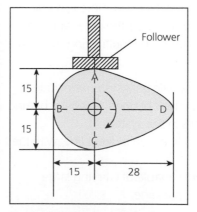

Figure 6.68 A pear-shaped cam

Epicyclic gears

A simple epicyclic gear train consists of three main components.

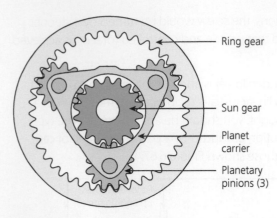

Figure 6.69 Epicyclic gear train

In an epicyclic gearing application, one of the three main components must be held stationary. Epicyclic gear trains can produce different gear ratios depending on which part is used as the input, which part is used as the output and which part is held stationary. There are several combinations, and therefore several gear ratios that can be obtained from a single system. However, in practice, only certain combinations are usually used and these are summarised in the table below:

Table 6.8 Epicyclic gear train combinations

Input	Output	Stationary	Gear ratio
Sun	Planet carrier	Ring	Speed reduction
Planet carrier	Ring	Sun	Speed increase
Sun	Ring	Planet carrier	Reverse and speed reduction

If power is fed in to any two components at the same time, the entire gear train rotates as one, providing a 1:1 gear ratio.

As with ordinary gear systems, two or more epicyclic gear trains can be placed one after the other and the overall gear ratio can be found by multiplying together the gear ratios from each stage.

The benefits of using epicyclic gear systems include:
● several gear ratios can be obtained simply by braking different components in the system
● the output shaft is co-linear (in a direct line) with the input shaft
● they are very rugged, capable of carrying large torques in a compact package.

For all these reasons, epicyclic gear trains are used in cars with automatic transmission, and they can also be found in turbine engines, electric drills and as a compact gearbox on some miniature DC motors.

Bearings and lubrication

A bearing is a component which supports a moving part and allows it to move only in the desired motion, with little friction.

Bearings are most commonly used to support a drive shaft so it can rotate freely. A drive shaft needs to be supported at a minimum of two points along its length (sometimes more in high-load applications) so that the shaft is held accurately in place, ensuring that gears and other components stay in precise mesh with each other. The forces acting on a shaft can be radial forces which try to push the shaft sideways, or axial forces which try to push the shaft along its axis.

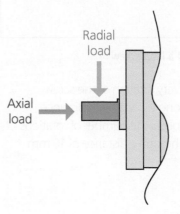

Figure 6.70 Forces acting on a shaft

A plain bearing is made by simply passing the shafts through a hardwearing material which also offers low resistance as the shaft rubs against it. Several polymers make popular plain bearings, such as nylon, PTFE or polyacetal. A lubricant such as oil or graphite may be used, or the bearing material may be self-lubricating, such as bronze which has the property of releasing low-friction material as the bearing wears away.

Ball bearings consist of a series of metal balls running between two races. The inner race turns with the shaft; the outer race is held stationary within the bearing housing. The balls roll (rather than slide) offering very little friction and low wear. Ball bearings handle radial loads well but cannot deal with axial loads; a variation is the roller bearing which can handle even higher radial loads.

Figure 6.71 Ball bearing

Both types of bearing are lubricated with grease and may have seals covering the exposed bearings, in which case they are described as sealed bearings. Seals keep the lubricant in and grit and other contamination out.

Figure 6.72 A sealed ball bearing

Tapered roller bearings, due to their cone shape, are able to support axial as well as radial loads. A thrust bearing uses balls or rollers in a slightly different configuration and is used to support axial loads only.

Figure 6.73 A tapered bearing can support axial and radial loads

For very high-performance, high-speed applications, fluid bearings can be used to support the shaft within a sleeve of high pressure oil or air. Such bearings provide exceedingly low friction and almost no wear.

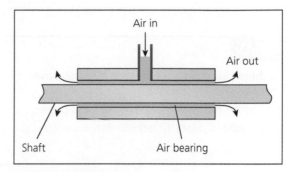

Figure 6.74 Fluid bearing

Efficiency in mechanical systems

An ideal mechanism transfers power without adding to or subtracting from it. This implies that it does not lose energy due to friction, or by generating sound or heat, or by flexing the parts or wearing them away. Of course, even with lubrication, friction is always present in real mechanisms and this leads to the output power being less than the input power. The **efficiency** of a mechanical system is:

$$\text{efficiency} = \frac{output\ power}{input\ power}$$

which is then multiplied by 100 to get a percentage.

Throughout this section we have been referring to the basic ideal machine principle:

$$(\text{input force x input speed}) = (\text{output force} \times \text{output speed})$$

which stems from the fact that:

$$(\text{output power}) = (\text{input power}).$$

However, if energy is lost in the mechanism, the output force (or torque in a rotary system) will be less than the above principle suggests.

If the input and output forces (or torques) are measured experimentally, then the actual mechanical advantage can be calculated:

$$\text{actual mechanical advantage} = \frac{measured\ output\ force}{input\ force}$$

The efficiency of the mechanism can then be calculated by:

$$\text{efficiency} = \frac{actual\ mechanical\ advantage}{ideal\ mechanical\ advantage}$$

which is multiplied by 100 to get a percentage.

The efficiencies found in typical gear systems are:
- epicyclic gear train 97%
- three-stage compound gear train 94%
- worm drive could be as low as 20%.

6.5 What forces need consideration to ensure structural and mechanical efficiency? (Design Engineering only)

LEARNING OUTCOMES

By the end of this section you should have developed a knowledge and understanding of:
- static and dynamic forces in structures and how to achieve rigidity, including:
  - tension, compression, torsion and bending
  - stress, strain and elasticity
  - mass and weight
  - rigidity
  - modes of failure.

Static and dynamic forces and how to achieve rigidity

A structure is a collection of parts which provide support and rigidity for the range of static and **dynamic forces** that may be applied. Within a design solution, a structure may be a separate system or it may be integrated into other parts within the design.

Force, mass and weight

The SI unit of force is the newton (N).

The **mass** of an object describes the amount of material the object contains. Mass is measured in kilograms (kg).

The **weight** of an object is the force of gravity pulling down on the object. Since weight is a force, weight is measured in newtons (N).

There is sometimes confusion between mass and weight. 'My weight is 75 kg' is a scientifically incorrect thing to say, but we all know what is meant by this. When solving problems in Design Engineering it is important to be specific – are we dealing with mass (in kg) or weight (in N)?

The link between mass and weight is gravity. On Earth, gravity pulls down on each kilogram with a force of 9.8 newtons, so a 9 kg dumbbell will weigh 88.2 N.

weight = mass × gravitational field strength

On Earth, gravitational field strength = 9.8 N kg^{-1}

So:

weight of an object (in N) = mass of the object (in kg) × 9.8

Static and dynamic forces

A wide range of loads (forces) will be exerted on a structure during use. For example, consider the flagpole in Figure 6.75.

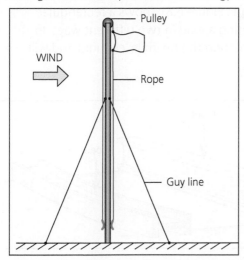

Figure 6.75 Consider the static and dynamic forces on this flagpole

A **static force** is a load which is always present and does not move or change in magnitude. An example of a static force is the weight of the flagpole and the downward forces applied by the guy lines. It is relatively easy to predict and calculate the magnitude of static forces. Dynamic forces are loads which may move and may change in magnitude or direction during use. For the flagpole, dynamic forces would include wind loading and the intermittent force applied when a flag is raised up the flagpole using the pulley. Dynamic forces can be hard to predict, so a design engineer must sometimes use their 'best guess' based on research of similar structural systems. A safety factor is usually applied, which means that the structure would be designed to safely take dynamic loads much greater than predicted. A safety factor of 2 is normal; for safety critical systems a safety factor of 4 is used. Aircraft structures are designed with a safety factor of 6+.

ACTIVITY

Identify the static and dynamic forces involved with the following structural systems:
- a chair
- a cordless drill
- a flying drone.

Types of forces in structural members and modes of failure

The different members within a structure are there to support different kinds of forces, which could be tension, compression, torsion or shear forces. Each kind of force will affect a structural member in a different way, causing the member to flex due to the elasticity of the material used. If the force becomes too great, the member will fail and the mode of failure depends on the type of force in the member, as explained below. If the member is in equilibrium (i.e. it is stationary within the structure) the forces will be balanced and will often act in pairs.

A tie is a structural member which resists tension. Tensile forces try to stretch the member. As the tension increases, the tie will extend slightly until it ultimately fails by snapping. A tie will extend less if it is thicker or made from a stiffer material, and its ultimate strength depends on its thickness and the tensile strength of the material. As they are always in tension, ties are often made from wires or cables.

Figure 6.76 A tie resists a tension force

A strut is a structural member which resists compression. Compressive forces try to squash the member. As the compression increases, the strut will shorten slightly, but it will also tend to buckle. Struts nearly always fail by buckling. To minimise buckling, struts can be made shorter or they can be braced along their length to prevent sideways movement. A strut can also be made more rigid by changing its cross-sectional shape, or by using a stiffer material.

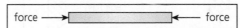

Figure 6.77 A strut resists compression

A beam resists bending. When a beam is loaded, as in Figure 6.78, it bends and the top surface is in compression while the bottom surface is in tension. A beam will eventually fail, usually by the bottom surface cracking under tension. The amount a beam deflects under load depends on how it is loaded and how it is supported. Shorter beams will bend less and a beam can be made more rigid by changing its cross-sectional shape, or by using a stiffer material.

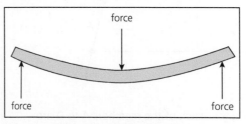

Figure 6.78 A beam under load

Torsional forces cause a member to twist. Just as with compression and bending, the amount of twisting can be reduced by increasing rigidity, which can be done by reducing the length of the member, changing its cross-sectional shape, or by using a stiffer material.

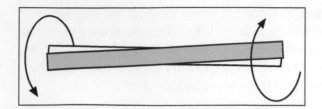

Figure 6.79 Torsion

Shear forces are scissor-like forces which try to slice through a component. The strength of an object in shear depends on its cross-sectional area and the shear strength of the material.

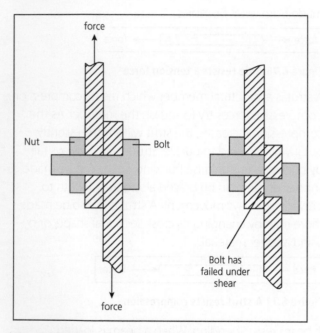

Figure 6.80 Shear forces acting on a bolt

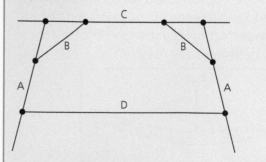
Increasing rigidity

For compression, bending or torsional forces, the rigidity of a structural member depends on its cross-sectional shape. Figure 6.81 shows a rectangular beam supporting a load in two different ways. In the second method the beam is less rigid and will deflect more.

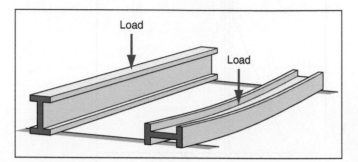

Figure 6.81 An I-beam in its stiffer and its more flexible orientation

Structural members can be made stiffer, using less material, by careful choice of cross section and consideration of how the member is loaded. This means the structure can be made lighter and cheaper. A selection of commonly used material cross sections is shown in Figure 6.82. Some will withstand bending evenly in any direction, others have a stiffer axis and a more flexible axis. Try to determine an order of stiffness for the materials; it is impossible to do this properly without dimensions, but trying to develop a 'gut-feeling' is a useful exercise.

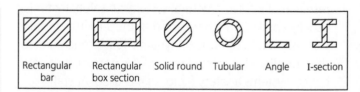

Figure 6.82 Commonly used material cross sections

The choice of cross section might also be influenced by other factors such as:
- how the structure is joined together
- other parts which may attach to the structure
- aesthetic implications.

Stress, strain, elasticity and Young's modulus

It is frequently necessary to calculate the amount by which a part will extend or compress when loaded. Consider a length of material under tension as shown in Figure 6.83.

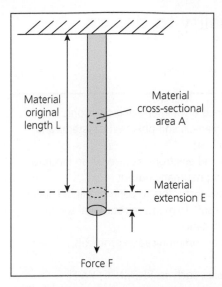

Figure 6.83 A material extending under load

The stress in the material is defined as:

$$stress = \frac{force}{cross\text{-}sectional\ area}$$

The units of stress are pascals (Pa). 1 Pa is equal to 1 N/m^{-2}.

This stress will cause the material to undergo strain (stretching):

$$strain = \frac{extension}{original\ length}$$

Strain has no units as it is a ratio of lengths.

Stress and strain are linked by **Young's modulus**:

$$Young's\ modulus = \frac{stress}{strain}$$

The units of Young's modulus are Pa.

Young's modulus describes the stiffness of a material, i.e. the opposite of its elasticity.

Young's modulus example

Mild steel has a Young's modulus of 200 GPa.

A steel wire 1.8 mm diameter and 1000 mm long has a force of 500 N applied to it. Calculate the extension of the wire.

The wire will have a circular cross section, and the cross-sectional area is:

$$area = \frac{\pi d^2}{4}$$
$$= \frac{\pi \times 0.0018^2}{4}$$
$$= 2.54 \times 10^{-6}\ m^2$$

$$stress = \frac{F}{A}$$
$$= \frac{500}{2.54 \times 10^{-6}}$$
$$= 1.97 \times 10^8\ Pa$$

Using the Young's modulus of steel we can now calculate the strain in the wire:

$$strain = \frac{stress}{Young's\ modulus}$$
$$= \frac{1.97 \times 10^8}{200 \times 10^4}$$
$$= 9.85 \times 10^{-4}$$

And the extension can then be calculated:

$$extension = strain \times original\ length$$
$$= 9.85 \times 10^{-4} \times 1000$$
$$= 0.985\ mm$$

KEY TERMS

Dynamic force – a force that might change position, size or direction.

Mass – the amount of material the object contains (kilograms).

Static force – a force that does not change.

Weight – the force of gravity acting on an object (newtons).

Young's modulus – the resistance of a material to elastic deformation under load.

KEY POINTS

- The role of a structure is to provide support and rigidity in a design solution.
- A structure will need to withstand a range of types of static and dynamic forces.
- Mass and weight are linked by gravitational field strength.
- The rigidity of a structural member depends on a number of factors, including material choice, member size and shape and the nature of the applied force.

6.6 How can electronic systems offer functionality to design solutions? (Design Engineering only)

LEARNING OUTCOMES

By the end of this section you should have developed a knowledge and understanding of:
- how electronic systems provide input, control and output process functions, including:
  - switches and sensors to produce signals in response to a variety of inputs
  - programmable control devices
  - signal amplification
  - devices to produce a variety of outputs, including light, sound and motion
- the function of an overall system, referring to aspects, including:
  - passive components: resistors, capacitors, diodes
  - inputs: sensors for position, light, temperature, sound, infra-red, force, rotation and angle
  - process control: programmable microcontroller
  - signal amplification: MOSFET, driver ICs
  - outputs: LED, sounder, solenoid, DC motor, servo motor (A Level: stepper motor, piezo actuator, displays)
  - analogue and digital signals and conversion between them
  - open and closed loop systems including feedback in a system and how it affects the overall performance
  - sub-systems and systems thinking

- what can be gained from interfacing electronic circuits with mechanical and pneumatic systems and components, such as:
  - the ability to add electronic control as an input to mechanical or pneumatic output
  - the use of flow restrictors to control cylinder speed
  - the use of sensors to measure rotational speed, strain/force, distance
- networking and of communication protocols, including:
  - wireless devices, such as: RFID, NFC, Wi-Fi, Bluetooth
  - embedded devices
  - smart objects
  - networking electronic products to exchange information
- the basic principles of electricity, including:
  - voltage
  - current
  - Ohm's law
  - power.

Electronic systems

You will have gained a basic knowledge of electricity from GCSE science. You should recall that the voltage at a point in a circuit is a measure of the electrical 'pressure' trying to cause a current to flow and current is a measure of the actual electricity flowing. The units of voltage and current are volts (V) and amps (A), respectively, although milliamps (mA) are commonly measured in electronic systems:

$$1\,A = 1000\,mA$$

Ohm's law links **voltage**, **current** and resistance in the following formula:

$$voltage = current \times resistance$$

$$V = IR$$

Resistance is measured in ohms (Ω).

The power dissipated (given out) in an electrical circuit is given by the formula:

$$power = current \times voltage$$

$$P = IV$$

Power is measured in watts (W).

By substituting the Ohm's law formula into the power formula, we can produce two further formulae for power and both of these are very useful when working with resistors in circuits:

$$P = \frac{V^2}{R}$$

$$P = I^2R$$

Passive electronic components

Whilst many advanced electronic components are available for use in A Level Design Engineering, it is important to understand the use of the basic components. Resistors, capacitors and diodes are called passive components, which means that they cannot amplify signals or control other signals.

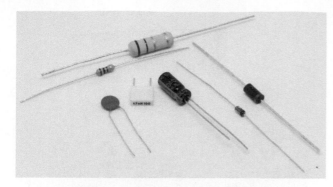

Figure 6.84 Resistors, capacitors and diodes

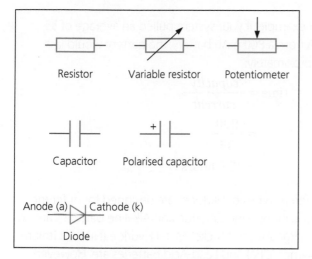

Figure 6.85 Circuit symbols for R, C and D

Resistors are used to control current flowing in parts of a circuit. By doing this, they will produce different voltage levels at different parts in a circuit, according to Ohm's law. Most resistors have a series of coloured bands on them to indicate their resistance (in ohms). The resistor colour code is shown in Figure 6.86.

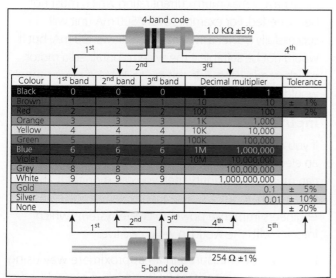

Colour	1st band	2nd band	3rd band	Decimal multiplier		Tolerance
Black	0	0	0	1	1	
Brown	1	1	1	10	10	± 1%
Red	2	2	2	100	100	± 2%
Orange	3	3	3	1K	1,000	
Yellow	4	4	4	10K	10,000	
Green	5	5	5	100K	100,000	
Blue	6	6	6	1M	1,000,000	
Violet	7	7	7	10M	10,000,000	
Grey	8	8	8		100,000,000	
White	9	9	9		1,000,000,000	
Gold					0.1	± 5%
Silver					0.01	± 10%
None						± 20%

Figure 6.86 Resistor colour code chart

Resistors have a maximum power rating, and exceeding this can damage the resistor. The power formulae are used to check that the rating has not been exceeded. For example, a 470 Ω 0.25 W resistor is connected across a 12 V power supply. The power it dissipates is:

$$P = \frac{V^2}{R}$$

$$P = \frac{12^2}{470} = 0.3 \text{ W}$$

This resistor is therefore under-rated and it should be replaced by one with a higher power rating, for example 0.5W.

Capacitors are components which store electrical charge. Capacitance is measured in farads (F), but units of μF or nF are more likely to be used in electronic components. As a design engineer, it is unlikely that you will be required to decide when to use capacitors, or to select a particular value. However, during your research and your design iterations you may find advice or instructions on how to use them to help supress the electrical noise generated from DC motors, or to help smooth out fluctuations in a power supply.

Different types of capacitor are available. The smaller values (<1μF) are often 'ceramic' or 'polyester' type, and these can be connected either way round in a circuit. The larger values (1μF and above) are usually 'electrolytic' type and these are polarised, meaning they have a positive and a negative lead and must be connected the correct way round, or they are likely to explode. The negative lead is sometimes shorter, and is indicated by '-' signs on the capacitor case.

Diodes are 'one way' components in that they only allow current to flow through them in one direction. They have an anode (+) lead and a cathode (-) lead. The cathode lead is indicated by a band or strip on the diode's case. Current can flow from an anode to a cathode, but not in the other direction. In Design Engineering, diodes have a few different uses:

- A diode inserted in the battery lead will prevent damage to the main circuit by blocking the current if the user accidentally connects the battery the wrong way round. This could prevent expensive damage to the microcontroller or other key components.
- Diodes prevent current flowing backwards. For example, in Figure 6.87 pressing switch A will light the red and green lamps, while pressing switch B will only light the green lamp. If the diode is replaced by a wire, pressing switch B will also light both lamps.
- Diodes remove 'back emf'. (See later in this section.)

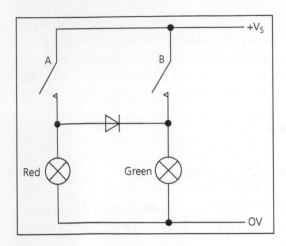

Figure 6.87 Diode circuit diagram

Figure 6.88 Power supplies used in electronic systems

Power supplies for electronic systems

An electronic system needs a power supply which can be provided by batteries or a mains-derived power supply such as a variable voltage bench **power supply unit (PSU)** or a 'plug-top' power supply. In some designs there are also options to use renewable energy sources such as solar panels or a wind generator, but these would probably be used in conjunction with rechargeable batteries, so for practical purposes they can be treated in the same way as a battery supply. During the experimental iterative stage of project development, the use of a bench PSU is strongly advised as these provide a reliable and stable supply voltage and also include safety features such as current-limiting which can protect you and your circuit if you make a wiring mistake.

The job of a power supply is to provide a steady voltage while pushing current through a system. If an electronic circuit is not working properly, the first thing you should do is monitor the power supply voltage, using a multimeter, and check that it stays stable as the circuit operates. Many faults are down to an unstable power supply which would need correcting before the system will work reliably.

If you do intend to eventually use batteries, you should research the options early on because their voltage, capacity, size and method of recharging/replacing all need to be considered.

Battery capacity is measured in milliamp-hours (mAh). This figure is roughly the number of milliamps of current you can draw from the fully-charged battery for one hour:

$$\text{battery capacity (mAh)} = \text{current (mA)} \times \text{time (h)}$$

For example, if your system pulled an average of 35 mA from a 800 mAh battery, the battery would last approximately:

$$\text{time} = \frac{capacity}{current}$$
$$= \frac{800}{35}$$
$$= 22.9 \text{ hours}$$

Different types of batteries are designed for different applications. Zinc-Chloride and Alkaline batteries are not rechargeable, but Nickel-Metal Hydride (NiMH), Lithium Polymer (LiPo) and Lead-Acid batteries are. However, you must research how to recharge your chosen battery and decide whether the battery will be removed from the product to be recharged in a dedicated charger, or whether you wish the battery to be recharged within the product, in which case you will need to include specialised charging circuitry within your system.

'Plug-top' power supplies (integrated into a 13 A mains plug) have a maximum current rating which must not be exceeded. For example, a 12 V 500 mA unit will successfully power a system which draws 150 mA, but it would not be suitable if the system contained a motor which needs 750 mA; in this case a 12 V 1 A power supply would be needed.

Inputs

If you can think of a physical quantity, there is probably an electronic sensor available to measure it, and if you can measure a quantity then you can monitor it and take action about it. Consequently, sensors give design engineers immense power to create systems which interact with the world in which we live.

Force can be measured in an approximate way using a force-sensitive resistor (FSR), which is placed under the load to be measured. As the load increases, the resistance of the FSR will decrease. The FSR is, therefore, an analogue resistive sensor and would be connected to a microcontroller as described below.

For more accurate measurement of force, a load cell would be used. These devices contain carefully designed metal shapes which bend when the cell is loaded. Strain gauges attached to the metal will change in resistance when it bends and this resistance change can be measured. A load cell is used in conjunction with an instrumentation **amplifier** which is a sensitive electronic module designed to register the tiny change in resistance and give an output signal which can be read by a microcontroller. Different sized load cells are available for measuring different ranges of forces. Load cells are used in electronic scales or weighing machines.

It is often useful to be able to detect the presence (proximity) of a nearby object, especially in robotic systems. A simple proximity sensor can be made with a switch, such as a microswitch, but this requires the object to actually press against the switch to be detected. Touchless detection can be achieved with sensors such as a magnetic reed switch, or a Hall effect sensor but, since both these devices detect magnetism, they would require a magnet to be placed on the object being detected. The range of these magnetic sensors is quite small (a few millimetres) so they can only detect at short range.

Figure 6.89 Force sensitive resistor

An opto-switch is a digital sensor containing a light source and a light receiver. A reflective opto-switch will give a high output when a reflective object moves close to the sensor (within a few millimetres). A slotted opto-switch (shown in Figure 6.91) contains a small gap within the sensor and an output signal is generated when an object moves into this gap.

Figure 6.91 Slotted opto-switch

Rangefinder distance sensors use ultrasound or infra-red to produce output signals which are calibrated to measure the actual distance to an object, rather than just indicating its presence. The range of these sensors is usually in excess of 1.0 m so they can be useful for detecting fairly distant objects, but they are also prone to false triggering by other unintentional obstacles such as bumps in the floor or nearby chair legs.

Figure 6.92 Ultrasonic and infra-red rangefinder sensors

It is sometimes necessary to measure rotation and rotational speed, and this can be done in a number of ways.

- A slotted opto-switch can be used to detect the gaps in a slotted disc which rotates with a shaft, as shown figure 6.93.

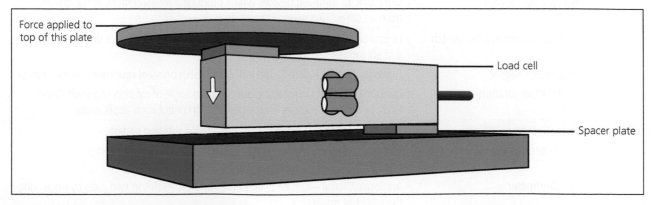

Figure 6.90 A load cell measures force

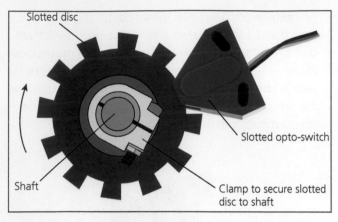

Slotted disc

Shaft

Slotted opto-switch

Clamp to secure slotted disc to shaft

Figure 6.93 Measuring the rotation of a shaft with a slotted opto-switch

- A reflective opto-switch can be used to detect the passing of a reflective patch which is attached to a rotating wheel.
- A device known as a quadrature rotation sensor produces a pulse signal when a shaft has rotated through a specific angle. The sensor can also indicate the direction of rotation. Quadrature sensors are commonly used as rotary controls in household products (for example, a volume control on a radio, or a time input knob on a microwave oven); they have no end-stops and can be rotated continuously, clicking as they do so. Every click produces a pulse, which is read by a microcontroller.

Table 6.9 Types of sensor

Quantity	Sensor	Type/Interface
Light	Light dependent resistor (LDR)	Analogue resistive
Temperature	Thermistor	Analogue resistive
	Analogue sensing IC	Analogue voltage
	Digital temperature sensor	I2C
Sound	Microphone	Analogue output (pre-amplifier required)
		Digital output (indicates yes/no presence of sound)
Infra-red	IR sensor	Digital output for detecting the presence of a warm object (e.g. a hand sensor for an automatic tap)
	PIR sensor	For detecting a moving warm object (e.g. intruder detector)
	Photodiode	For receiving IR serial data (e.g. from a remote control unit)
Strain/Force	Force sensitive resistor (FSR)	Analogue resistive – for an approximate indication of applied force
	Load cell (containing strain gauges)	For accurate measurement of force/weight (instrumentation amplifier required)
Position/ Distance	Switch (e.g. microswitch, reed switch, float switch)	Digital switch
	Hall effect sensor	Digital or analogue types indicating the presence or strength of a magnetic field
	Proximity sensor	Digital output – different types indicate the presence of a range of objects/ materials
	Reflective/slotted opto-switch	Digital output – to indicate the presence of a reflective object or an object that has moved to block a light beam
	Potentiometer	Analogue resistive – linked mechanically to a moving system
	Rangefinder sensor	Ultrasonic or infra-red devices which produce a measurement of the distance from an object. Various interfaces available
Rotation	Reflective/slotted opto-switch	Digital output – can detect a reflective patch or the gaps in a slotted wheel as a shaft rotates
	Gear tooth sensor	Analogue/digital – detects the individual teeth on steel spur gears as they rotate
	Quadrature rotation sensor	Dual digital outputs indicate the direction and angle of rotation of a shaft. Known as an incremental encoder. Frequently used in control knob applications
Angle	Potentiometer	Analogue resistive – mechanically linked to a rotating system
	Angle sensor	Analogue/digital – produces a signal indicating the absolute angle (0–360 degrees) of a shaft
	Accelerometer	Analogue/digital – can be used to measure tilt angles in two axes by measuring the effect of gravity

New sensors

New sensors are continually being developed for cutting-edge technologies such as smartphones or drones. Understandably, these sensors are often physically very small and almost impossible to handle and use directly. Fortunately, many such useful sensors are now available mounted on a **breakout board** which increases the scale of the component up to a reasonable size to handle and allows it to be integrated into a prototype system. Consequently, a huge number of exciting, modern and easy-to-use components are available for use within A Level projects.

Interfacing sensors with a microcontroller

There are four different ways that sensors may be **interfaced** to a microcontroller.

Digital switch sensor

A variety of sensors are actually just switches which open/close when a physical quantity changes. Examples include a float switch, magnetic reed switch, BCD (binary coded decimal) rotary switch and, of course, a push switch. All switches are connected to a digital i/o pin on the microcontroller, along with a resistor (which, when used in this way, is referred to as a 'pull-down resistor') as shown in Figure 6.95.

Figure 6.94 **Magnetic reed, tilt and vibration switches**

Note the +5 V and the 0 V power supply lines; these are the same power lines that the microcontroller is connected to.

A multimeter can be useful to trace faults in electronic circuits. The multimeter should be set to a voltage range. The black (common) probe is always connected to the 0 V rail; this means that all voltages in circuits are measured relative to the 0 V rail. Never move the black probe from the 0 V rail otherwise your voltage measurements will not make sense. In this simple circuit, the red multimeter probe is connected to the digital i/o pin, and the meter should read 0 V when the

switch in open, and approximately 5 V when the switch is closed.

Analogue resistive sensors

Many simple sensors are analogue resistive sensors, such as a light dependent resistor (LDR) or a thermistor (for measuring temperature). These are connected similarly to the switch, using a resistor in an arrangement called a 'voltage divider'.

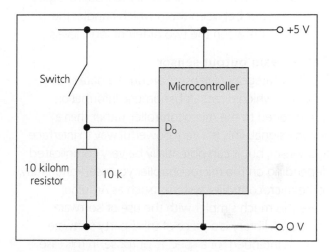

Figure 6.95 **Connecting a switch sensor to the input of a microcontroller**

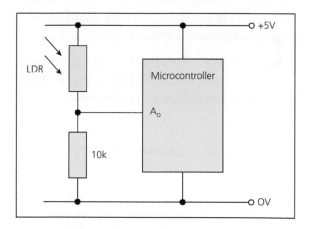

Figure 6.96 **An LDR connected to an analogue input**

Analogue/digital output sensor

Some sensors directly produce an analogue or a digital output voltage and these are easy to use with a microcontroller.

An analogue output sensor will connect to an analogue input pin on the microcontroller. The program will need to read the analogue input voltage and store the result in a variable, available for use by the rest of the program. Consideration will need to be given to the sensor's calibration, so that the value measured can be related to the actual physical quantity.

A digital output sensor will connect to a digital i/o pin. When the program reads this pin it returns a logic 1 if the sensor output is high, or logic 0 if it is low. Some digital sensors work from a 3.3 V power supply. If the microcontroller is running from a 5 V supply then the logic levels between the two devices will not be compatible and it will be necessary to install a logic level converter between them which handles the conversion between the two different logic levels.

Data stream output sensor

The final category of sensors produce a data stream output in which direct alphanumeric information is transferred to the microcontroller, rather than a voltage signal. This is a very powerful way to interface to a sensor, but it can potentially be very complicated depending on the microcontroller you use. Fortunately, some microcontroller systems (such as Arduino) make this much simpler with the use of 'software libraries'. A library is a piece of third-party software which is intended for a specific purpose, in this case to communicate with an input device. The library is imported into your own program and gives you some extra commands to help communicate with the sensor. One huge benefit of this is that the data transferred is often directly usable without further processing.

There are a few different interfaces in use for these types of sensors and you will need to look up further technical details should you need them for project work.

Outputs

Output devices are used to provide light, sound, motion or information.

Light

LEDs are available in a huge range of colours, sizes, shapes and variations in brightness. Larger 'power' LEDs are increasingly being used to replace filament lamps for room illumination, bringing improvements in efficiency and reduced energy consumption. Multi-colour LEDs are available, including red-green-blue (RGB) which can be made to light up in any colour, including white. LED strips consist of several miniature

LEDs on a flexible self-adhesive roll. LED strips are available as single-colour or RGB.

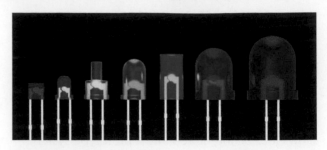

Figure 6.97 A variety of LEDs

LEDs require a resistor to limit the current flowing through them, otherwise they will be damaged.

Ohm's law is used to calculate the value of series resistor for an LED:

$$R = (V_s - V_{led})/I$$

where:

V_s is the power supply voltage

V_{led} is the voltage drop across the LED

I is the current the LED needs to light up

The values for V_{LED} and I can be looked up for the LED you intend to use.

Sound

Microcontrollers can easily produce sound signals and this feature is useful for creating alarm noises or for providing a simple 'click' feedback to indicate that a button has been pressed. Some microcontrollers can play music files (such as MP3) or record and play voice files, allowing a product to speak information to the user, such as in a satnav.

To convert the electronic signals into sound waves, a piezo-electric sounder is used. These small and cheap devices can generate surprisingly loud tones, especially at higher frequencies. For louder volumes, an amplifier circuit will be needed and a loudspeaker will be used.

Motion

Controlled movement is very important in engineered systems. A range of actuators are available and each type is useful for different applications.

Table 6.10 Types of actuators

Actuator	Type of motion	Method of control	Example application
Solenoid	Reciprocating Short stroke	Simple on/off using a **driver**	Door lock release Pneumatic valve
DC motor	Rotary High speed	Speed control using PWM (pulse width modulation) output from microcontroller Direction control using a motor driver IC	Very common in a wide range of applications, e.g. cooling fans, robots, automatic doors.
Servo motor	Oscillating, turns to an angle then stops	PWM control from a microcontroller	Robotic systems
Stepper motor	Rotary, turns through precise angles, step-by-step	4-wire control from a microcontroller through a driver IC	CNC machines
Brushless motor	Rotary High speed, high torque	PWM control from a microcontroller through an electronic speed controller (ESC)	Flying drones
Piezo actuator	Reciprocating Very short stroke	Varying a voltage produces small linear movements	Focusing systems in miniature cameras

Figure 6.98 Motors – DC, servo, stepper and brushless

For A Level Design Engineering projects, the 'radio-control type' of servo motor is very useful. Unlike ordinary DC motors, a servo does not rotate continuously but has an output disc which will rapidly rotate to a specific position and then stop. The disc can rotate through a maximum of 180°. In order to control the servo, a microcontroller is used to generate a pulse-width-modulation (pwm) output signal. As the pulse width varies from a duration of 1 ms to 2 ms, the servo will rotate from one end position to the other. Therefore, a pulse width of 1.5 ms will place the servo disc in its central position. Servos are essential components in robotic systems and they are connected through mechanical linkages to provide repeatable, accurate positioning of robot arms, for example.

A stepper motor produces a rotary output which is divided into a number of steps. It has several control wires through which the current must be switched on and off in a particular sequence to make the motor rotate to its next position. It can rotate clockwise or anticlockwise, one step at a time. A typical step angle is 1.8°, so such a motor would have

$$\frac{360}{1.8} = 200 \text{ steps per revolution.}$$

Stepper motors can be controlled directly from a microcontroller, but it is often more convenient to use a stepper motor driver between the microcontroller and the stepper motor. The microcontroller tells the driver how many steps to rotate and in what direction. The driver takes care of producing the correct sequence of pulses through the control wires, and it handles the high currents needed for the motor.

Information

LEDs are available in packages known as displays. Two types are commonly available: the 7-segment and the dot matrix display.

The 7-segment display is used to display digits. These devices are bright and colourful and popular for use in clocks or displays on household appliances such as microwave ovens. It is necessary to light up the appropriate segments and this can be done using a suitably programmed microcontroller. The 7-segment display can be connected directly to the microcontroller output pins, or through a dedicated 7-segment driver IC.

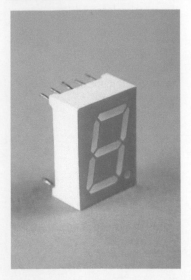

Figure 6.99 7-segment display digit

A dot matrix display is harder to interface, but permits the display of letters, digits and other characters. Several dot matrix displays can be connected to create a scrolling message display.

When several digits or several characters are needed, displays can be multiplexed. This technique reduces the potentially large number of wires needed by displaying only one character at a time and then switching this off before displaying the next character, and so on. If this is done very rapidly, the eye sees all the characters displayed simultaneously. **Multiplexing** also reduces the power consumption of large displays.

Alphanumeric displays are useful when it is necessary to display detailed information and messages. These displays are either liquid crystal displays (LCD), which are often used with a backlit panel, or LED which are bright and emit their own light. Alphanumeric displays are available in a range of character sizes and varying lines of text. The easiest way to control them with a microcontroller is using the serial interface method.

Figure 6.100 Dot matrix display unit

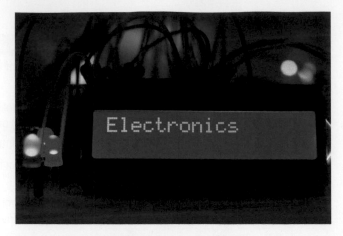

Figure 6.101 Backlit LCD alphanumeric display

Converting between analogue and digital signals

Each analogue input pin on a microcontroller contains an **analogue-to-digital converter (ADC)**. When an analogue voltage is read, the microcontroller converts the voltage into a number which is proportional to the voltage. The resolution of the ADC will determine the number of different levels that can be read and, therefore, the range of numbers that can be produced.

- An 8-bit ADC will produce $2^8 = 256$ levels, so the number range is from 0 to 255.
- A 10-bit ADC will produce $2^{10} = 1024$ levels, so the number range is from 0 to 1023.

The full-scale input voltage (the voltage which produces the maximum ADC number) is usually equal to the power supply voltage.

The result of an analogue-to-digital conversion is given by:

$$\frac{input\ voltage}{supply\ voltage} \times full\ scale\ value$$

So, for a 10-bit ADC running from 5 V, an input of 3.7 V will produce a result of:

$$\frac{3.7}{5.0} \times 1023 = 757$$

In practical applications, it is usually discovered that the range over which an analogue sensor is used does not produce the full range of ADC outputs. Under these circumstances it is often necessary to scale the result of the ADC to produce a result between two wanted values. For example, in a system where a temperature sensor controls the speed of a fan, the temperature value from the ADC could be scaled so that a temperature range of 20°C to 30°C results in a fan speed ranging from 0 per cent to 100 per cent.

Some microcontrollers can produce an analogue output signal which can be used to control the brightness of LEDs or the speed of motors, etc. However, this output is not a true analogue signal, but rather a rapidly pulsed on-off signal known as **pulse width modulation (PWM)**. The ratio of on-time to off-time determines the average analogue effect produced. For example, output pulses which are on for 0.25 ms and off for 0.75 ms will produce an effective 25 per cent analogue output. 0.80 ms on and 0.20 ms off produces 80 per cent output, and so on. PWM works for many applications, but it is important to understand that it is a pseudo (not genuine) analogue output.

Signal amplification

Sometimes it will be necessary to amplify the voltage of small signals from sensors before putting them into the ADC so that they produce a good range of output values. Microphones produce very small signals and will require a pre-amplifier. A load cell (used to sense force/weight) will need connecting to an instrumentation amplifier to produce an analogue signal large enough to be registered by the ADC.

The microcontroller output pins cannot supply much current, a few mA at most. This is sufficient to light a small LED or sound a buzzer but, for most output devices, a driver is used to amplify the output current to a level where the device works properly.

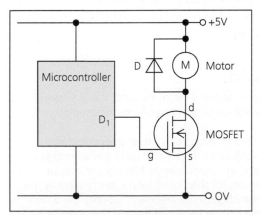

Figure 6.102 Using a MOSFET as a driver

The simplest driver is a component called a MOSFET (metal-oxide-semiconductor field-effect transistor). It has three leads, named drain (d), gate (g) and source (s) and it must be connected correctly as shown in Figure 6.102.

This example shows a MOSFET driving a DC motor, but the motor can be replaced with any output device, such as a buzzer, LED or solenoid, etc. The diode (D) is included in this circuit as a protection device for the MOSFET. The diode is only needed if an electric motor or solenoid is used as these devices generate high voltage electrical 'spikes' (called 'back EMF') which can damage the MOSFET. If LEDs or buzzers are being used, the diode can be omitted.

MOSFETs are available with different current ratings. Two common types are shown in the table below:

Table 6.11 MOSFET information

MOSFET type	ZVN2106A	IRF510
Pinout		
Maximum drain current	0.45 A	4.0 A

A MOSFET is used when simple on/off control is required. It also works with the PWM analogue output from a microcontroller.

In motor applications where reversing is required, a motor driver IC will be needed (sometimes called an H-bridge driver) in place of a MOSFET. These ICs connect to two microcontroller outputs, as shown in Figure 6.103.

The outputs are used to control the motor as shown below:

Output A	Output B	Motor
Low	Low	Stop
Low	High	Forward
High	Low	Reverse
High	High	Stop

Motor driver ICs can also be connected to two PWM outputs so that motor speed and direction control can be achieved.

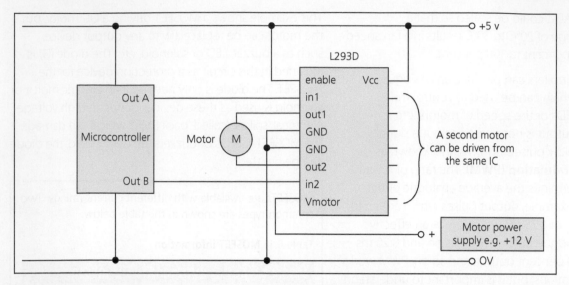

Figure 6.103 A motor driver IC allows the microcontroller to reverse the motor

Other specialist driver devices are available for controlling specific types of motors, such as stepper motors or brushless motors.

Open and closed loop systems

In many products and systems, electronics are used to control something, for example, in:

- an oven, to control the temperature
- a cruise control, to maintain the speed of a car
- a laser cutter, to move the cutting head
- a cordless drill, to control the motor speed.

In an open-loop control system, the microcontroller controls the output device but it has no way of knowing whether the desired result has been achieved. The cordless drill is an example of an **open-loop system**; when the user pulls the trigger part way, the motor turns at a reduced speed, but if the drill is then applied to a workpiece it will slow down. The user needs to pull the trigger harder to maintain the speed because the electronic system does not realise that the drill has slowed down under load.

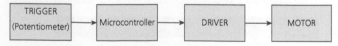

Figure 6.104 Open-loop system of a cordless drill

In a **closed-loop system**, a sensor is used to monitor the output. Information from the sensor is fed back to the microcontroller which can then adjust the output device until the required result has been achieved. An oven is a closed-loop control system; the microcontroller will switch on a heater and a temperature sensor will monitor the oven temperature. When the desired temperature is reached, the microcontroller will cycle the heater off-on-off to maintain that temperature. The electronic system knows the temperature of the oven so it knows whether the heater needs to be on or not.

Closed-loop systems use feedback to provide precise control. Open-loop systems do not use feedback; instead they predict (or guess!) the output needed to achieve the desired result. Closed-loop systems generally provide more accurate control, but their increased complexity is sometimes not needed in simple systems, or in situations where the output device produces a predictable result. In CNC machines, for example, stepper motors are used to move the cutting head through precise distances; feedback is not needed because the controller knows that ten steps of the motor moves the head the required distance.

ACTIVITIES

1   Consider whether a car cruise control and a laser cutter are examples of open-loop or closed-loop systems.
2   Identify one example each of an open-loop and closed-loop control system in products around your house and in school/college. Identify the output devices and, for the closed-loop system, identify the sensor which provides feedback. Identify the datum in the open-loop system, if there is one. Draw a system diagram for each example, identifying any other input devices, and explain how the whole system operates.
3   Consider the differences between digital and analogue closed-loop control. Look up the terms 'lag' and 'hunting', and try to explain the problems these cause in control systems and the methods of reducing them.

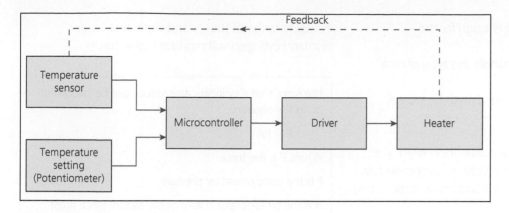

Figure 6.105 **Closed-loop system of an oven**

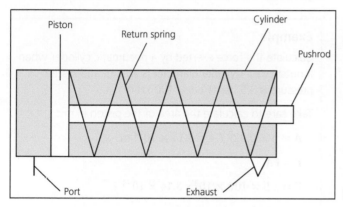

Figure 6.106 **Schematic symbols for a single-acting pneumatic cylinder**

Open-loop systems sometimes require a datum which is a known position that they move to at start-up. All other movements are subsequently made from this datum. The datum point may be achieved with a digital sensor such as a microswitch; the system will move the output until it touches the switch and then it will stop, in the known position.

Sub-systems and systems thinking

When a Design Engineer begins to develop an electronic system for a particular application, they will follow a logical process which analyses the needs of the application and identifies the physical quantities that need to be monitored, and the devices that need to be controlled. This is called systems thinking, and the outcome of this will be a system diagram showing the interconnections between the sub-systems, and the signal flow through the system as a whole.

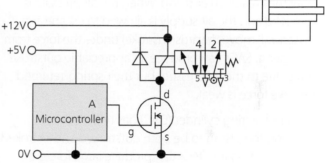

Figure 6.107 **An example of an electronic system diagram**

The stages in producing a system diagram are:

1 Identify the output devices:
   ● Are outputs to be controlled digitally or in an analogue way?
   ● Are there any display or sound outputs?
   ● Determine whether any of the output devices require a particular type of control signal, for example a servo motor requires a PWM output signal.
   ● Decide whether a driver is required for each output.

2 Identify the input devices:
   ● Determine what sensors are to be used.
   ● Consider issues of calibrating the sensors.
   ● Determine what kind of signal the sensors produce and how they will be interfaced to the microcontroller.

3 Decide on the microcontroller to use:
   ● How many I/O pins are required?
   ● Will the microcontroller interface satisfactorily to the input/output devices chosen?
   ● Are you familiar with the programming language?
   ● What are the power supply requirements?

- Is the microcontroller fast enough for the application?
- Can the microcontroller handle the type of data being processed?
- Does the application require data to be stored?

Pneumatics

Pneumatic cylinders are linear actuators in which the driving force is provided by a supply of compressed air. The simplest pneumatic cylinder is called a single-acting cylinder (SAC).

When compressed air is supplied to the port on a SAC, the piston and pushrod will move outwards (outstroke) until they reach an endstop. When the air pressure is removed and the air supply is allowed to escape, the piston will move inwards (instroke) under the force from the spring. SACs can produce an appreciable outstroke force due to the air pressure but their spring-returned instroke force is weak.

A double-acting cylinder (DAC) has two air ports, allowing the piston to be both outstroked and instroked under air pressure. DACs can produce large forces on both outstroke and instroke.

The distance that the pushrod can travel is called the stroke of the cylinder. Pneumatic cylinders are available in a wide range of strokes and diameters. Larger diameter cylinders will produce higher forces.

The force that a cylinder can produce can be calculated using the equation:

$$F = PA$$

Where F is the force.

P is the compressed air pressure.

A is the cross-sectional area of the piston. Since most pistons are circular, $A = \pi d^2 / 4$ where d is the piston diameter.

Example

Calculate the force exerted by a pneumatic cylinder when it outstrokes. Cylinder diameter is 20 mm and system air pressure is 3.5 bar (1 bar = 100 kN m^{-2}).

Take care to calculate the area of the piston in m^2:

$$A = \pi \times 0.02^2 / 4 = 3.14 \times 10^{-4}\,\text{m}^2$$

$$F = PA$$

$$F = 3.5 \times 100 \times 10^3 \times 3.14 \times 10^{-4}$$

$$F = 110\,\text{N}$$

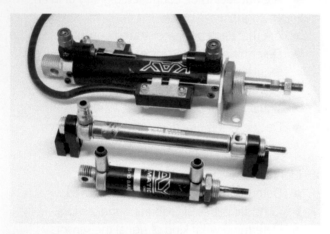

Figure 6.108 Pneumatic cylinders

Since the pushrod prevents the air from acting across the entire face of the piston when it instrokes, the instroke force is slightly less than the outstroke force.

The interface component between an electronic control system and a pneumatic system is a solenoid valve. These components control the flow of air to the cylinder and they ensure that air is released to an exhaust when necessary. A 3/2 solenoid valve has

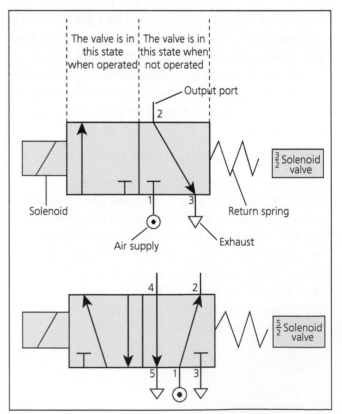

Figure 6.109 **Schematic symbols for pneumatic solenoid valves**

three air ports (air supply, exhaust and output) and is used to control a SAC, while a 5/2 solenoid valve with five ports is needed to control the air flow to both ends of a DAC.

Figure 6.110 shows how a 5/2 solenoid valve is connected as an interface between a microcontroller and a pneumatic cylinder. Notice the use of a MOSFET driver and notice the schematic symbols used to represent the compressed air supply and the exhaust.

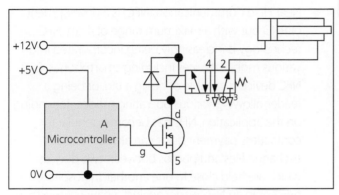

Figure 6.110 Schematic for 5/2 solenoid value as in interface

When the microcontroller output A goes high, the solenoid valve will operate and the cylinder will outstroke. When output A goes low, the solenoid valve will switch off and its spring will return the valve to the original state, causing the cylinder to instroke.

Pneumatic cylinders can move very rapidly and it is often necessary to slow down their motion to provide a controlled movement in a system. A one-way flow restrictor is a component which is placed in the air line of a double-acting cylinder such that it restricts the flow of air out of the cylinder, while allowing the air to flow in unrestricted. Reducing the flow out produces a smooth and controlled motion, the speed of which is controlled by adjusting the screw on the flow restrictor.

Pneumatic cylinders produce reciprocating motion, but they are often coupled to mechanical systems to create oscillating motion, such as opening a hinged door or operating the jaws of a clamp, as shown in Figure 6.112 on page 268.

In such circumstances, the pushrod coupling needs to be a pivot which allows for the rotation of the mechanism as the cylinder outstrokes. In addition, it is often necessary to mount the cylinder itself on a pivot, allowing it to rotate slightly to maintain the geometry of the mechanical system.

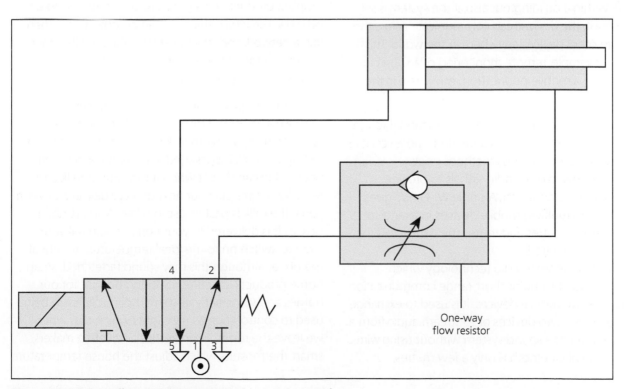

Figure 6.111 Using a flow restrictor to control outstroke speed

Figure 6.112 Pneumatic cylinders operating a clamp

Networking

Electronics designers now focus much of their attention on building systems which are compatible with other systems so that they can exchange information when required. Electronic systems which are part of a network can share data and tasks, control each other and receive updates when required.

For networking to be successful, all devices on the network must agree on a communication protocol. The most popular of these is the Internet Protocol (IP); each network device has a unique IP address, which is a number that identifies that device to others on the network. This allows devices to share data on a local network within a building but also, if the system is set up to allow it, the devices can communicate across the internet to other devices anywhere in the world. This allows, for example, remote monitoring of an IP security camera using a mobile phone, from anywhere in the world.

Networked devices can exchange information through cables, but wireless communication is far more versatile and several different wireless methods exist:

- Wi-Fi uses microwave radio signals to provide high-speed data transfer. A single Wi-Fi hub gives coverage to multiple mobile devices in a range of a few tens of metres and routes the data across the network as required.
- Bluetooth is another radio technology which provides point-to-point short-range communication between two mobile devices. It is used to exchange files between two devices or to stream audio from a music player to a sound system without using wires. The range of Bluetooth is only a few metres.

- Radio Frequency Identification (RFID) allows data to be read from passive tags or ID cards. When the tag is held near an RFID reader, radio energy from the reader 'energises' the tag which responds by transmitting its pre-programmed data to the reader. The tag is then identified by the reader which responds as required. RFID tags are used in a huge number of applications including security 'swipe cards', keyless entry systems, cat flaps, passports and shop security tags. The range varies depending on application.
- Near Field Communication (NFC) is a development of RFID but with a maximum range of 4 cm. NFC technology is increasingly being incorporated into various mobile devices including smartphones. Each NFC device is capable of being a tag, or being a reader, allowing various operating modes depending on the application. NFC has become popular for contactless payment but it can also be used to exchange files and photos between two devices which are held close to one another. In hospitals, nurses tap an NFC reader against a patient's wristband to identify the patient, but also to read medication dosage and when the last dose was administered.

The massive growth of electronics and the ability to miniaturise electronic controllers and produce them at extremely low cost has led to the concept of embedding networked electronic systems into the simplest products, creating what are known as **smart objects**. The idea is that these smart objects can then join a network and share their information with other devices on the network, leading to what has become known as the Internet of Things (IoT).

The IoT offers powerful opportunities for devices to sense and control their surroundings and to act collaboratively with other devices and controllers. For example, if milk is packaged in a smart carton then your fridge can detect when it runs out of milk and re-order the milk in your next grocery delivery, or warn you if the milk is past its use-by date. A smart pizza box, as it is delivered to your house, could tell your oven to switch on to the ideal temperature to reheat the pizza. Although this may sound farfetched, smart home products are currently being bought for our homes and voice-activated smart controllers are being used to control them: smart lights which dim when we leave the room, smart kettles and coffee makers, smart thermostats which adjust the house temperature according to who is at home (by detecting their mobile phone) and by referring to the weather forecast. The range of smart objects will grow, along with the novel ideas for using them.

ACTIVITY

Carry out internet research to find a range of smart products which are currently available to buy. Explain how they function through networking.

KEY TERMS

Amplifier – a subsystem to increase the size of a small signal.

Analogue-to-digital converter (ADC) – converts an analogue voltage into a numeric value.

Breakout board – a method of mounting miniature components which allows them to be easily used in prototypes.

Closed-loop system – a system which uses feedback to provide accurate control.

Current – a measure of the actual electricity flowing.

Driver – a current-amplifying subsystem which ensures an output device operates correctly.

Interfacing – connecting two parts of a system together to allow signals to flow between them.

Multiplexing – a method of driving a large display to reduce the number of connections and reduce power consumption.

Open-loop system – a control system which does not use feedback.

Power supply unit (PSU) – provides a steady voltage while pushing current through a system.

Pulse width modulation (PWM) – a method of producing a pseudo analogue output.

Smart object – an item which features an embedded, networked electronic system.

Voltage – a measure of electrical pressure.

KEY POINTS

● A wide range of input devices are available to sense various physical quantities.
● Output devices can produce light, sound, movement or information.
● Signal amplification is required with some sensors.
● A programmed microcontroller is used to enhance the function of a system.

● A driver is required with many output devices.
● Control systems can be open- or closed-loop.
● Electronic systems can be interfaced to mechanical or pneumatic systems.
● Electronic systems can be networked to create versatile solutions.

Further reading on technical understanding

Design Engineering
● Brain, M. (2003) *How Stuff Works*, Hungry Minds, ISBN: 978-0764567117
● Duncan, T. (1997), *Electronics for Today and Tomorrow*, Hodder Education, ISBN 978-2091910734
Microcontrollers
● Arduino: www.arduino.cc
● PICAXE: www.picaxe.com
Mechanical and electronic simulation
● Animated mechanisms: www.mekanizmalar.com

General information and resources
● Cool components: www.coolcomponents.co.uk
● Howstuffworks: www.howstuffworks.com
● Instructables: www.instructables.com

Fashion and Textiles
● Kettley, S. (2016), *Designing with Smart Textiles*, Bloomsbury Publishing, ISBN 9781472569158
● Pailes-Friedman, R. (2016), *Smart Textiles for Designers: Inventing the Future of Fabrics*, Laurence King Publishing, ISBN 9781780677323

PRACTICE QUESTIONS: Technical Understanding

Design Engineering

1 (a) Explain what is meant by the structural integrity of a product or system.

   (b) Discuss the factors which affect the structural integrity of a product or system.

   (c) By giving specific examples, identify three ways in which structural integrity has been achieved in a product or system.

   (d) Identify two ways in which a specific product has been designed to be resilient in operation.

2 (a) Give three examples of smart materials and identify a practical application for each one.

   (b) With reference to a specific example, explain how the use of a smart or programmable technology has enhanced its operation.

3 (a) State and explain the significance of the basic machine principle.

   (b) Use annotated sketches to explain the operation of a compound lever and a compound gear train. Write down equations for the mechanical advantage of each.

   (c) Describe the benefits offered by a worm drive system.

   (d) Describe the operation of a lead screw and identify an application.

   (e) Identify three benefits offered by an epicyclic gear train.

   (f) Identify the types of bearings which are suited to axial and to radial loads.

   (g) Explain why the efficiency of a mechanical system is less than 100 per cent, and describe ways of improving the efficiency.

4 (a) Explain the relationship between the mass and the weight of an object.

   (b) Explain what is meant by a structure and explain ways that the rigidity of a structure can be increased.

   (c) Give examples of the static and dynamic forces acting on a specific product or system when it is being used.

   (d) Write down equations to define stress and strain. Explain how they are linked by Young's modulus.

5 (a) Describe, giving specific examples, the different kinds of electrical signals produced by different sensors.

   (b) Identify three different kinds of electric motor, explaining how each is controlled and giving an example application.

   (c) Explain the function of an ADC in a microcontroller.

   (d) Explain the difference between open-and closed-loop control systems, giving an example of each.

   (e) Draw a schematic diagram to show how a 5/2 solenoid valve can be used to electrically control a double-acting pneumatic cylinder.

   (f) Give three examples of wireless technology.

   (g) Describe the advantages of networking electronic devices.

Fashion and Textiles

1 Explain the difference between whole garment and fully fashioned panel knitwear and the advantages and disadvantages of each.

2 Many sportswear manufacturers have incorporated hi-tech fabrics within their products. Discuss how these can benefit the user and give examples to support your answer.

3 Discuss how the structural integrity of fashion and textiles products can be improved. Give examples of products to support your answer.

4 Discuss how the use of colour-changing technology within smart materials has been incorporated within the fashion and textiles industry.

5 Explain the benefits of digital printing within the fashion and textiles industry.

Product Design

1 Explain what is meant by a structure and explain ways that the rigidity of a structure within a piece of wooden furniture can be increased.

2 Discuss how the structural integrity of consumer products made from polymers and metals can be improved. Give examples of products to support your answer.

3 Many product designers have incorporated hi-tech fabrics and materials within their products. Discuss how these can benefit the user and give examples to support your answer.

4 Discuss how the use of smart materials has been incorporated within the product design industry.

5 Explain the benefits of digital printing within the design of consumer products.

6 Discuss the factors that a manufacturer must consider before deciding on the most appropriate surface finish for metal components used in outdoor products.

7 Consider the reasons why a surface finish is necessary on metal and wood products used within a domestic environment.

8 Find examples of wooden and metal products used outdoors and identify the material and what surface finish has been used.

Manufacturing processes and techniques

PRIOR KNOWLEDGE

Previously you could have learnt:

- the materials, methods and processes, equipment and specialist techniques used when making models, mock-ups and prototypes relevant to your chosen material area

- the methods used for manufacturing at different scales of production
- the manufacturing processes used for larger scales of production appropriate to your material area.

TEST YOURSELF ON PRIOR KNOWLEDGE:

1 Using annotated sketches, explain the main stages of the following processes:
   - laminating timber to form a curved component
   - permanently joining two pieces of acrylic
   - joining two pieces of aluminium sheet by pop-riveting
   - soldering electronic components onto a printed circuit board
   - adding a lining to a pencil skirt
   - screen printing a two colour design on cotton.

2 Produce a worksheet aimed at GCSE students on the following processes:
   - forming of thermoplastics
   - heat treatment of metals
   - 3D printing
   - developing and downloading a control program for a microcontroller
   - etching designs into denim with a laser cutter
   - the importance of efficient pattern cutting.

3 Explain and give examples for each of the following levels of production:
   - one-off, bespoke production
   - batch production
   - high-volume production.

7.1 How can materials and processes be used to make iterative models?

LEARNING OUTCOMES

Product Design and Design Engineering
By the end of this section you should:
- understand that 3D iterative models can be made from a range of materials and components to create block models and working prototypes to communicate and test ideas, moving parts and structural integrity
- demonstrate an understanding of simple processes that can be used to model ideas using hand tools and digital tools, such as rapid prototyping or digital simulation packages to support the creation of iterative developments.

Fashion and Textiles
By the end of this section you should:
- understand that iterative models can be made from a range of materials to create samples, toiles and other modelled concepts to communicate and test ideas, fit and structural integrity
- demonstrate an understanding of how to develop iterative models using pattern making, pattern drafting and toiles to be able to test garments and forms of other textile products
- understand the use of both hand tools and digital tools such as rapid prototyping or digital simulation packages to support the creation of iterative developments.

Creating iterative models from a range of materials

When designers are developing products, it is crucial that they are tested throughout the process to ensure they are fit for purpose. Once a designer starts using materials and fabrication techniques they are able to refine their ideas to make them work. Models complement drawings as a communication tool and enable designers to work out and visualise their ideas before committing to a design. They also enable clients and stakeholders to gain an instant overview and understand the product fully.

Modelling and testing is a key process in the development stages and leads to feasibility decisions before production. Although prototyping can be a costly process, in the long term it can save money as the further along the production line the manufacture goes, the more expensive it may be to make alterations.

Models can be 2D or 3D and can be a quick model of any size or a full-sized mock-up. The finish of these early models is not important as their purpose is to test factors such as functionality and proportions, and to allow a visual representation to present to clients.

More detailed models help to refine and develop the aesthetic and technical aspects of the design.

DESIGN ENGINEERING AND PRODUCT DESIGN

Block models are created to give an accurate physical representation of the appearance of a product being designed. Consisting of shaped blocks, they tend not to contain any moving or working parts, but are finished to a high standard to give a realistic indication of the overall form of the product.

During the iterative process, several block models may be produced before the final form and appearance of a product is decided. In the commercial world, modelling is also a key process as it enables manufacturers and designers to test ideas and modify products and processes. The process may involve a number of iterations but in the long run it saves time, reduces manufacturing costs and avoids mistakes such as the product being uncomfortable, awkward to use or not functioning as intended.

Models to test mechanisms and components being considered in a design are also important. These may start simply, but where complex and critical functions are involved, these models may be highly sophisticated to fully test performance and reliability, using materials and methods that will potentially be used in the final product.

Materials used for modelling

Commercial model makers use a variety of materials, depending on the type of model they are making or the requirements of the design solution. Two processes are used; the additive process where models are created by adding material to form a model, for example in 3D printing, and the subtractive process.

The subtractive manufacturing process involves the machining, carving or sculpting of a model component. A part can often be made in a number of pieces and assembled to create the final product. Once a part is roughed out, hand finishing is carried out. For some models, colours, textures and graphics are applied to create a part that closely mimics the desired future product. However, producing a complex model in this way can be time consuming. CAD/CAM programs make the production of these parts much simpler and provide high tolerances for parts and components. Model makers commonly use laser cutting technology to precisely cut materials like foam board, high-density papers and card and other materials to create panels used in the construction of structural models. Product design model makers also use moulds and castings or create vacuum formed pieces.

Sometimes less is more; using simple materials and bought-in components to create models can allow you to test with users quickly and move onto the next iteration. James Dyson famously created 5,127 prototypes of his first machine before developing one that he considered worked perfectly, the DC01.

Unless you have easy access to a workshop and a reasonable level of experience with machinery, it would be best to work with card, foam board, corrugated cardboard or similar easy-to-cut materials such as balsa or styrofoam that you can cut with either a craft knife or junior hack saw. If you are cutting with a craft knife or a scalpel, it is better to use several light passes rather than trying to cut all the way through in one go. You will get a cleaner cut and you are less likely to slip and cut your finger.

- Clay, plasticine or Polymorph can be used to quickly make 3D parts that can then be drilled and used within models to test ideas and mechanical parts.
- Electronic circuits can be built and tested using kits and breadboards to quickly test ideas.
- Styrofoam or hard wax can be used for 3D block models as detail can be carved into shapes. Thin sections of styrofoam can be used to create curves easily by rolling it around formers.
- **Calico** or rip-stop nylon can be used to model fabric parts of a product.

Folding techniques and the use of nets are quick ways to make 3D forms from 2D sheets of material. A variety of materials such as cardboard, fabric, plastic or metal can be used and then applied to a range of products.

Other useful modelling materials include:
- metal wire for modelling structures
- aluminium wire, which is easy to bend and form complex shapes
- various profiles and sections of materials are available, for example ABS tubes, rods, bends and rings for modelling pipework and tubular steel
- high impact polystyrene (HIPS), which can be easily joined. A 1 mm sheet can be cut with a craft knife and it can be easily joined with liquid solvent cement. Vacuum forming and dome blowing can also be used to create models
- corriflute board, corrugated card and foam board, which can be used to quickly create rigid shapes with walls. Cutting a series of parallel lines will make foam board easier to bend and form.

To join models together, fasteners designed for use with Corriflute and other plastic screws/nuts, rivets and clips and other specialist components can be used. A hot glue gun or a strong all-purpose adhesive can be a quick and easy way to join pieces. It can be useful to use small pegs or string, rubber bands or masking tape to clamp and hold parts of models together when gluing.

Mechanical and structural parts such as linkages, levers, cams or chassis components can be designed on CAD and then laser cut out of card. This allows you to check that the dimensions are correct, including features such as bearing-mounting hole sizes and separation of shafts to ensure the correct meshing of gears. 3D printing provides an alternative method of modelling mechanical components.

During the modelling stages you can start to experiment with bought-in components such as motors, gearboxes, and sensors, and the fixed sizes of these parts will begin to determine certain dimensions of the overall design. Kits such as Lego, Meccano or Vex Robotics can be used to model mechanical systems and structures.

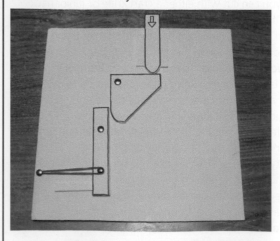

Figure 7.1 **A linkage mechanism modelled in card**

Foam board is another useful sheet material for modelling mechanical components and is easy to cut using a modelling knife. A useful method for testing the movement of linkages is to make a 2D model out of foam board and to use paper fasteners as pivots. Such models can be used to test that the full range of required motion can be achieved and to prove that parts do not collide or lock up.

A breadboard (or prototyping board) is used to model electronic systems without the need for soldering. Breadboards consist of holes into which component wires can be inserted. The holes are electrically joined in rows of five and wires sharing the same row will be connected. Breadboards take a little practice to use effectively, but they are by far the most effective method of modelling electronic circuits.

Figure 7.2 **An electronic system modelled using a breadboard**

Computer-aided engineering (CAE) software provides a very effective way of modelling electronic circuits, especially in the stages before breadboard prototyping. Components can be quickly connected and disconnected and the designer can easily modify a component's value (such as the resistance of a resistor) and observe the effect. This provides a rapid method of optimising component values by trial and error.

Integrating the full programmable functionality of microcontrollers is a problem for some circuit simulation software. Certain software packages will simulate only specific microcontrollers, while others cannot simulate a microcontroller program at all. It is, therefore, important to understand the limitations of CAE circuit simulation.

Throughout your NEA project, the use of models and prototypes will be an important part of your designing, to test iterations with users and stakeholders and move towards an optimised design solution. Do not try to make the model show every detail of your design or you will not have time to finish it. It is better to use quick and easy materials to model with and focus on learning from it and moving on to the next iteration. The use of colour is an area where models can go wrong; unless it is a vital part of what your model is trying to show, do not worry about the final appearance. It is more important to be able to test how a design works or feels than worry about how well your model is finished.

Figure 7.3 **The images show a design student recording iterations and using a variety of modelling techniques in her university project when designing an Eye Tracker**

It is impossible to make a perfect sphere from a flat sheet of cardboard. Cardboard can curve in one direction, but cannot curve in two directions at the same time. There are many examples of 3D shapes made from card on the internet. It's worth trying some of these techniques to develop your skills. Cylinders and curves can be quickly made by scoring or bending card. The more scores, the smoother the finish of the curve. Simple contours can be created by using repeated patterns cut out of card that can then be covered to make a solid shape or base.

Tents and arches can be formed using thin wooden sticks or wire and fabric or acetate can be sewn or pulled into place. To give the appearance of a dome, you could stack circles together through cutting circles smaller each time or use a petal system. The circumference of the circle is divided it into equal segments to produce a petal template. Different materials will give you different looks for whatever model you are trying to create.

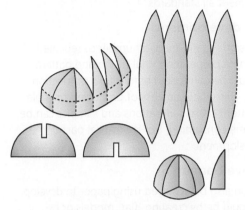

Figure 7.4 **This is a basic net shape for making a sphere; simple strips stuck together at the ends with pins, staples or glue can quickly create shapes**

FASHION AND TEXTILES

The initial idea is usually 2D, drawn by hand or using CAD software. This allows the designer to continually refine ideas following feedback from stakeholders and users.

Particular benefits of CAD for fashion and textiles designers include the following:
- Texture mapping and virtual reality mean the CAD models give very realistic results.
- 3D virtual garments refer to the real garment design and the programmes can be linked to produce the 2D patterns, which saves time.

- The simulated garment is visible from all sides so the customer can see the design from all angles and adjust if necessary.
- The designer can use a substantial fabric library to offer solutions and many options.

If testing is required, the following methods could be used:
- Prototype patterns can be developed from **block patterns**.
- A **toile** made from calico or other suitable fabric may be used to test the fit, drape and construction.
- A sample garment could be produced from the intended fabric to test the material, planning and costing.

Texture mapping to create realistic fabric draping

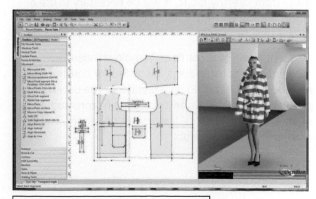

Final CAD garment on virtual model

Figures 7.5–6 **CAD examples of garment representation**

For upholstery, for example, templates may be made of fabric or of firmer material such as paper, card, plywood or polymer. The template must allow for a sufficient tolerance for the type of fabric and purpose for which it will be used.

Sometimes testing is not required. For example, with soft furnishings like cushions, the sizes are standard. The fabrics will have been tested by the manufacturer to ensure they meet all standards.

Prototype development

Once the design is confirmed, a number of repeated cycles of sample preparation, trial fitting and pattern alteration must be conducted before final production commences. This whole process of producing multiple sketches, pattern drafts, alterations and samples can be costly and time consuming, but is a critical part of the concept development.

Pattern development

Modelling can also be achieved using paper to develop ideas. This could be by creating 'flat' models or by manipulating the paper on a mannequin. The paper or card can be folded, scrunched, glued or stitched. By using this technique, the designer can visualise how the actual material will need to be cut and shaped to achieve the desired effect.

When the design has been finalised, the idea needs to be developed into a 3D **prototype**. The first stage will be developing the patterns. The designer or pattern developer will start with a basic block to adapt or, if the idea is totally unique, use a dressmaking mannequin to pin paper to for the desired look. For other textile products the same process could be used without a mannequin.

Toile

The toile is a mock-up of the product and can be fitted and tested until the desired result is achieved. The toile is usually constructed from cotton calico as it is cheap and easy to construct with. However, if the product is to be made from stretchy fabric, it will be necessary to use a fabric with similar performance characteristics for the toile to give realistic results.

Garments are tested for fit on sample standard-size models and textile products are tested for construction method, suitability or to gain feedback. The development process may just involve testing different construction methods or components. It does not necessarily involve making the whole product.

Figure 7.7 **A toile made from calico**

How to develop iterative models

The use of patterns, toiles and other forms of prototyping support an iterative approach and, along with illustrated concept developments, help to show the designer's thinking process. When developing designs, a designer may end up with many versions of the same pattern that will have been adapted in some way or another and a number of toile models. You may use patterns and toiles in a similar way in your NEA project.

Pattern making

The methods for developing patterns include:
- basic block patterns – this is a basic shape usually made out of card from which many different styles can be developed.
- **working patterns** – this is used for marking out design features and can be adapted many times
- **production patterns** – this includes all pattern markings and information to make up the product.

Figure 7.8 **Example of an A Level student's product development and toiles**

Figure 7.9 **Examples of basic block patterns**

One important factor to consider when developing patterns is to include the seam allowance (usually 1.5 cm).

If a pattern needs modifying to create improved fit or adapting to add fullness it is important that you understand the techniques used. You may need to use techniques known as **slashing or spreading** to add fullness (see page 330 for further information).

Pattern drafting

Pattern drafting takes measurements from a person or a form to produce a pattern which is used as a starting point for manufacturing a design. In industry, pattern drafters are employed to create fashion and textile products for the designers. They are experienced and knowledgeable on how fabrics can be manipulated.

The stages to draft a pattern include the following:
- The pattern drafter starts with a sketch of the design.
- Using measurements from a form or person, the patterns drafter starts to 'break down' the garment or product into individual sections that will make the whole product.
- The pattern is developed and shaped using pattern drafting techniques.
- Fabric is cut from the pattern draft.
- A toile is produced and then further modifications are made to perfect the pattern according to the design.

When drafting your own patterns, there are a wide range of techniques you can use to achieve the desired effect. To take a piece of paper and manipulate it to fit a form will involve many trial and error techniques. For example, to allow a flat piece of fabric to fit a curve and give a smooth result, darts or curved seams are used.

When developing your own models, try to select a fabric that has similar properties to the one that will be used for the finished product. For example, if the end result needs to have excellent drape, stretch or stability, select a similar inexpensive version for producing the toile.

It will be necessary for you to practise with a range of techniques. The best way to achieve this is to select

printed and textured fabrics and experiment to see how best to mimic them.

Figure 7.10 **Printed fabrics created using colouring pencils and fibre tipped pens**

3D modelling

Draping is one of the most traditional methods for creating a model. It is still widely used today by many designers and is a more free and experimental approach. The principles behind it include:

- draping the fabric, wrong side visible, on the required form.
- pinning and drawing on the fabric to indicate where the seams, etc. will be
- once the fabric is removed from the form, copying each component onto paper
- adding allowances to the paper. These may be seam allowances, pleating, gathering, etc.

Figure 7.11 **Moulage techniques being used on a toile with pins and markings**

Moulage is the French term for **draping**. An advantage of using moulage is that prototypes can be mocked up quickly, allowing the designer to experiment with different techniques and silhouettes before committing to the final shape. This allows the designer more freedom than working with 'flat' paper, as the fabric can be repositioned quickly.

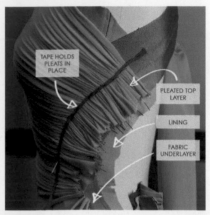

Figure 7.12 Moulage techniques using final fabric, pins and tape

You can also create a toile as follows:
- You will have created a pattern using the draping method or drafted it using a range of measurements.
- Use the pattern, following pattern markings accurately, to cut out the fabric.

- Depending on how quickly and accurately you need the model, you can tack, machine stitch or even staple the pieces together. Remember to follow the seam allowance to get an accurate model.
- Adjust the toile accordingly. At this point the pattern can be modified. This may mean adding or removing fabric.
- Once the fit of the model is accurate, finishing techniques such as piping and binding may be added so that the presentation prototype represents the final product as closely as possible.

Sampling

Sampling is one of key elements of the pre-production processes in the fashion and textiles industry. Before a manufacturer produces bulk orders, a prior sampling of styles are produced to obtain approvals from stakeholders and users before starting the manufacture of products. Once the toile has been finalised and the patterns adapted accordingly, a sample product will be manufactured. This will be made from the intended fabric and components, using production type machinery. Once the sample product has been given the go ahead, the production pattern is finalised.

ACTIVITIES

1 Explain why modelling is such an important part of the **iterative design** process for fashion and textiles products.
2 Explain how modelling is carried out in the workshop and within commercial practice for the fashion and textile products.
3 Take an existing design idea for a fashion and textiles product.

a) Model it in 2D by producing sketches of the pattern templates required to produce it.
b) Model the same product in 3D by producing the pattern templates and pinning together.
4 Use draping techniques on a mannequin using fabric to create a prototype garment. Record each stage.

DESIGN ENGINEERING

While a design engineer may need to consider the physical or aesthetic requirements of a design, they will often need to focus more on creating models to test the functional aspects when developing a solution which is fit for purpose. As such, design engineering models will frequently be used to test and develop mechanical, structural, electronic and software systems.

Making mechanical and structural models

Mechanical and structural systems are quite difficult to design without actually building and testing them. Forces, mechanical advantages and gear ratios can be calculated, but it is often not clear whether a system will work properly until it is built. For example, when using electric

motors and gearboxes it is difficult to predict how fast the motor will turn when it is under load.

Making electronic models

Electronic models are usually constructed in order to check:
- that the interfacing works between the input/output devices and the microcontroller
- that the input/output devices achieve the desired effect
- the functioning of the flowchart program.

Microcontrollers designed for experimental work allow for electrical connections to be quickly made using crocodile clip leads (see Chapter 6). For more advanced designs, which require the use of other components such as resistors or driver ICs, a breadboard may be needed.

FASHION AND TEXTILES

Table 7.1 highlights a number of hand tools that may be useful when creating prototypes. This list is not exhaustive and it may be necessary for you to use other tools to achieve the desired effect.

Table 7.1 **Hand tools used for iterative developments of fashion and textile products**

Tools	Purpose
Tracing wheel	To transfer pattern markings quickly onto the fabric
Dressmaker's shears/scissors	To cut out fabric. You can use inexpensive scissors for paper
Craft knife	To cut thick fabrics and card
Iron	To flatten the seam allowance and to attach interfacing and bondaweb if required
Tailor's chalk	To draw onto fabric temporarily
Seam ripper	To unpick mistakes or to make a quick opening for buttonholes

PRODUCT DESIGN

Laser cutters are used to cut materials like foam board, high-density papers and card and other materials precisely, to create panels used in the construction of structural models. Product design model makers may use moulds and castings or create vacuum-formed pieces. CNC routers or milling machines are widely used to machine parts out of different materials.

However, producing a complex model in this way can be time consuming. CAD/CAM programs now make the production of these parts much simpler and provide high tolerances for parts and components.

Additive modelling process and rapid prototyping

3D printing is known as **additive manufacturing** because it is usually created layer by layer. This process is sometimes referred to as **rapid prototyping**. It allows designers to produce innovative designs that are complex to manufacture that can be produced at a fast pace. The low cost of producing numerous models lends itself to iterative designing and rapid prototyping. This makes it feasible to produce products on demand, rather than in expectation of demand. The parts can be printed in a variety of finishes, colours and textures. Sometimes the parts will require additional hand finishing afterwards, especially if a high-quality model is required, but often the parts come off the machines ready for study and testing.

3D printing electronically 'prints' a part in three dimensions in a base material of powder or liquid (depending on the type of process). The parts can be printed in a variety of finishes, colours and textures. Sometimes the parts will require additional hand finishing afterwards, especially if a high quality model is required but often the parts come off the machines ready for study and testing. A design is created on a 3D computer modelling package and then this data is used to three-dimensionally print the data into a base material. The processes used in rapid prototyping include: SLA (stereolithography), MJM (multi-jet modelling), SLS (selective laser sintering) and FDM (fused deposition modelling).

One of the main benefits of rapid prototyping is the ability to program and build a part quickly. The ability to build a component with complex details makes this type of model-making valuable to the design process and minimises cost and time to production.

Digital simulation

With advances in virtual reality applications, designers are increasingly designing products digitally. The designer can use computer simulation technology to get quick decisions from stakeholders without having to make physical prototypes. The advantages of this include the following:

● Design alterations made in 2D or 3D are immediately visible.
● 3D simulated products refer to the real design; the 2D drawings can be plotted and used for the production of the product.
● The simulated product is visible from all sides so you can adjust it if necessary. This means that errors can be corrected sooner.

- Amendments or alterations are made relatively fast.
- New ideas can be tried quickly and changes can be visualised.

Mathematical modelling and computerised simulation software can be used to test circuits and mechanical devices without the need to physically build them.

As they don't use physical components, money isn't wasted on expensive parts and this can speed up the production process. In the design of structures and moving parts, stresses on individual components can be predicted and parts can be strengthened before physical prototypes are built.

DESIGN ENGINEERING

Making microcontroller software models

The microcontroller program is crucial to the successful operation of the whole system and must play a part in the ongoing iterative design process. It is very tempting to rush this aspect, to leave it to the end, or to assume that the first version is going to be satisfactory.

Microcontroller programs are modelled and tested to remove bugs. Some software packages for programming microcontrollers allow the program to be run and simulated before downloading it to the microcontroller and this can be a useful way to iron out obvious bugs. Running the program on a computer also allows system parameters to be monitored; the program can usually be run one step at a time, giving you a chance to identify exactly where a problem is arising.

Many products have a user interface (UI) that is directly influenced by the microcontroller and a poor UI will make the product difficult or even impossible to operate properly. Allowing real users to test a microcontroller program is essential because they are likely to use the product or system in an intuitive way, and the first attempts at writing the software may not provide a suitable user experience. Iterative software testing is also needed to fine tune program parameters or to calibrate the system. One such example would be in a cooling fan system, where program values are iteratively adjusted until the fan switches on at the desired temperature.

FASHION AND TEXTILES

Advantages of digital simulation for fashion and textiles designers include the following:
- Fabric drape can be accessed and adapted from a fabric library.
- Fast and exact positioning and scaling of prints, logos and other details is possible.
- New ideas can be tried quickly and changes be visualised.
- Complex garment shapes can be realised.
- Fabrics and component availability are unlimited.

PRODUCT DESIGN

Other digital simulation programs such as SolidWorks, Autodesk Inventor and Rhino are all used to create models to show what products visually look like, to test stresses on parts and to explore new shapes.

ACTIVITY

Read about how an algorithm for Rhino (a 3D parametric CAD package) can be used in footwear design at http://blog.rooy.com/post/160701339125/former-yeezy-developer-shares-some-secrets-of-the.

KEY TERMS

Additive manufacture – uses equipment to electronically print a part in three dimensions in a base material of powder or liquid (depending on the type of process).

Block patterns – basic patterns from which patterns for many different styles can be developed.

Calico – a plain-woven textile made from unbleached and often not fully processed cotton.

Draping and moulage – the process of positioning and pinning fabric on a form to develop the structure of a garment design. Moulage is the French term.

Iterative design – a continual and cyclical design development process to refine and perfect the product. Ongoing testing of models and prototypes, incorporating improvements and progressing towards an optimum solution for all stakeholders.

Modelling – the preliminary work or construction used to communicate and test design ideas and which forms the basis for further design iterations and the design of the final product.

Production pattern – the pattern that includes all pattern markings and information to make up the product.

Prototype – a working prototype is a working model made to test the function and feel of a design before production commences. It represents all or nearly all of the functionality of the final product and helps to identify faults and enable last-minute changes to be made.

Rapid prototyping – the technique used to quickly fabricate a scale model of a physical part or assembly using 3D digital technology (CAD). Construction of the part or assembly is usually 3D printing or additive layer manufacturing technology.

Sampling – this refers to the first product to be made from the actual production line, so that operators know what they are going to make. This sample is made with actual fabric, trims and accessories and made by sewing line tailors.

Sketch models – quick models, made from easy-to-work and low-cost materials such as cardboard, calico or foam.

Slashing and spreading – the techniques used to develop models and patterns to add or remove fabric to create the desired fit.

Subtractive modelling processes – these can involve use of clay and wood or styrofoam, to shape, carve, sculpt or machine a model component.

Toile – an early version of a finished garment made from inexpensive material so that the design can be tested and perfected.

Working pattern – this is used for marking out design features and can be adapted many times.

KEY POINTS

- Modelling is a key stage in the design and manufacturing process, enabling testing to be carried out in various formats to assess the viability of the proposed idea.
- Virtually any material can be used to make a model, depending on the purpose of the model, mock-up or prototype, from **sketch modelling** (very quick, sometimes 'rough' models) to show early iterations to stakeholders, to a full-scale prototype of an idea or component that could be fully tested by users, for structural strength, for example.
- The level of 'finish' of models and prototypes is dependent on the purpose of the model/prototype. An iterative model may have the purpose of obtaining views on finishes and aesthetics, in which case a high quality and realistic finish is important.

- In fashion and textiles, toiles can be produced from pattern blocks or freehand manipulation of fabrics.
- Prototyping gives the designer and manufacturer knowledge of design features and potential manufacturing approaches that won't work, enabling them to focus on the ones that do.
- CAD and digital simulation software is used to model and test ideas.
- CAD software enables the designers and manufacturers to coordinate and link production from the design stage to pattern development and cutting.
- The use of models, prototypes and sampling is cost effective in the long run as money is not wasted on production runs that may not be profitable.

7.2 How can materials and processes be used to make final prototypes?

LEARNING OUTCOMES

Design Engineering
By the end of this section you should:
- understand how to select and safely use common workshop tools, equipment and machinery to manipulate materials by methods of:
  - wasting/subtraction processes such as cutting, drilling, turning, milling
  - addition processes such as soldering, brazing, welding, adhesives, fasteners
  - deforming and reforming processes such as bending, vacuum forming.
- demonstrate an understanding of the role of computer-aided manufacture (CAM) and computer-aided engineering (CAE) to fabricate parts of a final prototype, including:
  - additive manufacturing (3D printing) to fabricate a usable part
  - subtractive CNC manufacturing such as laser/plasma cutting, milling, turning and routing.

- demonstrate an understanding of measuring instruments and techniques used to ensure that products are manufactured accurately or within tolerances as appropriate
- understand how the available forms, costs and working properties of materials contribute to the decisions about suitability of materials when developing and manufacturing their own products.

Fashion and Textiles
By the end of this section you should:
- recognise the order of assembly for different fashion and textiles products, including:
  - assembly of fabric pieces including lining
  - addition of working parts such as zips and fastenings
  - reduction of fullness according to the design, using darts, gathers, elastic, pleats
  - adding embellishment
  - adding functional details, such as pockets and quilting.

- demonstrate an understanding of the tools, processes and machinery required to accurately manufacture fashion and textiles products in a workshop environment, including:
  - dyeing processes
  - hand and digital printing processes, such as screen, roller and transfer printing methods
  - transferring pattern markings using tailor's chalk, tailor's tacks and tracing wheel
  - cutting fabrics using fabric shears or a cutting wheel
  - joining fabrics using a sewing machine, overlocker, needles and pins
  - finishing fabrics and garments using a steam iron.
- understand how digital technology, including the use of computer-aided design (CAD) and computer-aided manufacture (CAM), can be used in the making of final prototypes
- understand how the design of **templates** and patterns can ensure quality and accuracy when making a final prototype
- understand how the available forms, costs and working properties of materials contribute to the decisions about suitability of materials when developing and manufacturing their own prototypes.

If you are studying at A Level you should also:
- demonstrate an understanding of the principles of pattern cutting, including:
  - pattern sizing
  - pattern symbols and instructions
  - how to manipulate patterns for different applications.

Product Design

By the end of this section you should have developed a knowledge and understanding of:
- methods of joining similar and dissimilar materials within products to fulfil the following functions:
  - permanently joining materials to include constructional joints
  - temporarily/semi-permanently joining materials
  - adhesion and heat
  - using standard components and fixings
- an understanding of a variety of processes used to manufacture final prototypes in the workshop made from wood, metal and polymers, including:
  - wasting/subtraction techniques such as drilling, sawing, shaping, abrading
  - moulding methods such as **thermoforming**
  - milling metals and turning woods
  - casting of metals such as lost wax casting, sand casting, low temperature and resin casting
  - forming and lamination
  - bending, rolling and forming sheet material
- how digital technology, including the use of computer-aided design (CAD) and computer-aided manufacture (CAM), can be used in the making of final prototypes
- how the design of jigs, formers and moulds ensure quality and accuracy when making a final prototype
- how the available forms, costs and working properties of materials contribute to the decisions about suitability of materials when developing and manufacturing your own prototypes.

A final prototype is a one-off, pre-production version of an intended product. It is manufactured to represent as closely as possible the intended product so that it can be thoroughly tested and evaluated by stakeholders and users. Modifications will be made to the design following testing, before commercial or industrial production begins in quantity.

The following sections describe the techniques, processes and equipment used to complete prototypes within a workshop or design studio environment. You are likely to use several of these processes in your NEA project in the making of prototypes, both as part of the iterative design process and in the final prototypes.

Many of the techniques used when developing a prototype will differ from those used in industry. The prototype should evolve following various methods of modelling and testing. The development of prototypes will allow you to understand the factors that dictate the order of assembly. While the choice of tools and equipment may be restricted within a school workshop or design studio to achieve professional results, they can still be used effectively to create a functional prototype that will allow the designer to make fundamental decisions about further developments and costs. In some cases it might be more appropriate to make a series of prototypes to best demonstrate and test the intended design solution, rather than trying to make one prototype that achieves everything.

Manufacturing tools, equipment, methods and processes for prototypes

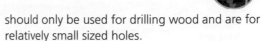
DESIGN ENGINEERING AND PRODUCT DESIGN

Processes, tools and machinery used to accurately manufacture prototypes in a workshop

Wasting/subtraction techniques

Wasting or subtraction techniques remove or cut away material to leave the desired shape, and involve drilling and milling, turning and sawing. The term 'wasting' is used because the material that is removed is usually thrown away. Any type of material can be shaped by wasting.

Drilling

When drilling holes in a material, a number of factors should be considered, including:

- the material being drilled
- the hole diameter and depth needed
- whether a through or blind hole is being drilled
- the rotation speed/feed speed required
- the capacity of the drilling machine.
- the need for coolant
- the method of holding the work piece.

Holes can be drilled, and also reamed – a sizing process by which an already drilled hole is slightly enlarged to a desired size. They may also be counterbored or countersunk, which both involve the enlarging of one end of a hole to accommodate a bolt head or screw head so that it will be below or flush with the work surface.

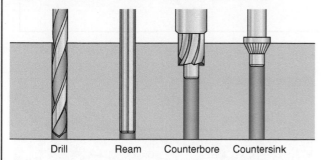

Figure 7.13 **Drilling holes**

For accuracy and safety, the drilling speed is important; for example, aluminium can be drilled at a faster speed than mild steel. Tables giving the correct drilling speeds can be found in many engineering reference books. Drilling jigs are often used to position holes accurately on more than one workpiece. The **jig** locates against a datum position on each workpiece and incorporates a guide hole for the drill.

- Twist drills are probably the most common drilling tools. They can be used on timber, metal, polymers and similar materials, although HSS (high speed steel) twist bits should be used for drilling metals.
- Spur point bits are also known as wood or dowel bits. The bit leaves a clean-sided hole. Spur point bits

should only be used for drilling wood and are for relatively small sized holes.

- Flat wood bits are for power drill use only. The centre point locates the bit and the flat steel on either side cuts away the wood. These bits can be used to drill fairly large holes. Sizes range between 8 and 32 mm.
- Hole saws have interchangeable toothed cutting rings and can be used to cut small or large holes in thin sheet metals as well as wood or polymer. They are best used in a power drill at low speed as the blade saws its way through the material.
- Forstner bits are used to form holes with a flat bottom, such as for kitchen cupboard hinges. These are best used in a pillar drill. If used freehand, the positioning is difficult to control as there is only a very small central point.
- Wood auger bits are ideal when drilling deep holes in wood or thick man-made boards. Auger bits can be used in a hand brace but are now mostly used in portable and cordless power drills. The bit will cut a clean and deep hole.
- Expansive bits are similar to auger bits, but the distance of the cutting edge from the screw point of the drill can be adjusted.
- Adjustable tank cutters are used to cut large diameter holes in thin sheet materials.

Safety procedures when drilling

Drilling takes place in many different practical situations. In some, such as using a battery-powered hand drill to cut a hole in a car body panel, the work is large enough and has enough mass to stay still while being worked on. When a smaller workpiece is to be drilled, however, it must be held securely – both to assure accurate placement of the hole and to prevent it binding to the drill bit and spinning around with the rotation of the drill. This can obviously be a safety hazard, especially if the work is a piece of sheet metal with sharp edges. There are several methods of work holding available, from clamping the work in a vice to be drilled with a portable drill, to fixing it in a machine vice attached to the table of a pillar drill or clamping it to the machine or a work bench with G-cramps if the work is larger. As with any process of this nature, appropriate guarding and PPE (personal protective equipment) should be used.

Sawing

A wide range of saws are available, all varying in size and with different sized teeth that are set according to the material to be cut. Saws are divided into groups known as back saws, frame saws and hand saws.

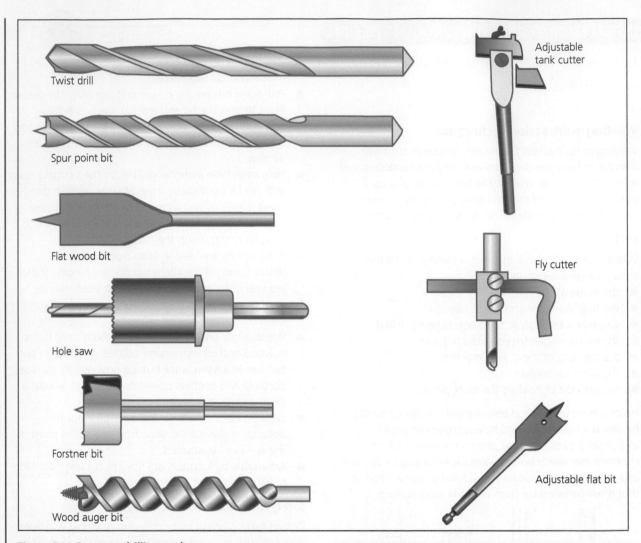

Figure 7.14 Common drilling tools

Twist drill

Spur point bit

Flat wood bit

Hole saw

Forstner bit

Wood auger bit

Adjustable tank cutter

Fly cutter

Adjustable flat bit

Back saws

Frame saws

Hand saws

Figure 7.15 **Types of saws**

Table 7.2 **Descriptions of saws**

Saws	Uses
Back saws	
Tenon saw	General cutting out in wood
Dovetail saw	A smaller version of the tenon saw used for finer, more accurate work; the back of the saw limits the depth of the cut
Frame saws	
Coping saw	Cutting curves in thin sectioned wood
Hacksaws and junior hacksaws	Cutting out metal, hacksaw has a finer blade, a junior hacksaw can be used for cutting thin section of metal and tubes
Piercing saw	Has a fine blade for delicate work in metals; often used by jewellers
Hand saws	
Cross cut saw	Used on large sections of wood when cutting across the grain
Panel saw	Cutting panels in large sheets of wood
Rip saw	Cutting or ripping down the grain on large sections of wood
Other types of saw	
Bandsaws and circular saws	Used by trained teachers and technicians to cut and prepare wood; widely used in the manufacturing industry
Scroll saws or Hegner saws	These are fixed saws that are useful for cutting intricate shapes on thin sheet materials
Jig saws	Jig saws are portable power tools that can be used to cut around curved shapes in sheet material
Power hacksaws	Power hacksaws are also available for cutting through bars of metal and polymers.

The width of the cut that the saw makes is known as the 'kerf' and this must be wider than the blade itself to avoid the saw sticking when the operation is carried out. To create the necessary gap, the teeth are set by turning them to the left or right. Smaller saw blades tend to be set in a wave form.

Snips and shears can also be used to cut thin metal sheet and soft polymers. These may be hand held or bench mounted. When bench mounted the lever greatly increases the force that can be applied.

ACTIVITY

With a magnifying glass, look closely at the shape and set of the teeth on a range of saw blades. Look at both hand-held saws and machine saws. Can you see a difference between those designed to cut metals and those designed to cut wood?

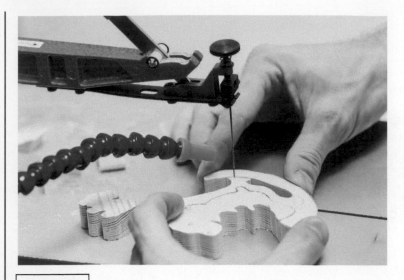

Scroll saw

Jig saw

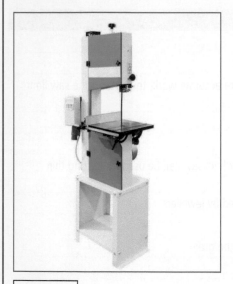

Bandsaw

Circular saw

Figure 7.16 Power/machine saws commonly used in schools and small workshops

Tin snips and pliers

Bench shears

Figure 7.17 **Tin snips and bench shears**

Turning

Timber can be turned using a wood lathe in two ways:

- using a faceplate and holding it on the lathe headstock. This method is used to create items such as bowls, plates and dishes
- between centres when slimmer, long items such as turned table legs, candlesticks or stair rails are produced. Timber used in this process needs to be held between the headstock using a driving dog centre and supported with a fixed or revolving centre in the tailstock.

Although almost any timber can be turned, those particularly suitable for turning include ash, beech, cherry, elm, sycamore and teak.

Metal can be turned using a centre lathe, the most common type of lathe found in workshops and manufacturing industries. The material is machined by cutting tools held in a tool post which can be moved in and out, and along the bar. Longer pieces of metal can be supported in the centre by the tailstock.

Furniture manufacture and the construction industry require large batches of turned components, for example legs, banister rails and so on. Copy lathes can be used to either manually or automatically trace the profile of the required shape and save, store and repeat it. Alternatively, CNC machines that use a profile generated by CAD software can be used.

Figure 7.18 **A typical wood working lathe**

The main function of the turning process is to produce parallel cylindrical shapes or tapered cylindrical shapes to a very high degree of accuracy.

The basic turning process involves the use of a single point cutting tool in which material is removed from the outside diameter of the workpiece, or a twist drill used to bore a hole into (or through) the centre of the horizontal axis of the workpiece. There are many types of lathe cutting tools that are available to the engineer. A wood working lathe is shown in Figure 7.18.

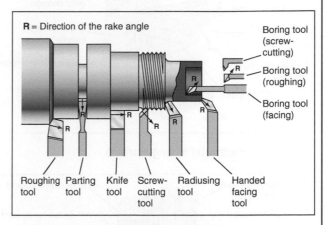

Figure 7.19 **Metal turning lathe with turning tools**

The removal of material is achieved by clamping the material that is to be worked on (the workpiece) firmly into a work-holding device.

There are two main forms of chucks. One is a three jaw self-centering chuck and the other is a four jaw independent chuck. A third work-holding device is called a faceplate. Their uses depend on the initial form of the workpiece.

A face plate allows irregularly-shaped casting or forgings that are too big for a standard four jaw chuck to be securely held in place. It also enables diameters and faces to be machined on workpieces that are parallel or perpendicular to a surface which is pre-machined to a flat finish. The pre-machined face is placed against the face plate allowing the parallel surface to be machined.

Milling

Milling allows for the rapid removal of material using multi-tooth cutting devices. The workpiece is secured to the machine work table and is fed under the cutter. There are two basic types of milling machines:

- horizontal miller – the tool spindle axis is in the horizontal plane
- vertical miller – the tool spindle axis is in the vertical plane.

A universal miller is similar to the horizontal miller but has a cutting head that can be swivelled through a prescribed angle.

Milling machines are used to machine flat surfaces, slots and steps. They have a work table that can be raised or lowered and also moved horizontally in two perpendicular directions. On most milling machines it can also be set to traverse automatically beneath the cutter.

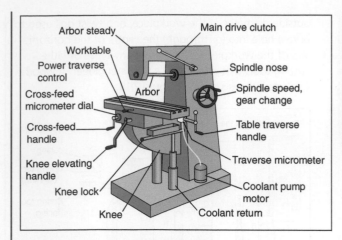

Figure 7.20 **Horizontal milling machine**

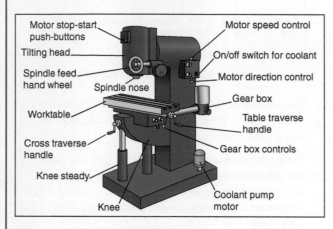

Figure 7.21 **Vertical milling machine**

The spindle, which holds the cutting device, can be driven either by a variable speed drive motor or through a gearbox so that the correct cutting speed can be selected depending on the nature of the material being worked.

For milling operations, there is a wide variety of cutting tools, all of which comprise of a series of wedge-shaped teeth that have been ground with suitable rake and clearance angles.

The most common cutting tools used in horizontal milling operations are:
- Slab cutters – sometimes known as slab mills or roller mills. They are used to produce wide, flat surfaces.
- Side and face cutters – these have their cutting teeth around the periphery and the side faces of the tool. They are generally used for light facing operations and for cutting slots and steps in a workpiece.
- Slotting cutters – these are somewhat thinner than either of the two above cutters and have cutting teeth on the periphery only. They are used for cutting narrow slots and keyways in shafts.
- Slitting saws – the thinnest of all milling cutting tools. They are used to cut very narrow slots and also to cut material to size (parting off).

CNC lathes and milling machines are commonplace in industry and also schools and colleges.

Fig 7.22 **A CNC milling machine**

Shaping

Shaping of materials can be carried out with tools such as files, rasps and surforms.

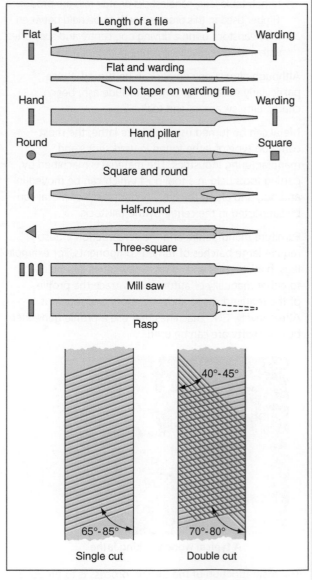

Figure 7.23 **File types**

Files are available in a variety of shapes, lengths and grades of cut. The shape of file – flat, round, half round, three square and knife – can be selected according to the profile of material being cut. A swiss or needle file is used for more delicate work.

The selection of cut will depend upon the amount of material to be removed or the surface finish required. Rough cut and bastard cut files are used to remove material quickly. The use of these files can be followed by a second cut file and finally smooth or dead smooth cut files can be used to produce a surface ready for finishing with abrasive paper or emery cloth.

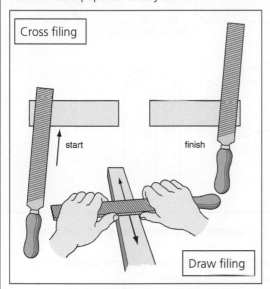

Figure 7.24 **Filing is normally carried out by pushing the length of the work, which is called cross filing; surface finishing is done by draw filing**

Rasps are similar to files but they have coarser teeth and are more suitable for use on wood.

Surform tools are available in a range of shapes and sizes with replaceable blades. Due to surforms having a range of blades, all with different cutting edges, they can be used on several material groups including hardwoods, softwoods, soft metals, nylon and polymer laminates.

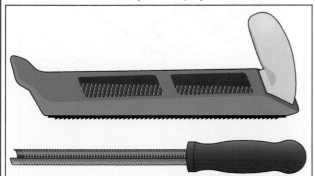

Figure 7.25 **Surforms**

Abrading

An abrasive is any substance that wears down a surface by rubbing against it. Abrasives are available in many forms, including powders, compounds, papers, wheels or disks, brushes, belts and more.

- Material can be removed from woods, metals and polymers using abrasives. Disc and belt sanders are commonly used.

- Grinding of metals can be carried out using a disc or angle grinder, surface grinder or off-hand grinder. All the grinders make use of discs that have been made from abrasive grit that has been bonded together. The size of the abrasive grit used determines the grade or coarseness of the discs.
- Glasspaper is an abrasive sheet that can be used manually to smooth wooden surfaces. Emery cloth can be used on metals and silicon carbide paper can be used on polymers. Aluminium oxide is another grit commonly used in abrasive sheets.
- Disc sanding machines use abrasive sheets mounted on a backing disc. Belt sanders use a continuous abrasive band over two rollers and workpieces are pressed against it. These machines are used for cleaning up and finishing work. You may also have used portable power sanders in your workshop.
- Polishing or buffing machines use abrasive compounds such as Tripoli applied to cloth mops.

Addition processes

Permanently joining materials

The properties of some materials make it necessary to create a mechanical joint to join components. This is particularly true of timber where the internal structure means that any joint that relies on contact of the end grain is unlikely to be successful. Joints provide a greater surface area for gluing and sometimes require an additional fastener or fixing to increase strength.

Joining timbers (Product Design only)

Timber structures are usually frame or carcase construction. A number of joints are used in frame construction – for corners, 'T' joints and cross joints where pieces cross over.

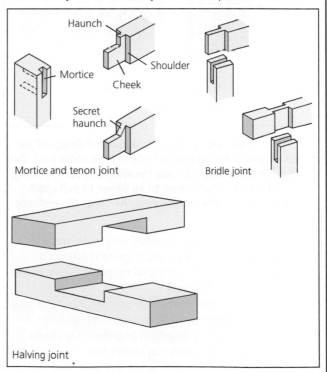

Figure 7.26 **A selection of joints**

Carcase construction methods are used for boxes, shelves, etc. The aim of this construction is to provide mechanically strong joints such as dovetails or large gluing areas such as dowels and comb/finger joints. A dowel can be a through hole or a blind hole that hides the joint completely. These joints lend themselves to machine manufacture and take advantage of the advances in modern adhesives. Finger joints are particularly suited to mass production. Accurate machining allows simple assembly and produces a strong joint that requires no additional fittings.

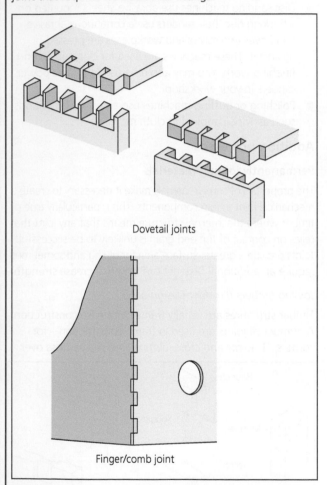

Dovetail joints

Finger/comb joint

Figure 7.27 **Compare the angular shape and design of the dovetail joint with the parallel sides of the comb joint. One joint is ideally suited to be load bearing and resist pulling forces; the other would seem to be easier to pull apart. However, we may choose one over the other for aesthetic reasons**

Sometimes it is necessary to obtain very wide material for such things as table tops and cupboard sides or to create longer lengths for structural purposes. Some of the following methods are used:

● tongue and groove joint – only one piece has a groove cut in it; the other is cut to provide the matching tongue. This joint is easy to fit together. It is usually glued but in some cases can be left free

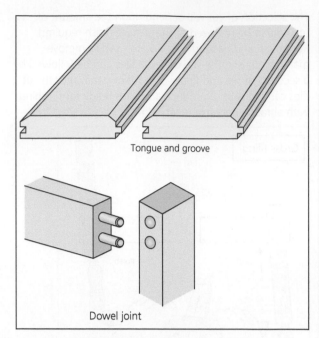

Tongue and groove

Dowel joint

Figure 7.28 **Tongue and groove joint and dowel joints**

● dowel joints – created with matching holes and dowels forming the joint
● biscuit joint (sometimes called a flat dowel joint) – a matched profile joint produced by the router cutter
● scarf joint – a type of finger joint used by manufacturers to produce long lengths of timber for production. It minimises waste because short lengths that would otherwise be unusable can be joined to make usable lengths of material.

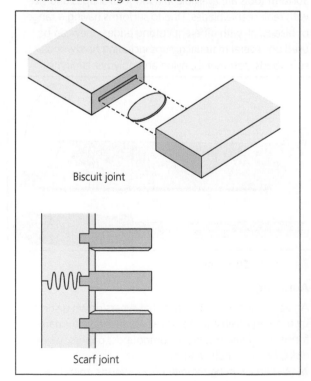

Biscuit joint

Scarf joint

Figure 7.29 **A biscuit joint and a scarf joint**

Riveting

Riveted joints are usually associated with joining flat sheet metal components, but the principle can be applied to most resistant materials and even fabrics. The stages involved in making a basic riveted joint are shown in Figure 7.30.

A riveting set is used for setting or pressing together two metal plates and ensuring the rivet is pulled all the way down the hole. The hole in the set is the same size as the diameter of the rivet. The rivet snap is used to support the head of a round head while riveting, and to finish a round head rivet to the correct shape. It has a concave hole the same size and shape as the rivet head, and two are required when using a round head rivet.

Both set and snap can be combined in one tool and are available in various sizes to fit different rivet diameters. In practice a complete riveted joint may be made up of many individual rivets to give the required strength and the diameter of the rivet can be from 3 mm to 50 mm.

A significant drawback to the standard riveted joint is that access to both sides of the joint is needed. This is overcome with the use of pop (or blind) rivets. These tend to be relatively small in diameter (2 mm to 5 mm). Blind rivets usually consist of a hollow aluminium alloy rivet (rather like a top hat) and a steel pin. They are passed through the hole in the components.

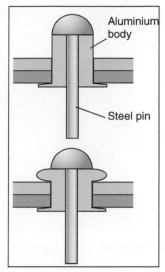

Figure 7.31 **Stages in pop riveting**

There are a range of technologies that are specifically designed to join dissimilar materials. Inserts used in plastic mouldings provide features such as threaded anchorages and electrical terminals. Metallic inserts can be added at the moulding stage or by pressing into moulded or drilled holes. Alternatively, an interference fit joins two parts by friction. They are usually brought together by pressure, or expansion or contraction of one part, using heat, cooling or shape memory properties. Other mechanical methods include crimping and stitching.

ACTIVITY

Find four different wood joints used in furniture and household items. Sketch the joints in their context and state why each joint is suitable for that particular product or situation.

Soldering

Soldering involves two or more metal components that are joined together by melting and flowing a metal filler material into the joint. The metal filler material has a lower melting point than that of the materials being joined. Traditionally an alloy containing lead and tin was used as the filler but concerns about the health and safety implications of the fumes produced during heating resulted in lead-free alternatives becoming more common.

For a bond to be successful, the metals to be joined must be extremely clean so, in addition to clean working conditions, a 'flux' is used to clean the surfaces and prevent oxidisation. Soldering is most commonly used for electrical connections (because of its low electrical resistance) and can be automated for the production of printed circuit boards.

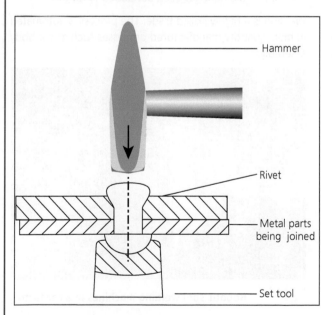

Figure 7.30 **Stages in riveting**

Figure 7.32 **A typical light-duty soldering iron**

Brazing

In principle, brazing is similar to soldering but takes place at a higher temperature (>450°C) by using a different filler material. The filler materials are typically either an alloy of silver (when the technique is sometimes called silver brazing or silver soldering) or brass or bronze-based alloys.

As with soldering, a flux is applied to the joint before heating to clean the surfaces of the components to be joined to allow the filler metal to bond effectively with their surface layers.

Typically, the filler materials are an alloy of silver (when the technique is sometimes called silver brazing or silver soldering), brass or bronze.

In most cases a gas torch is used to provide the heat needed for brazing – butane, propane and oxy-acetylene can all be used successfully.

It is important to remember that for both soldering and brazing, the metals to be joined do not melt, and that usually the temperatures involved in the process are considerably lower than the melting points of the metals that are joined.

Welding

Welding is usually associated with the joining of metals but in principle can be applied to many thermoplastics. The materials are joined by heat. The pieces to be welded are placed together, heated, usually with the addition of a filler material, brought to melting temperature and then left to solidify on cooling. This produces an extremely strong permanent joint. It is essential that the materials to be joined and filler material are similar.

In principle many metals can be joined by welding but, because of their metallurgical structure, some are extremely difficult to weld. For example, mild steel is considered one of the most easily welded metals yet stainless steel, which

contains a higher level of carbon, is considerably more difficult to weld successfully. Types of welding include:

- Friction welding uses mechanical energy (for example a rapidly rotating component pressed against a stationary component)
- Gas welding usually uses a mixture of oxygen and acetylene
- Electricity is used in arc-welding (MMA), metal-inert-gas (MIG) and tungsten-inert-gas (TIG) welding and resistance (spot) welding
- Ultrasonic energy (sound waves) are commonly used in the welding of polymers

While the metal is molten during welding it is likely to absorb impurities from the atmosphere that will weaken the material once it solidifies. In gas welding this is prevented by increasing the flow of oxygen to the flame, which effectively excludes the ambient atmosphere.

In MMA welding the molten metal is protected by a shield of inert gas produced by the vaporisation of a powdered coating around the welding rod. In MIG and TIG welding the shielding gas is introduced directly from bottled sources. The higher temperatures involved in welding (approximately 3000°C is typical) can cause significant problems of distortion. For some joints, where ultimate strength is not necessary, this problem is reduced or eliminated by spot welding, that is, making the weld in isolated spots rather than a continuous seam, which generates much more heat. Specialist equipment has been developed to do this.

Welding is a highly skilled trade. The process is automated in high-quantity manufactured processes such as car body assembly.

Figure 7.33 **Robotic spot welding machines in a car factory**

Adhesion

In some situations, adhesives offer significant advantages over other joining methods. They:

- can produce discrete joints with little visible evidence of how the joint is secured
- reduce the need for mechanical joints or other complex joining processes, therefore reducing manufacturing time, level of skill required and costs

- make permanent joints that, when correctly made, will not fail or become loose under normal design loading
- can act as sealants by filling the gap between components.

Increasingly, adhesives are used for joining dissimilar materials in manufacturing. The use of adhesives involves additional processing, for example, preparing surfaces to be joined and waiting time for setting or curing, although high-performance, fast-acting adhesives are becoming more common. Increasingly, adhesive tapes are used, reducing the need for setting time.

Synthetic adhesives are often stronger than the materials they are joining. Natural adhesives are generally made from non-toxic substances and present few hazards in use. Synthetic adhesives can contain substances requiring careful use.

Commonly used adhesives:

- Polyvinyl acetate (PVA) is a rubbery synthetic polymer usually sold as an emulsion in water. It works very well with porous and semi-porous materials to which it bonds by penetrating the surface. To work really effectively, pressure must be applied to the joint, usually through clamps or other devices that can be attached temporarily. Drying time is 30 to 60 minutes.
- Hot melt adhesives become a viscous liquid at fairly low temperatures. They are applied to the joint while hot and the joint is closed immediately. The main advantage of this type of adhesive is the speed of curing, but they seldom result in a high-strength joint; the most commonly used is EVA.
- Polyurethane resin (PU) is a multipurpose adhesive that works with almost any porous material. It is formulated to begin the curing process when the resin comes into contact with moisture (normally contained naturally in the materials to be joined). This is known as a 'reactive adhesive' and usually these have a shorter setting period. Another well-known example of a reactive adhesive is cyanoacrylate or Super Glue. Other reactive adhesives include epoxy resins (a thermosetting plastic). These are usually supplied in two parts that are mixed to initiate curing. Epoxy resins form a strong bond with the surface layer of the materials to be joined and so work well with both porous and non-porous materials. They are useful because they can join materials that are quite dissimilar to each other and a joint made under ideal conditions will have great strength. Other synthetic resins used as adhesives include urea formaldehyde polymer, sold as a high quality waterproof wood glue under the trade names of Cascamite and Extramite.

- Silicone adhesives and silicone sealants are based on tough silicone elastomeric technology. Silicone adhesives have a high degree of flexibility and very high temperature resistance (some are flexible between -40°C and 1200°C) when compared to other adhesives, but they lack the strength of other epoxy or acrylic resins. The most common silicone adhesives and silicone sealants are room temperature vulcanising (RTV) forms, which cure through reaction with moisture in the air and give off acetic acid fumes during curing.
- Pressure-sensitive adhesive (PSAs) are very common in everyday life in the form of adhesive tapes such as masking tape or sticky tape.
- Solvent adhesives are a niche group of adhesives used solely to join some polymers, particularly those containing styrene. They dissolve the surface of the materials to be joined into a semi-liquid state. The surfaces then flow together until the solvent evaporates, leaving the components permanently joined.

Contact adhesives are either based on natural or synthetic rubbers in a solvent solution. This is applied to both surfaces to be joined and allowed to dry. During this process the rubber forms a bond with the surface of the material and, when the two items are brought together, the faces of the adhesive bond rapidly to each other. This joins the two materials together quickly and without the need for clamping or support. They are most effective where the area to be glued is large.

ACTIVITY

Explore the strength of adhesives. Use a range of different adhesives and join a range of different materials, following the manufacturer's instructions closely (for example, allowing time for the glues to fully set). Devise and carry out some simple tests (safely and sensibly!) and record your findings.

Temporarily/semi-permanently joining materials

Mechanical fastenings are important in the production of many products. They allow similar and dissimilar materials to be joined and many allow the joint to be disassembled for maintenance and/or adjustment.

Self-tapping screws cut their own thread and require only a pilot hole before use. They are easily identified by a fairly coarse open spiral thread with a tapering diameter – sometimes reaching a sharp point. The most common self-tapping screws are wood screws. Wood is a material in which it is difficult to produce a threaded hole ready for a machine screw. Self-tapping screws are also available for other materials, but the angle of the thread and point are not the same as those for wood. Self-tapping screws are used when repeated disassembly and assembly of the product is unlikely.

Machine screws can be used with a nut or a tapped hole. They can be identified by their evenly pitched thread and constant (non-tapering) diameter. It is essential that the thread of the screw and hole correspond exactly. Historically there were many different types of thread but in recent years ISO metric screw threads have become the manufacturing standard. Diameters typically range from 2 mm to 20 mm, with the thread denoted M2 to M20. When it is likely that a joint will be taken apart, for adjustment or maintenance, a machine screw will be used.

Because they tend to damage the material into which they are screwed, self-tapping screws are used when repeated disassembly and assembly of the product is unlikely. When it is likely that a joint will be taken apart, for adjustment or maintenance, a machine screw will be used, even though it is likely to be more expensive to manufacture.

Over the years many different drive systems have been introduced to supplement the familiar slotted and Pozidriv® screws. In some cases, these have been developed to allow greater torque to be applied to the screw and in others they are intended to prevent the end user dismantling the product. Hex-drive (Allen key), Torx and Triwing are popular types of drive system for screws used in engineering applications. Screws are also available with a number of different types of head styles to suit different applications – these include countersunk head, pan head, button head, flange head and socket cap.

Nuts, bolts and washers

A bolt is similar to a machine screw. The difference is that a machine screw is threaded over its entire length whereas a bolt is threaded at the end and has a plain shank running between the thread and the head of the bolt. A bolt is inserted into a hole that is slightly larger in diameter than itself and the screw threads at the end of the bolt are secured into a nut or threaded component. There are several different types of nuts available for use with bolts. The most common is a plain hexagonal nut but other types are frequently used, for example, locking nuts (with a nylon locking ring) that prevent accidental loosening of the joint and wing nuts that allow the joint to be tightened without the need for tools.

Washers are an essential part of a bolted joint. Several different types of washer are available for different applications. They help to prevent damage to the components being joined and some are developed to prevent loosening of the connection. Most common are plain and spring washers.

Figure 7.34 **Washers help prevent damage to components that are being joined together**

Designers often find that a joint is required between different materials. One example is a joint between a timber panel (for example, a shelf) and metal tubing. Mechanical fasteners such as screws and threaded fasteners or 'knock down' fittings can often be used.

ACTIVITY

Make a study of the use of mechanical fasteners in products. Look around your school, college or home. Find as many different types, different materials and different finishes as you can. Use images from the internet to produce a poster for your Design and Technology area.

Using standard components and fixings (Product Design only)

Many mechanical joining systems have been developed for the manufacture of flat-packed furniture and products made from multiple materials. They are manufactured in huge numbers, which results in low unit costs. They allow furniture to be assembled at home without the need for specialist equipment. The use of manufactured boards such as chipboard, MDF and plywood in modern furniture, and the growing interest in self-assembly, has led to the need for a whole new range of joining methods. These are known as knock down (KD) fittings. They allow components to be assembled or taken apart as many times as required, without weakening the joint or the material.

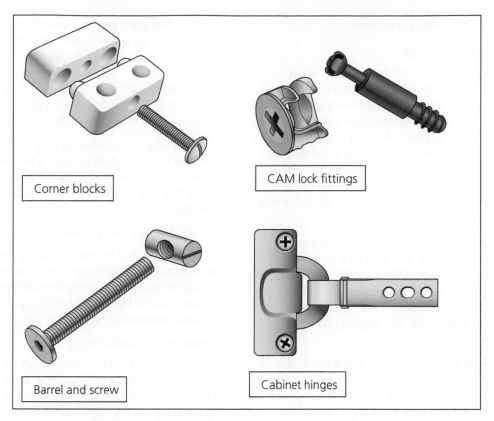

Corner blocks

CAM lock fittings

Barrel and screw

Cabinet hinges

Figure 7.35 **Commonly used KD fittings**

Some of the most common types are:
- Corner blocks. A matching set of nylon blocks are screwed to the inside of the joining sides. A central screw locks them together. They work quite well but look unsightly.
- Barrel and screw. A nylon or metal barrel is fitted into a drilled hole in one piece, the other is drilled to take a machine screw. The screw is tightened into the barrel. A dowel should be used alongside the screw to prevent twisting.
- Self-locking plates. Each plate is screwed to the joining sides. Then, as they are pushed together, a springing action locks them together. They produce a strong joint commonly used in bed frames but also in a wide range of products in similar and dissimilar materials; they are often used in conjunction with dowels, which may or may not be assembled with adhesive.
- CAM lock fittings. These are used extensively in flat-packed furniture; they are quick and simple to assemble and, like barrel and screw fittings, are often used together with dowel joints.
- Corner plate and screw bolt. A central screw bolt is fitted into the corner piece or leg. The side pieces or rails are slotted for the plate and have normal joints cut on the end. The joint is fitted together dry, the plate located over the bolt end and a wing nut tightens the whole unit together.

- Cabinet hinges. These are ideal for hanging doors to cupboard carcasses made from man-made boards. They fasten onto the faces (rather than the fragile edges) of the boards and provide easy adjustment for correct door alignment.

Although all KD fittings are simple to use, their success relies on the amazing accuracy of modern manufacturing methods that utilise CNC machinery to pre-drill holes ready to receive the fittings. If KD fittings are to be used in the school/college workshop, jigs and templates are invaluable to position holes as accurately as possible.

Other components used in furniture include screw cups and caps, many types of hinge mechanisms and fittings, drawer runners, sliding door fittings, latches, catches and locks. Many were once only available for commercial manufacture and for use by tradespeople. All are now readily available for use in small workshops and in the making of prototypes and help to give a commercial feel to products.

ACTIVITY

Examine a range of products in different materials. Identify the fasteners and fittings used in their assembly and make sketches and notes of how they have been used. Explain the benefits they bring to the product and the advantages to the manufacturer and user.

Deforming and reforming processes

Polymer moulding methods

Thermoforming is a manufacturing process in which a polymer sheet is heated and then formed to a specific shape using a **mould** or former. Many thermopolymers can be thermoformed, they include polystyrene, polypropylene and PVC. Thermoforming is one of the most common methods of producing polymer components and vacuum forming is one of the most common as it is accessible to both schools and colleges and large scale production. Other types of thermoforming are drape forming and plug-assisted forming.

Thermopolymers in sheet form can be heated gently to between 160°C and 180°C and then bent into shape. If the sheet is held in position while it cools it will remain in its new shape while it is at room temperature or below. If the polymer sheet is reheated, it will try to return to its original shape of a flat sheet. This property is known as plastic memory.

To bend a straight line a strip heater or line bender can be used. The heating element heats the polymer only along the area held above it. When the polymer in the heated area becomes soft it can be removed and held over a former, or a simple jig, to hold it in shape until it cools.

Vacuum forming is a process that is commonly used to make simple trays or containers. A mould is produced, usually from plywood or MDF. The polymer is heated and a vacuum is used to remove the air and allow the polymer sheet to take the form of the mould. This is then cooled and trimmed to shape. The process of vacuum forming is outlined in more detail later in Section 7.3.

Press forming is also known as plug and yoke forming. A two-part mould is used to shape a heated sheet of polymer. In school this mould can be made from plywood or MDF. The shape that is the plug is smaller than the hole in the yoke.

Casting (Product Design only)

Complex shapes may be required during the making of a prototype; this is not always possible using standard material removal techniques so the prototype is made in one piece, usually by means of a casting process. There are several methods of casting that can be used.

Investment/Lost wax casting

Investment casting is also known as lost wax casting. It is used where a complex component is needed that requires a high degree of accuracy and is difficult to machine after casting. The moulds used in this process are made from fine refractory materials that can withstand wide heat

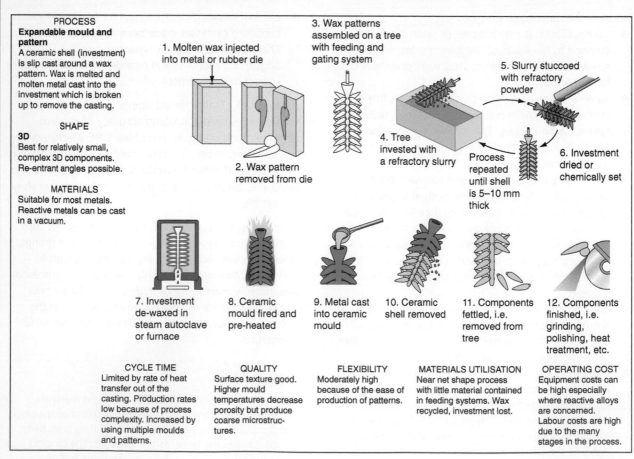

PROCESS
Expandable mould and pattern
A ceramic shell (investment) is slip cast around a wax pattern. Wax is melted and molten metal cast into the investment which is broken up to remove the casting.

SHAPE
3D
Best for relatively small, complex 3D components. Re-entrant angles possible.

MATERIALS
Suitable for most metals. Reactive metals can be cast in a vacuum.

1. Molten wax injected into metal or rubber die

2. Wax pattern removed from die

3. Wax patterns assembled on a tree with feeding and gating system

4. Tree invested with a refractory slurry

5. Slurry stuccoed with refractory powder

Process repeated until shell is 5–10 mm thick

6. Investment dried or chemically set

7. Investment de-waxed in steam autoclave or furnace

8. Ceramic mould fired and pre-heated

9. Metal cast into ceramic mould

10. Ceramic shell removed

11. Components fettled, i.e. removed from tree

12. Components finished, i.e. grinding, polishing, heat treatment, etc.

CYCLE TIME	QUALITY	FLEXIBILITY	MATERIALS UTILISATION	OPERATING COST
Limited by rate of heat transfer out of the casting. Production rates low because of process complexity. Increased by using multiple moulds and patterns.	Surface texture good. Higher mould temperatures decrease porosity but produce coarse microstructures.	Moderately high because of the ease of production of patterns.	Near net shape process with little material contained in feeding systems. Wax recycled, investment lost.	Equipment costs can be high especially where reactive alloys are concerned. Labour costs are high due to the many stages in the process.

Figure 7.36 Investment casting

ranges and which can provide fine dimensional accuracy. The process involves the following procedure:

- A wax pattern is made of the component being manufactured; if there are a large number of these components being made, then a mould would be made to produce the wax pattern.
- The wax pattern is then coated with a refractory slurry (either by dipping or spraying); as the slurry dries, it forms a hard, brittle shell which now forms the 'die'.
- The die is placed in a preheated furnace or autoclave and the wax is melted out; this process leaves a cavity and also assists in the setting of the refractory mould.
- Molten material is then poured into the cavity and allowed to solidify; when solid, the refractory lining of the mould is broken off.

This process allows castings to be made from virtually any material that can be melted, commonly metals and polymers, and gives a high degree of accuracy and complexity. It is an expensive, slow process because of the number of stages that need to be followed.

Sand casting

Sand casting is a process in which an impression is made in sand using a pattern of the required product. The pattern is slightly oversized to compensate for the shrinkage of the final product during the cooling stage. The process is commonly used for machine components, car engine blocks and turbines.

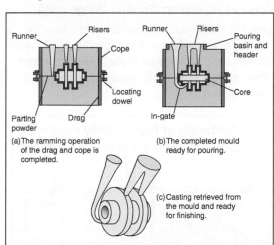

Figure 7.37 The sand casting process

Wood is quite often used to make the pattern, particularly for small batch work, as it is a relatively cheap material that can be easily worked into the desired final shape. Sand is used (usually green sand or Petrobond) because it is strong enough to withstand the pressures of the molten metal and permeable enough that it can allow the escape of the hot gases produced during the casting process.

Depending on the complexity of the final product, the pattern may have to be made in two halves, which are joined together using locating dowels as shown in Figure 7.37. This is called a split pattern.

- One half of the pattern is placed face down on a 'turnover board' and the lower half of the two-part steel mould, called a drag, is placed around it.
- Parting powder is then sifted over the pattern until it is completely covered (this is to assist the final removal from the mould) and the whole is then packed tightly with green sand.
- The drag is then inverted, the turnover board removed and the second half of the pattern located onto the first using the locating dowels.
- The top half of the mould, known as the cope, is then placed in position and the process is repeated. The only difference is that two extra items are inserted before packing with sand: these are known as a runner, which allows for the pouring of the molten metal into the cavity of the mould, and the riser (which is always placed in the highest part of the cavity), which allows the gases to escape and also shows when the mould is full. These two pieces also act as reservoirs for the molten metal so that the casting can draw down additional metal as it cools.
- To prevent cohesion of the two parts of the mould a parting powder is dusted onto the sand in the drag before the cope is placed in position.
- Once the mould has been packed, it is carefully separated and the pattern removed to leave a mould cavity. At this stage, if required, ready-formed sand 'cores' can be placed into the mould cavity to produce holes in or through the finished casting.
- The molten metal is then poured into the assembled mould. During the solidification stage, the metal in the mould contracts slightly, hence the need for a slightly larger pattern at the outset.

This process allows for the manufacture of complex-shaped engineered products from virtually any metal that can be melted. Although the sand can be reused, the moulds cannot and have to be remade each time. This makes it a time-consuming process and suitable for one-off and batch production only.

The major disadvantages of the basic sand casting process are as follows:

- The mould has to be remade for every component manufactured.
- The overall accuracy of the finished product is relatively poor.
  These disadvantages are overcome by the use of die casting (see page 333)

Resin casting (Product Design only)

Resin casting is a method of casting where a mould is filled with a liquid resin which then hardens. It is primarily used for small-scale production and prototypes. It requires little initial investment and is used in the production of collectible toys, models and figures, as well as small-scale jewellery production. The synthetic resin for such processes is a monomer or plastic thermoset. A curing agent is mixed with the resin (often referred to as a catalyst, usually up to 5 per cent is added).

Typical uses start with traditional sculpting processes where a clay sculpture or wooden mould is made, although 3D printed parts can be used. From this, a flexible mould is made from a (RTV) silicone rubber. The mould can be made of one or two parts depending on complexity.

- Position the part in a moulding box usually made from wood. Boxes made from LEGO also work well and can be reused.
- Mix a quantity of silicone rubber and pour the silicone rubber into the moulding box
- Allow the silicone to cure
- Remove the original part from the silicone mould
- Mix the casting resin and pour the mixed resin into the silicone mould.
- Allow to cure and then de-mould.

Figure 7.38 Models made from resin casting

Low temperature casting (Product Design only)

Some metals, such as pewter, which melts at approximately 245°C, can be cast easily in the school/college workshop. Moulds can be made from many materials, including heat resistant silicone rubber, MDF, wood or cuttlefish. Moulds can easily be made by hand or using CNC milling or laser cutting. The mould usually has two parts, which are held together with small cramps or sometimes wire. The metal ingot is cut and placed in a ladle; a brazing torch flame can then be used to heat the metal, which will melt and forms a molten liquid. The molten metal is then poured into the mould.

Once it has cooled, the mould is split open and the casting removed. Waste material, such as the sprue, is easily removed using hand tools. A soldering iron can be used to remove blemishes after casting and the material can be finished easily with files and wet and dry paper.

ACTIVITY

For the manufacture of the following components, explain which casting process would be the most suitable:
- A motor vehicle engine block
- A model toy car
- A turbine blade for a jet engine
- A piece of bespoke jewellery
- The housing for a small electric motor
- A water pump for a motor vehicle

Figure 7.39 Pewter casting in a workshop

Forming and lamination (Product Design only)

The term 'laminated' refers to components where the grain of the laminates (layers) is uni-directional. Curved components are created when multiple layers of wood and adhesive are assembled in moulds or formers and pressure is applied.

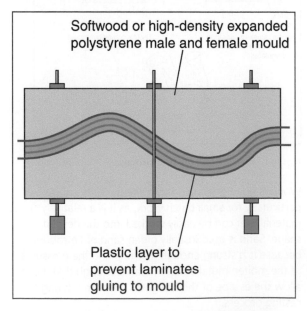

Figure 7.40 Salad server moulds using simple male and female formers

Producing an item using the process of **laminating** greatly increases its strength. Plywood is manufactured by laminating an odd number of thin layers of wood veneers together, each layer glued at 90 degrees to the previous layer. An odd number of veneers is used so that the grain on the outer surfaces runs in the same direction. This method creates a strong product, especially when compared to natural timber of an equivalent thickness.

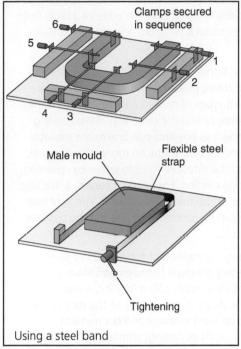

Figure 7.41 **Laminating larger structures**

In 1830, Michael Thonet began producing laminated and steam-bent chairs in Europe. There are many design and economic advantages of "bentwood." Designers and manufacturers can build their own formers or purchase a range or standard shapes/components.

'Male' and 'female' formers are typically used for production of laminated or formed-plywood parts. The contact surfaces must be high quality. A plastic film can be used to protect the product and moulds from glue spillage. A sealed, pressurised rubber tube system is preferred for more complex shapes.

Larger structures, such as table supports, can be produced using a male **former** secured to a base. The laminates are glued and held using clamps with protective blocks. The clamps are tightened in sequence.

Vacuum press systems are commonly used in commercial furniture manufacture and are increasingly used in schools and colleges. Large moulds can be built quickly from expanded polystyrene, with a thin laminate glued to

its surface to ensure a high-quality finish and strengthen the mould. The laminates are glued, placed on the mould and put inside the heavy-duty clear PVC bag. Air is withdrawn from the bag using a vacuum generator. Atmospheric pressure (up to 8250 kg per square metre) ensures uniform and even compression.

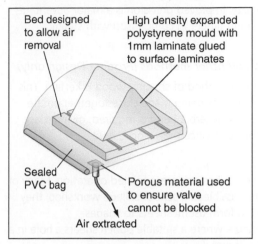

Figure 7.42 **A vacuum press system**

Bending, rolling and forming sheet material

Steam bending (Product Design only)

Because of the grain structure of timber, it can only be bent significantly in one direction, namely, at right angles to the grain flow. Bending along the grain gives a limited curvature, so in cases where a tighter curve is required the timber is first steamed to make it more pliable.

As timber is dried following the felling of the tree, it loses water from the cell boundaries which increases its stiffness. Immersing the timber in steam reverses this process and makes the timber more workable. If the timber is held in the bent position it will keep this shape as it dries out.

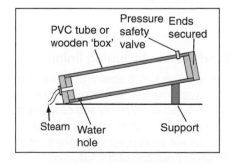

Figure 7.43 **A simple steam chamber for bending wood**

Kiln-dried wood is often unsuitable for steam bending because the lignin in the wood is set during the hot, dry kiln process. Air dried woods (known as seasoned wood) such as ash, elm, cherry and oak can all be easily steam bent. The amount of time needed for steam bending varies according to the timber selected. The drying timber is, however, prone to twisting and bending and this can lead to a lack of accuracy and repeatability. Improved precision is a major advantage of laminating timber in comparison with steam bending.

Kerfing and other methods (Product Design only)

A traditional method of shaping wood is kerfing. This involves saw cuts being made close together on the inside surface where a bend is required, giving space for compression once the wood is bent.

Pressing

Pressing describes any press-related operation. Many of these are possible in a school/college workshop; they are used to form useful sheets to shapes:

- piercing – where a suitable punch shears a hole in a piece of metal
- blanking – where the punch shears the required shape from the metal
- notching – where a punch shears an open-sided hole in the metal
- cropping – where the punch shears or cuts a plain or shaped length from the metal
- bending – where a suitable punch shapes the metal by a folding process
- forming – where the metal is forced into the shape of the surface contours of a die.

The above terms cover the majority of the shaping and manipulative techniques used in sheet metal work and, to a lesser extent, on plate. Metal up to about 3 mm in thickness is classified as sheet metal; above 3 mm it is called metal plate. Virtually all presswork operations are carried out on sheet metal as the majority of engineered and manufactured components are made from metal that is less than 2 mm thick. In a school/college workshop, sheet metal notchers, guillotines, folding bars and sheet metal rollers are common.

The easiest way to bend metal is using a former or bending the metal over an edge or shape. You can push a flat piece of metal over the edge of a table (with your hands) to make a bend but a former will ensure accuracy.

A sheet metal bending machine is the most common way to make clean, precise bends in metal. There are many different types of metal folder available but they all work in more or less the same way. Metal folding machines known as box-and-pan brakes are available that clamp and fold the metal on more than one side, after bending the sides are fixed together by soldering, welding or pop riveting. For bending curves, a machine that has three adjustable rollers is used. The tightness of the curve can be controlled by altering the position of each roller.

Metal pressing is a method of forming sheet steel into a shape by pressing it between two shaped dies. The pieces of sheet steel, called 'blanks', are pressed into the same shape as the surface of the dies. Metal pressing can be used to shape and cut metal in different ways such as cutting, embossing, bending, and flanging. You could also use a hammer and shaping dolly to bend the metal or by placing metal on a leather sand bag and shaped using a bossing mallet. Often the metal is annealed first to allow it to be worked more easily.

FASHION AND TEXTILES

Order of assembly for prototyping different fashion and textiles products

Assembly of fabric pieces including lining

The order of assembly is essential to ensure the product is manufactured to the quality desired and to minimise errors during manufacture. Product disassembly and commercial patterns are extremely helpful to understand the stages of manufacture. This order will be dependent on the product but will usually involve any number of the following:

- **fusing** onto fabric, if required
- reducing fullness
- assembly of the fabric pieces
- addition of working components; for example, zips
- decorative stitching; for example, top stitching
- lining construction (this is usually made from a different fabric from that of the main product and

made up in a similar way. This is added to the product to improve its appearance, drape or warmth)

- functional details; for example, pockets (these can be added at various points throughout the assembly process and are dependent on the design)
- finishing – this could be finishing hems, adding components like buttons
- **pressing** – ironing should be carried out throughout the making of the prototype. The seams, darts and so on should be ironed at each stage as they are completed.

When deciding the order of assembly, the basic principle is to think about the finished prototype. Visualise it and try to work out the assembly in reverse. This can help as it can establish what will not be accessible once the lining has been added, for example.

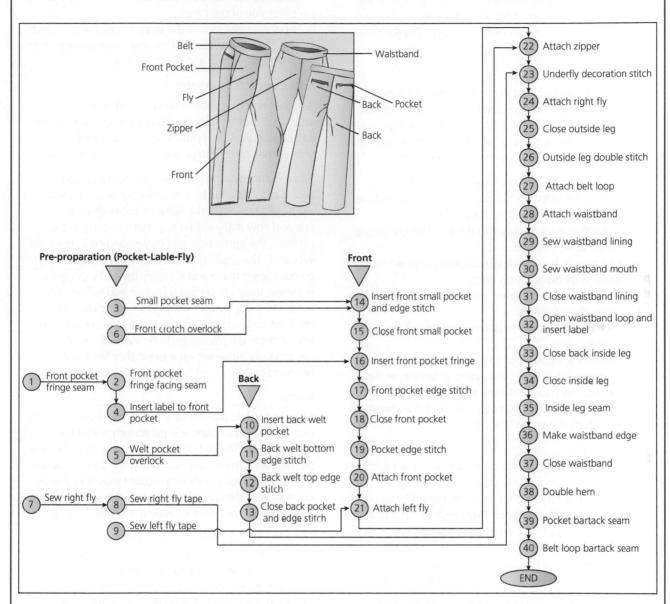

Figure 7.44 Order of assembly

Fusing

Fusing generally refers to applying another layer to the main fabric to reinforce it. This can be done by hand with an iron to fuse the two layers together; for example, by applying interfacing or using Bondaweb for a hem. Interfacing is used in sewing to add structure or firmness to certain areas of a garment – such as cuffs, collars and waistbands. The main thing to ensure is that the interfacing is a similar weight to the fabric so that it adds firmness but doesn't make it too firm.

Any type of fusing onto fabric tends to be the first stage carried out once the fabric has been cut and before sewing, as the fabric is still a flat surface and therefore easy to iron interfacing on to.

Table 7.3 Applying iron-on interfacing to fabric

Stages	Method
1	Cut the interfacing to the same size as the pieces it is to be applied to, remembering that the sticky side is applied to the reverse side of the fabric.
2	Trim off the seam allowance from the interfacing, if required.
3	Place the interfacing onto the wrong side of the fabric and cover with a fine cloth or fabric (this protects the iron and the interfacing).
4	Gently press with a hot dry iron for a few seconds until the interfacing is attached to the fabric.

Addition of working parts

Working parts are explained in further detail in Chapter 6. The emphasis here is on how to include the working parts as part of the assembly process. You will be able to use this section when constructing your own prototypes.

Zips

Zips are usually included in the assembly of a product once the elements of the design that are to be 'closed' have been part assembled. For example, if using a zip in a skirt, trousers, dress or jacket, it would be inserted once the main templates are stitched together

To insert a zip, regardless of style, requires a sewing machine to be fitted with a zipper foot. This allows the machinist to sew close to the zipper teeth.

- Stitch the seam together to the point where the zip needs to be inserted.
- Machine or hand tack the seam together where the zip is to be placed. Iron open.
- Place the zip, face down, over the wrong side of the seam. Pin into place.
- Tack the zip into place down each side of the zipper tape.
- Attach the zipper foot to the sewing machine.
- Machine stitch down the centre of the zipper tape. Pivot at the base of the zip to stitch across the end.
- Unpick to tacking down the centre of the seam until you reach the end of the zip.

Tips

- Select a zip that matches the colour of the fabric, unless you want it to contrast.
- If the zip is too long it can be shortened by doing a few stitches at the point you wish to cut it to and then cutting off the excess. The stitches will prevent the zipper from coming off.
- The top end of the zipper tape should match the edge of the fabric to which it is to be stitched.

For open-end zips, the steps are similar but the seam is not stitched together before attaching the zip. The zip can also be 'sandwiched' between the outer fabric and lining.

Invisible zips are particularly useful when making fashion and textile products where the zip should be hidden and not spoil the aesthetics of the product.
- Finish the edges of the fabric using zigzag stitch or using an overlocker.
- Iron the zipper tape to make it flat (on an invisible zip the edge curls slightly and ironing it helps to sew close to the teeth).
- Place the zip to the edge of the right side of the fabric. Place the pins at right angles to the edge so the sewing machine can stitch over them.
- Attach the invisible zipper foot to the sewing machine.

- The invisible zipper foot has two little grooves at the bottom – slot the zip teeth into the left groove. Stitch all the way down one tape. Repeat for the other side of the tape.
- Now we need to sew the seam below the zip. Close the zip. Change the foot on your sewing machine to a regular zip foot and stitch the seam.

Buttons and buttonholes

Buttons are used for decoration as well as for fastenings. Buttonholes are usually completed near the end of manufacture. Buttons are also attached at this time as they can interfere with the assembly process.

When manufacturing your own prototypes, if you decide to include buttons as a functional closure you will need to establish the distance between each one and how many will be required. On commercial patterns, the buttonhole and corresponding buttons are indicated. The position of the buttons depends on the product being made and is largely up to the designer. However, there are certain rules that you will need to adhere to if the buttons are to work as a closure down the front (or back) of a shirt or jacket. For example, on a shirt, if there is a placket (as in Figure 7.45), the buttons run vertically. However, on a jacket they tend to run horizontally.

Velcro

Attaching Velcro is a relatively simple and a cost effective method of manufacture. It is particularly useful for children's wear, quick-fastening requirements and where garment adjustments are required. It can be used when making prototypes to show a product closed to give an impression of the completed product rather than going to the expense and time of adding the 'real' fastener.

Some Velcro is self-adhesive and is either pulled off the backing paper and placed in position or heated to bond it to the fabric. For a more permanent method of attachment, Velcro can be machine stitched into place. Due to its thickness it is quite difficult to hand stitch.

Figure 7.45 **Buttons and buttonholes on a shirt placket**

Other fasteners

There are a wide range of other fasteners that can be selected depending on how they are to function as part of a product. (Many of these are featured in Chapter 6.) Due to the fact that many fasteners can obstruct the machine presser foot, they are generally added at the end of the assembly process.

- Poppers are used on the front of jackets, and are both functional and decorative.
- Press studs are used for concealed extra fastenings; they are not very strong on their own and are used where there will be little strain.
- Parachute clips tend to be added to straps and webbing as a secure method of fastening. They often feature on bags and rucksacks.
- Hooks and eyes are used in positions of strain; a good example is at the top of a zip.
- Laces and ties give a decorative way of fastening; rouleau loops are fastened with laces or used with buttons.
- Buckles are used on belts and bag straps.
- Toggles are used on sports and outdoor pursuits clothing and equipment.
- Magnetic fasteners are used as a hidden closure on textiles products, particularly bags.

Reduction of fullness

Textile products need to have shape; this is achieved not only by the cutting but by techniques that dispose of fullness. Sometimes the disposal of fullness is visible as part of the aesthetics of the product. When developing prototypes, different disposal techniques can be trialled to establish which would be the most effective.

You need to be able to describe all of the different techniques listed with the aid of annotated drawings and recognise why they are selected.

- Reducing fullness involves mathematical calculations. It is useful to know how much fullness needs to be reduced, the measurement it needs to be reduced to and by which method.
- Elastic and gathers allow for more **tolerance** but pleats and darts must be measured accurately to get the perfect fit.

Darts

Darts are folds of fabric that end in a point, giving a smooth shape. Stitching a dart involves folding and removing a triangle shape in order to give a flat piece of fabric a 3D form. They are used extensively in garment design and are usually completed as one of the first stages of the product assembly because if the dart is in the seam line, stitching the dart results in the seam being the correct size before attaching to the corresponding side, and if the seam is in the middle of a garment that

is to be lined, the dart would be less accessible after attaching the lining.

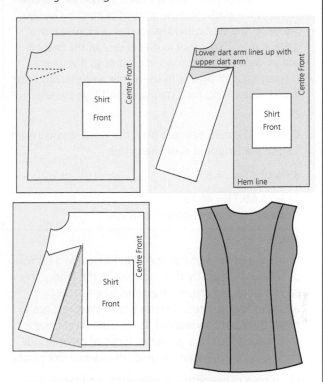

Figure 7.46 Disposal of fullness in garments using darts

There are three main types:

- single pointed, as used in the waist of skirts or trouser
- double pointed, generally used on dresses where there is no waist seam
- dart tuck, which is an opened dart used as a feature on blouse shoulders or for waists.

Gathers

Gathering can be done by hand or machine. This technique is widely used in many textile products. Two or three parallel lines of even, long stitches are pulled up to evenly draw in the fullness to the required amount. The gathers are distributed evenly within the fixed length required. This technique is usually one of the first stages of assembly.

Elastic

Elastic is available in many different widths that are suitable for a multitude of applications. It is a particularly useful method for reducing fullness as there is an increase in flexibility. It is often used in the waist of children's garments where no other fastening is required. The main options for adding elastic are:

- sewing the elastic direct to the fabric
- creating a casing to thread the elastic through (this is favourable when the elastic is to be hidden)
- using shirring elastic to create an elastic panel.

Calculating the amount of elastic

> The amount of elastic you will need is dependent on the width of the elastic and the amount of stretch. Usually, the wider the elastic, the less you need. However, as a general rule, 25 per cent to 40 per cent of the finished actual length is required. So, for example, if you were making a pair of pyjama trousers with an elasticised waist (using wide elastic) for a standard size 12, you would subtract:
>
> Waist = 71 cm 35 per cent from the waist measurement to give the length of elastic required:
>
> 35% of 71 = 24.85
>
> 71 – 24.85 71 = 46.15 cm
> (For ease, round this down to 46 cm)

Pleats

Pleats are used to give shape with fullness in a fashion or textiles product by forming folds in the fabric. This can be done in synthetic thermoplastic fabrics by applying heat to press a creased line firmly in place to make the pleats permanent. Unpressed pleats give a soft, folded look to a garment. Types of pleat include knife pleats, box pleats, inverted pleats and kick pleats.

Pleats have to be constructed accurately and can be time consuming in their preparation. However, accuracy is crucial to the end product to ensure the pleats hang correctly. Due to the extra fabric required and the time taken to complete, pleats can increase the cost of the product.

Figure 7.47 Examples of pleats

Calculating fabric required for pleats

> Pleats may be an integral part of your design and as such you will need to develop a pattern to calculate the amount of fabric required. This can only be achieved using careful measurements. The design and type of pleat selected will dictate this, but below is an example of how to calculate the fabric for a size 12 simple knife pleat skirt. The same principle can be used for other fashion and textile products.

> Waist measurement = 71 cm
>
> Depth of each pleat = 2 cm (this is then doubled as the fabric is folded to create the pleat)
>
> Number of pleats = 12
>
> So the first calculation would be:
>
> 71 + (4 × 12) = 119 cm
>
> You would then need to add the distance between each pleat (for this example we use 1.5 cm).
>
> 119 + (1.5 × 12) = 137 cm
>
> So the total length of fabric required would be 1.37 metres.
>
> Note that this does not take into account seam allowances or the length of the skirt.

Varying forms of pleats are also used in curtain construction. The pleats can be achieved by hand or curtain heading tape can be purchased that creates the desired pleat effect once it has been stitched to the curtain.

Tucks

Tucks are narrow folds held in place by a line of machining on the tuck. Very narrow tucks are referred to as pin tucks and are often used as a decorative finish as well as to dispose of fullness.

Figure 7.48 Examples of tucks

Adding embellishment

Appliqué

Appliqué means applying one fabric on top of another. Appliqué was traditionally carried out by hand but manufacturers now use industrial machines. Appliqué **embellishments** can also be purchased by the manufacturer in the form of a pre-manufactured component. This is then ready to include in the assembly stage, which speeds up the manufacturing process.

EMBELLISHMENT

One of my objectives at the start of the project was to focus on embellishment. I will explore different methods of embellishment by creating samples and by sketching designs. The embellishment will act as a decorative feature of the dress to blend in with the theme "Nautical".

Experimenting with appliqué

Appliqué is one of the embellishment techniques I researched in my trend board. It was one of my main trends I discovered in wedding dresses. Most of the shapes were very floral as this too was another major trend. However instead of experimenting with floral shapes, I will develop this to Nautical related shapes such as shells and create small samples of appliqué.

Appliqué is method of placing a fabric piece on top of the main fabric and sewn decoratively either hand stitched or using the machine. I used a decorative stitch on the machine that zigzagged around the piece. I cut the fabric piece in a spiral as this was inspired from my moodboard. I used a different fabric to show the contrast.

From my experiment I discovered that appliqué is a technique to portray to people the different fabric and shape of the appliqué. However if I used this technique it will give texture and some relief to the garment. However as I showed and discussed with my client, so commented that the decorative stitch may be too much and may look to messy for a wedding dress. In addition most of the time on the day people will be looking at the back of the dress compared to how much they will look at the front of the dress as the bride will be facing the back to the audience and moving around to greet numerous guests.

Experimenting with negative appliqué

Negative appliqué is the same principle as normal appliqué but the fabric that wants to be shown is sewn underneath the fabric using zigzag stitches. When it is sewn in the desired shape, using scissors I cut the area that was sewn around showing the fabric underneath (white). Unlike normal appliqué this technique can be seen as messy if the fabric is not cut accurately as small amounts of fabric that are unable to cut with sharp scissors can ruin the overall look. In addition This will cost more that appliqué as I would need to buy double the amount of fabric to create this sample for my final garment.

Experimenting with quilting

Quilting is another decorative technique where padding is placed between two layers of fabric. It is then sewn diagonally using a machine creating diamond shapes. Looking at the sample it was very attractive and I could add small beads or gems where the lines intersect, however as it was very thick compared to one layer of fabric, this might make the garment uncomfortable for the bride as it is very heavy and as padding is used for warmth, make the bride hot. I could use padding in some areas however this may still make the client warm. The client agreed with my opinions, therefore I experimented with other techniques.

YOUNGJU IM #1065 Centre 14723   Page 29

Figure 7.49 Examples of a student work using extensive appliqué

The stages in appliqué are:
- using patterns or markings to cut out the appliqué design in fabric
- applying interfacing or Stitch-N-Tear to the underneath of the background fabric to strengthen it if required
- positioning the fabric design by pinning and tacking, or by ironing Bondaweb onto the right side of the background fabric (the heat melts the Bondaweb and holds the pieces in place)
- machine stitching in place using a zigzag stitch or, if hand sewing, using a blanket or herringbone stitch
- trimming and pressing.

Another example of appliqué is mola. This is where the fabric is layered and the outline of the required image is stitched through all the layers. The fabric is then cut away in layers to access the required layer to make the image.

Free machine embroidery

Free machine embroidery has developed from the traditional hand processes and is carried out as follows:
- The presser foot is removed from the machine and the feed teeth are lowered.
- The fabric is stretched tightly in an embroidery ring, which is used upside-down to keep the fabric flat on the bed of the machine.

- The fabric is positioned under the needle and the presser foot is lowered.
- The fabric is moved around under the needle to fill in the design.
- Straight or small zigzag stitches are used for small areas and wide zigzag for larger areas.

Figure 7.50 Examples of free machine embroidery

Computer controlled stitching

Industrial embroidery machines are usually controlled by computer software and are used to embroider on a large scale. The most common use of these machines is to produce a logo on fashion and textiles products. This can be done at varying stages of product assembly and occasionally the completed product is sent to a specialist company to carry out this process. This can be more cost effective.

Computerised embroidery is, however, usually completed before assembly for a number of reasons:
- If an embroidery design is to added to a pocket, for example, it would need to be added before the pocket is attached to the main product. If it were attached afterwards, the pocket would not function.
- The fabric that the embroidery is going to be added to must fit flat into the frame that is attached to the machine. After a product has been assembled this is difficult to do.

Luxury embellishment

As fashions come and go, the use and popularity of embellishing fabrics follows suit. However, certain embellishments remain popular, particularly in bridal wear and luxury gowns, in the form of sequins, beading or crystals. The addition of these forms of embellishment to any fabric or textile product increases the cost. At the bespoke level, embellishing fabrics for one garment can take weeks or months to complete.

Figure 7.51 **Luxury embellishments of fabrics**

There are many ways to attach beads and crystals to fabric and several factors should be considered when deciding how you are going to attach them.

- Some forms of embellishment cannot be washed by the same method as the fabric to which they are applied.
- Many fashion and textile products that have beading, crystals and sequins attached have to be dry cleaned which can deter some of the target market due to cost.

Methods of attaching embellishments include:
- Sew on – using a conventional needle and thread or a beading needle for very fine holes. This method is time consuming.
- Glue on – using specialist fabric glues. This method can be surprisingly strong and definitely quicker than sewing.
- Hot fix – glue on the reverse is activated when heat is applied. Crystals can be bought on a transparent film that is simply laid onto the fabric and ironed on. This is a quick method of application; hundreds can be set at one time.

If planning on doing a placement of crystals or beads on a product, it can be useful to plan the design first on paper and then use a hole punch to indicate where each embellishment should be positioned. The mark can then be transferred onto fabric using tailor's chalk.

Functional details

Pockets

Pockets come in many shapes and forms to fulfil certain requirements. The main ones are for storage and security. However, they can be for warmth or simply for decoration. The style of pocket usually dictates at which stage of the product's assembly the pocket is added.

The order of assembly when considering pocket application is that when the pocket is to be included in a seam, it is added before the seam is sewn. If the pocket is a welt version, it is constructed pre garment assembly as this is the most complex style of pocket. Patch pockets tend to be added before the main assembly and are usually the simplest to construct.

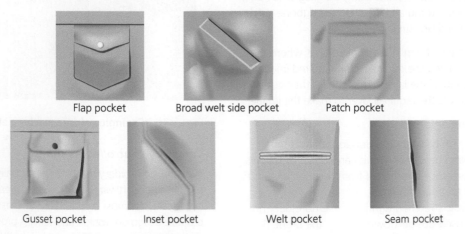

Flap pocket   Broad welt side pocket   Patch pocket

Gusset pocket   Inset pocket   Welt pocket   Seam pocket

Figure 7.52 **Examples of different pockets**

Table 7.4 **Assembly of a seam pocket**

Stages	Image
Stitch pocket facings to pockets.	
Stitch pockets to side seams and press.	
Lay front and back pieces together and pin so they match. Stitch the seam above and below the pocket opening.	Pivot at turn points
Reinforce the opening above and below the seam.	Before you turn to the right side, secure-stitch 2 cm from both sides of pocket opening Pocket facing is needed to prevent lining being visible from pocket openings

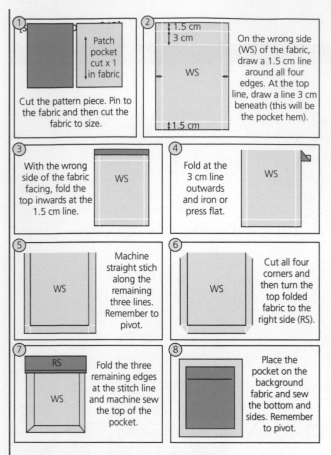

Figure 7.53 **Assembly of a patch pocket**

The diagram panels contain the following text:

1. Patch pocket cut x 1 in fabric. Cut the pattern piece. Pin to the fabric and then cut the fabric to size.

2. 1.5 cm, 3 cm, WS, 1.5 cm. On the wrong side (WS) of the fabric, draw a 1.5 cm line around all four edges. At the top line, draw a line 3 cm beneath (this will be the pocket hem).

3. WS. With the wrong side of the fabric facing, fold the top inwards at the 1.5 cm line.

4. WS. Fold at the 3 cm line outwards and iron or press flat.

5. WS. Machine straight stich along the remaining three lines. Remember to pivot.

6. WS. Cut all four corners and then turn the top folded fabric to the right side (RS).

7. RS, WS. Fold the three remaining edges at the stitch line and machine sew the top of the pocket.

8. Place the pocket on the background fabric and sew the bottom and sides. Remember to pivot.

Quilting

Quilting involves the sewing together of two fabrics with a layer of wadding in between. The fabric can be bought ready quilted, which may be a sensible option when developing prototypes. However, this fabric can be expensive. If cost is a crucial factor it may be sensible to create your own quilted fabric. This will also give you the option of greater fabric choice and quilting pattern. The quilting can be done at one of two stages:

- at the first stage of assembly before cutting out the pattern templates. This is usually the preferable option as quilting after the templates are cut out reduces the size of the product
- after the templates are cut out.

Quilts are usually filled with polyester wadding, which is warm, washable, lightweight and easy to work with. Quilting is essentially functional, to add insulation to the product, but it adds decoration as well. Quilting is used in clothing and other textile products.

Figure 7.54 **Example of quilting**

Tools, processes and machinery required to accurately manufacture fashion and textiles products in a workshop environment

It is important that you understand the different techniques and processes required for making fashion and textile products within the workshop environment. You should be able to differentiate between the tools, machinery and processes required to produce similar end results whether in the workshop or industry. This section concentrates on the workshop.

Dyeing processes

Fabric is primarily dyed for aesthetic purposes. The fabric content can dictate how successful the dyeing will be and the fabrics have to be prepared prior to dyeing to remove any finishes that may hinder the dyeing process.

Natural dyes, in the form of pigments from vegetables and plants, have been used for many years. However, for convenience, modern dyes are produced from chemicals and the ones normally purchased to dye fabrics in the workshop are in powder form. This powder has to be mixed with water, either in a bowl or in the washing machine. It is crucial that the dye remains colour fast. This means that the dye will not 'run' or fade with future washes. To achieve this, a mordant is added to act as a fixative. In the workshop, salt is often used. The basic process of dyeing fabric in a workshop is shown in Table 7.5 on page 309.

Table 7.5 **Dyeing fabric in a workshop**

Stage	Process	Tools/machinery
Preparation	The fabric must be washed before dyeing to remove any surface finish. Tip: Weigh the fabric to calculate the amount of dye required.	The fabric can be washed in the machine or by hand. Use scales to weigh the fabric.
Preparation	Mix the dyes. The dyes need to be mixed according to the instructions on the pack as there are many variations. Some dyes require salt to be added.	Equipment required depends on the method chosen but if hand dyeing you will need a deep bucket. Rubber gloves must be worn.
Dye	Place the fabric in the diluted dye and leave. Dyeing in the washing machine takes one wash cycle to complete. If dyeing in a container, leave the fabric in the dye for the required time.	Equipment required depends on the method chosen.
Finishing	Set the machine to the rinse setting. If hand dyeing, rinse the fabric numerous times until the water is clear.	Equipment required depends on the method chosen.

Tips

- Cotton, linen, wool, silk and viscose usually dye to the colour advertised on the pack.
- Polyester and other synthetic fibres dye to a much paler variation than the one advertised on the pack or not at all.
- The original colour of the fabric will affect the finished result. To get a true colour, the fabric should be white.
- Dyeing can be done after assembling the products but components and thread will not dye to the same colour.

Hand and digital printing processes

Printing differs to dyeing as it is the placement of colour onto the surface of the fabric. There are various methods for printing onto fabric in the workshop. Some are traditional techniques and some are simpler versions of the method used in industry.

Block printing

This is one of the oldest methods of printing onto fabric. Traditionally, engraved wooden blocks are dipped into dye and stamped onto the fabric. Similar effects can be achieved in a workshop by using styrofoam. The designs can also be created using a laser cutter. This is a laborious method of printing and subsequently is an expensive method when larger quantities of printed fabric are needed.

Figure 7.55 **Block printing**

Screen printing

Screen printing is a common method of printing onto fabric, as it is quick. Variations of screen printing are used in industry but the same effects can be achieved in the workshop. The traditional method was to use silk but today polyester can be used. Screen printing involves stretching fabric over a screen. Dye is then pushed through the screen using a squeegee and a stencil blocks off the areas that do not require colour. The same screen can be used multiple times after washing. One colour is printed at a time, so several screens can be used to produce a multi-coloured image or design. Detailed and intricate designs can be created by using numerous colours. Figure 7.56 shows the process:

1 Stretch fabric over a wooden screen.	**6** Use a squeegee to evenly spread the dye over the stencil.
2 Attach the fabric.	
3 Create a stencil.	**7** Check for an even spread.
4 Place the stencil on the screen.	**8** Remove the screen.
5 Apply dye.	

Figure 7.56 **Screen printing in the workshop**

Roller printing

Roller printing is a technique of applying print designs to fabric, similar to block printing. The main difference is that engraved copper rollers are used instead of wooden blocks. A different roller is required for each colour. The ink is fed through to the rollers from a reservoir. This method of printing is expensive due to the rollers required and is unsuitable for small runs. However, it is fast. To create the same effect in the workshop for a prototype, block printing would be a more appropriate method.

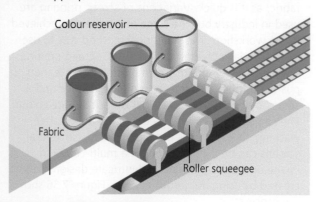

Figure 7.57 **Roller printing**

Transfer printing

The term 'transfer printing' refers to the methods used to transfer a print from a non-textile substrate onto fabric. The advantage of using these methods is that small runs or individual prints can be achieved. This is particularly beneficial if you want to create a placement print or a bespoke printed fabric for a prototype at a relatively low cost.

Sublimation transfer

The design is first printed onto a special type of paper with sublimation inks. The pattern is then transferred to the fabric with the aid of a pressurised, heated press. The temperature is very high and causes the dye to pass into the vapour stage (sublimation). The vapour diffuses into the fabric. This method of printing is particularly suited to 100 per cent polyester fabric.

If you have access to a sublimation printer but are restricted by print size, create a print that can be positioned repeatedly onto the fabric to create an all-over print.

Film release transfer

The design is printed onto heat transfer paper. The ink is 'held' in a layer which is transferred completely to the fabric from the paper using heat and pressure. The ink film layer is bonded onto the fabric by heat. The paper is then peeled away. The image remains on the fabric.

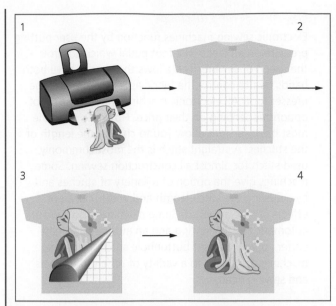

Figure 7.58 **Heat transfer design in a T-shirt**

Digital printing

Digital printing works much like a desktop inkjet printer. During the printing process, the fabric is fed through the printer using rollers and ink is applied to the surface as tiny droplets. This is an ideal method for producing samples or one-off designs.

There are a number of different types of digital printing ink technologies. Each works on specific fabric types and is best suited to particular purposes. Fabric choice dictates the ink selection. These inks are divided into two categories:

● dye
● pigmented.

Table 7.6 shows the types of ink used for different fabrics and their applications.

Table 7.6 **Fabrics used for digital printing and their application**

Ink	Fabric	Application
Acid	Silk, wool, nylon	Ties, scarves, swimwear
Reactive	Cellulosic	Fashion
Pigment	Cotton, cotton/polyester	Furnishings
Disperse/sublimation	Polyester	Fashion, sportswear, automotive

The fabric should be pre-treated before printing. This assists the print quality and is known as padding.

Once the fabric has been printed it must be 'fixed'. This is achieved by steaming the fabric so the printed ink bonds to the fibres of the fabric.

Transferring pattern markings using tailor's chalk, tailor's tacks and a tracing wheel

All patterns have markings on them that have to be accurately transferred onto the fabric. The markings on the pattern are there to indicate the positioning of details such as darts or pockets. However, these markings are only required on the fabric temporarily and should be easy to remove when no longer required. There are a number of ways to transfer the marking.

● Tailor's chalk can be purchased in a number of forms but they all perform the same function. The chalk temporarily marks the fabric where required. The markings can then be brushed off.
● Using tailor's tacks is the most traditional method for transferring the marking and is still used for bespoke tailoring, but it is a time-consuming process. The advantage of using tailor's tacks is that they don't rub off and are easy to see.
● The tracing wheel is a more modern method for transferring markings and is excellent for transferring the position of darts, etc. The wheel has small spikes on it which make minute holes in the fabric where

the markings are required. This is a faster method for transferring the pattern markings but is not suitable for delicate fabrics.

Thread a needle and take a stitch through the pattern and all the layers of fabric (1). Take another stich on the same spot, leaving the tail of the thread long with a short loop (4). Gently pull the pattern off, then seperate the fabric and snip the threads (5).

Figure 7.59 **How to sew a tailor's tack**

Cutting fabrics using fabric shears or a cutting wheel

When producing prototypes or bespoke products, the fabric will be laid out as a single or double layer and the patterns are pinned onto the surface. Alternatively, weights can be used. The pattern and fabric should be kept as flat as possible. Once the pattern markings have been transferred to the fabric, the next stage is cutting out. This can be done using either fabric shears or a cutting wheel. Fabric shears are the traditional method. These are scissors that are larger than usual, as the fabric needs cutting out in long clean cuts. A cutting wheel is accurate but takes more skill to use.

Figure 7.60 Tips for cutting out fabric

Note the following when cutting fabric:
- Do not use too many pins as they can distort the fabric.
- Make sure the fabric shears are sharp.
- Do not lift the pattern or fabric off the cutting surface as everything needs to stay as flat as possible for accuracy.
- Keep the lower blade of the scissors against the cutting surface.
- Weights are better to use than pins for knit fabrics as pins can stretch the fabric before cutting.
- For fine, delicate fabrics that have a tendency to slip when cutting, place the fabric between tissue paper while cutting.
- Always cut out from the wrong side of the fabric so markings can be placed easily without having to move the patterns.

Joining fabrics using a sewing machine, overlocker, needles and pins

Once the fabric has been cut out, the next stage is assembly. It may be necessary to hold the fabric pieces together temporarily before progressing to the sewing machine by using pins or by tacking the fabric together along the seams or hems. This will prevent the fabric from slipping when sewing on the machine and will give a more accurate result.

Sewing machines

There are a huge variety of sewing machines available but in a workshop producing prototypes, the sewing machine will be mainly used for sewing seams and hems. The main types of sewing machine are electronic, computerised and overlockers. A combination of these is more than sufficient to complete most prototypes.

Electronic sewing machines

Electronic sewing machines function by the user putting pressure on an electronic foot pedal which controls the speed of sewing. This allows the user to have both hands free to control the fabric as it passes under the presser foot. All electronic machines have a variety of options depending on their price. However, even the most basic versions allow you to change the length of the stitches. A straight stitch is the most commonly used stitch for almost all construction sewing. Some machines give the option of a variety of stitches and by adjusting the stitch length and width, different effects can be achieved. Some machines also have an automatic tension setting and an automatic thread cutter, as well as a set buttonhole stitch. Electronic machines are suited to a variety of different projects and sewing needs.

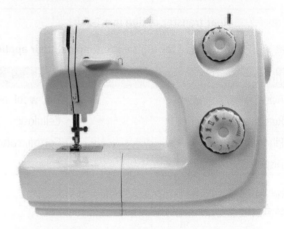

Figure 7.61 Electronic sewing machine

Computerised sewing machines

Computerised sewing machines have an LED display, LCD display or large touch screen, rather than the dials and buttons used on an electronic version. The most basic version will memorise the most frequently used stitches and the tension will be set automatically. The more expensive and complex machines allow the user to select designs already programmed in the machine. This can range from simple lettering to complex embroidery designs that require different coloured threads. Some machines also have a USB port available, which enables the user to create embroidery designs on a computer using specialist software and then save it to the USB to load onto the sewing machine. While computerised sewing machines are more costly than electronic versions on the whole, they offer a huge selection of decorative and embroidery stitches as well as automatic tie-offs and thread cutting.

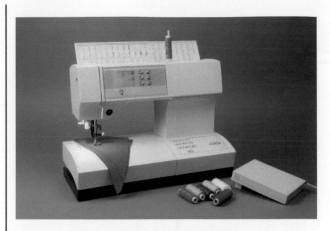

Figure 7.62 **Computerised sewing machine**

Presser foot

The presser foot holds the fabric against the feed dogs and guides it in a straight line as you sew. Every machine is supplied with the standard all-purpose presser foot but some techniques can prove difficult or impossible with this foot. There is a large selection of presser feet available that can be attached to the sewing machine to perform distinct operations. Selecting the correct foot can result in completing specific tasks quicker and with a more professional result.

Sewing machine needles

It is important to select the correct size of needle for the fabric being sewn. Needles are sized by numbers. The smaller the number, the finer the needle. The size of a needle is calculated by its diameter; for example, a 90 needle is 0.9 mm in diameter. A general rule when selecting the needle is to use the higher numbers for thicker fabrics. It is important that you understand the names of the different needles available to help you complete successful prototypes.

- Sharps – used for natural materials; the most common needle used.
- Ballpoint needles– used for man-made materials such as polycotton, polyester and viscose. Sometimes ballpoint needles are referred to as jersey needles.
- Stretch needles – one common fault when stitching stretch fabrics is missing stitches. Stretch needles are coated to enable them to slide easily through the fibres.
- Leather needles – these are spear shaped at the end to allow it to pass easily through leather.

Seams

The type of seam selected when assembling fashion and textile products will be dependent on the fabric being used and the function of the product. If completing toiles or prototypes, a plain seam can be used to assess the fit of the product. The majority of seams use a 1.5 cm allowance as standard.

- Plain seam – this is the most common way of joining fabrics. The seam edges will need to be finished to avoid fraying.
- French seam – this seam tends to be used on sheer fabrics and those that have a tendency to fray easily. This seam encloses all of the raw edges.
- Flat fell or double stitched seam – this seam is commonly used on jeans. It is flat and very strong. This is the best choice if the seam is going to be subjected to a lot of strain.

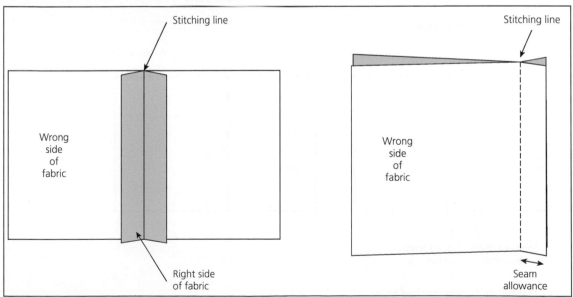

Figure 7.63 **Plain seam**

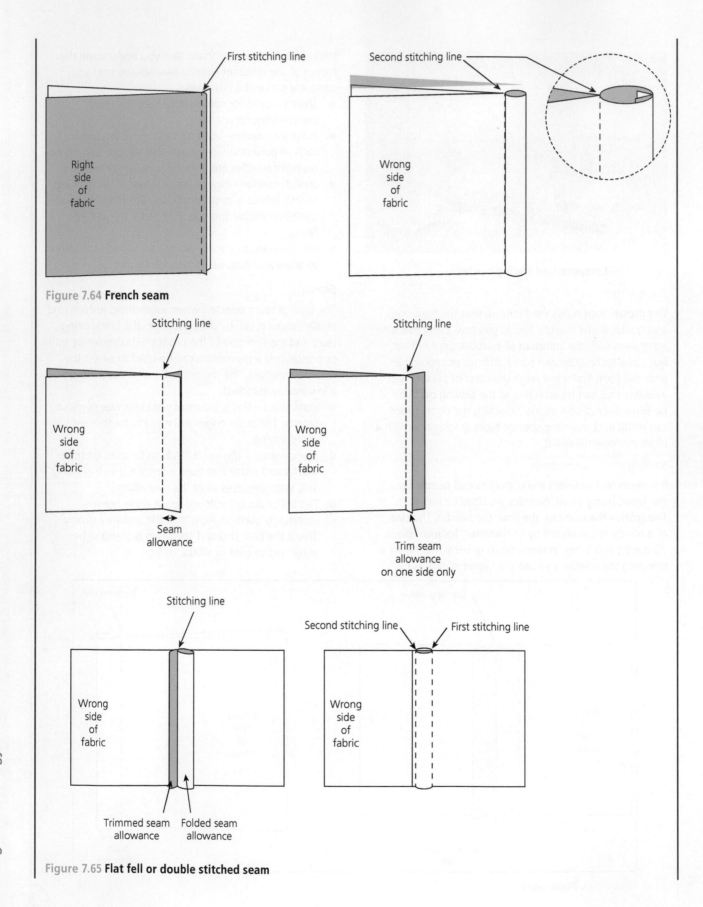

First stitching line

Second stitching line

Right side of fabric

Wrong side of fabric

Figure 7.64 French seam

Stitching line

Stitching line

Wrong side of fabric

Wrong side of fabric

Seam allowance

Trim seam allowance on one side only

Stitching line

Second stitching line

First stitching line

Wrong side of fabric

Wrong side of fabric

Trimmed seam allowance

Folded seam allowance

Figure 7.65 Flat fell or double stitched seam

When completing curved or angled seams, for example, when joining a collar to the neckline, it is necessary to clip into the curve. This stops the fabric from puckering and allows the seam to lie flat. It also reduces the bulk, giving a neater finish.

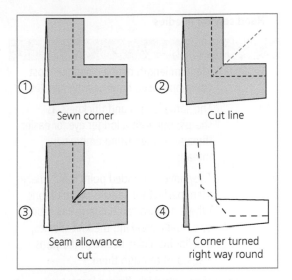

① Sewn corner   ② Cut line
③ Seam allowance cut   ④ Corner turned right way round

Figure 7.68 **Snipping a corner**

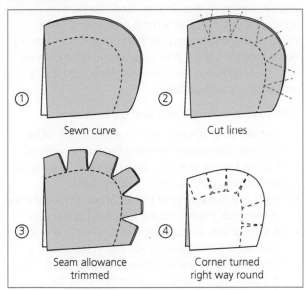

① Sewn curve   ② Cut lines
③ Seam allowance trimmed   ④ Corner turned right way round

Figure 7.66 **Snipping a concave curve**

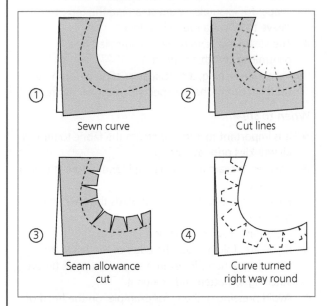

① Sewn curve   ② Cut lines
③ Seam allowance cut   ④ Curve turned right way round

Figure 7.67 **Snipping a convex curve**

Overlocker

Overlockers can join, finish and trim in one operation, and are better than sewing machines at joining fabrics that stretch. If a garment has to stretch on the body and the seam will be under strain, the stitches would snap if joined on a sewing machine using a conventional straight stitch as they do not stretch with the fabric. However, the stitching on an overlocker stretches with the fabric.

When fabrics have been joined it can be necessary to finish the edges. This prevents the fabrics from fraying and gives a neater finish. The method used is largely dependent on the fabric used, as some fabrics are more prone to fraying than others. The quickest and most professional way to finish the edges is by using the overlocker, as it finishes the edges and trims the excess at the same time, giving a neat finish.

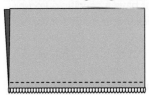

Figure 7.69 **Overlocked edge**

Needles

We have already looked at the importance of using the correct needle when using the sewing machine. The same principle should be used with needles for hand sewing.

Hand sewing needles vary in thickness, length, point shape and size of the needle eye. Select the type of needle according to the project and the size of needle according to the weight of fabric and thread you are using. The basic options are listed in Table 7.8.

Table 7.8 Hand sewing needles

Type	Purpose
Sharps	Medium length needles and the most commonly used hand sewing needles.
Embroidery needles	The same length and thickness as sharps, but with a longer eye for easier threading when using embroidery threads.
Ballpoint needles	They have a rounded point which makes them perfect for sewing on knit fabric. Where a sharp can damage knit fabric by actually breaking the knit stitches, a ballpoint glides between the yarns instead of through them.
Chenille needles	These are thick, have a large, long eye and are sharp. They are used for heavy embroidery as the eye can easily accommodate several strands of embroidery thread. The sharp point makes it easy to stitch through coarse fabric.
Beading needles	Long, thin, flexible needles that come in a range of sizes to fit even the tiniest of seed beads.

Pins

Pins are generally used to hold fabrics together temporarily before sewing either by hand or on the sewing machine. They are also invaluable when cutting out fabric to hold the paper patterns to the fabric. The most commonly used type is the straight pin, which is also known as the hemming or basting pin. Some have decorative pin heads and most have a specific purpose.

- Longer pins tend to be thicker and made for coarser fabrics.
- Dressmaker pins are the most common and are designed for light-to medium-weight fabric.
- Pleating pins are a bit shorter and extra fine, suitable for delicate fabric.
- Short pins, often called appliqué and sequin or sequin pins, are some of the shortest pins available.

Tips:
- Use the minimum number of pins required to hold the fabric together as too many can pucker the fabric.
- Position the pins at right angles to the direction you are sewing so the machine needle sews over the top of them.

- The thickness of the pin tends to relate to the thickness of the fabric with which it will be used.

Finishing fabrics and garments using a steam iron

The finishing of fabrics is a crucial part of the quality assurance process. Pressing the fabric during manufacture is important as it ensures a more professional finish. When completing prototypes in the workroom, a steam iron is used. This is not only used for the more obvious ironing of seams but also for adding definition to stitching lines and for shaping darts.

How to steam iron fabrics and garments

- First test your iron setting on a small swatch of fabric. Different fabrics can tolerate different amounts of heat and steam and it is vital to test first. Too high a temperature can scorch or melt the fabric.
- You can place a piece of muslin or cotton over the top of the piece you are ironing for added protection.
- It is important to understand that when assembling your fabric pieces, you do not iron in the usual by way, moving the iron backwards and forwards over the fabric. You should place the iron onto the fabric, hold it still for a few seconds and steam if required.
- Steam can be used to soften the fabric and add shape; for example, when steaming the top of a sleeve to give it a rounded effect.
- The steam inlet and outlet is controlled by a controlling switch on the iron.
- For steam ironing, a special air suction bed can be used to control the unexpected crease on garments.

When to steam

- It is important to iron and steam the fabric to remove all wrinkles prior to cutting.
- Press each seam after stitching it and before stitching across it.
- Press seams either open or towards the back of the garment.
- Press darts towards one side.
- If you are folding over the edge of the fabric, to stitch it as a hem, for example, give it a press before stitching to flatten and neaten it.
- Avoid pressing gathers, for example, on the head of a set-in sleeve or around the waist of a garment, as it will flatten them.
- Steam and press pleats into shape.
- Steaming fabrics can manipulate them into shape.

The use of digital technology (including CAD and CAM) in the making of final prototypes

Computer-aided design (CAD) software allows a designer to define and test a prototype component before manufacture (see Chapter 4). Computer-aided manufacturing (CAM) software is then used to translate the CAD part into a set of manufacturing commands or instructions which will be used by a machine to manufacture the component. The manufacturing machine acts under the control of a computer, known as computer numerical control (CNC). A wide range of CNC machines are available, many of which are used in schools and colleges, including laser cutters, 3D printers, vinyl cutters and computerised sewing machines.

CAD/CAM parts are manufactured to high tolerances and repeated parts will all be identical. It is quite common for a prototype part to be designed using CAD software and then a model created by rapid-prototyping. This model may be cut from card using a laser cutter, machined from polystyrene modelling foam using a CNC milling machine, 3D printed in a polymer material or embroidered by a computerised sewing machine. The model part will be used to check dimensions and functionality in the prototype. The CAD design may then be modified before the part is manufactured by the same CNC machines in the final chosen material. Time spent in the iterative CAD designing stage pays off because the actual manufacturing of the perfected part is usually fast.

When designing parts for manufacture using CNC machines, it is important to understand the limitations of the machines used, in respect of the materials they can handle, and the methods they use to process the material. It is also important to understand that many CNC machines require a great deal of 'setting-up' and adjusting before they can be used.

CNC laser cutter

Laser cutters are 2D manufacturing machines which cut and engrave a variety of thin sheet materials such as polymers, wood-based sheets, card, textiles, etc. Some school/college laser cutters will cut through thin sheet aluminium.

Laser cutters work by burning and vaporising the material. The laser power and speed of cut needs to be matched to the thickness and type of material. Different colours/opacities of acrylic may also demand different laser cutter settings.

The laser beam needs to be correctly focused onto the top surface of the material to ensure the finest cut and the lowest power settings. The sheet material must be completely flat otherwise it is impossible to achieve focus across the entire sheet.

The laser itself is located underneath the cutting bed and the beam is directed to the cutting head through a system of mirrors and lenses. If the mirror system is not adjusted properly, or not kept clean, then poor cutting results can be experienced.

Figure 7.70 **Laser cut fabric**

3D shapes can be built up from layers of materials glued together. In this way, it is possible to produce complex 3D designs which would be difficult to manufacture by other methods. One advantage is that it is not necessary to clamp the material when cutting, although lightweight material (such as thin card or paper) risks being blown about by the cooling air jet that most laser cutters use. Incorrect settings also risk the material catching fire.

Figure 7.71 **Plasma cutting sheet metal parts**

CNC plasma cutter

A CNC plasma cutter is essentially the same as a laser cutter, but the cutting method uses an accelerated jet of hot, electrically conductive gas, known as a plasma.

Plasma cutters can only be used to cut electrically conductive materials (metals such as steel, stainless steel, aluminium, brass and copper). They are particularly useful for cutting thick metal sheets (up to 150 mm thick in some industrial cases), but laser cutters are generally preferred in industry for thin sheets because they produce a more accurate, well-defined cut.

CNC router and CNC milling machine

A CNC router is a 2D cutting machine which uses a rotating tool to cut out profiles (shapes) from a sheet material, usually wood, polymer, foam, etc. A router can also engrave, drill precise holes and cut out intricate shapes from the material, as well as cutting only part way through the material to produce slots and steps.

A CNC milling machine also uses a rotating tool, similar to a router, but the machine will generally be intended for creating a greater depth of cut on its z-axis (the vertical axis), often at the expense of less movement on the x and y axes, compared to a router. CNC milling machines can be used for producing 3D parts by cutting away material from a solid block (called a blank). It has the capability to create a wide array of shapes, holes, slots and other profiles.

Some CNC milling machines have multiple tools and they can be programmed to automatically change tool during the manufacturing process to suit the type of cut being made. Many CNC milling machines in schools/colleges are capable of machining metals such as aluminium, brass and mild steel.

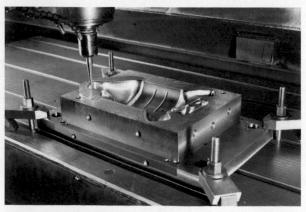
Figure 7.72 CNC milling a metal blank. Note the material clamps

CNC lathe

A CNC lathe is a machine for producing cylindrical shapes under computer control from a CAD part drawing. The CNC lathes found in schools/colleges would typically be capable of machining wood, polymer, wax, foam or metals such as brass, aluminium or mild steel.

3D printing

3D printers are increasingly being used by schools and colleges. This opens new design possibilities and solves many of the problems with other forms of manufacture. By fully understanding the limitations of the technology and by becoming familiar with individual machines, excellent results can be achieved.

Some thought must be given to the orientation of the part when it is being 3D printed. There are two reasons for this:

- The layering method of manufacture results in parts which have an inherent weakness between the layers which could cause structural parts to snap under load. By re-orientating the part for manufacture, it is possible to shift the direction of weakness in to a less critical plane.
- Overhanging parts may require the 3D printer to produce a scaffolding structure to support the part when it is printed. This structure adds to printing time and increases material costs. It also takes time to remove the support structure after printing is complete. By rotating the 3D build, it may be possible to reduce or eliminate the need for the support structure.

Rapid prototype models can be used immediately after FDM printing. However, you can visibly see the printed layers, so ABS and PLA prints will need finishing to achieve a quality finish. Pieces can also be spray painted.

- Sanding the ABS print is simple and straightforward; wet and dry or glass paper removes stepping lines using small circular movements evenly across the surface of the part will result in a smoother finish and epoxy filler (or a filler primer paint) can be used to fill gaps and sand before final painting.
- Sanding PLA is more difficult because of how soft it becomes if you try to sand too quickly; be patient and again sand in small circles evenly across the surface.
- Dremel wheels can be useful when sanding and finishing 3D printed parts and heat guns can help smooth surfaces and use of plastic polishes can avoid the need to spray paint.
- Acetone vapour can be used to smooth and polish the surface of an ABS 3D print. Acetone fumes are harmful, and very flammable but if care is taken, this method can produce finished parts that look as good as injection moulded pieces.

Further information can be found in the following published articles:

- http://makezine.com/projects/make-34/skill-builder-finishing-and-post-processing-your-3d-printed-objects/
- www.fictiv.com/hwg/fabricate/ultimate-guide-to-finishing-3d-printed-parts
- https://3dprinting.com/filament/finishing-3d-printed-parts/

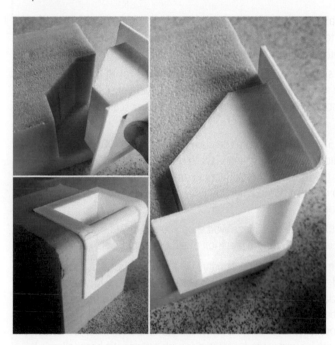

Figure 7.73 **3D printed parts used in an NEA project**

Ensuring quality and accuracy when making a final prototype

Tolerance

Prototype parts need to be closely monitored during manufacturing to ensure that they are fit for purpose. The precision or tolerance to which a part needs to be made depends on the function that part plays in the prototype.

The tolerance of a part specifies the range of sizes over which the part would be acceptable. Tolerance may be expressed as a percentage or as an actual quantity.

DESIGN ENGINEERING

Mechanical components such as gears or bearings usually need to be accurate, known as close tolerance, otherwise problems such as excessive friction prevent the parts from moving efficiently. For example, the diameter of a drive shaft may be specified as 5.00 mm ± 1%, which is the same as 5.00 mm ±0.05 mm.

FASHION AND TEXTILES

When including tolerances in your patterns and templates, be aware that if too many are allowed it will affect the overall fit.

For example; the seam allowance is 1.5 cm with a tolerance of ±2 mm. This would obviously not create a problem to the fit of a prototype if there are only two seams. However, if the prototype has eight seams this could make the overall difference of ±16 mm. This would result in a poor fit and poor quality.

Further advice can be found in the following published articles:

- http://yourwardrobeunlockd.com/images/stories/pdf/easypatterndrafting.pdf
- www.thecuttingclass.com/pattern-notches-alexander-wang/

Measuring instruments

Various measuring instruments are used to check the dimensions of prototype manufactured parts.

ACTIVITIES

1 Calculate the percentage tolerance of a dimension specified as 250 mm ± 5 mm.
2 A 10 mm diameter hole needs drilling to a tolerance of 2%. Calculate the tolerance in mm.
3 A part being turned on a lathe needs to be 17.0 mm diameter with a tolerance of 5%. The part is measured to be 16.4 mm. Calculate whether this is within tolerance.

PRODUCT DESIGN

The method of manufacture will have a huge impact on the accuracy of a part. Where wastage methods are used, it is sometimes advisable to aim to make the part slightly oversize and then carefully file or sand the part down to its accurate size. Parts manufactured using CNC machines will generally be accurate, but dimensions can still be wrong if the machine was not set up properly, so it is important to check dimensions after manufacturing. For example, suppose a designer requires a laser cutter to produce 5 mm diameter holes in acrylic sheet to receive 5 mm LEDs. The actual holes cut will be slightly larger than 5 mm due to the width of the laser beam (the kerf). Consequently, the LEDs may be a loose fit. Reducing the diameter of the CAD circles would result in more accurate 5 mm holes. It's always important to test these features before finalising designs.

Table 7.9 Measuring instruments

Instrument	Description
Rule/tape measure	Simple rules are ideal for larger measurements. They generally have markings at 1 mm intervals, giving them a precision of ±0.5 mm. This precision arises because any attempt to judge a measurement between the mm markings would be subjective (down to the opinion of the user) and the definitive measurement would only be precise to the nearest mm mark which, in the worst case, could be up to 0.5 mm above or below the judged value.
Vernier caliper	A vernier caliper is a versatile instrument for taking measurements up to about 150 mm to a precision of typically ±0.02 mm
	The caliper has jaws for measuring external and internal measurements (such as cutouts). Traditional calipers have a vernier scale which needs to be read correctly to take a measurement. Many modern calipers have a digital readout which greatly simplifies their use.
	The caliper consists of a main scale and a sliding vernier scale. The measurement is taken by reading the whole number of mm from the main scale as indicated by the zero point on the vernier scale. The fractions of mm are then measured off the vernier scale by finding which mark on the vernier scale lines up with any line on the digital vernier caliper.
Micrometer	Measurements are taken by adding the 0.5 mm readings from the main scale to the 1/100th mm readings from the rotating scale. When closing a micrometer against an object to be measured, the ratchet should be used to rotate the spindle as this stops the user from overtightening onto the object, causing possible damage. Modern micrometers may have a digital readout which simplifies their use as well as providing 'HOLD' and other useful measuring functions.
	There are different types of micrometer, but in a school/college workshop you are most likely to find a type capable of taking external measurements up to 25 mm to a precision of ±0.005 mm.
Dial gauge	A dial gauge is a precision instrument for measuring small movements between two parts. It consists of a large dial (which could be analogue or digital) and a probe which is placed in contact with the surface to be monitored. The dial gauge is firmly clamped in place so that it cannot move and any movement of the surface is transferred through the probe and displayed on the dial. These gauges typically measure a movement range of 0–10 mm to a precision of ±0.005 mm. Movement in either direction can be detected. Dial gauges are often supplied complete with a sturdy stand, sometimes with a magnetic base, allowing for easy attachment to machinery.

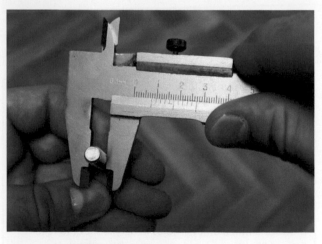

Figure 7.74 **Vernier caliper in use**

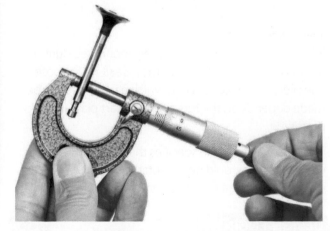

Figure 7.75 **Micrometer in use**

Figure 7.76 **Measuring run out with a dial gauge**

Electronic measurements

In addition to taking distance measurements, design engineers sometimes need to take measurements in electronic systems to ensure that prototypes are manufactured and set up correctly.

A multimeter measures voltage and current (DC and AC) and resistance. Many multimeters also indicate continuity, which means that they produce a 'beep' if there is a good electrical connection between the probes. Continuity checks are useful to quickly check tracks on a PCB (printed circuit board), or the leads connecting a sensor to a microcontroller, for example. Remember that continuity tests can also be used to check that two points are not connected if they shouldn't be. The use of multimeters in the development of electronic systems is covered in Chapter 6.

An oscilloscope is another useful instrument for displaying and measuring signals and waveforms in advanced electronic systems.

Templates and patterns

The use of a pattern that has been correctly drafted will ensure the final fashion or textiles product is made accurately, which assists in the overall quality.

It is important that all measurements are accurate. If they are not correct it can result in the pieces not fitting together when assembling. For example, if the armhole is too small, the sleeve will not fit in. In a final product this will result in poor fit or puckers where an attempt has been made to 'stretch' the fabric piece to fit the required space.

In industry, basic blocks are used and there are standardised measurements for specific sizes. There can be a slight tolerance either side of the measurement. When making prototypes within a workshop, it may be necessary to fit to individual requirements. Therefore when developing your own patterns and templates, keep checking measurements and ensure pattern markings are aligned.

Depending on the prototype being made, there are certain steps to follow when developing patterns and templates to ensure accuracy. As well as measuring accurately, other factors include the following:
- Use of a pattern will ensure all fabric pieces are cut out correctly before assembly.
- Using the same pattern template, for example, when creating complex patchwork, will ensure the fabric pieces are all cut to the same size and therefore fit together accurately.

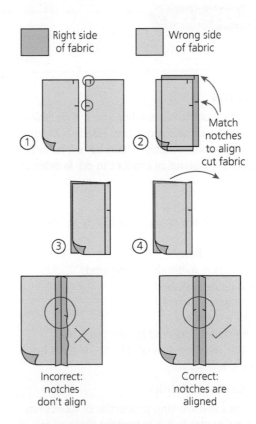

Figure 7.77 **Transferring the pattern markings to the fabric and then aligning them is crucial to ensure pieces fit together accurately**

321

- The use of the grain line on the pattern ensures the fabric is cut in the correct direction of the weave or knit. This helps with the fit and also the drape.
- If using pile fabrics, the pattern should indicate which direction to cut to ensure the pieces of the fashion or textiles prototype look the same colour when assembled. Pile fabrics can look a different colour depending on the angle from which they are viewed.

- Pattern markings (notches) are transferred onto the fabric to ensure all pieces align correctly.

Further advice can be found in the following published articles:
- http://yourwardrobeunlockd.com/images/stories/pdf/easypatterndrafting.pdf
- www.thecuttingclass.com/pattern-notches-alexander-wang/

PRODUCT DESIGN

Jigs, formers and moulds

When making products with more than one common component, jigs, formers, templates and moulds can be used to ensure consistency and speed up the process.
- A jig is used to hold and position work and/or locate or guide a tool to ensure accuracy and repeatability.
- A former is used to make sure that parts are formed or bent to the same shape; It can be used to laminate wood or form metal or plastic parts.

- A template can be drawn around to mark a shape onto material, so that it can be cut or shaped. A template can be made from cardboard, MDF or thin aluminium sheet.
- A mould is a hollow shape used when casting. It can be made from sand in the case of metal casting or in plaster or rubber for casting resin.

See Section 7.3 for more on jigs, formers and moulds.

FASHION AND TEXTILES

The principles of pattern cutting (A Level only)

Selecting fabrics

One of the key decisions to make when you are designing and manufacturing your prototypes is to establish which materials will be the most suitable. It may be helpful to list what you expect the material or product to do, in order of priority.

Ask yourself questions such as:
- What conditions is the final product likely to have to stand up to?
- What budget do I have?
- What construction methods are required?
- Is the fabric easily available?
- Is there a minimum order?
- What width is the fabric?
- Is special machinery required to sew the fabric?
- Will the fabric require a specialist finish or can it be applied after assembly?

Working properties of fabrics

The characteristics and working properties of fabrics has been covered in Chapter 5. It is important that you are familiar with a wide range of fabrics and recognise why they are selected. This will help you to make informed decisions when developing your own prototypes.

Budget

The perfect fabric for your design is one that not only looks and feels good but also fits within your budget. Remember that you will also need to factor in the cost of any components required and these can make a significant difference. You will need to think about a realistic budget and stick to it.

Construction methods

When selecting materials it is important to think about the construction methods that will be required to complete the prototype. Will the fabric require specialist tools or machinery, which may only be available in industry? Some fabrics are notoriously difficult to sew and it would be prudent to carry out tests to ensure the equipment available to you will be sufficient.

Fabric availability

Once a fabric has been selected it is important to make sure the quantity required is available and what the minimum order is. Some fabric suppliers have a high minimum order and while this takes advantage of economies of scale, when only a few metres are required for a prototype it is not helpful. Some companies charge extra for small orders.

Fabric widths

The majority of fabrics are sold by the metre but some are sold by weight. The width of the fabric may seem to be insignificant but if the fabric is too narrow it will cost more to make the prototype. At the same time, if the fabric is very wide there could be unnecessary waste. This comes into play even more if the fabric has one direction or a print that requires a placement or needs matching.

Fabrics are generally available in the following widths:
- 90 cm
- 115 cm
- 150 cm.

Some fabrics are sold at much narrower and some at much wider widths but these tend to be for specific fashion or textile markets. For example, many delicate sari and hand-loomed fabrics are available in narrow widths and some upholstery fabrics are available in wider widths. Table 7.10 illustrates how much fabric would be required to complete the same prototype with differing widths.

Table 7.10 **Fabric required according to width**

Fabric width	90 cm	115 cm	150 cm
Metres required	1.6 m	1.3 m	1.0 m
	2.3 m	2.0 m	1.5 m
	3.5 m	2.7 m	2.1 m
	4.6 m	3.6 m	2.7 m

Finishing

Some fashion and textile prototypes may require specialist finishes. These could include coatings or finishes to make the prototype weather proof or fire retardant. In industry, these finishes will be applied to meet specific requirements and the cost will be budgeted into the product. However, when completing your own prototypes, this service will probably not be available. There are some excellent fabric manufacturers that do supply fabrics that have specialist finishes and a quick search on the internet will reveal suppliers.
- Request samples before purchasing large amounts. This will enable you to carry out any tests required.
- Fabrics can be made water proof in the workshop with silicone spray, a mixture of white spirit and soybean oil or with wax. While these techniques would not be ideal in industry they are more than sufficient for a prototype.
- Fire retardant fabric sprays can be purchased.

The principles of pattern cutting

The production and use of pattern pieces

As we have seen earlier in this chapter, production patterns are developed from basic blocks and working patterns.

Basic block patterns based on a British Standard Institute (BSI) standard sizing are used in the industry as the basis for designs. The basic block will not include any seam allowances as these are added to the final pattern once complete.

From the basic block, a designer will develop a working pattern to trial the garment. The final production pattern will include all the seams, hems, grain lines, pattern markings and information needed to make the product.

When developing your own prototypes, it may be useful to create a block pattern or use one of the commercial ones as a base for creating modelling variations throughout the iterative process.

Commercial patterns

There are numerous companies that produce patterns. The patterns give a range of sizes ready printed onto pattern paper to indicate which cutting line to follow depending on size. The patterns have all the relevant symbols printed on them, as well as a list of suitable fabrics and components required to manufacture the product. The instructions also give numerous lay plans to ensure economical cutting dependent on fabric width. There will be clear step-by-step instructions on how to complete the product. The advantage of using a commercial pattern is that there is no need to develop a pattern from a basic block, which saves a lot of time. However, the patterns are in standard sizes which will not always be a perfect fit.

Pattern sizing

Patterns may require adjustments to achieve a perfect fit. Some are supplied with a range of sizes printed on the same template and the line is cut relevant to the chosen size.

On the reverse of the pattern envelope will be a sizing guide. This will help you to decide what size pattern to cut out, depending on how much ease you prefer. Ease is the amount of 'give' you prefer in garments to make them comfortable to wear.

It is possible to take a commercial pattern and grade it to fit individual requirements. This means increasing or decreasing the size but keeping all the styling and pattern markings in the correct places according to the size. Today, pattern companies take a middle-sized pattern (typically a size 12) and grade it up for larger sizes and grade it down for smaller sizes.

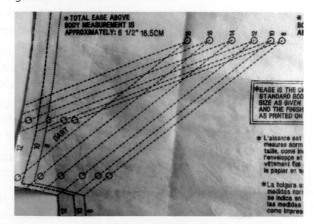

Figure 7.78 **Range of sizes printed on a commercial pattern**

There are three basic methods of **grading**.

Cut and spread method

This is possibly the simplest method of grading and involves cutting the pattern and either spreading it to grade up or overlapping it to grade down.

Pattern shifting

This is a process of increasing the overall dimensions of a pattern by using a specially designed pattern grading scale ruler to move the lines and points a measured distance up or down and left or right.

Figure 7.79 **Pattern grading using the cut and spread method**

Computer grading

In industry, pattern grading is now carried out using computer software that stores the basic blocks and allows new designs to be digitised and developed on screen. Once a design has been developed it will need to be altered to fit a range of sizes. This grading means the master pattern can be used to create larger or smaller sizes without changing the design, shape or appearance. Computer-aided grading systems do this calculation quickly.

For bespoke products, the alteration of patterns will be done manually to ensure the garment fit is accurate and individual to the customer requirements.

Pattern symbols and instructions

It is necessary for you to recognise the range of pattern symbols and their purpose. Whether you are adapting a commercial pattern or drafting your own, you will need to understand the range of pattern symbols. Most pattern symbols are standardised so they are easily recognised.

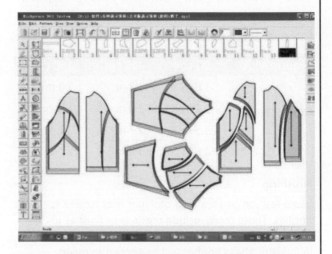

Figure 7.80 **Pattern grading of a garment using CAD**

Table 7.11 **Pattern symbols and their meaning**

Symbol	Meaning
←——————→	Grainline – used to position the pattern, usually on the warp of the fabric. If the fashion or textiles product is bias cut, the grainline is placed at a 45° angle from the warp or weft.
↓ FOLD ↓	Place on fold – position the fabric on the fold of the fabric. This is used when the pattern is symmetrical. It indicates which edge of the pattern piece is aligned with the fabric fold. The pattern piece only indicates half of the fabric piece, so this line is very important as it ensures the creation of a whole piece.
1.5 cm {	Seam allowance – the seam allowance is the measurement between the cutting line and the stitching line. On commercial patterns this is usually 1.5 cm. In industry it is 1 cm as this means less fabric is required, which reduces cost.
△△	Notches – used to ensure fabric pieces are lined up correctly. They indicate joining points on seams. They can be single or in a group and should be lined up with the corresponding piece that is to be sewn together.
⊢——⊣ ✕	Buttonhole and button symbol – the buttonhole will be marked to the correct length. The cross indicates the position of the button.
● ■	Dots and squares – these are used to indicate the position of another component; for example, a pocket, collar joining or dart points.

	Lengthen or shorten lines – these lines indicate where on the pattern the garment can be shortened or lengthened according to individual requirements. If the product is to be shortened, a fold can be created at this point. To lengthen, the pattern is cut at these lines and then the two pieces are positioned an equal distance apart according to the measurement required.
	Stitching line – this line shows where the product should be stitched.
	Cutting line – this indicates where the fabric has to be cut.
	Darts – these usually have dots on them at certain points to transfer to the fabric. The darts can be single ended or double ended.
Tuck	Tucks and pleats – these are indicated by solid and broken lines. The solid lines are to indicate that one line needs to be placed to the other following the direction of the arrow. The broken line indicates the centre of the fold.
	Centre front and centre back line – this indicates the centre front or centre back position. This is particularly helpful when positioning collars and facings around a neckline.

Tips:
- Pin all pattern pieces to the fabric before cutting out to ensure the most economical pattern lay.
- Double check the grain lines are positioned correctly and fold lines are on the fold.
- Make sure that any alterations to the main pattern pieces have been applied to corresponding pieces such as facings.

Manipulating patterns

Designers use many different techniques to put their ideas into practice but at some stage a pattern will be adapted or created to enable the fabric to be cut. Pattern development is a complex process that involves a good understanding of garment technology. This will include using various of techniques to produce the desired effect, for example, reducing fullness, using one of the methods described previously.

The process will normally start with the basic pattern which will then be developed. This could involve adding sections to or removing sections from the pattern, or completely redesigning it.

KEY TERMS

Embellishment – a decoration or feature that is added to the product to enhance its appearance.

Former – used to make sure that parts are formed or bent to the same shape.

Fusing – adding a layer of interlining to the fabric to strengthen it or give it shape.

Grading – proportionally increasing or decreasing the size of a pattern, while maintaining shape, fit, balance, and scale of style details.

Jig – used to hold and position work and/or locate or guide a tool to ensure accuracy and repeatability.

Laminating – assembling multiple layers of material and adhesive for decoration or to achieve a form.

Mould – a hollow shape used to pour material into, like a jelly mould.

Pressing – a lifting and setting motion using an iron. It is used to flatten an area that has been stitched, such as seams, darts, pleats and hems.

Template – can be drawn around to mark a shape onto material.

Thermoforming – a manufacturing process in which a plastic sheet is heated and then formed to a specific shape using a mould or former

Tolerance – the permissible range of variation in a dimension of an object. Tolerances may also refer to other characteristics such as weight, capacity, quantity or hardness. Sometimes known as allowance.

KEY POINTS

- Most products are produced as an assembly of a number of different components. The methods used to join components depend on the materials being joined and how permanent the join is – whether the joint needs to be dismantled for maintenance or adjustment.
- Riveted joints are usually associated with joining flat sheet metal components, but the principle can be applied to most resistant materials and fabrics.
- Mechanical fastenings are important in the production of many products. They allow similar and dissimilar materials to be joined and can allow the joint to be disassembled for maintenance and/or adjustment.
- Adhesives and heat can be used to join materials permanently. They can be invisible and provide a watertight seal.
- The increased use of manufactured boards (chipboard, MDF, etc.) in furniture and the growing interest in self-assembly, has seen a big increase in the need for specialist fittings. These have become known as KD or knock-down fittings. They allow the work to be assembled and taken apart as many times as required, without weakening the joint or the material.
- Wasting or subtraction techniques cut away material to leave the desired shape; they involve drilling and milling, turning and sawing. They are so called because the material which is removed is usually thrown away.
- Shaping by wasting can be done on any type of material. Milling and turning removes material with cutting tools and can be CNC controlled.
- Complex shapes can be moulded in one piece of material, usually by using a casting process – a mould is made, then molten metal or resin is poured into the mould and cooled or cured.
- Steaming or laminating wood allows curved components to be created. In laminating, multiple layers of veneer or thin wood and adhesive are assembled around formers and pressure is applied.
- A variety of pressing and forming operations are carried out on sheet metal as the majority of engineered and manufactured components are made from metal that is less than 2 mm thick.
- The iteration of ideas and subsequent prototyping is a crucial stage of the product development process as it is key to making the correct manufacturing decisions.
- There are many factors that dictate the manufacturing processes used for a final prototype. This can be dependent on materials, budget, method of manufacturing, and tools and equipment available.
- Fabrics can be shaped and manipulated in a number of ways to ensure the correct fit is achieved.
- Basic pattern blocks can be developed and altered to meet individual needs.

MATHEMATICAL SKILLS

Design Engineering and Product Design

Pressing and stamping is often associated with high-volume production. A great deal of time is spent in planning the stamping (blanking) layout for a particular component: the orientation of the component on the metal strip is important as it affects the economic use of the material. Tessellation can be used to plan the most effective layout in terms of waste and cost.

For maximum efficiency of material usage, it is beneficial to have more than one punch carrying out the stamping operation and that the production rate is increased by the number of punches being used. The increased cost of the punches is more than offset by the increase in the production rate and associated economy of material. The logical conclusion is to arrange for multiple blanking layouts, the limit being governed only by the size of the component being produced and the width of the steel sheet supplied.

Most steel producers today will accurately slit the steel coils to any desired width, so multiple layouts should not present a problem.

The only limiting factor is that, when planning for multiple stamping layouts, one must bear in mind the minimum 'land' (space) between adjacent components and the outer components – the edge of the strip cannot be less than the thickness of the material being used.

In the past, arranging the layout for multiple stamping was a skilled and time-consuming task for the engineer or manufacturer. A computer program is now used to ascertain the maximum usage of the metal strip (referred to as a nesting process or tessellating).

MATHEMATICAL SKILLS

Fashion and Textiles

- To ensure efficient manufacture and assembly of fashion and textiles products, each process is timed to establish the total manufacturing time for product completion. This is then factored into the cost for manufacture.
- When reducing fullness of fabric, the measurement of excess fabric is calculated and distributed evenly, depending on the method chosen and taking body measurements into account.
- The size of a repeat pattern is taken into consideration when designing and manufacturing fashion and textile products. For example, the amount of fabric required to match the repeat is calculated when making soft furnishings
- Ratios are used when dyeing fabric, according to the weight of fabric and the amount of dye required.
- Maths skills are used to calculate the cost of manufacturing a product.
- Body measurements are used to develop basic block patterns and draft patterns to know where to add or subtract fabric to make a garment fit well.
- Geometry is used to understand how a flat pattern shape will make up into a 3D garment as well as what part of the flat pattern to change to fix the fit.

7.3 How can materials and processes be used to make commercial products?

LEARNING OUTCOMES

Design Engineering

By the end of this section you should:

- demonstrate an understanding of the industrial processes and machinery used for manufacturing component parts in various materials, including:
  - polymer moulding methods, such as injection moulding, blow moulding, compression moulding and thermoforming
  - metal casting methods such as sand casting and die casting
  - sheet metal forming methods using equipment such as punches, rollers, shears and stamping machines.
- demonstrate an understanding of the benefits and flexibility of using computer-controlled machinery during industrial production, such as:
  - automated material handling systems
  - robot arms to stack, assemble, join and paint parts
- understand the necessity for manufacturers to optimise the use of materials and production processes, such as:
  - economical cutting and costing, ensuring cost effective production for viability
  - working to a budget through efficient manufacture and making the best use of labour and capital throughout the design and manufacturing process.

Fashion and Textiles

By the end of this section you should:

- recognise the tools, processes and machinery required to complete a range of textiles products in industry, including:
  - dyeing processes
  - hand and digital printing processes, such as screen and roller printing methods
  - transferring pattern markings using thread markers, drills and hot notchers
  - cutting fabrics using multi ply fabric cutting, computer-controlled knives, lasers, water jets, plasma or ultra sound to cut fabric and prevent fraying
  - joining fabrics using lockstitch, overlocker, seamcover, linking, automatic buttonhole and computer-controlled sewing machines
  - finishing fabrics and garments using pressing units, ironing and sleeve boards, steam dollys, tunnel finishers and flatbed presses for trousers
- understand the necessity for fashion and textiles manufacturers to optimise the use of materials and production processes, such as:
  - economical lay plans and costing; ensuring cost effective production for viability
  - working to a budget through efficient manufacture and making the best use of labour and capital throughout the design and manufacturing process.

Product Design

By the end of this section you should:

- understand commercial production processes and machinery used to manufacture products to different scales of production, including:
  - moulding methods such as injection, rotational, compression, extrusion and blow
  - thermoforming and vacuum forming
  - die casting and sand casting
  - sheet metal forming and stamping
  - automated material handling systems
  - robotic arms to stack, assemble, join and paint parts
- understand how the design of jigs, fixtures, presses, formers and moulds in commercial production is used to ensure consistent accuracy and quality, and different scales of production methods
- understand the necessity for manufacturers to optimise the use of materials and production processes, such as:
  - economical lay plans and costing; ensuring cost effective production for viability
  - working to a budget through efficient manufacture and making the best use of labour and capital throughout the design and manufacturing process.

Methods for assembling electronic products

Surface mount technology

Prototype printed circuit boards (PCBs) manufactured in schools/colleges will generally use through-hole components, where the components sit on the top side of the PCB and their wires pass through holes drilled in the PCB to be soldered to the pads and tracks on the reverse side. The PCB will probably be single-sided, meaning that the copper tracks exist on one side only, although double-sided boards can be produced in schools/colleges with some skill and care. Assembling and soldering a through-hole PCB is labour intensive, fiddly and difficult to mass produce.

Figure 7.81 **Through-hole components on a single-sided PCB**

A look inside any modern electronic product will reveal surface mount technology (SMT) PCBs. This is an assembly system in which most of the electronic components do not have wires and, instead, are placed straight onto the tracks of the PCB and soldered directly to them. SMT systems allow PCBs to be assembled at high speed by robotic pick-and-place machines, then all the components are soldered simultaneously in an oven. Surface mount devices (SMDs) are much smaller than their through-hole counterparts which allows a higher component density which results in smaller, more complex electronic products. SMDs are manufactured in a range of standard sizes which are designed to be handled by pick-and-place machines using vacuum cups or grippers to pick up the component.

Figure 7.82 **A SMT PCB, showing an IC, resistors, capacitors and vias**

Manufacturing a SMT circuit board

A PCB is designed and optimised on CAD and CAE software. Industrial PCBs will almost certainly be double-sided, having a pattern of tracks on the top and bottom sides of the board. This allows more complex designs to be achieved, without the problem of tracks crossing each other. Links between tracks on each side can be achieved through a copper-plated hole passing through the board, known as a via. Some modern PCBs are multi-layered and these can be thought of as several thin single-sided boards laminated together. Multi-layered boards allow highly complex circuits to be constructed in a very compact space and they are invariably used in products such as smartphones.

At the same time as the PCB is manufactured, a solder stencil is also produced. The stencil is a laser-cut thin stainless steel sheet which contains holes in the positions where solder is wanted on the PCB. The stencil is used in the first stage of the PCB assembly process.

Figure 7.83 **SMT components on ribbons being fed into a pick and place machine**

DESIGN ENGINEERING AND PRODUCT DESIGN

Polymer Moulding Methods

Injection moulding

Injection moulding is the most commonly used method for mass production of thermoplastics, due to high production rates, dimensional accuracy and quality of finish. Polythene bowls, nylon gear wheels and polystyrene casings can be manufactured using the injection moulding process. Products can be manufactured with features such as finger grips, surface texture and other intricate details such as fastening clips and reinforcing ribs/webs as an integral part of the moulding.

The main steps in the injection moulding process are outlined below:

● Polymer granules are placed into the machine through a hopper
● The injection moulding machine consists of a heated barrel equipped with an Archimedean screw (driven by a hydraulic or electric motor), which feeds the molten polymer into a temperature-controlled split mould via a channel system of gates and runners
● The screw melts (plasticises) the thermoplastic pellets to the point of melting and becoming liquid, and also acts as a ram during the injection phase
● The liquid plastic is injected (forced through a nozzle) into a mould tool that defines the shape of the moulded part
● Water is pumped around the mould chambers to rapidly cool the moulding
● The mould opens and the part is ejected by ejector pins. Once removed from the machine the item will need to be finished or 'fettled' by removing the sprues and any flashing.

Injection moulding machines are similar in some ways to extruders, except that where an extruder continually forces material through a die, an injection moulding machine uses force to push the required amount of liquid plastic into a mould to produce the specified product.

The pressure of injection is high and depends on the material being processed. Tools (moulds) tend to be manufactured from tool steel or aluminium alloys. The mould may have more than one cavity.

The high costs associated with tool manufacture means that injection moulding lends itself to high-volume manufacture; this is the main limitation of the process.

Injection moulded components can be complex, as can be seen in Figure 7.84

Once a designer has decided to injection mould a product, they then need to produce a design that favours this process. Some of the considerations would be:

● Make the wall thickness of a product uniform and to a minimum without compromising strength. It is better to use ribs than increase thickness; the thicker the walls of a product the longer the product cycle and the higher its cost
● Provide generous fillet radii, avoiding sharp corners where possible
● Avoid using blind holes as more complicated moulds will be needed. Holes are produced using core pins
● A draft angle of at least one degree is needed so that the product can be ejected easily. Undercuts should also be avoided as they require sectional moulds that can be withdrawn after moulding to release the component
● Moulded components will have visible gates – this is the point at which the polymer is injected into the mould, and sometimes you will notice surface marks created by ejector pins.

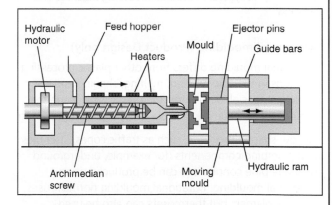

Figure 7.84 **A typical injection moulding machine**

7 Manufacturing processes and techniques

Figure 7.85 injection moulding can be used to make the casings of toasters and food mixers

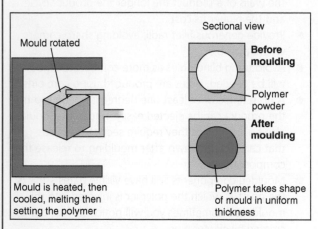

Figure 7.86 A typical rotational moulding machine and process

Rotational moulding (Product Design only)

Rotational moulding differs from other plastic processing methods in that the heating, melting, shaping and cooling stages all occur after the plastic is placed in the mould and no external pressure is applied during forming.

Hollow-shaped products, such as traffic cones, litter bins and plumbing components (for example, underground drainage pipe connectors) can be produced using rotational moulding. Rotational moulding normally uses thermopolymers but thermosets can also be used. The main steps in the rotational moulding process are outlined below:

- A pre-determined amount of polymer powder is placed in the mould. With the powder loaded, the mould is closed, locked and loaded on to the arm of the machine and into the oven
- Once inside the oven, the mould is rotated around two axes, tumbling the powder. As the mould becomes hotter the powder melts and sticks to the

inner walls of the mould. The polymer gradually builds up an even coating over the entire inner surface of the mould.

- When all the polymer has been used to form the required shape, the heater is switched off and the mould cooled either by air, water or a combination of both. The polymer solidifies to the desired shape of the mould with a uniform wall thickness.
- When the polymer has cooled sufficiently to retain its shape and be easily handled, the mould is opened and the product removed. At this point powder can once again be placed in the mould and the cycle repeated.

The process has a relatively long **cycle time**, but it enables complex parts to be moulded with low-cost machinery and tooling. The advantage of this is that large products can be produced economically with relatively low mould costs.

Compression moulding

Compression moulding is mainly used to process thermosetting polymers. Common products made by this process include light switches, plug sockets and knobs for appliances.

The process involves inserting a pre-measured amount of polymer into a closed mould and subjecting it to heat and pressure until it takes the shape of the mould cavity and solidifies.

The mould in this process is made up of two parts and is usually made of highly polished, high-carbon steel. Each part of the mould has a heater plate attached to it.

- The measured polymer granules (known as charge) are placed into the bottom part of the mould; the second mould is attached to a hydraulic ram, which helps to apply the pressure once the mould is closed.

- Once the mould is closed the polymer is heated until it reaches the required temperature. The pressure is applied and the plastic takes up the form of the mould. The heaters are turned off and the job allowed to cool before the mould is opened. The base resin, being a thermosetting material, cures and hardens, cross linking the plastic.
- The part is then ejected and removed.

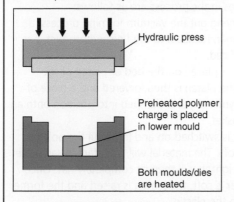

Figure 7.87 **The compression moulding process**

The cycle time for compression moulding is very long compared to a process such as injection moulding but it has its advantages. These include low capital cost, as the tooling and equipment is simpler and cheaper. There is no need for sprues or runners, which reduces waste. However, there are limitations upon the size and complexity of products that can be manufactured.

Transfer moulding (Product Design only)

Transfer moulding is a modified version of compression moulding and is aimed at increasing productivity by accelerating the production rate. The process involves placing the charge in an open, separate pot. A hydraulic ram then squeezes the softened polymer into a mould where it sets into the defined shape. It means that sprues and runners are needed. The surfaces of the sprues and runners are kept at high temperatures (approximately 150°C) to promote curing of the thermoset. The entire shot including the sprues is then ejected.

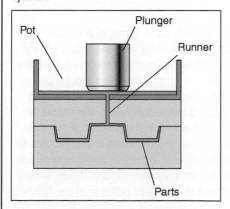

Figure 7.88 **The transfer moulding process**

Extrusion (Product Design only)

Extrusion is a continuous process used to produce both solid and hollow products that have a constant cross section. Polythene pipes and nylon curtain rails are produced by extrusion and uPVC window and door frames are extruded in a semi-automated process to meet the high-volume demand in the construction industry.

The thermopolymer is heated and extruded out of a die, a continuous process capable of forming an endless product that is cooled by spraying water and then cut to the desired length. The shape of the die used during this process will determine the cross section of the extrusion. The speed of the process varies according to the cross section being manufactured: thicker sections are extruded slowly as time is required for the initial heating and subsequent cooling of the larger quantities of material involved.

- Polymer granules or powder are loaded into a hopper on the extrusion machine.
- They are fed from the hopper into the heating chamber by a rotating screw. As the screw turns the granules pass through the chamber, and as the temperature increases the polymer becomes molten.
- The screw continues to turn compacting the polymer and forcing it to leave the chamber through a pre-shaped die in a continuous stream.
- As the polymer emerges from the die, it is hot and will lose its shape unless it is cooled fairly quickly. Water baths or jets of air tend be used for this purpose as the item is supported on a moving belt.

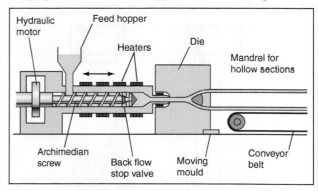

Figure 7.89 **The extrusion process**

Blow moulding

Blow moulding is a process in which a softened thermopolymer tube inflates against mould walls and hardens by cooling, typically to form a hollow vessel or container. Mass produced drinks bottles as well as other hollow items, such as large water containers, can be produced using blow moulding.

Blow moulding is fast and efficient. The hollow products manufactured in this way usually have thin walls, but range in shape and size.

Although there are different versions of the blow moulding process, for example, injection blow moulding and extrusion blow moulding, they basically involve blowing a tubular shape (parison) of heated polymer within the cavity of a split mould. Air is injected into the parison, which expands to the shape of the mould in a fairly uniform thickness.

Injection blow moulding is used for the production of hollow objects in large quantities. The main applications are bottles, jars and other containers. The injection blow moulding process produces bottles of superior visual and dimensional quality compared to extrusion blow moulding. The process produces them fully finished with no flash.

- The process requires a split mould to be accurately produced prior to the production system being put into action.
- A parison is produced by injection moulding or extrusion and may consist of a fully formed bottle/jar neck with a thick tube of polymer attached; it is transferred to the blow mould while still hot.
- The process commences with the mould open and a hollow length of heated thermopolymer placed between the two parts of the mould.
- The mould is then closed and compressed air is blown inside the hollow tube, forcing it against the walls of the mould. As the walls of the mould are cold, this causes the polymer to cool.
- Once the polymer is cool the mould can be opened and the product removed; excess material is trimmed off and the product is ready for use.

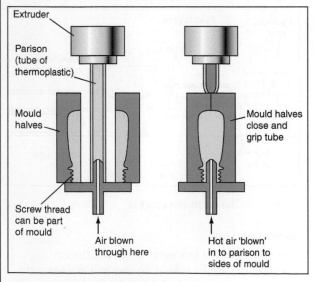

Fig 7.90 **The blow moulding process**

Thermoforming

Thermoforming is a manufacturing process in which a polymer sheet is heated and then formed to a specific shape using a mould or former. For more details see Section 7.2.

Vacuum forming

Vacuum forming is a process that is commonly used to make simple trays or containers. In industry, products produced can vary from polystyrene food trays used in supermarkets to acrylic sinks and baths. Products are manufactured from thermoplastic sheets by a sequence of heating, forming, cooling and trimming.

The main stages in the process are as follows:
- Before carrying out the vacuum forming process, a quality former, identical to the finished product, has to be produced.
- The former is placed on the bed or platen of the machine. The platen is then lowered and a piece of thermopolymer sheet is clamped into position onto an air-tight gasket.
- The heater is switched on and left until the polymer becomes soft. The material will first start to sag under its own weight. It will then sit completely flat. Once the polymer is soft, the platen is raised and the former pushes into the plastic.
- The heat is then removed and the vacuum pump switched on to remove the air. The polymer is forced against the former by atmospheric pressure. Where a deep draw is required, a top 'plug' may be used to push material into the former during the forming process.
- The material is allowed to cool. The cooling process may be shortened with blown air or even a fine water spray.
- The component is then released from the mould by introducing a small air pressure.
- After forming, any finishing may be performed (trimming, cutting, drilling, polishing, decorating and so on).

Moulds are usually made of aluminium because of its high thermal conductivity. For low-volume production, wood-based materials or plaster of Paris are often used. Draft angles are needed for easy removal. As the process involves the forming of sheet material over or into mould, intricate details and features are not possible except when using very thin sheet material, where surface details added to the mould will be reflected in the surface finish of the product.

The limitations of the process in terms of thickness of material, the depth of moulding, and the thorough and even heating of the polymer must be observed. Webbing, where the polymer creases as it stretches around or into the mould, can be a problem; draft angles usually of 5 degrees need to be added to the mould. Very small air extraction holes are required in the corners of complex moulds to aid removal of the air. While 'male' moulds for the polymer to form around are much easier to produce, 'female' moulds, for the polymer? to form inside, ensure a more constant wall thickness of polymer in the finished component.

Sophisticated machines and moulds are used for continuous automated production of high-volume items like yoghurt pots, disposable cups and sandwich packs. Bathtubs can be made at a rate of one per minute. Multiple parts can be made using a single multi-former. The main materials used in vacuum forming are high impact polystyrene, ABS and polycarbonates. Similar processes are now beginning to be used to inflate and superform metal.

Polymer domes that may be used as food covers can also be produced by blow moulding in a similar way to that already described in vacuum forming. A shaped ring will be clamped over the thermoplastic on the vacuum forming machine. The polymer will be heated as described in the vacuum forming section but once it reaches the required temperature, instead of removing air out and causing a vacuum, the motor is reversed and air is blown into the chamber causing the heated polymer to rise; a shape will be produced according to the clamped ring positioned on the surface.

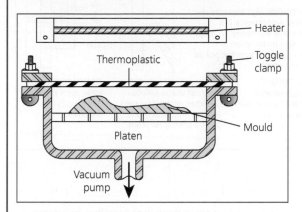

Fig 7.91 **The vacuum forming process**

Metal casting methods

Die casting

Die casting is used to commercially produce engine components (large and small) and alloy wheels for cars. Many components found inside domestic appliances, such as washing machines, are die cast.

The main materials used are aluminium alloy, zinc, magnesium and copper.

Molten metal is poured into reusable steel moulds called 'dies' and the resultant casting is, in general, left in its finished state with little or no further machining required. This is a fast process and subsequently leads to a reduction in production costs per unit once initial investment in the die is made. Die casting is usually used with non-ferrous metals as the higher temperatures needed for casting steel and iron damage the expensive dies, which are usually made from steel, leading to premature loss of accuracy and finish of the product

being cast. Highly intricate detail can be achieved. Tiny toy cars, now volume produced in polymer by injection moulding, were first made in large quantities by die casting. The die casting process is not dissimilar to injection moulding.

Figure 7.92 **The making of a steel mould for use in die casting**

Gravity die casting uses a permanent metal die; it closely resembles the processes of basic sand casting in that the die is filled with molten metal by the natural force of gravity. If cores are needed, they are often made so that they can collapse in order to allow them to be withdrawn from the casting. Where this proves to be a problem then, just like the sand casting method, sand cores are used. This process is not suitable for zinc-based alloys as it tends to promote a fairly coarse grain in the finished product.

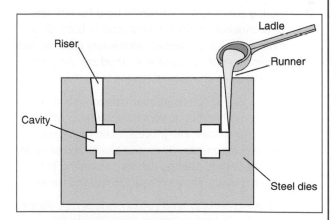

Figure 7.93 **Gravity die casting**

Pressure die casting utilises pressure. The pressure is applied to the molten metal as it is fed into the dies (moulds) using a simple plunger. This technique ensures that the molten metal makes good contact with the walls of the casting dies and, by maintaining the pressure throughout the cooling stage, gives a sharp and well-defined casting that requires little or no additional machining.

If the dies are cooled when filled, the components will solidify quicker and the casting can be removed while solid but still hot, enabling the process to be repeated again, giving a much faster turnaround of components.

One disadvantage of this type of process is that the cost of producing the dies is quite high. It only becomes economical if a large number of components are being made.

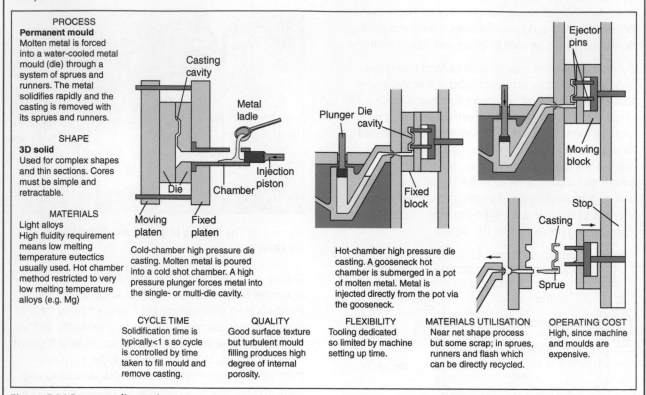

PROCESS
Permanent mould
Molten metal is forced into a water-cooled metal mould (die) through a system of sprues and runners. The metal solidifies rapidly and the casting is removed with its sprues and runners.

SHAPE
3D solid
Used for complex shapes and thin sections. Cores must be simple and retractable.

MATERIALS
Light alloys
High fluidity requirement means low melting temperature eutectics usually used. Hot chamber method restricted to very low melting temperature alloys (e.g. Mg)

Cold-chamber high pressure die casting. Molten metal is poured into a cold shot chamber. A high pressure plunger forces metal into the single- or multi-die cavity.

Hot-chamber high pressure die casting. A gooseneck hot chamber is submerged in a pot of molten metal. Metal is injected directly from the pot via the gooseneck.

CYCLE TIME
Solidification time is typically <1 s so cycle is controlled by time taken to fill mould and remove casting.

QUALITY
Good surface texture but turbulent mould filling produces high degree of internal porosity.

FLEXIBILITY
Tooling dedicated so limited by machine setting up time.

MATERIALS UTILISATION
Near net shape process but some scrap; in sprues, runners and flash which can be directly recycled.

OPERATING COST
High, since machine and moulds are expensive.

Figure 7.94 **Pressure die casting**

Sand casting

Sand casting is a process that can be used for different levels of production. It is a low-cost process but is time consuming and labour intensive. Aluminium patterns and a programmed sand compaction method are also used. You can read more about sand casting in Section 7.2.

Castings can vary in size from small components that would fit in your hand up to very complex products weighing many tonnes. The process is commonly used for machine components, car engine blocks and turbines. Although the casting method can vary according to the product being made, the basic process remains the same.

Recent developments in sand casting include CO_2 silicate casting which provides greater accuracy due to the sodium silicate making a tougher mould, and shell casting where the sand mould is coated in resin giving a closer tolerance and improved surface finish.

Sand casting is a primary process to produce a raw component normally requiring secondary processes to complete the component. In addition to the removal of the sprues resulting from the pouring of the molten metal into the mould, secondary processes often include the machining of critical surfaces.

Sheet metal forming and stamping

Stamping (also known as pressing) describes any press-related operations, which can be categorised as follows:

- piercing – where a suitable punch shears a hole in a piece of metal
- blanking – where the punch shears the required shape from the metal
- notching – where the punch shears an open-sided hole in the metal
- cropping – where the punch shears a plain or shaped length from the metal
- bending – where a suitable punch shapes the metal by a folding process
- forming – where the metal is forced into the shape of the surface contours of a die.

The above terms cover all of the shaping and manipulative techniques used in sheet metal work and, to a lesser extent, on plate (metal up to about 3 mm in thickness is classified as sheet metal; above 3 mm it is called metal plate).

Virtually all presswork operations are carried out on sheet metal as the vast majority of engineered and manufactured components are made from metal that is less than 2 mm thick.

The main users of these manufacturing techniques in industry are the automotive, aeronautical, heating and ventilation industries, along with canning and container making, and domestic appliance industries. Examples include:

● car body panels
● aircraft panels
● air conditioning units
● washing machine and tumble dryer carcases.

More than half of the production of metal products in the western world involves the use of sheet metals. It should be noted though that large presses can be used to fold and form materials up to about 50 mm in thickness and presses exist that can cater for capacities exceeding 45,000 tonnes, which can be used for large items.

The majority of press-related operations are confined to large batch production work due to the high costs and lengthy tool-setting involved. The press tool costs for producing a motor vehicle body part, for example, can run into tens, sometimes hundreds of thousands of pounds.

A great deal of time is spent in planning the stamping (blanking) layout for a particular component: the orientation of the component on the metal strip (the roll of sheet metal) is important as it affects the economic use of the material. Computer programs are used to ascertain the maximum usage of the metal strip (referred to as a nesting process).

Pressing and sheet metal forming operations are virtually all automated now (or, at the very least, semi-automated) with only very small batches of products being suited to manual press operations. Often a product will go through a number of stages. Jigs enable complex shapes to be formed with no maximum size, allowing the creation of precise components. Multiple operations can be carried out on sheet metal as it progresses through a series of press tools (for example,. piercing, drawing then blanking) to create a complete component.

ACTIVITY

Explore some of the many online videos showing pressing and stamping processes used in the commercial production of sheet metal components. Make a list of some of the unusual components you discover.

Types of presses

Presses can be manual or power operated to suit the working conditions. Manually operated presses are obviously limited in the magnitude of the forces that can be applied to the metal, with large fly presses being capable of operating at just a few tonnes at best. Tooling costs are relatively low in comparison to other processes.

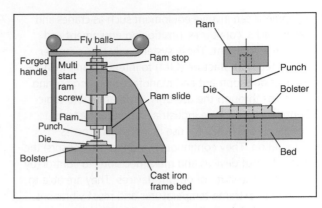

Figure 7.95 **Fly press for stamping or blanking**

Commonly used presses include:

● Arbor presses – these are usually operated by a handle working through a rack and pinion system or pneumatically, with the air supply operating the ram directly. They are used for punching, bending and cropping operations on relatively thin gauge material or for light assembly work.
● Fly presses – these are relatively cheap to install and the tooling costs are low. They consist of a heavy cast iron frame with a large diameter square thread operating the ram. As the operator swings the handle, the rotational kinetic energy imparted to the cross bar (and the balls) is stored with the downward movement of the ram and is expended as the punch makes contact with the metal workpiece. Fly presses are often used to prove and check the alignment of larger tool sets for use on large power presses.

Developments in manufacturing processes

As materials and processing techniques develop there is more overlap between processes. Superforming of aluminium alloy uses a process similar to vacuum forming, it involves heating aluminium sheet to 450-500 degrees celsius in a pressurised oven and forcing it into or over a single surface tool to form a 3D shape.

Designers such as Ron Arad and Mark Newson have made use of it to create furniture. Some alloys can be formed in minutes and a high surface finish is achieved. British designer Stephen Newby has recently introduced a way of inflating stainless steel which is similar to blow forming of glass.

Using computer-controlled machinery during industrial production

Automated material handling systems

Material handling systems use mechanical equipment for the movement, storage, control and protection of materials, goods and products throughout the process of manufacturing, distribution, consumption and disposal. The different types of material handling equipment can involve equipment such as cranes and pallet trucks, conveyors, positioning equipment and storage equipment. These systems rely on Computer Integrated Manufacture (CIM) to integrate and manage product development and manufacturing engineering functions through the use of IT. Programmable Logic Controllers (PLCs) are industrial computer systems designed specifically for use in industrial process automation. They continuously monitor (by sensors) the state of input devices and make decisions to control (by actuators) the state of output devices. They are able to operate in extreme temperatures and resist vibrations, impact and electrical noise.

The flow of materials in, around and between production facilities is an important consideration.

Material flow diagrams are used to map and efficiently plan the movement of materials and components between the various activities and stages in the manufacture of a product. Industrial trucks, mobile robots, that use a variety of systems to navigate known as Automatic Guided Vehicles, and overhead conveyors are widely used. The aim is to reduce manual handling to a minimum.

Robotic arms

Robots are used in industry to carry out material lifting and the handling and placing of components. They are also used to carry out repetitive tasks that workers may find tedious or operations that could be classed as hazardous, such as working with corrosive liquids or processes that may cause fumes that may have an effect on workers' health. Industrial robots are built to copy human movements; they are expensive to produce but over time these costs are recouped by increased productivity and efficiency. Robots are adaptable as they can be reprogrammed to carry out new tasks once an operation is completed.

The introduction of robots into the workplace may have had an effect on the number of workers employed but some human input is still necessary in order to carry out routine maintenance, quality control checks of work and for programming. In manufacturing applications, they can be used for assembly work, processes such as painting and welding and for material handling. More recent robots are equipped with sensory feedback. Through vision and tactile sensing, they can check production quality and self-reset as necessary.

Other advantages of using robots include:
- decreased cost per unit – throughput speeds increase, which directly impacts production rates and results in lower costs per unit
- improved quality – tasks are performed with accuracy and precision every time and the amount of raw material used can be reduced, decreasing costs on waste
- increased safety – robots increase workplace safety and can perform dangerous applications in hazardous settings.

Figure 7.96 **Robots in action in a car factory**

Table 7.12 **Benefits and drawbacks of automated systems**

Benefits of automated systems	Drawbacks of automated systems
- Low labour cost - Low product cost through **economies of scale** - No production loss through disputes - Consistent quality - Very efficient stock-control system enables JIT manufacture - System checks and modifications easy to carry out	- Expensive initial layout - System breakdown costly - Specialised workforce requires regular training - Protocols of all systems may not communicate - Some flexibility in processing is required to enable manufacture of different products

ACTIVITIES

You should be able to find plenty of examples of SMT PCBs around your school/colleges workshop or inside electronic products.

1 Take photos of these and try to identify the different components (ICs, resistors, capacitors, etc.). Use online resources to help you.

2 With the aid of a magnifier, try to read the markings on the resistors and capacitors and work out their values. Use online resources for reference.

3 If you are able, try to rework an old SMT PCB. This involves melting the solder and removing components, then replacing them with new ones. Various online videos will guide you.

Processes used to make commercial products

Dyeing

Prior to dyeing, fabric has to go through certain stages of preparation to remove oils or impurities that could affect the end result. Fashion and textile products can be dyed at the fibre, yarn or fabric stage of manufacture.

Direct dyeing

Direct dyeing involves the dye being applied directly to the fabric without the aid of a mordant. In this method the dyes are either left to ferment or are chemically reduced before being applied to the fabric. Direct dyes are water soluble and are mainly used for dyeing cotton. The colours produced are bright but have very poor wash fastness. However, treatments can be applied to the fabric to improve this.

The dyeing process is normally carried out in a dye bath, at or near boiling point. The dyes are mixed with cold water to make a paste. The paste dissolves once it is added to the boiling water. The solution is constantly stirred to ensure that no lumps form and the dye fully dissolves into the boiling water.

Stock dyeing

Stock dyeing refers to the dyeing of the fibres, or stock, before it is spun into yarn. There are two methods of stock dyeing:

- The older method involves removing the packed fibres from the fibre bale to loosen them and then immersing them into large vats containing the dye. The liquid is then heated to the appropriate temperature required for the dye application and dyeing process. The liquid dye is forced through the fibres.
- The newer method involves placing the whole bale in a specially designed machine and forcing the dyes through the entire fibre bale. This method saves time and labour. However, the disadvantage of this method is that the dye may not penetrate some areas of the bale. But following subsequent blending and spinning operations to convert the fibres into yarns, the undyed areas become unnoticeable and an even colour is achieved. This method of whole-bale dyeing is suitable for wool and all types of man-made fibres.

Top dyeing

This method of dyeing is also done at the fibre stage. The short fibres are removed before the dyeing process. The word 'top' refers to the long fibres of wool from which the short ones are removed. The fibres are then wound onto spools and the dye solution is passed through it. This method results in very even dyeing.

Figure 7.97 **Bale stock dyeing**

Yarn dyeing

Yarn dyeing involves dyeing the yarns before the refers to fabric production stage. There are several methods of yarn dyeing. Some of these include:

- skein dyeing – involves immersing loosely wound skeins of yarn into the dye bath
- package dyeing – the yarn is wound onto a spool (or package) and multiples of these are immersed into the dyeing machine
- warp beam dyeing – an entire warp beam is wound onto a cylinder which is then placed in the beam dyeing machine
- space dyeing – the yarn is dyed at intervals along its length.

Yarn-dyed fabrics are usually deeper and richer in colour. This method can achieve interesting effects, which is one of the main reasons the dyeing is done at the yarn stage. The different colour yarns can be woven to create checks, stripes and plaids.

Piece dyeing

Piece dyeing refers to the dyeing of constructed fabrics. The fabric can be dyed in batches according to the manufacturer's requests, which avoids wastage and saves on costs. There are several methods of piece dyeing and these include:

- beck dyeing – used for dyeing long yards of fabric. The fabric is passed in rope form over a reel and through the dye bath. This process is repeated many times until the desired colour is achieved
- jig dyeing – the fabric is held on rollers as it is passed through the dye bath
- pad dyeing – the fabric is passed through a dye trough and then between two rollers which force the dye into the cloth and squeeze out the excess dye. Heat is then applied to set the dye before it is washed, rinsed and dried again.
- jet dyeing – the fabric moves along a heated tube where jets of dye solution are forced through it at high pressures.

Solution pigmenting dyeing (dope dyeing)

This method is used to dye synthetic fibres. The dye is added when the synthetic fibre is still a solution and before it is extruded through the spinneret. This gives a colour-fast fibre.

Garment dyeing

Garment dyeing refers to dyeing the finished textile product. A number of garments are packed loosely in a nylon net and put into a vat filled with dye that has a motor driven paddle. The dye is thrown upon the garments by the moving paddles. However, the disadvantage of this method is that the dye may not completely penetrate all the fibres and, in particular, the components used, for example, the threads, zips and lining. For this reason, this method is mainly used for lingerie, socks and sweaters.

Figure 7.98 **Piece dyeing**

Fabric can be dyed by means of continuous or batch dyeing.

Continuous dyeing

The methods used for continuous dyeing include the pad steam process, pad dry process and thermosol process. The method used depends on the type of dye. The fabric is passed through a dye bath and then squeezed between rollers to spread the dye evenly and remove excess. This method is high speed and can dye between 50 to 250 metres per minute. Continuous dyeing is used for colours that do not need to change too quickly with fashion.

Batch dyeing

Batch dyeing involves fabrics being dyed to order in large batches according to the colours required. It is used for fabrics when colours have to be changed frequently because of fashion. A specific weight of fabric is treated in a machine containing a specific amount of dye. In the jigger process the smoothly spread fabric is led backwards and forwards through the dye bath giving uniform colour distribution.

Hand and digital printing processes

Screen printing

This is the most popular method of putting defined areas of colour onto a fabric. The process of screen printing is shown in Section 7.2.

Most fabric is rotary screen printed, which allows continuous production. Some fabric is flat screen printed, which involves lowering successive screens onto the fabric, pressing the dye through the screen with a squeegee blade then moving the fabric onto the next screen.

Roller printing

This is the oldest mechanised method of printing. The design is engraved onto copper rollers, one for each colour. The fabric travels round a main cylinder and the dye paste is in a trough. See Section 7.2 for further details of roller printing.

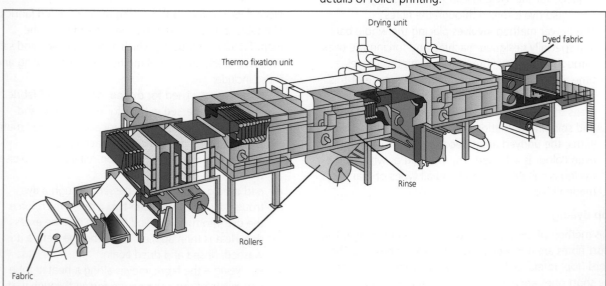

Figure 7.99 **Continuous dyeing**

Digital printing

Digital textile printing is becoming one of the fastest growing industrial print methods. One of the main attractions to the fashion and textiles industry of using this method is the quick turnaround of producing the printed fabric due to inkjet production with its reduced set up times. Speed is increasingly a priority as the fashion segment embraces multiple mini-seasons and print-on-demand delivery models.

Transfer printing

The industrial method of transfer printing does not differ from the workshop method apart from the size, speed and heat of the process. Transfer printing involves printing onto a substrate (usually specialist paper) that is then transferred to a textile by heat. On an industrial scale the method is faster to meet demand. The image is printed onto large sheets which are fed through a machine that has heated rollers and transfers the image from the paper to the fabric.

Transferring pattern markings

In industrial production, once the fabric has been spread ready for cutting, the **pattern markings** need to be transferred onto the fabric. Conventional methods of using tailor's chalk would be time consuming and not cost effective. There are three main options to choose from:

- Thread marker – a machine that provides a method of short-term marking when a permanent mark is not desired. The machine has a large needle which pulls a thread through the material. This allows the machinist to determine where to place pleats, darts, buttons or pockets.
- Drill marker – drills holes through numerous layers of fabric, indicating buttonholes, darts, pockets and other attachment points. The hole is minute and does not damage the fabric.
- Hot notcher – transfers the notches from a pattern to the fabric by burning a notch into the edge of the fabric. The temperature is controlled, so as to leave a brown burn mark without melting or doing excessive damage to the fibres. A hot notcher can also be used to fuse together loosely woven fabrics temporarily to ensure precise alignment for sewing.

Figure 7.100 Drill marker to transfer pattern markings temporarily onto fabric

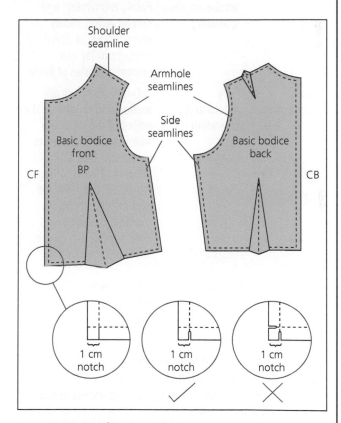

Figure 7.101 Notches on patterns

Cutting fabrics

All fabric, regardless of product, has to be laid out before cutting. This is called fabric spread. It is a very important part of the production process. The type of spreading and cutting system is dependent on the level of production. For mass production, the fabric is laid out using a mechanical spreader; it rolls out the fabric back and forth in a large stack. The layers of fabric are referred to as plies. The number of plies is dependent on:

- capacity of the cutting machine
- volume of production
- type and thickness of the fabric.

See Table 7.13 to understand the different fabric spreading methods used.

Table 7.13 Spreading the fabric

Level of production	Fabric spreading	Method
Bespoke or small batch	Manual	Fabric is pulled from the roll by hand and cut to the length required. The fabric is usually single or double ply.
Large batches	Semi-automatic spreading	Fabric is unwound and spread automatically using a manual driven carriage to lay and smooth the plies of fabric.
Mass production	Fully automated spreading	Fabric is automatically loaded and unloaded. The edges of the fabric are automatically levelled to make sure they align. At the end of each cutting run, there is automated end-of-roll cutting. An automatic tensioning device is used to control the fabric.

The cutting of the fabric depends on the level of production. The methods used take into account factors that will reduce cutting time to meet demand and therefore save on costs.

Manually operated power knives

Types of manually operated knife include:

- straight knife – used by many cutting rooms as it is versatile and accurate
- band knife – the narrow blade of this machine allows the finest of shapes to be cut very accurately
- round knife – this is a very fast machine, excellent for cutting straight lines or gradual curves
- die cutting – this process involves the use of a hydraulic press which forces a shaped metal cutting die through a pile of material and is mostly used when large quantities of small components have to be cut very accurately.

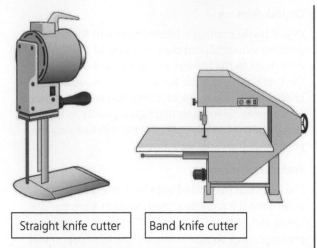

| Straight knife cutter | Band knife cutter |

Figure 7.102 Manually operated power knives for fabric cutting

Computerised methods

The use of CNC machines for cutting fabric is usual for mass produced items. This step of the manufacturing process can be fully automated. There are machines that incorporate 'intelligent' knives that can adjust to the weight of the material.

- Computer-controlled knife cutting – automated blade cutting machines are the most highly developed and widely used computerised cutting systems. Numerically controlled knives cut multiple plies with great accuracy and speed.
- Laser cutting – a powerful beam of light is projected onto a minute area to cut the fabric. Lasers cut with incredible speed (twice that of automatic knives), accuracy and multidirectional ability. Another benefit is that lasers never need sharpening.
- Water jet – a water jet cutting machine is another computer-operated, multidirectional method. Water jet cutting is performed by propelling a tiny jet of water through the fabric at very high pressure. This method is particularly effective for cutting fabrics that are usually difficult; for example, leather. It is a quiet method and no excess heat is produced. Another advantage is, as there is no knife, there is no need for sharpening.
- Plasma torch cutting – fabrics are cut by blowing compressed, high-temperature, ionised argon gas through a nozzle. One or more fabric plies can be cut, but this system is more useful for cutting single ply fabric.
- Ultrasonic cutting – this method of cutting involves mechanical oscillations created by an ultrasonic generator. These oscillations are amplified, creating a hammer-like action which cuts through the fabric and seals the edges of the cut to eliminate any fraying. This method is very clean as there is no burning or colour change.

Figure 7.103 **Computer-controlled knife cutting**

Joining fabrics

In industry, a range of different sewing machines are used for different purposes. Whereas in the workshop environment you may only have access to a domestic sewing machine and possibly an overlocker, in industry, machines are manufactured that perform specific processes. The method for joining depends on the fabric, the design of the fashion or textiles product and its use. The correct technique is critical to the production of a successful and structurally secure product. Industrial machines are generally used to complete an operation efficiently to save on time and therefore costs.

Lockstitch machine

The lockstitch machine is the most common type of sewing machine and is similar to the domestic sewing machines used in small workshops. It uses two threads: upper and lower. The top thread runs through the tension mechanism and finally through the needle, and the bottom thread is wound onto a bobbin. The stitch is created by the upper thread 'catching and locking' the lower thread. The machine is called lockstitch due to the upper and lower thread being 'locked' together. This stitch is widely used as it does not easily unravel. The industrial lockstitch machine is available in many configurations depending on the application and fabric being used.

Figure 7.104 **Single needle lockstitch machine used for a wide variety of fashion and textile products**

Overlocker

The industrial overlocker performs in the same way as the domestic overlocker (see Section 7.2) and is used for either joining fabrics together or finishing the edges. The machine usually cuts the edge but some machines are made without the cutting blade. An overlocker with the cutting function is also known as a serger. Industrial overlockers use between one and five threads. The number of threads used depends on the product. However, the most common type is the two thread as it is used for seams and finishing edges, particularly on stretch and woven fabrics. The four and five thread machines tend to be used for seams that will be under a lot of stress.

Seam cover

A seam cover sewing machine is used with stretchy fabrics for longer seam life and to prevent thread breakage. There are many versions available but their primary use is to hem and finish stretch fabrics. They are used for sportswear, lingerie and swimwear, and for many other fashion or textile products that stretch. The visible stitch on the right side of the fabric resembles straight stitch and can be single, twin or triple needle. The underside resembles overlocking stitch and this is the one that creates the stretch.

Linking machine

Linking is a method of attaching pieces of a fashion or textiles product together after the pieces have been knitted. The linking machine joins the seams of knitted garments such as sleeves and can also be used for decorative linking. It works by placing the knitting on the circular point ring, which rotates and links the seams together.

Automatic buttonhole machine

Industrial buttonhole machines can be computer controlled to aid efficiency and consistency. The width and length of the buttonhole is selected and the machine automatically stitches and cuts the hole for the button. In fully automatic buttonhole operations, a pre-selected number of buttonholes can be sewn a pre-selected distance apart.

Computer-controlled sewing machines

The use of computers to control sewing machines improves efficiency and accuracy when manufacturing either the entire fashion or textiles product or just a part of it. For example, a CNC embroidery machine could be used to stitch the same logo on a batch of products.

One of the most common CNC machines used in the fashion and textiles industry is the integrated stitching unit (ISU). This machine can be programmed to carry out the same operation repeatedly. The bobbin holds more than usual bobbins and the machine will sense when the bobbin thread is running out and send a warning to ensure that the machinist knows to change the bobbin. This helps to avoid it running out of thread midway through sewing a seam, for example. Some of the main features of this machine include:

- automatic foot lift – after the programmed number of stitches, the presser foot will automatically lift
- automatic bartacking – the machine can complete a series of stitches used to reinforce areas of a garment that may be subject to stress or additional wear
- stitch counting – the machine can track a programmable number of stitches and then stop sewing
- automatic thread trimming – this eliminates the need to frequently cut the threads
- automatic fabric ply sensor – the machine can sense the thickness or the layers of fabric and adjust the presser foot pressure accordingly.

Figure 7.105 **Integrated stitching unit**

Finishing fabrics and garments

The finishing of any fashion or textiles product is a crucial step of the quality assurance process. Pressing improves the aesthetics of the product; it improves the stages of assembly throughout manufacture by neatening the seams before advancing to the next stage, and final pressing improves the presentation of the product.

Pressing takes place at different stages of the manufacturing process. These include:

- flat pressing – carried out to smooth fabric before cutting or on the finished product
- steaming – relaxes the fabric and removes stubborn creases
- under pressing – carried out on seams and darts during assembly and before moving onto the next assembly stage
- moulding – used to create and fix 3D shapes in a textiles product. This could be at the top of sleeves. This is most effective on wool fabric
- top pressing – final finishing of the product before selling.

Pressing can be done by hand or by using specialist machinery. As seen in Table 7.14, different pressing units are used for specific areas of a product. The pressing units can shape, stabilise and set seams using either dry heat or steam.

Table 7.14 **Pressing units and their functions**

Pressing unit	Function
Industrial ironing boards	There are many versions and options available. Some use an electrically heated ironing surface and vacuum motor which draws the steam through the garment, giving professional results. These can be used for flat, under and top pressing.
Sleeve boards	This small version of an ironing board allows sleeves to be 'pulled' over it to press the seams flat without creating creases in other areas. It is useful for difficult-to-access seams.
Collar, cuffs and sleeves	The flexible, adjustable forms on these pressing units allow pressing of different sizes. They have adjustable pressing forms for cuffs and flexible heating mats guarantee an even distribution of temperature and pressure over the whole surface. The temperature is set digitally. The pressure can be adjusted very sensitively for all different kinds of fabrics. A suction unit makes it easier to position collars and cuffs before pressing and optimises cooling down after pressing.
Steam dollys	There are many different versions but the principle of a steam dolly is to finish the completed fashion or textiles product. The completed product is placed over a form which is inflated by blowing with air and steam which removes any creases.
Tunnel finishers	The tunnel finisher is used for whole garments on hangers. The garments pass through a 'tunnel' where they are steamed and dried. The temperature can be adjusted to suit the fabric.
Flatbed press unit	A flatbed press is used for ironing garments or products that can be laid flat without getting creases elsewhere; for example, trousers or soft furnishings.

Ensuring consistent accuracy and quality through production

Jigs and fixtures

Jigs and fixtures are production tools that are used when a number of duplicate components or parts are to be made accurately and the correct alignment between a tool and the workpiece must be maintained. In order to provide this means of repetitive accuracy, a jig or fixture would be designed and made to hold, support and locate each part of a component to ensure that each part is drilled or machined in precisely the same way, within the limits or tolerances of the product's technical specification.

A jig is a special device that holds, supports or is placed onto a part to be machined. It is a production tool that not only locates and holds the workpiece but also guides the cutting tool as the operation is performed. A jig is usually fitted with hardened steel bushes/guides where the drill or other cutting tool enters. Small jigs are not necessarily fastened to the drill table; they can be hand held. However, if the holes being drilled are larger than 6 mm then the jig is generally clamped securely to the drill table.

A fixture is a production tool that locates, holds and supports the workpiece securely during a manufacturing operation. In a machining operation, set blocks and feeler gauges are used in conjunction with fixtures to reference the cutter in relation to the workpiece. Fixtures are always securely fastened to the table of the machine on which the work is being carried out. Although largely used for milling operations, they can be used on a wide variety of machine tools. They vary in design from very simple and inexpensive tools to quite complex, expensive devices.

Jigs are widely used to hold components together for assembly, for example in welding, where tubular steel frames for furniture, bicycle frames or car body panels are assembled. These are usually substantial steel frames with clamping devices to hold the components in the correct positions while welding is carried out. The clamping devices are sometimes mechanical, but in large scale production pneumatically operated clamps are often used as part of an integrated control system with robotic arms operating the welding gun.

Template jigs

These would be designed and made by a tool maker or fitter and are used for accuracy of production rather than speed. Template jigs are designed to fit over, on, or into the workpiece and are not always clamped to the work. They are the least expensive and simplest type of jig.

Presses, formers and moulds

Presses form material to shape when force or pressure is applied. They are usually associated with pressing of sheet material, such as metals or paper-based material. Most

Figure 7.106 Adjustable assembly jig for bicycle and similar frames. Note the backplate with holes to enable the clamping devices to be re-positioned

modern machine presses use a combination of electric motors and hydraulics to achieve the necessary pressure.

Moulding is the process of manufacturing by shaping raw materials using a rigid frame called a mould. The mould may take the form of a pattern or model of the final object. A liquid form of material might be poured into a mould and left to set or harden.

The design of presses, formers and moulds is important in commercial production, as they need to allow for consistent accuracy and quality. Press tools, formers and moulds tend to be made from hardened steel, alloy steels and materials that will not wear or scratch easily from repeated use.

Press tools, formers, and moulds in commercial production tend to be made from hardened steel, alloy steels and materials that will not wear or scratch easily from repeated use. The accuracy and quality of the component is only as good as the quality of the tooling – i.e. presses, formers, moulds used. Spark erosion and milling have been traditionally used to manufacture tools for injection moulding and die casting, although nowadays the efficient use of laser processes allows for accurate tool making. Laser deposit welding has established itself as a high quality method to make forming, punching or die casting tools. Additive manufacturing DMLS (direct metal laser sintering) produces strong, durable metal parts that work well as both moulds or as functional prototypes or end-use production parts.

Ford are exploring 3D printing to make tooling and prototypes, which can be tested just hours after their initial design. Manufacturing specialised moulds for a couple of hundred parts is really not economically sustainable, so 3D printing could be used.

Optimising the use of resources and production processes (A Level only)

Manufacturing in any company uses various resources, including raw materials, people and processing equipment. There are relationships between resources; for example, the most suitable process may depend upon the materials used or the availability of a suitably skilled workforce. For this reason, manufacturing resources must be considered in the designing stage for manufactured products to optimise production efficiency.

Resources must be available of appropriate quality and quantity and at the appropriate time. Time itself may be considered a resource when comparing alternative processes or sequences of manufacturing stages.

Electronic circuit boards are usually manufactured using robotic machines to place the components, and then they are soldered in place inside an oven. Circuit boards for electronic products are often very small, and it is more efficient and faster to assemble and solder several boards at once, all joined together in one larger panel. After assembly and testing, the panels are separated into the separate PCBs.

- Sourcing and processing raw materials – materials must be sourced to appropriate quality standards, in the right quantity and at the right time for production needs. Cost is not the only factor to consider when sourcing materials. Brands need to ensure their reputation isn't harmed through unethical or immoral choices, relating to labour conditions, corruption, child labour or sustainable resources, for example. Sourcing closer to the destination market increases efficiency and speed, but often comes at the expense of higher labour costs.
- Dyeing and printing (fashion and textiles) – selecting the most sustainable methods is very important as this sector of the industry uses and wastes the largest amount of water.

- Manufacturing – this area is similar to sourcing and processing of materials. However, to optimise production, the most effective method of assembly needs to be used. This could be **Quick Response Manufacturing** (QRM) to meet demand. The manufacturing stage can be the most expensive part of the whole process. **Design for Manufacture and Assembly (DFMA)** is one way of addressing this issue.
- Transport – this can include delivery of materials, components or completed products. Consideration should be given to selecting the mode of transport that reduces **lead times**, is most reliable and cost effective, and also limits emissions to reduce the environmental impact.
- Marketing – the manufacturer will need to decide upon the most effective method to market their products in terms of cost, while still reaching their target market.

Planning is crucial for efficient manufacture. Before detailed production planning can take place, it is essential to consider the timings and sequences of manufacturing activities in detail. Some stages of production can be carried out concurrently, for example, preparing different sub-assemblies; others must be carried out in a set order. The more complex a product is, the more likely it is to be more expensive to produce. Graphical methods are used in many organisations and computer software aids complex production analysis. Gantt charts are used in building and construction projects. Critical path analysis is used to identify key stages and critical points to aid project management and ensure projects keep to schedule.

FASHION AND TEXTILES

Lay plan

Fabric **optimisation** is an essential factor for every fashion or textile firm. To ensure this is achieved, marker making is done to avoid material wastage. Marking refers to the process of placing the pattern pieces to maximise the number that can be cut out of the fabric. Fabric width, length and type, and subsequent cutting method are all taken into account. This is all achieved by using computerised automated programmes. The marker will ensure the most economical **pattern lay** is created, taking into account various factors. For example, if the fabric has a one-way pattern (the pattern running in one direction), the pattern templates all have to face the same direction otherwise the pattern would be upside down on some areas. The same principle applies to pile fabrics. Today computer software and systems are used to optimise the process. One example of this is Lectra https://www.lectra.com/en.

The patterns will have been previously digitised and graded. This is done on a digitised table and a cursor is guided around the edge of the patterns and key grade points are recorded. This can also be done using photo digitisation. Pre-programmed grade rules for increase or decrease are automatically applied to the pieces of each grading location.

At this point, further modifications can be made to the pattern template to ensure a more economical lay.

The changes should not alter the finished product. One alteration could be to reduce the hem allowance and therefore the amount of fabric required. This may seem like a trivial amount, but when thousands of products are being manufactured, the cost benefit can be significant.

The aim of any business is to survive and, if possible, to make a profit. In general, any new product should contribute to the overall profitability of the company. Break-even analysis (a comparison of expected total costs and gross profit) determines the production level above which a profit will be made. If projected sales are not above this break-even point, it is unlikely that the new product will go into production.

Design for Manufacture and Assembly (DFMA)

DFMA is the integration of product design and planning for manufacture and assembly into one activity. The primary aim is to design products both economically and efficiently. By making use of standardised solutions it is possible to deliver many benefits. The main aims are to:

- minimise assembly costs by:
  - using standardised components
  - reducing the number of processes and parts
  - minimising the number of assembly operations
  - focusing on compatibility of materials and processes
  - using symmetrical parts for ease of insertion
- reduce the product development cycle
- manufacture high-quality products efficiently by:
  - ensuring the correct tolerances
  - focusing on sustainability through repair and replacement, reduction of waste and ease of disassembly.

Costing

Minimising production costs without compromising product quality is a key aim of production planning. Decisions made concerning preparation and processing of materials, assembly stages and the sequencing and timings of manufacturing stages are all linked to cost. For example, the majority of sewn products still rely on manual labour rather than automation. This is largely due to the complexity of assembling products. To ensure product assembly is efficient and therefore cost effective, the organisation of the production line is crucial. To put this into costs terms, the more efficient the worker, the lower the labour costs per unit of production and therefore the higher the potential profit.

There are many examples of software that can be used to optimise designs by highlighting areas for cost reduction and identifying components that can be changed or deleted while maintaining functionality. Concurrent costing software optimises costs by suggesting alternative tolerances, surface finishes and other details.

ACTIVITIES

1 Examine a designer/luxury fashion or textiles product and suggest strategies that could be used to manufacture it for a lower cost.
2 Examine a mass-produced fashion or textiles product of your choice and complete a production plan that could have been implemented in the factory for efficient manufacture. This could be completed using Excel.
3 Using the same or another product, suggest how **lean manufacturing** could have been introduced during the design and manufacture stages.
4 Examine an existing product and produce a table to list the stages of manufacture and the machines that would have been used to manufacture it for job production and also for mass production.

Cycle time – the time taken during the manufacturing of a product for the process to go through a range of steps and return to the same point.

Design for Manufacture and Assembly (DFMA) – the integration of design with planning for manufacture and assembly into one activity. The process includes simplification of design and reduction of costs while maintaining functionality, to achieve economy and efficiency throughout the whole production process.

Economies of scale – the greater the quantity of a goods produced, the lower the per-unit **fixed cost** because these costs are spread out over a larger number of goods.

Fixed costs – costs that remains constant regardless of the number of goods being produced.

Lead time – total time required for item manufacture.

Optimisation – a means of identifying the best choices from design alternatives in terms of optimum use of materials, manufacturability or ease of assembly, quality, performance, size, weight, design features, sustainability and so on.

Pattern lay – how the pattern templates are positioned on the fabric using the grainline to create the most efficient and cost effective cutting out.

Pattern markings – the markings on all pattern templates to indicate the position of crucial areas on the product.

Quick Response Manufacturing – the QRM process looks at how lead times across a company can be reduced to increase productivity.

Stock dyeing – dyeing a staple fibre before it is spun. This is the method commonly used for dyeing wool fibres.

MATHEMATICAL SKILLS

- Ratio calculations when dyeing fabric.
- Surface area calculations when creating pattern repeats.
- Calculating the fabric amounts using pattern lays and plies.
- Costing of prototypes and for volume production.
- Production planning timings.

KEY POINTS

- The stages of production for many fashion and textiles products follow a planned sequence. This sequence is dependent on the design of the product, level of production, materials and components and budget.
- Dyeing can be done at different stages: fibre, yarn or completed product.
- Fabric printing is carried out before product manufacture. Usually the more colours in a print, the more expensive it is to manufacture. Placement prints are usually done once the product has been constructed.
- The type of fabric spreading and cutting system is dependent on the level of production. The methods used in industry take into account factors that will reduce production time to meet demand and therefore save on costs.
- The finishing of any fashion or textiles product is a crucial step of the quality assurance process.
- The majority of sewn products still rely on labour rather than automation. This is largely due to product complexity when assembling.
- Manufacturers have to work to a budget which will restrict choices of materials and components and methods of manufacture.

7.4 How is manufacturing organised and managed for different scales of production?

By the end of this section you should:
- understand how and why different production methods are used when manufacturing products, dependent on market demand, including:
  - one-off and bespoke, batch and high-volume production systems
  - modular/cell production systems (A Level only)
  - lean manufacturing
  - just-in-time manufacturing
  - bought-in parts and components, standardised parts (A Level only)
  - fully automated manufacture (A Level only)
- understand how ICT and digital technologies are changing modern manufacturing (A Level only):
  - customised manufacture systems
  - rapid prototyping
  - additive and digital manufacture methods
  - stock control, monitoring and purchasing **logistics** in industry.

Production methods

The choice of production methods used to produce a product and/or system depends on market demand, the type of product and the materials it is made from. The main categories of manufacturing systems are one off, batch and high volume. Other factors that determine the selection of a production system are the availability of premises and workforce/skills, and capital and tooling costs. Large companies can produce goods at more competitive prices by spreading capital costs over a greater number of products and they can benefit from economies of scale. Large companies also benefit from bulk buying of materials and the ability to standardise parts across their products.

One-off/bespoke production

One-off/**bespoke** production refers to the manufacture of a single component or product, sometimes a luxury item. It is sometimes referred to as jobbing production. One product is usually fully completed before the next is started. Sometimes the product will be made to order, bespoke or customised.

Examples can include large-scale products such as ships, bridges and specialist constructions; or smaller-scale products such as jewellery, specialist furniture and bespoke clothing. Products are usually produced to a specific client specification or where there is a low level of demand. Manufacture often involves high skill levels and craftsmanship and can result in high labour and high unit costs. Capital costs will be lower and worker satisfaction is generally very high; one person will often be involved in every production stage from start to completion.

> **FASHION AND TEXTILES**
>
> One-off and bespoke production is a traditional method used in the fashion and textiles industry. **'Haute couture'** is the ultimate in one-off production. Bespoke is the British equivalent of the French term 'haute couture'. It means the product is made to order or custom made to the customer's specification. It involves a high degree of customisation and involvement of the user in the production of the good. These are usually products that are exclusive or specialist products.

Batch production

Batch production involves the production of batches of similar products. It refers to the scale of production, which can be a few items to several thousands, and to the type of production, where components are processed together in a planned sequence. Quality is controlled throughout manufacture to ensure each product is identical. Batch sizes can be increased and decreased as necessary according to demand. Examples include a batch of ten aeroplanes, a batch of fifty pieces of limited edition jewellery or a larger batch of several thousand pairs of training shoes or jeans.

Batch production methods often allow for more flexibility; a wide range of products can be produced and the system can be adapted to react to demand, stopping or increasing a production run.

The workforce is usually less skilled than in one off/ bespoke production; workers may operate one or two processes. A medium capital investment is needed for a range of machinery that can be set up for different operations.

Cars used to be mass produced with only one model manufactured. Modern car manufacturers produce batches of special editions or allow customisation with differing specifications, the quantities of each batch being decided by customer demand.

Batch produced products may be produced in different colours or styles to meet the demands of fast fashion; for example, clothing made for high-street fashion retailers. The quantity of products can vary from a set of six cushions to thousands for a department store.

High-volume production

High-volume production is often referred to as mass production. High-volume production systems usually operate 24/7 and are used to manufacture high-demand items. Examples include:

- **Continuous flow production** systems – these take in the raw material at one end of the factory and the finished product comes out at the other end. These factories seem like a complex, continuous, fully automated machine. Continuous flow production is used to manufacture products that are going to be produced over a long period of time. Glass, steel, paper and identical simple fashion or textiles items are produced in this way. The production is often highly automated and, as such, requires only a small workforce. However, it is expensive to shut down and restart production if something goes wrong.
- In-line production systems – these require the product to be moved from one process to another, usually by means of a conveyor system, with parts added in sequence. Cars and domestic products such as televisions and smartphones are produced in this way. The items are progressively assembled as they flow along a production line. The process is repeated constantly during the working day, with the assembly line only stopping in the event of a breakdown. Work carried out during the identified manufacturing stages can be done using a trained workforce, automation or a combination of the two. In-line production plants are very expensive to set up; however, as they are mainly used to produce huge quantities of items, the individual product cost is reduced a great deal compared to batch and one-off methods.
- **Repetitive flow production** – a large number of identical items are produced at a relatively low cost and production is broken down into sub-assemblies of smaller components. Set-up costs can be high,

but materials and components are bought in bulk and therefore take advantage of economies of scale. However, due to the large amount of stock, the **carrying cost** needs to be taken into consideration. Semi- or unskilled workers can be used which again keeps the costs down, but the tasks can be very repetitive and lead to job dissatisfaction.

- Fully automated production systems – pens, clothes hangers, buttons and paper clips are produced in high volume using machinery designed and built specifically for that product. There will be minimal, if any, variation in the product for which there is a regular, high demand. Initial capital investment is high for premises and machinery but a low unit cost is achieved. The workforce is often made up of low-skilled workers completing repetitive tasks or setting up machinery to run mostly fully automatic production systems. A highly efficient maintenance team is required, as stoppages to a production run can be costly. The method can result in low job satisfaction as the workforce is usually only involved in a small part of the production cycle.

Modular or cell production systems (A Level only)

In these systems, production cells or modules are grouped together to manufacture a component or sub-assembly of a larger product. The cells or modules usually consist of production machines and include inspection and assembly units. Very often the cells are operated by a small multi-skilled workforce, but they can be fully automated.

Some large manufacturing systems process large batches in sequence through several dedicated process or manufacturing sections. Sections usually have large, expensive machines designed to minimise unit costs by mass producing single identical components with minimal tool changes. This system largely requires advance orders and long production runs. It can be very wasteful as production can be held up if one section does not function correctly. Storage space is required for batches between processes.

In modular or cell manufacturing, the workstations are arranged in a logical manner to produce one complete item at a time, in a smooth and quick flow through the production process. The rate of production is decided by consumer demand. Production planning for the cells/modules must be accurately scheduled to ensure that the correct number of components/sub-assemblies is produced in time for the final assembly of the product. **Modular/cell production systems** require careful positioning of workstations to enable minimal and quick movement of parts from one operation to another and a multi-skilled workforce is required to offer maximum flexibility. The machines are often fitted with multiple tools and generally operate a rapid tool-changing system. Powered clamping systems, offering quick and easy location of the work piece, speed up production times. A standard size of manual locking method, usually in the form of a chuck key or spanner, is used to avoid operators wasting time searching for the correct tool.

FASHION AND TEXTILES

Sewing operators working as a team are an example of a production system. The operators do not carry out a repeated operation as in the PBU system and nor do they sew the whole product as an individual. Multi-skilled operators form a group and each carry out multiple operations. They help each other to complete a product and, as such, each is responsible as part of the team for the quality and production.

Lean manufacturing

Lean manufacturing offers companies a method to reduce costs, eliminate waste and increase productivity, while maintaining high levels of quality and still making a profit. It is a system that can be used not only in production but also any other business that is looking to streamline, such as retail, hospitals or offices. It requires top-down commitment and bottom-up involvement.

The method improves processes through continuous improvement (Kaizen) and elimination of waste, a philosophy derived mostly from the Toyota Production System (TPS). TPS is renowned for its focus on reduction of waste to improve overall customer value. The steady growth of Toyota from a small company to the world's largest car manufacturer has meant many companies have adopted its principles in the hope of duplicating its success. Lean manufacturing is a customer-focused approach used to continuously improve any process in manufacture and outputs through the elimination of waste in everything that is done. The aim of lean manufacturing is to get the right things to the right place at the right time, in the right quantity to achieve perfect workflow. In lean manufacturing, waste not only includes waste materials, but also any non-value-adding work. Questions that are asked include:

- Are products are being moved that actually do not need to be moved?
- Is this person walking further than they need to walk?
- Are there more processes than are needed?
- Can the product be produced with fewer components?
- Are there gaps between processes that can be closed to avoid waiting time?

Just-in-time manufacture

Many modern manufacturing companies, such as Toyota, Dell, Zara and Uniqlo, operate a **just-in-time** (JIT) **manufacture** system. It is a strategy used to increase efficiency and decrease waste by receiving goods only as they are needed in the production process. This has the knock-on effect of reducing costs, as the materials will be readily available to meet demand but not to the point that excess is ordered that would require extra storage.

Companies set up detailed arrangements with suppliers and distributors to ensure that advance orders are taken and regular deliveries of materials and components are made when required for manufacture, arriving at the appropriate point of the production line at the set time. Through the use of JIT delivery, the manufacturing company becomes more financially smart as parts are no longer stockpiled onsite and therefore a quicker financial return is made on the company's investment as delivered parts are used immediately. Computerised stock control systems ensure that production is continuous.

Wastage is reduced in terms of:
- storage space for materials, components and completed products
- defective products – all of the workforce have a responsibility for quality
- money invested in materials and components that will not be used and completed products that will not be sold
- movement of the product through the factory, which is kept to a minimum
- inefficient use of equipment – the system makes maximum use of production machinery and there is no waiting time between processing operations

- labour misuse – appropriately skilled workers are used
- product effectiveness – simplicity is a key feature of the system with the removal of product functions that are not necessary
- downtime with new product run up – detailed plans are made to ensure seamless flow from completed product to new product.

There are times where the system fails, for example, a delivery failure can result in a halt to production, which could be caused by natural disasters, road closures or fires. (Toyota production was halted as a result of a major fire that prevented supplies arriving.) Problems with workforce relationships may result in strikes or absenteeism, which could then hold up production. Frequent road transport of supplies and despatched items can have an environmental impact, unless major suppliers and customers are relatively close to the manufacturing plant.

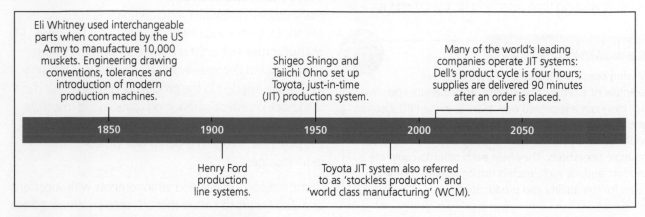

Figure 7.107 Lean manufacturing and JIT timeline

Bought-in parts and components (A Level only)
Many products make use of similar parts or components, for example:

- Different makes of computers often have the same make of hard drive or other internal components.
- Car manufacturers may set up sub-contractor arrangements for the supply of components such as headlamps and engine parts.
- Thread is usually bought in ready to be used in the assembly process.

Benefits of using bought-in parts and components
- No need for production space for the components
- Speeds up overall production
- Quality assured by the component manufacturer, using specified **tolerances**
- Specialist companies provide components, offering cost benefits through economy of scale
- Choice of suppliers if there are service/quality difficulties, offering cost benefits through price negotiations and loyalty contracts
- Reduces storage costs, components available when required
- Ensures consistency in weight, shape and colour (identical each time)

Disadvantages of using bought-in parts and components
- Need to have reliable suppliers
- More storage space may be required
- Longer time required for ordering and supplying
- No control over making/quality

Standardised components (A Level only)
Standardised components are the common items that are required in the manufacture of a wide range of products, such as screws, nuts and electronic components (batteries, resistors, capacitors, etc.), zips and buttons. They are usually small, simple items that are manufactured to guaranteed specifications and are of consistent quality. Hinges, locks, rivets, sequins and buckles are standard components.

- The benefits of bought-in components are relevant to standardised components.

- There are minimal interface and tolerance problems; standards are usually generated by an independent body, for example BSI or ISO.
- There is ease of maintenance; replacement components are readily available for consumers.

In the manufacture of a car, a chassis may be used as a standard component within that company and used in the production and development of several models of cars.

In the manufacture of soft toys, the same hollow fibre filling may be used as a standard component within that company but also be used by other manufacturers for other textile products.

ACTIVITY

Find three standard components or parts. Carry out research to determine their precise specification. Put together a short presentation for your group explaining the role of these parts in successful named products or systems. Include an explanation of the benefits to the manufacturer of using these standard parts in the products or systems you have chosen.

Fully automated manufacture (A Level only)

Lights-out manufacturing is a term used to describe fully automated manufacturing facilities that require no human presence. In most manufacturing systems, workers are necessary to set up systems, hold parts to be manufactured and often remove the completed parts from machinery. But as the technology necessary for lights-out production becomes increasingly available through developments in robotics, many factories are beginning to move towards a lights-out production system to meet increasing demand for products.

Robots play a key role in automated systems. They are used for material and component handling and can be programmed and used in a wide range of applications and manufacturing processes. They carry out repetitive tasks with consistency and precision. Most robots resemble mechanical arms and, in their simplest format, are used in 'pick and place' operations, for example, positioning components on a circuit board for a computer.

The use of process planning software ensures a rapid and consistent flow of operations with no down time. Software is also used to predict and check for tool wear in real-time and make adjustments accordingly, as well as scheduling periodic maintenance.

FASHION AND TEXTILES

New technology has advanced manufacturing processes in many fashion and textile areas. Industrial looms incorporate automated air-jets to weave at speeds of 2,000 picks per minute. There are a number of machines available in the market that help in the various stages of garment production, such as:
- autoconer – helps in the winding of yarn
- automated knitting and weaving machines
- robotic testing – for example, testing the strength of a fabric or how much stress a fastening can take before it breaks or falls off
- automated fabric spreading
- fabric cutting
- computer-controlled conveyor systems
- overhead travelling cleaners.

ACTIVITY

Examine an existing fashion or textiles product of your choice and list the manufacturing stages. Highlight all the areas where automation could have been incorporated into the process.

ICT and digital technologies (A Level only)

The last 40 years have seen a dramatic increase in the availability and use of ICT and digital technologies in manufacturing systems. The development of an increasing range of computer numerically controlled (CNC) machines signalled the single most significant contribution to the increase in manufacturing productivity. Ideas are shared and worked on in real time in different parts of the world, sent to manufacturing units within the same facility or, again, anywhere in the world. The development of 3D CAD tools enables designers to generate and explore ideas quickly, not just create life-like representations.

The final component designs are used to create tool paths for machining. CNC machines are used for a wide range of operations including cutting, pressing, stamping, embroidering, cropping, welding, forging, assembly and packaging. Digital systems, using visual measurement and/or weighing sensors, are used for inspection at key processing points to ensure quality control; for example, in modern car manufacture the tools used to insert the screws that secure seat belts have angle and torque sensors to ensure that the screws are fitted correctly and at the precise angle.

3D ideas can be realised using rapid prototyping technologies. The rapid development of these technologies allows manufacturers the flexibility to produce more complex products that are multi-material and multi-functional, with a reduced time to market. Over the past few decades, the manufacturing industry has needed to adjust to a rapidly changing business environment. The demand for fast fashion and constant updates of electronic devices has led to fast product turnaround and time to market. Many companies now manufacture abroad to keep costs down and to remain competitive. This has led to the development of quick response, rapid prototyping, and mass customisation strategies.

Customised manufacture systems

Companies such as Nike allow customers to customise and personalise products such as trainers and sports bags, furniture companies allow customers to choose fabrics and materials and car and bicycle manufacturers allow customisation of cars and cycles, and still achieve fast production. At the luxury end of the fashion and textiles market, Gucci has offered customers the option to customise their jackets at the flagship store in Milan. The development of affordable 3D printers, 3D scanners and CAD software has enabled small companies to offer bespoke customised designs. In a world of mass-produced products, modern technology has made it easier than ever for a single individual or small company to create and distribute items.

Company websites facilitate the customisation process through configurators that 'walk' customers through a process where they learn about the various options, visualise them on screen and then make their choices. As the process proceeds, the price is updated to reflect the choices made. At one time, only a selected number of options were available, but some companies are now able to offer almost limitless choices of colour and finish, product features, sizes and so on. **Mass customisation** is sometimes referred to as 'build to order' or 'made to order'.

> ### ACTIVITY
>
> Carry out a survey in your school or college, asking for reasons why students and staff like being able to personalise and customise products that they buy. Produce a chart of your findings.

Rapid prototyping

Rapid prototyping can be defined as techniques used to quickly fabricate a model. It gives designers and manufacturers the ability to turn ideas into reality at a much faster pace than using traditional techniques. For more on rapid prototyping see Section 7.1.

Rapid prototyping technology has been under research and development for over 30 years. The range of materials that can be printed is now wide, and includes polymers, paper, ceramics, metals, super alloys, wool and bio-materials. Methods of rapid prototyping include:

- 3D printing machines 'print' a thermopolymer material (ABS or PLA are the most common) in successive layers on top of previous layers to build up a 3D shape. Complex shapes often require an additional support material to be printed to support the object while it sets. Objects can be painted or electroplated for a higher quality finish.

- Paper-based rapid prototyping – allows any drawing or scan from a computer to be printed in paper. Software takes the drawing or scan and breaks it up into layers the same thickness as the paper. When the information is sent to the printer, it cuts each slice of paper to shape and layers them one on top of the other, using a water-based adhesive, to create intricate shapes.

- Stereo lithography – one of the best known methods of rapid prototyping. A 3D model of the object required is created using a CAD programme. The software is able to 'chop' the CAD model into thin layers with a typical thickness of 0.1 mm. A laser traces the shape of a layer onto a bath of liquid resin. This cures the resin. The platform is lowered and another layer is traced until the whole object is created.

- Laser sintering – works in a similar way to stereo lithography. The laser traces the shape onto fine heat-fusible powder (polymer, metal or ceramic). The powder becomes solid; another layer of powder is laid on top of the fused layer and the process is repeated until the object is completed.

- A new type of 3D printer which fabricates 3D objects from soft fibres (wool and wool blend yarn) is being explored by Carnegie Mellon University and Disney. This material is a form of loose felt formed when fibres from an incoming feed of yarn are entangled with the fibres in layers below it. This extends 3D printing from typically hard and precise forms into soft and imprecise objects and provides a new capability to explore the use of materials in interactive products and moveable parts/components.

3D body scanning can be used for rapid prototyping and customisation by creating a 3D model of the scanned body to create exact fit garments. The customer is measured and, through the digitised image seen on the computer screen, they can choose a garment that will be constructed to fit. Once the image is created, the measurement extraction software installed in the computer takes hundreds of individual measurements from head to toe. This data is then forwarded to the manufacturer to create the garment in a very short time with the exact measurements that matches the customer. It is an effective strategy for maximising customer satisfaction and minimising **inventory** cost.

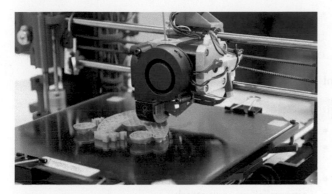

Figure 7.108 **3D printer in a school**

Additive and digital manufacture methods

There are a number of different types of additive manufacturing including 3D printing (see Section 7.1), rapid prototyping and **direct digital manufacturing (DDM)**. DDM refers to 'the process of going directly from an electronic digital representation of a part to the final product via additive manufacturing'. Some of these final products may be moulds or other items of tooling, however, in many industrial sectors including space, aerospace, automotive manufacture, healthcare, toy making, designer goods and fashion, we are starting to witness the application of 3D printing to directly fabricate final products or parts.

Ford are researching the use of DDM to manufacture parts that can be used in cars. In the future we might find ourselves downloading and printing parts or consumer products.

Advances are still needed to combine different families of materials, such as metals and polymers, in a single print cycle. Developments on this front are in the very early stages and it is likely to be more than five years before products are offered.

Stock control, monitoring and purchasing logistics in industry

Knowing the location, status, value and lead times of stock is essential to making the most effective business decisions. Stock is usually classified in three groups:

- materials/components – bought in from suppliers to be used in the manufacture of products
- work in progress – incomplete products currently being manufactured
- finished products – assembled products of desired standard ready for distribution.

Stock control enables production to flow without costly hold-ups and ensures that sufficient raw materials and components of acceptable quality are purchased and customer demand is met.

Computerised systems, including the use of barcodes and other digital recognition processes to monitor stock, are used. Links are easily made with the purchasing, marketing and sales departments. Accurate forecasts of predicted sales ensure that sufficient orders are placed to meet demand.

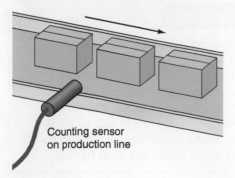

Figure 7.109 **A counting sensor on a production line**

The purchasing and logistics departments make decisions about suppliers who are reliable, competitively priced and able to provide the materials and components of the required standard.

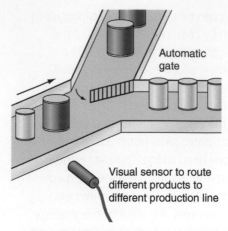

Figure 7.110 **A visual sensor to route different products to a different production line**

The system includes 'buffer' stock to cope with unforeseen problems or emergencies in the supply chain, or if there is a sudden and unexplained rise in demand.

Table 7.15 **Benefits and drawbacks of computerised stock control**

Benefits of computerised stock control	Drawbacks of computerised stock control
• Lower costs • Improved efficiency • Caters for fluctuating levels of demand • Efficiencies can lead to price reductions, improving competitive edge • Very quick system • No manual checks required • Detailed, accurate, well-presented data available for print out or distribution in electronic format • Data can be selective for presentation to different groups • Data easily stored	• Initial cost of set up and training • Sometimes only one person can operate system, so problems occur if that person is absent or leaves the firm • Possibility of manual input error • Software failure, virus attack • Computer breakdown can be costly; backup systems are essential • Digital data may be accessed without security system

ACTIVITY

1 Discuss how manufacturers meet consumer needs in rapidly changing product markets.
2 Discuss the workforce issues to be considered when selecting an appropriate production system.
3 Discuss the advantages and disadvantages of JIT manufacture.
4 Discuss the importance of using standardised parts and components in the manufacture of either domestic electronic products or textile products.

KEY TERMS

Batch production – the production of a specified quantity of a product. Batches can be repeated as many times as required. This type of production is flexible and is often used to produce batches of similar products with only a small change to the tooling.

Bespoke – describes a product that has been made specifically for an individual customer.

Carrying cost – the expense associated with holding inventory (stock) over a period of time.

Continuous flow production – highly automated production process where many thousands of identical products are made. The production line is kept running at all times to maximise production and eliminate the extra costs of starting and stopping the production process.

Direct Digital Manufacturing (DDM) – a process that produces parts directly from a CAD file.

Haute couture – the people and companies that create clothes that are very expensive, fashionable and frequently individual, using luxury fabrics and embellishments.

High-volume (mass) production – the manufacture of large quantities of standardised products.

Inventory – the raw materials, work-in-process products and finished goods that are considered to be the portion of a business's assets that are ready or will be ready for sale.

Just-in-time (JIT) manufacture – manufacturing system in which materials or components are delivered immediately before they are required in order to minimise storage costs.

Lean manufacturing – involves never-ending efforts to eliminate or reduce waste (activity that consumes resources without adding value).

Logistics – the careful organisation of a complicated activity so that it happens in a successful and effective way.

Mass customisation – a manufacturing technique that combines the flexibility and personalisation of custom-made products with the low unit costs associated with mass production.

Modular/cell production system – a contained, manageable work unit that includes an empowered work team. The teams may be used to perform all the operations or a certain portion of the assembly operations, depending on the organisation of the module and processes required.

Repetitive flow production – a large number of identical items are produced at a relatively low cost and production is broken down into sub-assemblies of smaller components.

Standardised components – parts that are usually produced in high volumes to the same specification and quality. Bolts, screws and fasteners are common examples.

Tolerance – the permissible range of variation in a dimension of an object. Sometimes known as allowance.

KEY POINTS

- With any manufactured product, there are different ways to organise the actual production of products, according to the type, the number and the diversity of products to be made. The type of production system is determined by the quantity of production and the required rate of delivery.
- Customised manufacture systems give consumers the option to personalise their purchases, allowing for individuality and self-expression.
- Many fashion and textiles manufacturers offer customisation options to remain competitive in the fashion and textiles world.
- Rapid prototyping gives designers and manufacturers the ability to turn ideas into reality at a much faster pace than using traditional techniques.
- Rapid prototyping allows for design iteration at a faster pace. It reduces waste and ideas can be communicated more quickly.

- 3D body scanning can be used for rapid prototyping and customisation by creating a 3D model of the scanned body to create exact fit garments.
- Additive manufacturing is revolutionising the way designers and manufacturers create prototypes in the industry. The methods accelerate the design process, reduce production time and waste, and lower costs.
- It is crucial for manufacturers to have reliable suppliers. In the ever-changing industry, retailers must manage stock continuously with the changing fashion, style and design.
- The implementation of inventory management applications has become a valuable tool for companies looking to manage stock more efficiently in a competitive world.
- In industry, the purchasing and logistics departments make decisions on the suppliers who can deliver the materials and components to the required standard, at a competitive price, as and when they are required.

7.5 How is the quality of products controlled through manufacture? (A Level)

LEARNING OUTCOMES

By the end of this section you should have developed a knowledge and understanding of:
- the processes that need to be undertaken to ensure products meet legal requirements and are high quality:

- quality control
- quality assurance
- Total Quality Management (TQM)
- European and British standards.

All products are designed and manufactured with consumers in mind and the end product needs to fulfil both the requirements and expectations of the customer.

These are likely to include appearance, performance, reliability, quality and safety. It is important that an organisation knows and understands its customer's needs and expectations and then puts in place the procedures and systems to ensure it meets them. Failure to meet the customer's expectations means a product does not sell as well as expected and this could cause damage to the reputation of the manufacturing company.

Quality control

Quality control (QC) is used by manufacturers to check quality against the required user and stakeholder requirements or standards or to make sure that items have been made within set tolerances.

QC involves using an inspection team who are looking for components or parts that are not up to the specific standards. Inspections can take place at identified stages from the initial sourcing of raw materials, through manufacturing, to the end of the process when the final item has been assembled. Inspections can be carried out on all, or a sample, of the products, depending upon the number of items being produced. You might notice stickers on products or garments you purchase, indicating they have been checked for quality.

Inspection checks can be carried out in a variety of ways, including:

- simple visual checks
- detailed data comparisons
- checking accuracy of dimensions
- checking calibration of sensors
- checking product functionality
- flammability tests
- checking weight
- electric circuit checks
- safety checks.

A 'quality control' approach on its own can be costly for a company if the numbers of defective products being rejected is high, especially if rejected products cannot be reworked and brought up to standard. Finding and dealing with the root causes of the failures has to be a priority. Beyond that, it is important to create a quality culture in a company, so that individuals care about and take responsibility for their own work. This is part of the quality assurance process.

Quality assurance

Quality assurance (QA) is carried out by a company to see that the product meets the quality standards set. A series of actions and procedures are set up to check the product before, during and after manufacturing operations have taken place. It is a proactive process aiming to prevent failure and to make sure that the quality of the product is right first time and every time. QA is considered part of the design development and testing processes, eliminating defects at the earliest possible stage.

NACERAP is a quality assurance term used in the fashion and textile industry. It provides a standard system for identifying faults and rectifying them. It provides a system for training everyone in the company to use the same words to describe, record, repair or prevent faults. NACERAP stands for:

- Name of the fault
- Appearance of the fault
- Cause of the fault
- Effect the fault has on the overall quality of the product
- Repair of the fault or equipment
- Action to be taken to correct the fault
- Prevention to prevent the fault reoccurring.

Total Quality Management (TQM)

Total Quality Management (TQM) is the aim of every company to achieve sustained levels of quality performance. It is based on the philosophy of perpetual improvement. In a TQM system, all the areas of an organisation and its suppliers use agreed specifications and quality control methods and quality is the responsibility of everyone, not just the quality control departments. An important consideration for the organisation of the company is how they can guarantee the manufacture of a quality product. It is because of a company's desire to gain customer satisfaction that TQM procedures are set up.

Companies that implement TQM are constantly seeking to improve the performance of their organisation and the quality of their products and services. TQM emphasises the importance of the whole manufacturing process, reviewing and monitoring every stage of management and manufacture across the company. Checks are made at every stage from the delivery of resources through to the final delivery of the product to the customer.

In order to be effective, TQM relies upon every employee within the workplace being responsible for their quality standards. Regular training of employees is a key element to support the process, encouraging employees to be pro-active in identifying and addressing quality-related issues. If any faults are discovered, they need to be corrected immediately; this allows repairs to be carried out or changes to the production methods to be made. If faults are identified at an early stage, the number of rejected items can be reduced.

European and British standards

A standard is an agreed way of doing something. Adopting a standard into a designing and manufacturing process ensures that the product is compatible with other similar products. The standards, testing procedures and quality assurance measures are set by British, European and International Standards Organisations. For more on these organisations see Section 8.1.

DESIGN ENGINEERING

Ensuring products are compatible is particularly important for standardised parts. For example, the ISO standard for metric screw threads ensures that screws and nuts from different manufacturers are all compatible with each other. Most standards are voluntary – it is up to a manufacturer whether they wish to adopt a standard, although there will be many good reasons for them to do so.

The International Electrotechnical Commission (IEC) publishes a number of standards you may be familiar with, including the IEC standard for electronic component circuit symbols, and the Ingress Protection (IP) rating standard which defines how effectively an enclosure is sealed against intrusions from fingers, dirt, water, etc.

FASHION AND TEXTILES

There are many British Standards relating to the fashion and textiles industry. The main ones relate to:
- fibre, yarn and fabric testing
- performance characteristics of the fabrics
- colour fastness, finishing and aftercare
- sewing machines and thread.

When providing information on a textiles label there are many regulations for manufacturers to follow. The guidance is issued by the BSI (British Standards Institute) textile labelling regulations department. The specific code relating to labelling is BS ISO 3758 Textiles – Care labelling code, using symbols.
- A textile product may be described as '100%', 'pure' or 'all' only if it is composed exclusively of one fibre type.

- Textile products containing two or more textile components of two or more different textile fibres must bear a label stating the fibre composition of each component.

With continual fibre and fabric developments in the textiles industry, legislation has to allow for new developments. New fibres will be considered for labelling if:
- the fibre is radically different from other fibres by chemical composition and/or by fibre properties
- a new generic name is justified, as the fibre cannot be classified into an existing generic name.

PRODUCT DESIGN

Sometimes small changes in products can make a big difference to how safe it is. The British Standard on writing and marking instruments (BS 7272) introduced in 1990 dramatically reduced the number of deaths caused by choking on pen tops. It specified that pen tops should have a hole in the top (or some other way to maintain airflow) to reduce the risk of choking. This standard is now used by all international manufacturers of pens.

Many products are covered by European and British standards and you should always check for labels to show products are safe and have been tested. Products used by children need to meet high safety standards, such as the BS EN 71 series that covers toy safety, from flammability to toxicity of materials and choking or injury hazards. This set of standards forms part of the EU Toy Safety Directive (2011). All toys supplied in the UK must meet a list of

essential safety requirements which are set out in these regulations. Manufacturers have a responsibility to make sure toys comply with safety requirements, conduct an assessment and use the CE marking to show compliance. BS EN 7 covers how toys should be manufactured and tested and the safety warnings they must carry. For example, all pushchairs need to be stable, durable, with good brakes and secure harnesses and be designed to minimise risk of injury when folding.

British Standards also cover inclusivity. Nearly half of over-65s, and one in six people under 40, find it hard to open everyday packaging items such as cartons or jars. The European technical specification for packaging covers ease of opening (CEN TS 15945) makes packaging easier to handle and open for everyone, including those with reduced hand strength.

ACTIVITIES – PRODUCT DESIGN

1 With help from your school or college, try to organise a group visit to a local small manufacturing company. Observe the production methods in use and prepare a list of questions in advance.
2 Watch several videos from the series *How It's Made* on YouTube to see how a range of products are manufactured, from bespoke items to high-volume production.
3 Describe the role of the British Standards Institute in relation to product quality.

ACTIVITIES – FASHION AND TEXTILES

1 Examine a fashion or textiles product of your choice and create a table that lists all the quality control checks that would have been carried out.
2 Examine the labels on a garment of your choice and carry out research to explain the standards that have been used to ensure the product is suitable for the consumer.

ACTIVITY – DESIGN ENGINEERING

Find out which products relate to the following British Standards. Choose one and prepare a short account of the key points about the standard.
  – BS 1363
  – BS 6102
  – BS 1852

KEY TERMS

NACERAP – a procedure used by fashion and textile manufacturers to provide a standard system for identifying faults.

Total Quality Management – TQM involves all members of an organisation in improving processes, products, services and the culture in which they work.

KEY POINTS

● Quality control in the fashion and textiles industry is implemented in terms of quality and standard of fibres, yarns, fabric construction, colour fastness, surface designs and the final finished garment.
● Manufacturers decide the acceptable standard that a product must meet to pass quality control checks. Any defects are highlighted as minor, major or critical.
● In a TQM system all the areas of an organisation and its suppliers use agreed specifications and quality control methods. It is the responsibility of everyone, not just the quality control departments.
● The standards, testing procedures and quality assurance measures are set by British, European and International Standards Organisations. The purpose of these organisations is to ensure that products meet the quality and safety requirements for their use.

8 Viability of design solutions

Across the creative industries, consideration of viability is continually planned and revisited at every level of the design process. Designs regularly fail to make it past the conceptual stages for numerous reasons, often relating to **commercial viability**, where stakeholder needs have not been fully considered.

They are many reasons why design proposals are not considered to be commercially viable. If a design idea does not meet the expectations of stakeholders it may not be considered viable at all. Over the course of this chapter you will learn how designers and manufacturers ascertain whether a design solution is **feasible**. You will learn how designers identify

key aspects and features of their intended concept to ensure the design solution is viable and meets stakeholder needs. This will help you to consider a broad range of requirements for your **iterative designs** and also better understand choices made for existing products.

As you work through this chapter, think about how and when you should consider viability in your own designing. The more often you do so, the better. Ultimately this will make you more likely to design products that people really want and that are truly commercially viable because they meet the needs of all the stakeholders impacted by the design.

8.1 How can designers assess whether a design solution meets its stakeholder requirements?

LEARNING OUTCOMES

By the end of this section you should have developed a knowledge and understanding of:
- how critical evaluation of a design solution is used to check if it has met its intended requirements, including:
  - functionality
  - ease of use and inclusivity of the solution
  - user needs
- the needs and methods for testing design solutions with stakeholders throughout the design development and when testing the success of a product

- the importance of testing the feasibility of getting a product to market, including considerations of cost, packaging and appeal.

If you are studying at A Level you should also have developed a knowledge and understanding of:
- the relevant standards that need to be met and how to ensure these are delivered, including those published by the:
  - British Standards Institute (BSI)
  - International Organization for Standardization (ISO) specific to the subject.

How can designers critically evaluate how a design solution has met its intended requirements?

At the start of the design process a range of core requirements should have been identified through feasibility studies, discussions with stakeholders and users and possibly focus groups. If the process has been undertaken thoroughly it will have included all relevant stakeholders and will have aided the creation of a detailed outline of the fundamental design needs. These requirements will be used by the designers and wider stakeholders to judge the success of the product and determine if it is viable for all concerned.

Designers will evaluate the viability of an intended concept from the outset of the design process right the way through to the final prototype and possibly beyond. They will utilise a range of tests to ensure the product meets all stakeholder needs through ongoing or final evaluations. The variety of approaches may include **qualitative** testing such as user feedback or detailed **quantitative** virtual stress loading or real-world destructive tests on physical prototypes.

Functionality

When designing any product one of the primary considerations is functionality. Fundamentally any product must do what it is designed to do and meet user expectations. For example, if a consumer purchased a pair of running trainers they would have an expectation that they would help running posture and foot strike, provide comfort in use and last for a considerable mileage. If the trainers fell apart quickly or were ill-fitting and caused blisters due to poor stitching or cut, they would not meet the functional requirements and would probably be returned. Designers strive to ensure the product meets its intended purpose in full, to safeguard against the manufacturer falling foul of regulatory standards or the Consumer Rights Act which lay out expectations of products; one of which is to be fit for purpose.

Designers engage in a variety of tests to ensure that the product functions correctly. These are used over the course of the design process, not just at the end. Within the NEA you are asked to undertake an explore-create-evaluate iterative design process which mimics industrial practice. An iterative design approach means that the product's function will gradually improve due to changes and adjustments being made in reaction to feedback gained. In industry, designers judge the feasibility of design iterations regularly against the original outlined intent and this may lead to more detailed or further refinement.

Figure 8.1 Car bodies are often shaped in clay prior to making a full prototype. Here a BMW prototype is ready for feedback from the design team

During the design process, designers may use strategies including rapid prototyping and virtual modelling to gather feedback. At early stages designers may simply use their own judgement and experience to determine the success of designs but they will then rapidly broaden the range of feedback to include more stakeholders. This approach is needed to avoid design **fixation** and issues associated with designing in isolation. Fixation from a design perspective means that a preconceived idea or approach is followed without taking on information from external sources or stakeholders. To avoid this, focus groups and intended users are asked to reflect on products. As design concepts progress, design professionals will aim to recreate the **environment of use** to ensure that feedback is authentic and highlights real-world practice. For instance, testing a wallpaper steamer in a large airy space such as a design office for a short period of time will probably not bring to light any issues associated with poorly fitting waterproof seals protecting electronic components; the location poorly reflects the real-life experience of using the product in a confined hallway over a prolonged time period where humidity levels increase significantly.

One way that designers aim to recreate a real-world product lifecycle is to use a design methodology called user-centred design (UCD) (see Section 1.2 for more on UCD). This close working relationship between designers and the intended user gives them excellent insight into problems, changes and also the merits of design solutions through timely feedback from the people expected to use the product.

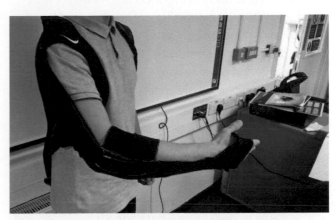

Figure 8.2 Here a model prototype design of a cycle armour design is evaluated to consider fit and range of motion

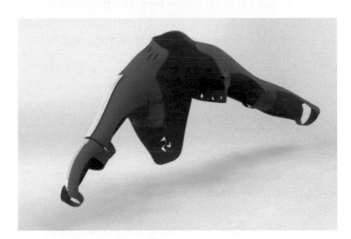

Figure 8.3 A final prototype has been determined for the armour and the design is created in CAD to allow pattern cutting and final feedback

Ease of use and inclusivity of the solution

Usability refers to how easy the product is to use – how clear and obvious the functions are. You should already be familiar with the term **inclusive (or universal) design**, meaning a product is accessible to (can be understood and be used by) everyone, regardless of age, ability and status, including those with some sort of disability, without special adaptation.

Consumers and users of products often interact with products without any training or knowledge of how they work. As human beings we are naturally geared to learn by trial and error. With simplistic systems this often works and given enough time we can even problem solve our way to understand how to use relatively complex equipment. However, for products to be successful they often need to engage users quickly. Users interact with products through the **human interface**. Typically this is the parts the user touches, presses, listens to or looks at. Consider, for example, your mobile phone – with relatively few buttons and a screen you are able to determine what happens on screen and understand what your inputs are doing. Overly complex systems with a daunting array of controls and displays will often put users off interacting with them. If this happens to a lot of users it could mean that the product in question is quickly rejected and fails to be viable against its competition.

Designers undertake market research with user groups to ensure that designs are intuitive to use, meaning that users can quickly understand how to use them. Through critical evaluation, user trials may highlight issues where users make common errors in using a product or find parts hard to identify. The trials consider the affordance of the design – the clues about how a product should be used. The feedback enables the designers to improve the **semiotics** of the design – the signs and symbols that direct the user. Door handles are an everyday example of affordance. If you see a handle which sticks out then you will assume you have to pull it and conversely if it is a flat plate you will push it. However, how many times have you been caught out and got it wrong? Is it you not understanding the product or is it bad semiotic design? See Chapter 10 for further details on affordance.

Figure 8.4 If affordance had been considered would the sign be needed?

To assess the ease of use or usability of a concept, designers aim to recreate the environment of use or a realistic situation. This involves thinking about the people who will use it (for example, their background and culture), different ways in which people interact with the product (such as opening, closing, operating, carrying, adjusting or switching) and the different situations (for example, bright or dark, outdoors or indoors) in which the product will be used. Designers aim to gain as much information on how users interact with prospective products as possible. This can be approached on a variety of levels:

- discussions with users
- group tests
- small-scale roll outs.

During the testing process, products sometimes perform well with users from a certain **demographic** (a group of people of a similar age, income or education) but not with others. Designers observe a range of users using a product and record key aspects of interaction, which may include:

- how often they are unable to proceed
- common mistakes in use
- areas where manual operation is difficult
- problems with understanding the interface.

By **critically evaluating** these issues, designers are able to understand where and how a current design fails to match ease-of-use or inclusivity requirements. In some cases these are quick to overcome by adding

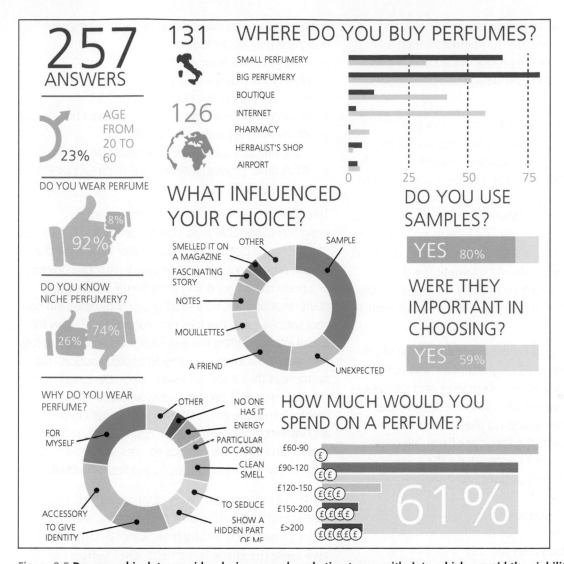

Figure 8.5 Demographic data provides designers and marketing teams with data which can aid the viability of their products. The data shown here allows designers to understand more about the consumers of perfume products and what influences their purchasing. This could allow for highly targeted advertising to take place

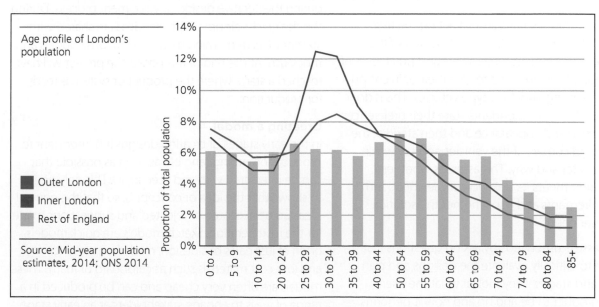

Figure 8.6 Demographic data can be used to identify users in certain areas. This data shows that a product aimed at people aged 20–40 could have a large customer base in Inner London

pictograms to bridge language barriers. Should the problems identified be more fundamental, then the current iteration of the product or approach may be rejected entirely and new design solutions sought.

User needs

The primary user is often the most important stakeholder as they are usually the person buying the product. Evaluating how a product meets the needs of its users is fundamental for it to be a viable product and to ensure a prominent place in the market. A common mistake when talking about user needs is to focus on just the person using the product. While the primary user interacting directly with the product will be the central concern during evaluating a design, we also need to think more broadly if we are to truly assess the impact of the product and its long-term viability.

The vehicles produced by the car industry are often focused on user needs. Designers plan how to fill niches within the saturated car market and use new innovations to promote sales to meet stakeholder requirements. When assessing the success of the products they look at the feedback from the driver as the primary user and contemplate their needs in detail. Some key considerations are:

- comfort
- adjustability of 'cockpit'
- safety of passengers
- ease of interaction with controls
- security of the vehicle
- aesthetics
- styling.

To ensure that they are critically evaluating the user needs, manufacturers hold 'car clinics' as part of their market research, where potential users and purchasers of a certain type of car are asked to give their opinions on a selection of existing and prototype vehicles. This is done with actual vehicles. Respondents state their preferences in terms of style and appearance and then stand at the ends, sides and corners of the vehicles and say which views they prefer and why. This assesses whether a design solution meets its stakeholders aesthetic and styling requirements and also whether the solution meets its user needs.

Designers assess user needs by asking questions that allow them to critically evaluate the designs and plan next steps and stages. They look at what the user wants the product to be and do and how it performs. Depending on the product and level of interaction the user may need many different things. Questions designers may ask include:

- Who will use this product?
- How will the user interact with the product?
- What will the product be used for?
- What type of environment will the product be used in?
- When will the product be used?
- Is the product used once or repeatedly?
- What other products/devices will it interact or interface with?

Throughout the design stages it is important for designers to consider **bias**. User feedback is a vital source of information but can be flawed. What happens if the person providing feedback has a very particular opinion or a preferred way of doing something? If a designer plans iterations around a small group of people who have very specific ideas, this may not match up with how the actual intended users will feel. Designers who do not fully take bias into account are likely to create products that fail to meet their intended users' needs. Designers must therefore also evaluate the quality and reliability of the feedback they are given. Often the designer's experience and instinct, along with a broad approach to gathering user feedback from numerous sources, will avoid fixation on issues which are only specific to a few individuals.

Demonstrate an understanding of the needs and methods for testing design solutions with stakeholders throughout the design development and when testing the success of a product

Pre-production testing can be undertaken at any stage during the iterative design development process. During the design development phase, ideas are shared with stakeholders using mind maps, simulations, models and mock-ups. At the end of this phase, the project will have reached a stage where the product or system is ready for production.

Creating a model

In the early stages of creating designs it is important to produce something tangible as soon as possible that can be interacted with and seen in 3D. This only needs to show what the idea or concept is, so that the general idea can be accepted or rejected and suggestions made for the next iteration. Sketch models are quick models, often of just parts of a design, made from easy-to-work and low-cost materials such as cardboard or foam. These models are often very cheap and can be produced in a range of ways to engage stakeholders at an early stage. You can read more about sketch models in Chapter 4.

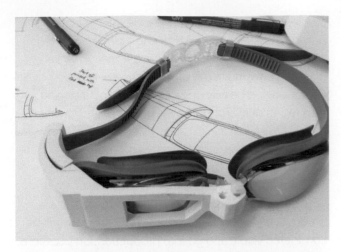

Figure 8.7 Designers gain feedback by making sketch models. Here we see a Swimar model that has been interacted with and improved

During the design process, designers are faced with the problem of giving their users and other stakeholders a real idea of what a product may be. Models give a good representation of what a product will look like and how it will feel and as such are used to test the relative successes of design solutions throughout and at the end of the development process. Design teams often use rapid prototyping to create models very quickly. These often produced from computer-aided design (CAD) files using additive printing or possibly through sketch modelling to help stakeholders see and interact with the product. Discussion, observation and pre-planned questions can give designers a good feel for the successes or failings of the concept.

Collaboration and multi-disciplinary working

Online collaboration is used extensively by design professionals, by means of web/cloud based software and networking. Designs can be developed concurrently with other designers and specialists around the globe. This often involves multi-disciplinary working – collaborating with those who have different areas of expertise to assess and ensure viability. These specialists aid the quick identification of problems and creation of alternative design solutions. (You can read more about online collaboration in Chapter 4.)

Delivery phase

The design development phase is followed by a delivery phase where the main activities are:

- testing of the final product with key stakeholders to confirm the product meets agreed requirements
- approval and launch – once all stakeholders agree the product is a success then it will be rolled out and launched to consumers

- setting targets – stakeholders will, for some products, agree new targets such as reducing production costs, emissions, delivery costs or waste
- evaluation and feedback loops – critical evaluation will be undertaken to determine improvements or failures and systems agreed to monitor the success of the product to inform future production.

Designers and manufacturers continue to work closely with stakeholders, sometimes for many years after a product has been released. The success of a product often comes from changes after launch, adjustments and new versions of products which reflect ongoing feedback and streamlining of production methodology or incorporation of new technologies to meet user needs and reduce unit price.

Demonstrate an understanding of the importance of testing the feasibility of getting a product to market, including considerations of cost, packaging and appeal

When bringing a product to a consumer market, designers should critically evaluate the steps needed, associated costs and the possible risks of a product to determine its viability. Smaller businesses are likely to have problems with funding and investment. James Dyson is renowned for having spent many years trying to sell or license his bagless vacuum cleaner concept to various manufacturers before eventually risking bankruptcy and going it alone to start his own business. Competing against larger firms with bigger budgets and established **brand identity** can be very difficult, so investors, design and marketing teams need to be reassured that the product will be successful. Likewise, larger firms have reputations at stake and a misplaced product that fails can be detrimental to future sales of other products, or even prevent the company from operating in certain markets.

Designers undertake a range of testing with consumers and stakeholders that aims to identify and address any potential issues. This allows them to give confidence to investors and to assure CEOs and other stakeholders leading the roll-out of a new concept that the product is viable and that ultimately it will be a success. There are many ways in which designers gather information and these methods of marketing research may include:

- CAD visual testing – using virtual mock-ups of both the product and its packaging to gather consumer opinions before the product is put into production

- User trials – inviting users to trial the product. This could be expert testers trialling the product against others or potential consumers chosen to test the product
- Focus group tests – gathering consumers together and discussing the pros and cons of design solutions
- Clinics (similar to focus groups) – using potential consumers who are often handpicked to represent specific demographic groups to assess, discuss and provide feedback on a range of possible outcomes to help the designers identify the most successful and pleasing outcomes (see User needs, page 364, for more detail).

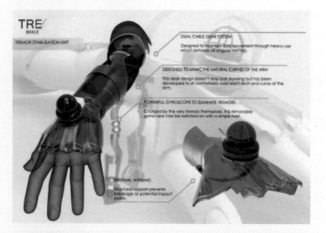

Figure 8.8 CAD models can help gather opinions about the likelihood of successfully bringing products to market. Here a proposed design for a 'tremor stability' device allows stakeholders to comment on the feasibility of the concept

Cost

Before bringing the product to market, designers and manufacturers must carefully consider all costs to determine if the product is viable. These costs may include:

- materials – the purchase (and reliable and ongoing supply) of raw materials or component parts
- design, research and development costs
- production – manufacturing costs including the purchase (and maintenance) of plant (machines, moulds, etc.)
- legal – costs relating to the protection of intellectual property
- marketing – costs surrounding product launch, advertising, branding and sales
- packaging – costs of printing and physical packing costs
- transportation – costs covering distribution, final delivery and supply chains.

Studies will be undertaken to gauge the expected number of sales (**forecasting**) and a **price point** will be judged, aiming to maximise profit margin while planning how much the product can be sold for without being heavily undercut by competition and thus losing sales. This price will be compared to the overall costs and forecast sales. The viability of the product hinges on selling a certain number of products at a certain price to match the costs. Once all costs are covered, the company meets a break-even point where further sales become profit. Normally only products that allow the company to break even will be considered viable and go into production. On occasions, companies will determine that the product will still go into production as a loss leader. These products rarely make profit and sometimes are produced at a cost to the manufacturer. However, it will have been identified that once customers have bought into a brand they will then buy additional items or consumables which make significant profit.

Examples of this are computer game consoles and razor handles. Consoles themselves attract little profit but once a consumer has bought one they will purchase games which make substantial profit. Similarly a reusable shaving razor system handle will often be sold below the price of production; however, the razor blades cost little to manufacture and will retail at relatively high costs and draw significant profit.

Consumer appeal

Once a product has been brought to market, its success often relies on its ability to capture the interest of its targeted consumers. During development stages designers plan testing of the product that ensures it attracts interest, while at the same time considering cost. As a consumer you may consider what drives you to buy a certain product rather than another. Is it simply the function of the product that influences buying decisions? The simple answer to that is no; decisions to buy a product are complex and cover a range of differing factors which designers must closely plan for, such as:

- personal taste
- fashion
- trends
- function
- brand loyalty
- aesthetic appeal.

Figure 8.9 Company branding is essential for consumer recognition. Here we see a range of tea brands. Different images, fonts and designs are applied to packaging to distinguish them from other brands selling a similar product

Research has shown that consumers often purchase items on impulse. These buying decisions are often swayed by the placement of products in shops, their branding, packaging and the way they are marketed. Designers aim to attract these purchases by planning how to engage consumers by meeting fashion trends and creating products with aesthetic appeal.

Companies aim to create brand loyalty to promote sales of their products. If users like a product they will often purchase more products from the brand due to their positive experience. Designers undertake feasibility studies on branding options to ensure the branding attracts the correct user groups and demographics and that it is also highly appealing.

Packaging
Packaging has numerous functions apart from simply containing the product and needs careful consideration when assessing the viability of many products. Considerations include:

- base material costs, including volume of material needed
- printed areas and range of colours
- production run size
- required level of protection afforded by the packaging
- additional functional requirements like **wet strength** or water resistance
- value added by packaging in terms of appeal
- environmental impact.

Packaging can significantly influence purchasing decisions and customers often buy products based on the packaging alone. Designers trial packaging options with consumers to see which are the most eye catching and appealing. In many cases, designers trial packaging which adds perceived value to the product. An example

of this is the packaging that comes with high-end perfumes and fragrances. The product itself often costs little to manufacture but retails with a high price tag. Designers utilise lavish packaging solutions which are expensive to produce but that they hope will lead to greater sales, making the product more profitable in the long-term and therefore improving its viability.

All extra packaging adds additional costs to the product and as such needs to be planned carefully. Small extra parts or additional printing rapidly grows to a significant sum if used on a large production run and can influence viability if it impacts heavily on potential profit.

Figure 8.10 Packaging helps identify very similar products and adds perceived value. Here, carefully considered packaging colours and bottle shape have been planned to make the consumer feel that they are purchasing something special and unique

Packaging also includes the shape of the product, not just the box, container or labels that accompany it. The shape, form and style of the product itself is very important to the viability of a product. Products such as electronics or electrical items are often designed to look stylish and appealing and careful consideration is given to the aesthetic qualities such as the lines and curves used within their physical forms. Despite how well a product might perform, if it were truly ugly would consumers want to have it on display in their homes? Exterior designers also ensure that product branding, including logos and product identification, is clearly visible to promote product recognition and reinforce brand image.

Understanding the relevant standards that need to be met and how to ensure these are delivered (A Level)

Standards are published documents that contain technical specifications or other precise information designed to be used consistently as a rule, guideline or definition. All designers need to be mindful of quality standards.

British Standards Institute (BSI)

BSI is an internationally recognised body whose policies and standards are used across the globe. There are five types of British Standard: specifications, methods, guides, vocabularies and codes of practice.

BSI standards on toy manufacture

The BSI standards on toy manufacture are outlined over 11 detailed documents. These standards promote the production of safer toys and aid manufacturers to meet the safety requirements set out in the Toy (Safety) Regulations 1995 and other directives. They are continually reviewed and updated to ensure they are current and encompass changes in the law.

The toy manufacture standards cover aspects including:

- mechanical and physical requirements
- flammability regulations
- organic chemical compounds restrictions, limits and tests
- electric toys
- children's play safety for playgrounds
- inflatable play equipment.

Before the 1995 Toy (Safety) Regulations, toys were manufactured to a variety of quality standards. Often cost or lack of awareness drove manufacturers to use parts or materials that were highly dangerous for the children using them. In the 1980s there was a spate of news stories about children being poisoned by toxic paint finishes or cut and injured by easily broken parts. Clearly these issues had to be addressed and, due in part to media pressure, new legislation was put in place in 1995.

BSI Kitemark®

The BSI Kitemark® was introduced in 1903 and has become a vital marketing tool to show that products are safe, reliable and are of an appropriate quality. Many organisations have made it mandatory for their suppliers' products to have been awarded the Kitemark® before they will place an order with them. As consumers become increasingly informed about their choices, conformity to recognised standards becomes a key purchasing decision.

Figure 8.11 The Kitemark is awarded to products to confirm they meet minimum standards. Consumers often look for this mark and avoid buying products that do not display it

To ensure the BSI standards are met, a variety of testing is done and manufacturers undertake a process called 'assessment of conformity'. This includes an initial assessment to check if the product meets the relevant standard and an assessment to see if the quality management system operated by the product manufacturer (see Section 8.2) is in place and fully documented. After they have been awarded a BSI Kitemark, licensees are subject to ongoing checks such as regular testing, assessment of production quality controls and possibly even mystery shopping.

Companies wishing to display the mark have to go through a quality assurance vetting procedure which not only looks at the product but also the manufacturing systems, tests, checks, record keeping and procedures dealing with identified failings. Designers plan the viability of their concepts and carefully consider the aspects the BSI may consider, which may include:

- durability
- fitness for purpose
- environmental impact
- ease of disposal
- waste
- product labelling
- energy efficiency and insulation
- material/surface finish.

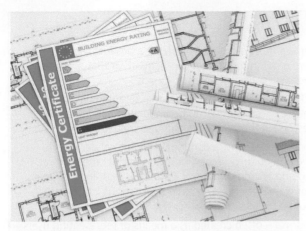

Figure 8.12 **Energy efficiency standards ratings are often graded A+++ to G. Here we see a certificate showing a good energy rating for a building**

International Organization for Standardization (ISO)

The International Organization for Standardization (ISO) is a non-government body which aims to promote the development of standardisation in order to facilitate international trade. Like the BSI, they also cover product-specific and services regulation.

ISO standards are broadly similar to the standards upheld by the BSI; indeed, in some cases, ISO standards are based upon documentation originally produced by the BSI or shared by both organisations. One difference is the geographical location where standards are recognised, with some countries choosing ISO above BSI or vice-versa.

ISO standards and products are regularly tested and revisited to meet current trends and practice. To attain an ISO standard, the company applying for it is required to go through a seven-point process which assesses:

- design and development planning – putting systems in place to ensure smooth data handling and that quality checks are in place throughout the design process
- design inputs – records that show all requirements from the customer or that are mandated by government or industry regulation are reviewed and taken into account for the current design proposal
- design and development outputs – these are all documents, CAD or machine data or assembly instructions, needed by financial and manufacturing teams to produce a product adhering to the original key criteria. In addition these will include details of the final product to ensure proper use
- design and development review – records that show all stakeholders within the company are considered through review systems to ensure processes are working correctly and that quality-review issues are identified and addressed
- design and development verification – documents showing that plans are in place to meet government regulations and that the output documents are taken into account to ensure a rigorous testing procedure that identifies pass/fail criteria
- design and development validation – records that show that upon completion of the first product, full testing to ensure fitness for purpose has taken place
- control of design and development changes – a planned process that ensures any changes to the design and development documentation are both necessary and agreed, while considering knock-on effects for related parts that could affect usability.

ACTIVITIES

1 In a group of four or five, adopt the **personas** of a range of relevant stakeholders and undertake a critique of one of your fellow student's design ideas. Remember to remain objective throughout the activity. Feedback can be provided on post-it notes, sketches or annotations separately or directly on the design sheets.

2 With the feedback you have gained, plan what changes are needed to improve your design further to ensure it meets a wider range of stakeholder needs. You may find that you have to rank order the requirements as some may be more pressing than others. What are your next steps going to be?

Affordance – the use or purpose of a product that is indicated to the user by the way it looks or is constructed.

Bias – a personal prejudice or inclination against a concept or approach.

Brand identify – the specific elements associated with a company or product such as name, logo, colours, typeface or tagline.

Commercial viability – the ability of a business to produce a product or service to compete against competition and make a profit.

Critically evaluating – closely examining a product or approach to identify significantly important issues or aspects and determine how important they are in relation to the success of the concept.

Demographic – a particular sector of the population. This may relate to age, education level, geographical location, etc.

Environment of use – the specific expected place and conditions in which a product will be used.

Feasible/feasibility – if something is feasible, it is capable of being achieved. It is possible. It is doable.

Fixation – a designer's preoccupation with a certain idea, or their fixed views on a solution to a problem, to the extent that they are unable to think of any other possibilities.

Forecasting – predicting or estimating what will happen in the future based on historical data.

Human interface – the part(s) of a product that the user comes in to physical contact with or that provides feedback, such as images or sound.

Inclusive (or universal) design – designing products that are accessible to (can be understood and used by) everyone without making changes or adaptations.

Iterative design – a continual and cyclical design development process to refine and perfect the product. Ongoing testing of models and prototypes, incorporating improvements and progressing towards an optimum solution for all stakeholders.

Persona – often created as a means of representing users or stakeholders. A persona is similar to a user profile. Information about users or stakeholders is used to create a collective persona, a 'typical' person, with their views, attitudes, preferences, lifestyle and skills and so on.

Price point – a chosen price at which something may be marketed and sold.

Qualitative feedback – feedback in the form of opinions, attitudes and thoughts.

Quantitative feedback – feedback that provides numerical data or information from sources which can be converted into numbers to aid analysis.

Semiotics – the study and use of signs and symbols.

Usability of a product – involves thinking about different ways in which people interact with the product and how easy it is to understand and use.

Wet strength – the ability of a material (normally card) to withstand loading while wet over a period of time.

MATHEMATICAL SKILLS

As a Desigen and Technology student you will need to undertake maths tasks which allow you to extrapolate data from graphs and calculate costs and percentages. These skills will allow you to create, read and interpolate data into your own NEA project and exam answers. For questions in this area you will need to be able to construct and use graphs to evaluate outcomes and present data.

ACTIVITIES

A company is checking the viability of its product packaging for a production run of 100,000 units. They have a budget of £110,000 for changes, but if they spend more than that, they will fail to meet their break-even point. Through feasibility studies they have identified that they need to make a range of changes to ensure the packaging is fit for purpose.

1 Fill in the gaps in the table and determine if the new packaging is viable.

Changes	Unit cost	Number of units per package	Cost of staff hours to manufacture and apply	Time in hours to complete tasks	Total cost
Corner reinforcements	£0.06	8	£7.00	0.01	£0.55
Additional panels of customer information	£0.05	2	£6.80	0.05	£
Additional tape to ensure seal	£6.50	0.02	£8.20		£0.171

Total additional cost per unit   £

Total cost for 100,000 units   £

2 Does this fall within the company's budget so they meet their break-even point?

Answer

Changes	Unit cost	Number of units per package	Cost of staff hours to manufacture and apply	Time in hours to complete tasks	Total cost
Corner reinforcements	£0.06	8	£7.00	0.01	£0.55
Additional panels of customer information	£0.05	2	£6.80	0.05	£0.44
Additional tape to ensure seal	£6.50	0.02	£8.20	0.005	£0.171

Total additional cost per unit   £1.161

Total cost for 100,000 units   £116,100

Does this fall within the company's budget so they meet their break-even point? **No**

KEY POINTS

- Products must be commercially viable to be a success.
- Testing takes place throughout the design and development stages to critically evaluate the strengths and weaknesses of iterations.
- Products must be proven to function as intended in their expected environment of use.
- Affordance promotes the viability of the product by ensuring that users understand how to operate and interact with it just by the way it looks.
- Detailed feasibility studies are vital to prove that the product will be competitive when brought to market.
- The BSI and ISO guide designers and manufacturers in meeting required standards of quality and safety.

8.2 How can design engineers, fashion and textiles designers and product designers and manufacturers assess whether a design solution meets the criteria of technical specifications?

LEARNING OUTCOMES

By the end of this section you should have developed a knowledge and understanding of:
- the methods and importance of undertaking physical testing on a product to ensure it meets the criteria it is meant to fulfil, including:
  – functionality
  – accuracy
  – performance.

If you are studying at A Level you should also have developed a knowledge and understanding of:
- how physical testing systems are integrated into the manufacturing process in the design industry to test functional feasibility, including:

Design Engineering
  – destructive and non-destructive methods
  – testing of materials for durability
  – testing models and prototypes for performance and fitness for purpose

  – testing products in use through different methods, such as:
    – consumer testing
    – virtual testing

Fashion and Textiles
  – testing of fibres and fabrics for durability and aftercare
  – testing prototypes, toiles and samples for performance and fitness for purpose
  – sampling garments and products through different methods, such as:
    – consumer testing, wearer trials
    – virtual testing

Product Design and Design Engineering
  – testing of materials for durability after care
  – testing models and prototypes for performance and fitness for purpose
  – testing products in use through different methods, such as:
    – consumer testing
    – virtual testing.

On completion of a final product, designers usually critically analyse the **viability** and feasibility of the final design solution through an intensive final evaluation called a **critique**. This detailed analysis of the design seeks to identify any weaknesses in the design for its stakeholder requirements and of course how it meets the final requirements detailed in the technical specification. A critique is undertaken with the specific goal of identifying issues, problems or faults in a product before it is put into full production. This is done in a very analytical way and the design team will look at every component and stakeholder viewpoint they can to ensure the product meets expectations.

Physical testing of a product

Due to the cost of production in large quantities, pre-production prototypes undergo rigorous physical testing before they are considered ready for final production. Pre-production models are used in their intended environment of use, analysed and scrutinised in detail by a wide range of stakeholders and professional testers. The data gathered is used to either confirm the manufacturing methodology is correct or highlight the changes that are needed. Only

once products have successfully passed this stage will they be considered viable and suitable to go into full production.

Figure 8.13 Pre-production prototypes are used to assess successes and failures. Here a team trials a wheel assembly prototype for an electric vehicle design to determine its fit and suitability for manufacturing in large quantities

Functionality

The requirement for a product to function as planned is a primary concern when assessing if a product has met its technical specification. Should the product not function in the way it was intended or serve its purpose to the desired level of quality, it is highly likely to be deemed a failure and non-viable. Function testing is planned for all products so that designers and manufacturers can critically evaluate how well they meet their intended purpose. To test function, designers and manufacturers carry out physical tests including destructive and non-destructive tests (see page 374) and user trials to assess how well products work during normal use.

Accuracy

The manufacture of components and assembly of parts to the required level of accuracy dictates how well they function or perform. For instance, if you purchased an electrical product but the manufacture of the plug was inaccurate, would you be able to plug it in to power the product? If there were inaccuracies in parts, could it pose a fire risk? A huge number of products, including fuses, plugs, credit cards, USB connectors and DVD discs, have standard sizes, with dimensional **tolerances** that they must be within to meet technical specification requirements. Small errors in manufacture would render them immediately useless for their intended purpose.

Designers and manufacturers ensure that the required standards are delivered by monitoring a range of different factors at different stages of development and manufacture. The key factors they assess are:

● tolerances
● differences in material quality
● performance.

Quality control (QC) inspection checks are carried out to monitor accuracy of parts and control the uniformity of production of products or components. How and when these checks are carried out varies but normally checks are made both during and after manufacture on some or all of the products. (See Chapter 7 for more information on quality control.)

Manufacturers often use bought-in parts and components. Larger manufacturers often use the just-in-time (JIT) method. For this method to work effectively, items are checked by the supplier or manufacturer of the specific part and delivered direct to the point in the production line where they are needed and at the right time. This reduces the costs of stocking and handling the parts which in turn increases the viability of the overall product.

To ensure that products conform to the technical specification, the checks that may be undertaken include:

● visual checks
● detailed data comparisons

● accuracy-of-dimensions checks
● weight checks
● quality-of-finish checks.

During the manufacturing process these tests usually include checking how accurately machines and processes are meeting tolerances. To do so the machines often have sensors which continually monitor that parts are being produced at a sustained level of quality.

For all products there is a need for accuracy and careful consideration and monitoring of tolerance to achieve viability. To be viable in a consumer market a product needs to be reliable and require minimum maintenance or repair. To achieve this, tolerances for both materials and manufacturing need to meet a minimum required standard. (For more on tolerances, see Chapter 7.)

Figure 8.14 Feasibility of use in the intended environment is vital to understand how a product will be used. Here the tolerances used in the manufacture of the ride-on toy have been carefully considered to ensure it will work despite being used in a number of different environments

Performance

Physical performance tests emphasise and are based on measurable performance characteristics as laid out in the technical specification. These may be used as proof of concept (that the product really works) or more often to ensure that the product will perform correctly in its intended environment of use. Testing carried out for performance is undertaken on completed products and component parts alike. Product performance tests are used for:

- subjecting products to stresses encountered in expected use and dynamic analysis (looking for errors or problems while a product is in use rather than in a laboratory setting)
- reproducing the types of damage to products found from consumer usage.

Quality control is used within performance testing. Here the checks are based around how the materials or components used react to given situations or stimulus. These can include:

- flammability tests
- electric circuit checks
- safety checks for components in terms of user interaction
- durability analysis of fabric/textile wear rates or the wear rates of bearing surfaces in mechanisms
- stress testing of key components.

Figure 8.15 Electric circuits are tested with a multimeter to confirm they perform to required standards. Here a laptop motherboard is tested with an amp meter

As part of performance quality control, manufacturers often consider not just components or parts but also the system being used to produce the products. If a product is to be made at a consistent quality then the system used to manufacture it must also be scrutinised. To ensure it is fit for purpose at the start of a production run and that it remains so throughout, checks must be made on an ongoing basis using Total Quality Management (TQM). (See Chapter 7 for more information on TQM.)

Integrating physical testing systems into the manufacturing process (A Level only)

DESIGN ENGINEERING

Destructive and non-destructive methods

It is important that designers are able to test parts and components as well as testing the manufacturing systems used to create the pieces. Often designers aim to understand what happens to a part if over stressed with the intention of understanding when it will fail. To do this they use **destructive testing** – breaking parts or materials to confirm what will happen. **Non-destructive testing** utilises other testing methodology to obtain data without damaging the product, often saving time, money and resources in the process. These tests allow designers and manufacturers to determine if they are meeting the technical specification requirements by:

- allowing them to understand the features and mechanical properties of the parts or materials
- determining comparative data between different material options or part sizes
- checking the operating characteristics of the material or part in controlled environmental test conditions
- simulating the ageing conditions of the parts, materials or fixings during their lifetime, in order to predict behaviour.

Table 8.1 **Methods of destructive and non-destructive testing**

Destructive	Non-destructive
Tensometer	X-rays
Vickers pyramid hardness test	Ultrasonics
Izod impact test	The Shore scleroscope hardness test
Brinell hardness test	

Destructive methods

Tensile testing using a tensometer – a test piece or component is subjected to tension through a worm drive gear system, which applies force through a spring beam. A force/extension graph is plotted using a stylus – this follows a mercury column, which magnifies the deflection of the beam. A connecting linkage like that used in a motor bike chain is likely to be subjected to this test to ensure that it will be able to perform as required and not stretch or break during use.

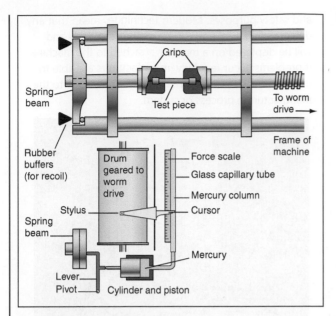

Figure 8.16 **Tensometer test rig**

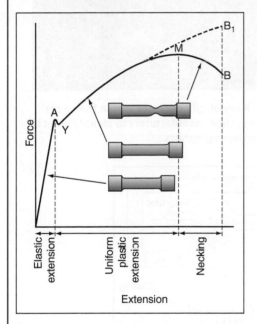

Figure 8.17 **Tensometer force/extension graph**

Note that Figure 8.17 shows a force/extension graph. This is different from a stress/strain graph. When necking occurs, the cross-sectional area reduces (s/s = B1).

Vickers pyramid hardness test – for very hard materials, a pyramid-shaped diamond is used to test for hardness.

Results are very accurate – a diamond will not deform. A microscope scale is used to measure the results of Brinell/Vickers indents (see below for the Brinell test). Products that must withstand high levels of wear such as gear components will often be tested to confirm they are hard enough to withstand abrasion during their working life.

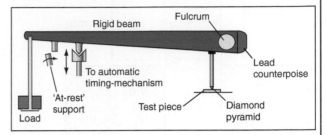

Figure 8.18 **Vickers rig**

Izod impact test for toughness/brittleness – a 'notched' test piece is fixed in the test vice. A heavy pendulum is released from a set position. As the pendulum breaks the notched test piece it absorbs energy. As the pendulum swings past, it drags a pointer with it. The pointer stops at the highest point of the swing. This indicates the amount of mechanical energy used. Comparative toughness/brittleness can then be ascertained and allow companies to plan specific material reactions to likely forces, loads or impacts.

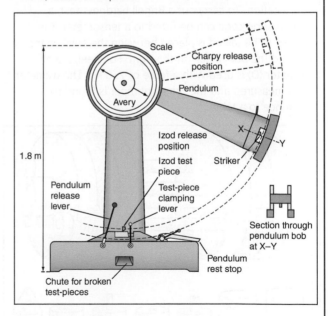

Figure 8.19 **Izod rig**

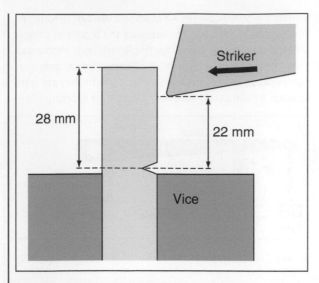

Figure 8.20 **Izod impact detail**

Brinell hardness test – it is essential for products to be suited to their environment of use. Imagine a mobile phone that quickly became scratched by being put down on hard surfaces like table tops. Customer satisfaction would be low and the phone could become unviable. If the designers have set out a specific hardness, they can carry out a Brinell test to confirm the material chosen will meet the expectations laid out in the technical specification. To carry out a Brinell hardness test, a compression cage can be fitted to a tensometer. The sample to be tested is placed between the anvil on the left and the Brinell ball on the right. For steel, a 5 m diameter (D) is used with a force of 750 kg. The diameter (d) is measured and the Brinell number is found by reference to a table.

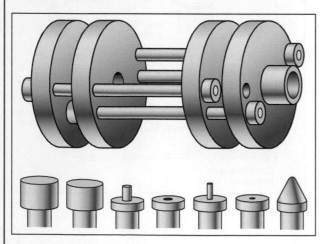

Figure 8.21 **Compression cage for carrying out the Brinell test**

Non-destructive methods

X-rays – an X-ray is a method of finding hidden internal defects in large castings before any expensive processing occurs. High-intensity X-rays necessary for this operation can be dangerous and a large shield is essential for health

and safety purposes. Modern techniques mean that any defect within a casting can be identified quickly and will be displayed on a monitor attached to the machine. If problems occur regularly, manufacturers are able to use the data collected to make relevant changes to the manufacturing process or system.

Figure 8.22 **X-ray of cylinder crank piston from a car engine. X-rays allow manufacturers to identify defects even in complete assemblies**

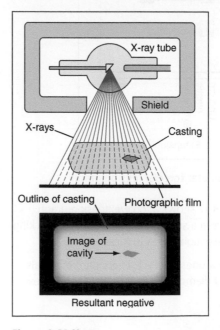

Figure 8.23 **X-ray**

Ultrasonics – this is another method of finding hidden defects. Ultrasonics work on the same principle as ultrasounds used to give images of unborn babies. Any defect within a metallic structure is picked up and recorded as an ultrasonic vibration. BMW and other car manufacturers use this method to check spot welds used in the manufacture of their vehicles. The test is completed by placing a probe on each weld and checking the ferocity of the weld. The process is often fully automated, being completed by the robots performing the welds and means that every weld is tested to confirm its quality.

Figure 8.24 **Welds are tested using ultrasonics. Here an angle probe allows the tester to scan for internal defects in the weldment**

The Shore scleroscope – prior to manufacture it is inconvenient to move very large samples of metal, so a portable method of testing hardness is invaluable. A glass tube can be held firmly above a large ingot or a very large casting, then a diamond-tipped hammer is released from a standard height. The height of the rebound measured against a scale gives a comparative hardness index.

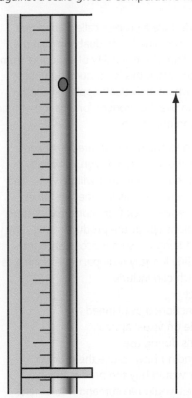

Figure 8.25 **Here we see the ball bearing rebounding up the tube allowing for measurement of hardness**

DESIGN ENGINEERING AND PRODUCT DESIGN

Testing of materials for durability and after care

Accelerated testing

A product's life can be planned by designers by looking at how a product is used and how often it is used. This will inform their choices of materials and parts to withstand these requirements. To understand if a product will meet its **lifecycle** targets, designers often implement accelerated testing. For example, manufacturers can test their washing machines by increasing the period of use. If the product is used continuously, the usage is accelerated by close to 17 times. Therefore a product designed for a life span of five years could be tested to this level in less than four months.

When required, designers perform real-world or virtual stress analysis tests or fatigue testing. These tests are vital to plan for a product that is capable of meeting its durability requirements. Virtual testing is normally based around mathematical modelling based on previous research (see below). Once the product is in production, similar tests will be carried out in real life on final products. These tests will be rigorous and completed with close attention paid to key tolerances and acceptable levels of performance.

Testing models and prototypes for performance and fitness for purpose

Designers and manufacturers put models and prototypes through a variety of testing to guarantee their performance in real life. IKEA uses their testing systems as part of their marketing and advertising campaigns both in store and for television advertisements. One example of this is IKEA's accelerated physical testing for their chair's. This test simulates a 20-stone person repeatedly sitting in the chair. This stresses the chair's parts repeatedly and is designed to highlight problems either in the design, materials or manufacture.

Testing products in use through different methods

Consumer testing

Companies often undertake a process called 'consumer acceptance'. This uses untrained individuals who aid the designers' understanding of the viability of a product from a normal user's stance. Often this is conducted through panels and websites where individuals apply to be testers of products and get to use the product for a period of time in its intended environment of use.

Consumer testing allows the manufacturer or design team to confirm that the intended design will meet consumer needs and that the product will be viable and not consistently returned as it fails to meet customer expectations. Negative feedback from consumer testers is likely to raise questions about the product's long-term viability. Users undertaking consumer acceptance trials normally complete fairly lengthy multipart questionnaires covering aspects which can include:

- quality of product
- if the product functioned as planned
- judgements made on visual appeal
- issues or problems during use
- ease of understanding how to use the product
- whether the user would buy the product
- whether the tester would recommend it to others.

Virtual testing

Virtual testing using CAD models is increasingly used to aid understanding of products and to confirm they are feasible and will work. CAD software packages utilising complex material mapping and comprehensive mathematical modelling systems allow products to be scrutinised virtually. An example of this is Solidworks 3D, an industry standard CAD modelling system which has the ability to virtually test materials and components within an assembly. Other packages include Autodesk Inventor and Autodesk Fusion. Tests can cover many different aspects and allow designers and manufacturers to understand how the product will perform under high compressive loads or when a shear load is applied. These simulations have visual outcomes which allow designers to identify potential areas of weakness and then adapt the product as required to ensure it meets design requirements or other standards.

Figure 8.26 Here we see a part being tested using Solidworks analysis to identify where stress will build up. If parts fail to meet tolerances then redesign will probably take place to thicken materials or reduce the stress the part is under

---

FASHION AND TEXTILES

Testing of fibres and fabrics for durability and aftercare

An item of clothing may have a specification stating that it must be washable at 40°C. To reflect the likely variance of washing machines the company may choose to test materials beyond the advertised washing temperature. For instance, some brands of washing machine may have a tolerance of 5 or even 10 per cent on their washing temperatures – a load set to wash at 40°C could actually wash at 44°C. The manufacturer is likely to plan for this in their controlled tests to ensure that the product can withstand 45°C as an upper limit to meet customer expectation.

British standards test for abrasion

A simple test is carried out to meet the British standard test for 'wear resistance'. Textiles are fixed to one block and abrasive paper is fixed to the other (see figure 8.28). For comparative tests, the pressure and frequency is kept constant.

Materials can also be tested in tension, with special grips fitted on to the tensometer: these spread the load and prevent tearing. The test checks how the material will perform under loads, simulating being stretched or pulled, and confirm that the material is suited to the garment or product.

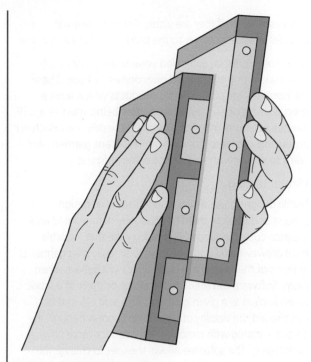

Figure 8.27 Abrasion test

Testing prototypes, toiles and samples for performance and fitness for purpose

Prototypes

Gore-Tex gloves are used extensively by climbers, skiers and snowboarders. Using the details set out in the technical specification, the gloves are put through numerous tests throughout the production process to ensure they are feasible and to confirm they are fit for purpose.

- Before production, the base materials (the sheets of Gore-Tex) are tested to ensure the cohesion of the laminate meets the company's own performance standards.
- Laboratory tests for abrasion, water resistance and **colour fastness**, among others, are undertaken to simulate the product in its intended application.
- Controlled performance tests are undertaken in a controlled environment to ensure repeatability and valid comparisons are possible. Judgements are made regarding comfort and feel to check the fit and quality of the glaves.
- The gloves are tested in real-world conditions to assess the experience of using them.
- Breathability test – to see how well the gloves let damp air out, steam is pumped into the gloves. Before and after the test, the gloves are weighed which allows the manufacturer to judge the amount of vapour that has escaped and calculate breathability.

- Water resistance test – water is sprayed onto the outside of the glove and the numbers of water droplets that remain on the surface are assessed. Fewer drops mean better resistance to water.
- Liner retention test using a tensometer – the inner of the glove is clamped and a force of 5 lb is used to pull on it to simulate a user removing their hand from the glove.

Toiles

During the design process, designers use toiles to understand the pattern needed to make a garment. The same is true for manufacturers when planning the final feasibility of the manufacturing process.

Checking samples for performance

During and after the manufacturing process, textile manufacturers assess the performance of the materials they are using to construct products. To do so, they test and check small samples of material to confirm that they are performing correctly. Manufacturers check raw or new material as it is delivered. Physical defects in the fabric or problems with consistency of colour are then identified. Should any problems be identified or the fabric be judged to be a poor match to the expected specified colour (such as that used in branded clothing) then it will probably be rejected. A wide variety of tests may be undertaken and can include:

- breaking strength
- tearing strength
- bursting strength
- air permeability
- wicking test (a test of absorbency and permeability)
- abrasion resistance
- pilling propensity.

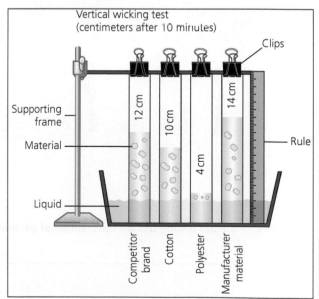

Figure 8.28 Vertical wicking test

Sampling garments and products
Consumer testing, wearer trials

Consumer testing is vital for manufacturers to confirm they are making products that will meet user expectations. In the textiles industry, wearer trials are often used to get feedback from consumers. In these trials, garments are provided to testers who usually wear the product, wash it according to the planned care instructions, assess the product and return it to the manufacturer. This allows the manufacturer to confirm that the intended design will meet consumer needs and be a viable product that will not be consistently returned as it fails to meet customer expectations. Wearer trials normally involve a multipart questionnaire covering important aspects, which can include:

● quality of comfort
● if the garment was a 'true fit'
● judgements made on visual appeal
● changes due to wear
● changes due to washing
● whether the user would buy the product
● whether the tester would recommend it to others
● if components, such as buttons, clasps, clips, zips or underwires, are correctly positioned.

The feedback is then assessed by the designers to confirm that the product is labelled, sized and manufactured to planned standards. With these aspects confirmed, the textiles manufacturer can be assured that the product is viable for its intended users and importantly will meet customer expectation. Products which have negative feedback are scrutinised to see if they are viable. The company will consider possible changes to the product or its manufacture.

For some products, extended wearer trials will also be undertaken in the intended environment of use. These are normally for performance products which have a specialist use or particular properties being used as a USP (unique selling point). For example, a textile manufacturer producing a waterproof or water-resistant garment will want to see how the product performs in use.

Virtual testing

Textile designers use virtual testing to aid the design process and to allow them to accurately determine fit and accurate cutting patterns. Technology like that available from Browzwear allows designers to virtually plan garments in terms of their sizing and how fabrics will behave when worn. Software can simulate the drape or sheer of a fabric when applied to a given garment. As a side effect this can also aid the viability of the design process through the costs associated with removing the need for more physical prototyping. The software allows the user to change the fit of a garment to match alternative body sizes and shapes quickly and efficiently and produce new patterns as required.

Virtual simulations can also be completed to indicate the viability of material choices by assessing damage through normal wear. This may look at aspects such as how a material will behave when stretched in both warp and weft directions. Companies may also plan the viability of materials by virtually testing new fabric mixes or construction.

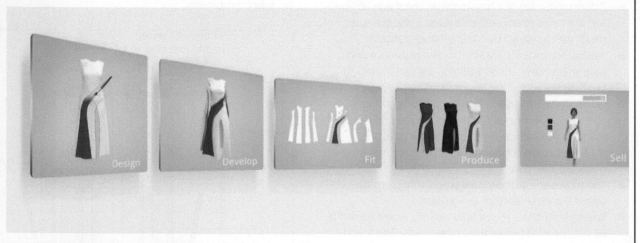

Figure 8.29 Virtual testing can aid the creation of garments

KEY TERMS

Colour fastness – how resistant the colour of a material is to fading or running when it is washed.

Critique – a document, chart or meeting of a small team which directly and intentionally looks for problems and faults in design ideas, prototypes or products, as well as recognising value and excellence.

Destructive testing – testing undertaken until the part or material reaches failure.

Lifecycle – a product's life from manufacture through to disassembly.

Non-destructive testing – testing which assesses the characteristics of a part or material without causing damage.

Tolerance – the permissible range of variation in a dimension of an object. Sometimes known as allowance.

Viability – the ability of a product concept to work successfully for its stakeholders.

KEY POINTS

- The process of critiquing products and prototypes helps designers to determine where designs fail to meet technical specification requirements.
- Quality is essential in all aspects of the product and the system used to manufacture it to ensure viability.
- Quantitative and qualitative data is gathered as feedback to confirm how well aspects of the technical specification have been met.
- Accuracy of parts is fundamental for most designs to ensure they correctly meet tolerance requirements when interacting or joining with other parts and products.
- Physical destructive and non-destructive testing of products is undertaken to confirm they meet required levels of performance.

8.3 How do designers and manufacturers determine whether design solutions are commercially viable? (A Level)

LEARNING OUTCOMES

By the end of this section you should have developed a knowledge and understanding of:
- the value of feasibility studies to determine the likely factors that influence the commercial viability of a product to market, such as:
  - the design solution's impact on user lifestyles
  - how well a product performs
  - technical difficulty of manufacture
  - stock availability of materials and components
  - costs and profit
  - timescales involved
  - promotion, brand awareness and advertising potential
  - balancing supply and demand
  - market analysis of similar products.

Feasibility studies

Both designers and manufacturers use a range of feasibility studies to ensure that their product can be brought to market quickly and efficiently while aiming to ensure sales and ultimately to make a profit. If completed correctly, feasibility studies influence the types of product being produced, ensuring they meet demand and also identifying flaws and problems before they impact on profit. In some cases an early **feasibility study** may mean that a product's development is halted completely as it has been identified that the product cannot make a profit in the current financial climate or that the concept will struggle due to competition already in the market.

To ensure a product is viable, designers and manufacturers normally undertake feasibility studies when:
- proposing a product needing major investment
- entering a new market place or product sector

- significantly expanding a product range or scale of production
- expanding overseas
- entering a saturated market place.

Bias

Studies are very useful to designers and manufacturers to understand the long-term viability of a design but when undertaking a feasibility study it is important for them to be mindful of bias. Often people personally invested in an idea become attached to their own ideas and ignore problems and issues raised in findings of studies. Designers often outsource studies to ensure that their results are non-biased and that the results are reliable. While this can bring about findings which are hard to hear, the independent assessment is often vital to unravel a concept and confirm that it is indeed viable and that further investment is warranted and likely to return profits.

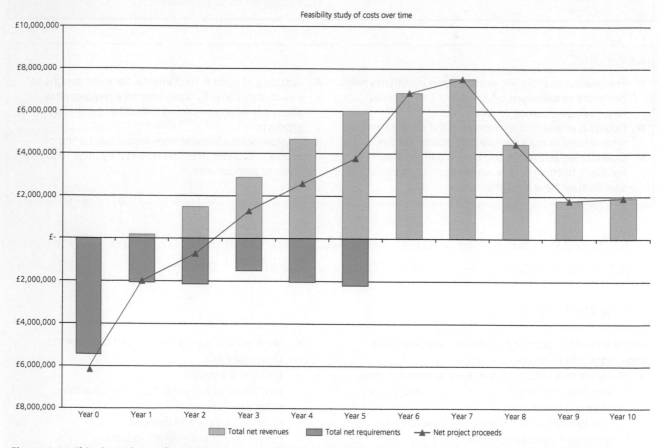

Figure 8.30 This chart shows the projected costs laid out in a feasibility study. Note the graph moves from costs to the break-even point midway through year 2. The product then moves into profit and peaks in demand by year 7

Scope

The scope of an independent study will depend on the product and the company's ability to invest both time and money in the feasibility study. A reliable study could take a significant length of time to return findings and cost a large sum of money. These factors also need to

be considered by manufacturers and designers as they can impact on the commercial viability of the product. If a feasibility study takes a long time to complete, it can jeopardise the feasibility of a concept because it gives competitors time to get a similar product to the market first.

Factors that influence the commercial viability of a product

The design solution's impact on user lifestyles

Designers aim to prove the viability of new products and show that they will have a positive impact on user lifestyles. Take for example a study that shows the benefits of a new way of completing a household chore. The development of the vacuum cleaner during the late 1800s and early 1900s meant that cleaning was made quicker, easier and more effective. It would have been clear from the outset that the product was feasible and would have a positive impact on the user's lifestyle.

Sometimes the feasibility of a product can be tricky to prove but with historical buying data, demographic information and solid results from user trials, products will often go into production. In the 1980s, mobile phones began to filter into the market. These phones were very big and, with a large battery, weighed in excess of 2 kg. However, designers were able to prove the concept viable by showing demand from business people. As technology moved forward, more companies entered the phone market and in the intervening years it has become one of the largest grossing consumer products ever produced with estimates showing that there are now close to five billion handsets worldwide. This product has changed modern society and has had a profound impact on social interaction and data sharing.

Figure 8.31 Mobile phone technology has advanced very quickly. Here we see an original 1980s phone and a later version from 2003. These show low levels of sophistication compared to present technology but were major leaps at the time

Negative impacts can often be hard to predict until a product is rolled out. Despite focus group feedback showing a product is likely be viable, external factors can play a part in a product's success. Language barriers, customer loyalty to other brands or simply poor traction in a market place due to a complication in the product's use can all cause problems. To counter this, designers will often have an ongoing improvement schedule in place for the product to review feedback, reviews and changes to standards legislation to ensure that the product consistently meets consumer expectation and remains fit for purpose and viable over an extended period.

How well a product performs

Feasibility studies confirming that a product will meet performance standards are vital if the product is to be viable. All user trials and feedback as well as virtual testing and assessments should show that the materials and parts will match the requirements of the product's intended lifespan. For a product to be truly viable it needs to perform as well as similar products on the market at the same price point while as a minimum, breaking even. Feasibility studies in the form of user or wearer trials, focus groups and clinics will need to have shown that the product will be easy to use and match the requirements of the intended end user. It would be relatively simple in many cases for designers to over engineer a product and make it function above its required performance standards over an extended life. However, it is highly likely that such an approach would lead to an excessively expensive product which would not meet its planned price point and would no longer be viable.

Companies want consumers to perceive a need to buy new products. Despite the ethical and environmental considerations this raises, companies often design-in planned obsolescence. This means planning for a part or system to fail after a given period of time. Clearly, to be viable the part must exceed warranty requirements and the customer must be satisfied with the performance of the brand to create a repeat sale. (For more on planned obsolescence see Chapter 4.)

Another viability consideration for designers and manufacturers is if the product's level of performance matches consumer demand. Consumer demand will often be based on a judgement of price against product life or quality. For instance, a consumer may buy a product to address a short-term issue or for a single job. If you were cutting a hole in a single cupboard, would you buy a market-leading branded pad saw with a lifetime guarantee or would you buy a significantly cheaper alternative product that would simply do the job? Feasibility study data allowing the accurate forecasting of sales will be essential to prove the product's performance is required and therefore viable in terms of demand.

Technical difficulty of manufacture

Feasibility studies must be performed to ascertain what the costs of production will be and to determine if these will fit within a certain budget. A major deciding factor in the production of a product is the costs associated with the difficulty of its manufacture. While in theory it is often possible to build highly complicated products, the level of complexity can mean that highly intricate mouldings and manufacturing solutions are needed to produce parts. This intricacy often means very expensive tooling which in itself can render a design unviable.

Designers and manufacturers aim to reduce the costs of manufacturing by adapting and simplifying products to make them easier to produce. During the development stages, designers will undertake a process called Design for Manufacturing (DFM). This process is in essence another form of feasibility study put in place with an aim to reduce manufacturing costs and improve the viability of the final product by identifying problems or lower cost options during the design stages. (See Chapter 3 for more on DFM.)

Often feasibility studies considering the scale of production will dictate how much of the manufacturing process can be completed by automated machines and robots. Often, for batch and one-off production, automation is not viable due to its cost. This means that parts of products that are hard to manufacture may have to be completed by workers and this can impact on the consistency of product quality. Labour costs are often a significant factor in the viability of a product and can impact on whether a product will go into manufacture at all. Careful analysis of the number of human hours needed to produce a product will be undertaken to determine if it can be made to meet budget considerations.

Figure 8.32 **A worker removes a large complex part from an injection moulding machine. Careful consideration of the time needed for this would be part of the feasibility analysis of the manufacturing process and design**

Stock availability of materials and components

The availability of stock or ease of obtaining materials can have a large impact on the viability of some products. Companies often buy materials or parts from other suppliers to reduce their own manufacturing costs or because they do not have the facilities to make them. Feasibility studies are undertaken to assess where parts can be obtained, their unit and transportation costs and the impact this will have on the overall manufacturing price of the product. Problems can occur if:

- a material is not available in sufficient quantities or availability is unreliable
- it will cost a significant sum to buy in
- there is a long lead time (the time between ordering and delivery is unreliable)
- there is strong competition for the material or part
- the materials have an impact on sustainability.

Manufacturing companies assess all of these points and use them to forecast stock availability and any trends in levels of stock becoming problematic.

Figure 8.33 **For a product to be viable, a reliable supply of raw materials and parts must be available. Here we see a range of metal extrusions ready for use. Note the colour coding that shows that they have been quality checked**

Feasibility studies highlight potential risks in supply chains and companies often outsource the risk of procuring parts or materials to other companies that form their supply chain. Often this approach is undertaken by larger companies manufacturing complex products like cars and other vehicles. BMW uses an extensive supply chain for its UK-based manufacturing operations. These suppliers produce parts and deliver them to BMW from all over Europe. The feasibility studies produced by BMW will have shown that they are able to produce their cars at a profit while buying parts from numerous other companies. BMW use the just-in-time (JIT) production methodology to organise the bought-in component parts.

Figure 8.34 JIT ensures that parts are delivered to car manufacturers shortly before being used. Here door panels are ready to be used on the assembly line

Costs and profit

The need for companies to generate profit means that feasibility studies carefully plan cost implications and the ability for a product to make a profit. Failure to meet this requirement (unless planned as part of a loss leader) could mean significant financial problems for the manufacturing company and its investors and stakeholders.

Feasibility studies calculate all expected costs and model the effects of issues such as underperforming sales figures or increased costs. This type of study allows the manufacturer to plan levels of production to meet demand. Companies often aim to reduce costs of production over time and calculate planned profits based on reduced costs from suppliers and improvements and reduction of waste in the manufacturing process due to effective use of TQM.

Costs fall broadly into two main types – fixed and variable – but on occasions there can be a mixed cost. Power usage is normally paid for monthly at a fixed cost. However, fluctuations in the price per kilowatt hour and potential changes in the amount used can lead this to change.

Table 8.2 Types of cost

Expense	Type of cost
Factory rent or mortgage	Fixed cost
Power	Mixed cost
Insurance	Fixed cost
Wages	Variable cost
Miscellaneous expenses	Variable cost
Machinery	Fixed cost
Materials	Variable cost
Tooling	Variable cost

Feasibility studies need to take into consideration that prices are not 100 per cent reliable. **Fixed costs** can vary over time; if a company extends its premises then rent may increase. In some cases **variable costs** such as materials can become fixed. For example, a planned quantity of materials can be bought in advance at a pre-agreed fixed price independent of the quantity of products actually being produced. Note that this is different from a negotiated price that varies with bulk buying in reaction to changing demand.

Timescales

The timescales involved in bringing a product to market are vitally important. Using feasibility studies to gauge the correct time to bring a product to market can dictate its success or failure. A product brought too late to a market where a competitor has a large market share can mean real problems in gaining sales and could mean a product and all its related investments will no longer be commercially viable. In some cases, the demand for a product is seasonal or time dependent. It is fairly obvious that bringing a range of winter coats to market in spring is unlikely to generate many sales but often subtle changes in demand due to fashion trends or even consumer confidence being affected by uncertain financial climates can have a major impact on a new product.

It is likely that any study will look at the likely time for manufacture and shipping times needed to bring a product to market. The study may bring the viability of the product into question if the times are overly long so that demand diminishes or the product itself is directly affected by the timescales involved. For instance, a consumable item may go off or past its best while in storage if sales are very slow. Studies will also show the long-term commercial viability of a product through forecasting of sales and analysis of market trends.

If the study takes too long then the product may take too long to reach its intended market and lose a considerable market share. Alternatively, if the study cuts corners so it can be completed quickly it could miss or underestimate essential indications that the product will fail in some other regard.

Another timescale that must be factored into a feasibility study is obtaining protection of design ideas through intellectual property rights, which ensures that the idea cannot be stolen by any of the other stakeholders or a competitor company. (See Chapter 3 for more on intellectual property.)

Promotion, brand awareness and advertising potential

Marketing often dictates the success of a product and its associated commercial viability. If consumers are unaware that a product exists they are very unlikely to buy it. The effective marketing of products to engage their intended target market is essential. Feasibility studies are used to determine the best way to engage the target market and to determine if this can be done at a viable cost. This analysis is often based on the marketing mix and aims to reach the maximum number of potential consumers at the most effective cost.

Companies often use **brand awareness** and **brand identity** to cross sell their products. As a consumer, you are more likely to buy another product made by a company if you have had a positive experience with another product in their range. Companies ensure that branding such as their logo and their identity (colours and specific design features) are included in all designs with an aim to increase sales. Feasibility studies help companies to understand the best way to approach consumers and may show that changes in tactics and rebranding are needed. Companies often use standard colours and fonts to aid customer recognition. Global brands like Coca Cola have a specific red colour to ensure that products in its cola range are immediately identifiable. To achieve this, Coca Cola will demand consistency of colour used on all of its products and in all of its advertising.

Figure 8.36 Consistency of brand colours aids sales across a range of products. Here Coca Cola utilise the same colours across three products to help consumers identify them as part of the same brand

Feasibility surveys look at buying habits and aim to understand why consumers are purchasing certain items. These studies aim to determine if the marketing strategies employed are working so that levels of demand will continue to ensure the product is commercially viable. Studies consider how widely advertising will be seen and the number of potential customers it will reach. The studies aim to show that the planned approach will access the right markets and forecast the sales that will result from it. They also aim to understand who and where their competitors are.

Balancing supply and demand

One of the main focuses for feasibility studies undertaken before going into full manufacture of a product is to determine if there is enough demand for a given product. Without demand then there is little likelihood of a product being viable. As consumers become increasingly aware of a product the demand for it will increase and this can be the reason for increases in production scale. In order for demand to be met there must be enough products to supply. Feasibility studies aim to confirm that the primary market is large enough to make it a profitable venture. If it is deemed as unviable then secondary or tertiary markets may be considered. Each expansion to a new market has a possible impact on the viability of the product due to the costs and possible risks involved with meeting the supply requirements.

Balancing both supply and demand is crucial and feasibility studies aim to provide data to allow this to happen efficiently. Too little supply can lead to consumers turning to alternative products and revenue being lost. Too much supply from numerous companies to a single market sector can lead to a saturated market where there may be a need for reductions in price to remain competitive. It is possible for a single company to

Figure 8.35 The marketing mix forms part of feasibility analysis so companies direct investment and time into the correct actions to ensure a viable outcome

supply too many units to a market. Providing there is a long-term demand this may not be a problem-providing, the product will not deteriorate over time and that it is feasible to store the products for later shipping.

Companies continually monitor the demand of products and can do so very accurately. Feasibility studies can consider demand at a very local level using feedback of sales through **Electronic Point of Sale (EPOS)**. This means that demand can be met by using historical data and trends to ensure supply is ready as it is needed.

Market analysis of similar products

During the product development and after its launch, companies often perform a range of feasibility studies looking at other competitor products, sales trends, marketing promotions and opportunities in new markets. These various studies allow the company to determine what opportunities there are for future sales and how competitor's actions will affect them.

During developmental stages, market analysis is primarily focused on similar products already available to consumers. Analysis may simply look at sales trends to show the popularity of a certain product range within a market to determine demand. A new product gaining even a very small market share of a large profitable market may be viable. Deeper analysis of competitor products is often undertaken to look at technology being used and manufacturing methodology. This can give designers and manufacturers useful insight into approaches to solve product-specific problems, competitor quality or to identify components or parts which are protected by copyright law. From these feasibility studies, companies may determine if their own ideas will be viable and if they will be able to sell their own product based on a USP.

Once the product is on sale, feasibility studies may also be carried out to monitor sales trends of products. These usually look at sales trends which may vary according to changing levels of demand.

ACTIVITIES

The consumer magazine *Which?* undertakes thorough reviews of products and compares them directly to other brands over a breadth of performance indicators.
1 Explain how, as a consumer, a positive review may influence your purchasing decisions.

2 Explain what impacts positive product reviews would have on a company.
3 Discuss the impact a negative review could have and what actions a company may take to address this to retain consumer confidence in the brand.

KEY TERMS

Brand awareness – the familiarity of consumers with the qualities or image of a particular company.

Brand identity – the specific elements associated with a company or product such as name, logo, colours, typeface or tagline.

Electronic Point of Sale (EPOS) – a computerised system used in retail shops to record sales of stock by scanning barcodes.

Feasibility study – this investigates the implications of a project before getting involved and investing resources. In the iterative design process, the feasibility of proposed designs can be assessed through experiments, trials, mock-ups, testing and modelling.

Fixed costs – costs that remain constant regardless of the number of goods being produced.

Variable costs – costs that fluctuate as output or demand change.

MATHEMATICAL SKILLS

Maths requirements for this section will primarily surround understanding data. In addition to this you should be able to understand tolerances and their importance, as well as create NEA work which shows your comprehension of them. Below is an example question that could be part of the examined content of the course. For questions in this area you will need to be able to calculate quantities, costs and sizes with a consideration of tolerances and percentages of profits.

Tolerance

Tolerance is often used to ensure that the possible range of size values for parts are acceptable to ensure fit.

Example:

A power cable for a mobile phone is required. The width of the connector is planned to be 6.8 mm wide.

The part has tolerance set as 6.82 ± 0.05 mm.

This means that the largest acceptable size for the connector is 6.82 + 0.05 = 6.87 mm.

The smallest acceptable size for the connector is 6.82 − 0.05 = 6.77 mm.

387

1 A 3D printer arm part is specified to be a total length of 82.1 mm when machined.
   The acceptable tolerance is ± 0.2 mm.
   What are the longest and shortest acceptable sizes?
   If the machined part is 82.4 mm long, is it acceptable?

2 The company produces large batches of arms. They accept a 0.01% failure rate.
   They are finding that in a batch of 12,000, on average 97 fail to meet tolerances. Calculate the percentage failure rate.
   Does this meet their standards?

Answers

82.3 mm and 81.9 mm

No

Failure rate is 0.008% so is acceptable

KEY POINTS

- Feasibility studies are essential to prove a product will be commercially viable.
- Designers aim to prove that a concept will have a positive impact on the user's lifestyle to help gain investment and demonstrate commercial viability.
- Truly commercially viable products are often based on ongoing long-term feasibility studies which confirm continuing demand.

- Awareness, and the demand for a product that this brings, are fundamental to ensure a product is commercially viable.
- The commercial viability of a product can be determined by the timescales involved in bringing it to market.

Further reading on viability of design solutions

- Papanek, V. (2005), *Design for the Real World*, 2nd edition, Thames and Hudson, ISBN 9730500273586
- Karis, C. (2017), *Feasibility Study: Startup and Sustainability*, CreateSpace Independent Publishing Platform
- Jackson, P. (2012), *Structural Packaging*, Laurence King
- Padayachy, M. (2015), *The Innovator's Method: Bringing New Ideas to Markets*, Cap Innovate

- Textile testing methods: www.slideshare.net/suniltalekar1/25-textile-testing-methods
- Testing: www.fira.co.uk/commercial-services/testing
- BRE: www.bre.co.uk
- Product testing: https://producttesting.uk.com
- BSI: www.bsigroup.com/en-GB/
- Which?: www.which.co.uk
- Pressure testing: http://forums.watchuseek.com/f2/pressure-testing-1356713.html

PRACTICE QUESTIONS: viability of design solutions

1   Identify the standardised parts that may be needed for the product to be viable. Discuss how the use of these parts would impact on the viability of the product in terms of accuracy and the technical difficulty to manufacture.

Design Engineering

1   An engineering manufacturer is building a new high-performance motor to drive its new product. The motor will be placed in a housing to protect it. Explain what testing could be undertaken to prove their material choices are viable.

2   Identify what tolerances may be needed for the product to be viable. Discuss how the application and use of these tolerances could impact on the viability of the product in terms of accuracy, performance and function.

Fashion and Textiles

1   A textiles company have identified a demand for a new range of ski jackets for beginners. Explain what range of tests they may undertake in order to judge the feasibility of their designs.

2   Identify what bought-in component parts may be needed for the product to be viable. Discuss how the application and use of these parts would impact on the viability of the product in terms of performance and commercial viability.

Product Design

1   A company is developing a new kettle which needs to be inclusive for a wide range of users. Explain which feasibility studies they may undertake and how these would help prove the validity of their proposed design solutions.

9 Health and safety

9.1 How can safety be ensured when working with materials in a workshop environment?

In any workplace, people need to work in a safe environment and yet accidents can happen. Accidents can be avoided by maintenance of machinery, training of workers and appropriate control measures.

United Kingdom (UK) legislation and guidance is intended to help minimise **risks**. Legislation currently includes the Factories Act (1961), the **Health and Safety at Work Act (1974)**, **COSHH (Control of Substances Hazardous to Health) Regulations (2002)** and the Management of Health and Safety at Work Regulations (1999). Legislation and guidance is regularly updated and it is always advisable to check for the latest updates.

Employers have a legal responsibility to assess the risks in a workplace and provide protective equipment and training. Employees have a duty to take reasonable care of themselves and others and not to misuse any equipment provided for safety purposes (for example, fire extinguishers or **personal protective equipment (PPE)** such as eye protection). Within school or college you have a duty to follow any health and safety guidance provided by your teachers.

Risk assessments

Health and safety law requires employers to assess the risks to health and safety and implement appropriate health and safety measures. In establishments where there are five or more employees, significant findings of the **risk assessment** and appropriate actions are recorded.

A risk assessment is a careful examination of what, in a workplace, could cause harm to people. Decisions have to be made as to whether enough precautions have been taken or if more could be done to prevent harm. Workers and others have the right to be protected from harm in the workplace. Failure to take reasonable control measures could result in injury or death.

In everyday life we all follow simple procedures that can control risks, for example keeping walkways clear so that people do not trip up. The law does not require all risks to be eliminated; it requires that all **reasonably practicable** measures have been taken to minimise the risk. In most cases, an employer will ask a suitably qualified member of staff to identify **hazards** and quantify the risk.

A hazard is seen as anything that could cause harm, such as working at height, using machinery or working with electricity.

The risk is the chance that someone may be harmed by a particular hazard. The risk could be high, medium or low, and an indication would be provided as to the chances of this risk occurring and what level of harm could be caused.

How to carry out a risk assessment

The Health and Safety Executive (HSE) is the organisation that oversees health and safety legislation and provides advice and information. HSE guidance describes five steps to carry out a risk assessment:

1 Identify the hazards. Inspect the workplace, carry out a tour with other suitably qualified personnel, ask employees their opinions, seek advice, use manufacturers' instructions or data sheets for chemicals and equipment, check accident and ill-health records – these often help to identify hazards.

2 Identify who might be harmed and the nature of the harm. Identify the groups or individuals who may be harmed, identify what type of injury or ill health might occur. Take into account special requirements, for example, expectant mothers and new workers.

Include others, for example, cleaners, visitors, etc. Consider the likely occurrence of the hazard – is it likely to be a very rare occurrence or high possibility.

3 Evaluate the risks and decide on control measures. The law requires that everything 'reasonably practicable' is done to protect people from harm. The easiest way to access this is to compare what is being done with best practice. Seek advice on good or best practice. Aim to get rid of the risk. If this is not possible, ensure that the precautions you take control the risks so that harm is very unlikely. Control measures could include:
   – training – update regularly
   – PPE (goggles, gloves, footwear)
   – guarding of machinery, fencing off areas.

4 Record findings and implement actions. Write down the results of risk assessments and publish them for staff (not required for establishments with fewer than five employees). Do not produce lengthy statements – clear, concise points are required. Accurately specify the locations of hazards.

5 Set a fixed period for review of risk assessment and update if necessary. Risk assessments should be reviewed annually. A review date ought to be fixed when the assessment is completed. When changes are made in the workplace, new hazards may occur. A review should be made as soon as changes are introduced. Encourage a culture where employees will inform of particular hazards.

Risk assessments are normally carried out using a table (see Table 9.1 on page 397).

The Health and Safety Executive website (www.hse.gov.uk) contains useful resources relating to risk assessment, including templates for conducting a risk assessment and case studies.

ACTIVITY

In a small group, select a machine and a process carried out on the machine. Think about the risks associated with using it. Identify the signage, safety equipment and PPE already in place, or that might be needed. For example, ears could be at risk from loud noise when using a circular saw – safety options include wearing ear defenders. Would different materials cause different risks?

9.2 What are the implications of health and safety legislation on product manufacture? (A Level)

The Health and Safety at Work Act (1974), COSHH (Control of Substances Hazardous to Health) Regulations (2002) and the Management of Health and Safety at Work Regulations (1999) are legislation and guidance intended to help minimise risks in the workplace.

Health and Safety at Work Act (1974)

The Health and Safety at Work Act 1974 (HASAW) forms the basis of British health and safety law. It states that employers have a duty to ensure, so far as is reasonably practicable, that employees and other visitors are protected at work.

'So far as is reasonably practicable' is a key phrase: employers do not have to take measures to avoid or reduce the risk if it is technically impossible or the time, effort and cost is grossly disproportionate to the risk.

The Management of Health and Safety at Work Regulations (1999) give detailed guidance on what is expected of employers. Employers are also required to keep everyone involved informed of health and safety issues. They are also required by law to display a HASAW poster or provide equivalent leaflets.

Key features of the Health and Safety at Work Act (1974)

Employers' duties include:

- making sure the workplace is safe and without risks to health by assessing risks
- ensuring plant and machinery are safe and that safe procedures of work are set and followed
- ensuring articles and substances are moved, stored and used safely by providing correct equipment and training
- providing adequate welfare facilities, including first aid arrangements
- providing the information, instruction, training and supervision necessary for personal health and safety
- making sure that work equipment is suitable for intended use and that it is properly maintained and used (annual inspections of machinery are often carried out by outside agencies)
- providing any necessary PPE equipment (see Figure 9.3)
- ensuring that mandatory and other appropriate safety signs are provided and maintained (see Figure 9.2).

Employees' duties include:

- taking reasonable care for their own health and safety and that of others who may be affected by their actions
- correctly using items provided by their employer, including PPE, in accordance with training or instructions (for example, wearing safety glasses or goggles when using machinery)
- using anything provided for health, safety or welfare correctly (for example, using guards or clamps when using machinery).

Figure 9.1 Examples of PPE from a construction site

SITE SAFETY

Hard hat must be worn

Protective footwear must be worn

High visibility jackets must be worn

Ear protectors must be worn

Warning
Construction site

Keep out

Danger
Demolition work in progress

No admittance for unauthorised personnel

Site safety starts here

Figure 9.2 Examples of mandatory signage from a construction site

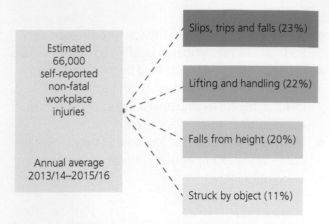

Estimated 66,000 self-reported non-fatal workplace injuries

Annual average 2013/14–2015/16

Slips, trips and falls (23%)

Lifting and handling (22%)

Falls from height (20%)

Struck by object (11%)

Figure 9.3 In 2015–16, 66,000 non-fatal workplace injuries were reported (www.hse.gov.uk/statistics/industry/construction/)

Control of Substances Hazardous to Health (COSHH) Regulations (2002)

Using chemicals or other hazardous substances at work can put people's health at risk. Employers have a duty to control the exposure to hazardous substances and to protect both employees and others who may be exposed.

Failure to control exposure to hazardous materials can result in harm, from mild eye irritation, skin complaints and fainting as a result of fumes, to chronic lung disease or, on rare occasions, death. Consequences for the individuals harmed are obvious. For the employer there could be lost productivity and liability for legal action, including prosecution under the COSHH Regulations and civil claims.

Hazardous substances include:
● substances that are used during work, for example, adhesives, paints, dye pigments, cleaning materials and developing materials
● substances that are created as a result of work activities, for example, fumes from soldering
● airborne particles, for example, dust from abrading materials or using dyes.

Asbestos, radioactive substances and lead are not included under this legislation as there is specific legislation relating to these materials.

For the majority of commercially available chemicals that we may use at work, for example soap and washing-up liquids, COSHH regulations are not applicable. COSHH regulations only apply to products if they have

a warning label, for example bleach, paints, solvents, fillers or dyes. The regulations cover safe disposal recommendations for waste from chemicals, solvents, batteries and oils that can occur within a workshop environment. Such waste must be separated and disposed of safely.

Complying with the COSHH regulations

HSE guidance suggests the following steps:

1 Look at each substance. Hazardous products come with a safety data sheet or will have information on the label.

2 Think about the task. Consider exposure by breathing in, contact with the skin, swallowing, eye contact or by skin puncture.

3 Carry out a risk assessment. Put in place control measures, control equipment, training, emergency procedures and arrangements for storage and waste disposal.

From HSE's *Working with substances hazardous to health* – a brief guide to COSHH. (See Further reading at the end of this chapter.)

Figure 9.4 Examples of international COSHH labels

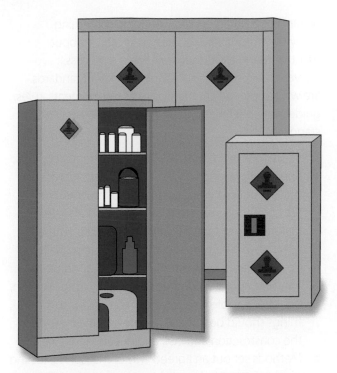

Figure 9.5 A typical storage solution for COSHH substances

Other important health and safety regulations

- Workplace (Health, Safety and Welfare) Regulations (1992) – covers issues including ventilation, heating, lighting, workstations, seating and welfare facilities.
- Personal Protective Equipment at Work Regulations (1992) – relating to protective clothing and equipment for employees.
- Provision and Use of Work Equipment Regulations (1998) – regarding the safe use of equipment and machinery provided for use at work.
- Reporting of Injuries, Diseases and Dangerous Occurrences Regulations (RIDDOR) (2013) – employers must notify certain occupational injuries, diseases and dangerous events.

Consumer protection

The **Consumer Rights Act (2015)** protects consumers by stipulating that all products must be of satisfactory quality, fit for purpose and as described. It helps consumers to obtain redress when purchases go wrong and also applies the same rights when buying digital content. See Chapter 3 for more information.

The **Trade Descriptions Act (1968)** makes it an offence for a trader to apply, by any means, false or misleading statements, or to knowingly or make such statements about services or goods. See Chapter 3 for more information.

The **Consumer Protection Act (CPA) (1987)** gives rights to consumers when buying goods and services; for example, manufacturers are legally required to put certain information on products, including health and safety information. If a defective product has caused damage, death or personal injury, the law gives consumers the right to claim compensation against the producer.

A dangerous electrical product can prove fatal or cause serious injury, so consumers should always look for labels and evidence of testing to meet the required **standards** for the product.

Product labelling

Product labelling is covered under the Trade Descriptions Act. Labels must include accurate information to ensure that products can be used safely and correctly. For example, motorcycle helmets must have a label attached to them informing you not to paint them or apply any kind of solvent. This could damage and weaken the helmet, meaning that it provides less protection in an accident. Similarly, aerosol cans must have a label warning the user to keep them away from heat. The British Standard BS EN ISO 3758:2012 textiles care labelling code using symbols, demonstrates how information can be passed to the consumer on the washing, bleaching, ironing, dry cleaning and drying of textiles. The symbols used on labels are consistent with those used on detergent packs, washing machines and irons.

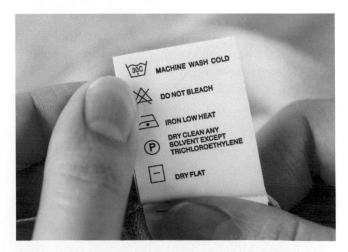

Figure 9.6 Example of a care label on a textile product

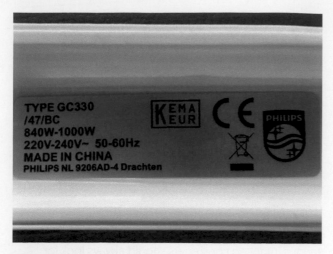

Figure 9.7 Example of labels on an electrical product

Examples of quality and safety assurance labelling

Some products, such as children's clothing and toys, are required by law to include safety warnings. All toys supplied in the UK must meet a list of essential safety requirements which are set out in the Toy (Safety) Regulations, 2011 (previously the Toy Safety Regulations, 1995 – updated in 2010). Currently all toys should also carry a CE Marking. There are seven British Standards which apply to toy safety and a separate standard which covers the electrical safety of toys. EN71 covers the safety standards for all toys for children below 14 and products that are not intended as toys but look like toys are also included. It covers aspects of safety such as flammability, toxicity and safety labelling. If any toy is found to be unsafe according to the regulations, then the producer can be found guilty of a criminal offence.

British Standards Institution (BSI)

BSI is the National Standards Body (NSB) for the UK and was the world's first NSB. It works with governments, industry, businesses and consumers to produce British, European and international standards. At the time of writing, it operates in 182 countries. (Source bsigroup. com press release, April 2016)

The most popular standard in the world – ISO 9001 Quality management systems (Requirements) – is used by over one million organisations in 178 countries.

The BSI Kitemark is the world's best-known product and service certification mark. Many organisations have made it mandatory for their suppliers' products to have been awarded the BSI Kitemark before they will place an order with them. As consumers become increasingly informed about their choices, conformity to recognised standards becomes a key purchasing decision issue.

A standard is an agreed way of doing something. Many standards relate to safety and can be about manufacturing a product, managing a process, delivering a service or supplying materials. Standards are written precise criteria for use as requirements, rules, guidelines or definitions. BSI has over 37,000 active standards.

There are four main categories of British Standard:

- Specifications are highly prescriptive standards that set out detailed absolute requirements. They are commonly used to stipulate important safety features in products or other applications where users need a high level of confidence and quality.
- Codes of practice recommend sound good practice. They incorporate a degree of flexibility in application, but offer reliable indicative benchmarks as to 'how things should be done'. They are commonly used in the construction and civil engineering industries.
- Methods set out an agreed way of measuring, testing or specifying what is reliably repeatable in different circumstances and places, wherever it needs to be applied.
- Guides are published to give less prescriptive advice that reflects the current thinking and practice among experts in a particular subject.

Standards help a company to:

- attract and assure customers that products are safe and fit for purpose
- demonstrate market leadership, creating a competitive advantage
- continue to develop and maintain best practice.

Some examples of British Standards include:

- BS EN 71-1:2014 Safety of toys. Mechanical and physical properties.
- BS EN 71-2:2011 +A1:2014 Safety of toys. Flammability.

Achieving these standards would be essential for any individual or manufacturer involved in the design and manufacture of toys.

A section of the BSI website is dedicated to standards for health and safety in the workplace: www.bsigroup.com/en-GB/industries-and-sectors/health-and-safety.

ACTIVITY

Look at some labels and packages for existing products in your material area. Which British Standards have the products been tested against? Use examples to explain how legislation can protect purchasers of domestic electric or textile products.

APPLICATION AND EXAMPLES FOR DESIGN ENGINEERING

Risk assessments are very important and should cover the use of machinery, tools, materials and chemicals in some components, such as mercury tilt switches, fumes when soldering and substances that could be subject to COSHH regulations.

ACTIVITY

Having looked at an example risk assessment, complete one for the school/college workshop for a typical range of equipment. The examples below might help but remember you need to think about your particular workshop and its users. Any risk assessment needs to cover the variety of materials and equipment used.

Table 9.1 Example risk assessment for engineering-type activity

Activity/Plant/ Materials	Hazard	Comments, existing controls, references used	Risk evaluation			Proposed controls/ Action and references required
			Severity	Likelihood	Rating	
TURNING: The production by turning, facing, boring, screw cutting, centre drilling, tapping and knurling of primarily cylindrical forms from both ferrous and non-ferrous raw material, possibly using oil based coolants.	Flying work piece – work pieces, chuck keys and tools can be ejected if not held properly or the machine starts unexpectedly.	Guards over hazardous areas should prevent anything flying towards the user. Students are trained to clamp work securely and to ensure the chuck key is removed from the chuck. All operators should wear eye protection. Horizontal mills require more extensive guarding. The level of training for staff and students required for milling is greater than for other operations. It is advisable to wear a full face visor.	Extreme	Unlikely	High	Staff and Students are trained in the use of this machine and are aware of risks. Staff should supervise students as necessary. Safety signs are in place.
MILLING: The removal of material to provide a finished form of close tolerance and good surface finish.	User injury	Human contact with rotating parts and swarf can cause cuts or abrasions. Guards around the hazardous area reduce the risk of hand or finger injury but training is essential. There is a high risk that students will put hands or fingers in hazardous places and experienced users may attempt shortcuts.	Moderate	Unlikely	Medium	Training to raise awareness of risks.
POWER SAWING: The cutting of work pieces to required length from stock bars or rods of both ferrous and non-ferrous material using a power saw. Cutting oils or lubricants may sometimes be required.	Trapping – closing movements between parts under power feed can present a trapping hazard.	Movement under power feed is usually slow, minimising this risk. The provision of adequate space around machines should also minimise the trapping risk.	Moderate	Unlikely	Medium	Training to raise awareness of risks.

	Entanglement – long hair, dangling jewellery or loose clothing can become entangled with rotating parts, dragging the user onto them.	Long hair should be tied back, jewellery should be removed or covered and loose clothing should be covered by a secure apron or overall.	Moderate	Unlikely	Medium	Training to raise awareness of risks.
	Manual handling – heavy parts, e.g. chucks, face plates or machine vices and stock bars, can present a manual handing hazard.	The risk is greatly reduced if more than one person handles any heavy items. Handling heavy components or awkward manipulation will not occur frequently but does present a real risk.	Moderate	Unlikely	Medium	Training to raise awareness of risks.

A number of BSI standards apply specifically to the engineering industry, covering areas such as technical delivery conditions. These are the standards relating to the appearance, labelling, quality and form of supply of engineering materials or products when delivered.

Ensuring machinery is well maintained

'PUWER requires that: all work equipment be maintained in an efficient state, in efficient order and in good repair; where any machinery has a maintenance log, the log is kept up to date; and that maintenance operations on work equipment can be carried out safely.'

Health and Safety Executive, Maintenance of Work Equipment, www.hse.gov.uk

In order to ensure workshop equipment does not deteriorate to an extent that it may put operators at risk, employers and others in control of work equipment are required by PUWER to keep it 'maintained in an efficient state, in efficient order and in good repair'. Many companies have annual maintenance contracts to service and check machinery.

ACTIVITY

Find examples of engineering products such as wearable technology, a lawnmower, garden shears, exercise equipment and temporary barriers for public places. Photograph and identify any labels that provide information or safety advice.

APPLICATION AND EXAMPLES FOR FASHION AND TEXTILES

All fashion and textiles products will be labelled in some way, giving information about the fabrics or materials used, the care instructions for washing or cleaning and any warnings, for example, of a fire risk.

The textiles and clothing industries employ around 189,000 workers in the UK, across 10,700 businesses (hse.gov.uk textiles). The industry can involve spinning, weaving, dyeing, printing, finishing and a number of other processes that are required to convert fibre into a finished fabric or garment. This means there are many safety and health issues associated specifically with the textile industry.

The major safety and health issues in the textile industry are:

- safety using machinery such as fabric cutting, sewing, spinning, weaving and warping
- exposure to cotton dust, other dust and chemicals
- exposure to noise
- ergonomic issues.

Some of these are looked at in more detail below.

Exposure to cotton dust

The workers engaged in the processing and spinning of cotton are exposed to significant amounts of cotton dust. They are also exposed to particles of pesticides and soil. Exposure to cotton dust and other particles leads to respiratory disorders.

Exposure to chemicals

Workers in the textile industry are also exposed to a number of chemicals, especially those engaged in the activities of dyeing, printing and finishing.

Exposure to noise

High levels of noise are evident in the textile industry, particularly in manufacturing units in developing countries.

Ergonomic issues

Ergonomic issues are often observed in units engaged in textile-related activities in developing countries. Many of these units have a working environment that is unsafe and unhealthy for the workers.

In America, the third largest producer of cotton, the OSHA (Occupational Safety and Health Administration in the US) has laid down a Cotton Dust Standard 2000, with a view to reducing the exposure of the workers to cotton dust. It has set up permissible exposure limits (PELs) for cotton dust for different operations in the textile industry.

For all of the identified risks above, consider what control measures would need to be in place to ensure the safety of machine operators/workers.

ACTIVITIES

1 Find examples of textile products such as clothing, soft furnishings and tents, and fashion accessories such as bags, costume jewellery or footwear. Photograph and identify any labels that provide information or safety advice. What extra advice might a product that uses e-textiles need?

2 Choose a specific step in a textiles manufacturing process, or an item of equipment for use with textiles, and produce a risk assessment.

Manufacturing Plan

Using existing pattern measurements and scale sketch of modifications, draw out pattern on pattern fabric with a seam allowance of 15mm.

Wash fabric to pre-shrink it, according to labeling, and then iron.

Pin pattern to fabric, making sure to tessellate patterns in order to minimise fabric wastage. In order to avoid injuries from pins, all the heads should be facing the same direction.

Trace over pattern lines onto fabric to mark it. Make sure the pen is machine washable so as not to indelibly mark the fabric.

Cut out fabric panels.

Cut line (using scissors) down front panel and sew in zip. Cut off excess zip at top and pin in place.

Insert magnet poppers to pockets and sew straight stitch around them to fix in place. Sew two fabric parts to each other. Overlock the sides of the pockets and then sew onto the back panel, folding over the hems and then using a straight stitch over the top.

Pin together seams inside out and then sew, using a zig zag stitch to allow for stretch, down the seam line. Then overlock to cut off excess and finish and finally iron the seam down flat. Start at shoulders and then continue around main body including mesh panels.

Hem sleeves, overlocking the edge and then folding it over and using a zig zag stitch (to allow the fabric to stretch). The sleeves can then be sewn onto the main body using the above method.

Sew on retroreflective panels, using a zig zag stitch along the edges so that it sits flush against the main body. These panels should cover the seams between the front and back panels and the mesh side panels.

Hem main body and neck, overlocking the edge and then folding over and sewing a zig zag stitch to finish. Sew on nylon loop to back. Snip off all excess thread.

Create light clip on Creo ProEngineer and then slice in Cura software. Place code on an SD card and then insert the SD card to the Ultimaker 3D printer for 3D printing.

Insert electronics, to be made using textiles components and connected using conductive thread tied through the terminals, to light clip and glue together with silicone based glue.

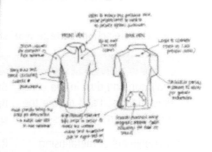

Figure 9.8 This section from a manufacturing plan was completed by a student during NEA work

Table 9.2 Example risk assessment completed as part of NEA

Process	Hazard	Who is at risk?	Severity of risk	Likelihood of risk	Controls
Pattern and fabric cutting	Risk of cutting fingers with scissors	User	Low	Low	Use blunt-ended scissors
Pinning	Risk of pricking fingers with pins	User	Low	High	Have pins all facing one direction
Sewing	Risk of fingers or clothing getting caught in machinery or under needle	User	Low	Low	Use needle guards Unplug when not in use or when changing settings Keep fingers away from needle

Care labelling code and symbols on textile products

Care labels on textile products give the consumer information about how to maintain the product and fabric. Labels provide details on the percentage of each fibre used to make a fabric, for example, 80 per cent cotton, 20 per cent polyester. This is a legal requirement. Symbols tell the consumer how to look after the product such as those in Figure 9.6 on page 395.

Flammability is a legal requirement for items such as children's fancy dress outfits, nightwear and soft furnishings.

Standard sizes are a standard measurement of the human body. For example, women's clothes come in sizes 10, 12, 14, 16, etc. or S, M, L.

APPLICATION AND EXAMPLES FOR PRODUCT DESIGN

The UK design and manufacturing industry is diverse and has an estimated 2.6 million workers. In manufacturing over the last five years there have been an average of 144 deaths a year through workplace accidents and reports of 3,100 major injuries. (Source: labour force survey, www.hse.gov.uk)

Risk assessments are very important and should cover use of machinery and tools as well as materials and substances that could be subject to COSHH regulations.

An example risk assessment for a factory employing 40 people can be found at www.hse.gov.uk/risk/casestudies/pdf/factory.pdf.

ACTIVITY

Having looked at the example risk assessment for the factory, complete one for a typical range of equipment in a school/college workshop. The examples that follow might help, but remember you need to think about your particular workshop and its users.

Table 9.3 Example school/college workshop risk assessment

Hazard	Control measures	Severity of risk	Likelihood of risk
Dust inhalation and dust in eye when finishing wood and tidying the workshop at the end of the lesson.	Brushes should be avoided when cleaning up dust. Handheld vacuum cleaner is provided.	Low	Moderate
Finishing polymers or metal using metal polish, or finishing wood with wax and stains.	Refer to COSHH and safety cards. Care should be taken when using polishes – students should be made aware of the risk of getting polish in their eyes or in mouth. Students should be encouraged to use products carefully, applying with a rag, etc. and avoiding prolonged contact with skin.	Moderate	Low

Hazard	Control measures	Severity of risk	Likelihood of risk
Gluing polymer using liquid solvent cement such as Tensol Cement. Gluing metal with epoxy resin. Gluing wood with PVA.	Refer to COSHH and safety cards. Care should be taken when using adhesives. Students should be made aware of the risk of getting adhesives in their eyes or mouth. Students should be encouraged to use them carefully, applying with a pipette or lollipop stick. Avoid prolonged contact with skin. Goggles must be worn when using liquid solvent cement or Tensol Cement.	Moderate	Low

Table 9.4 Example risk assessment completed as part of NEA

Process	Hazards	Person at risk	Controls in place	Risk evaluation		
				Severity	Likelihood	Rating
Using Hegner, reciprocating or scroll saw to cut straight lines.	Blade snapping, wood flying from machine, skin contact with blade.	User	Area around machine marked out to warn other workshop users. Blade well clamped in place and sharp.	Low	Medium	Low
Varnishing wood	Irritating vapours, harmful to eyes and lungs, flammable.	User, people in close surroundings	Sufficient ventilation. Varnish kept in flameproof cupboard when not in use.	Moderate	Medium	Medium
Hand tools with blades	Blades can cut skin; thin blades can snap.	User	Ensure blade is well attached and sharp.	Low	Low	Low

Product designers use a variety of materials and manufacturing methods, including textiles, so any risk assessment must take this variety into account.

ACTIVITY

Find examples of consumer products such as wearable technology, interior furniture, exercise equipment and outdoor play equipment. Photograph and identify any labels that provide information or safety advice. What extra advice might a product that uses textiles material need?

KEY TERMS

Consumer Rights Act (2015)/Trade Description Act (1968)/Consumer Protection Act (1987) – laws designed to protect consumers when purchasing and using products.

Standard – an agreed, repeatable way of doing something. It is a published document that contains a technical specification or other precise information designed to be used consistently as a rule, guideline or definition.

Further reading on health and safety

- Health and Safety at Work Act (1974) and all other legislation mentioned in this chapter – www.legislation.gov.uk
- BS 4163 2014 Health and Safety for Design and Technology in School – www.bsigroup.com
- Product safety for manufacturers 2015 – www.gov.uk/guidance/product-safety-for-manufacturers
- Health and Safety Executive – http://www.hse.gov.uk
- Working with substances hazardous to health – a brief guide to COSHH – www.hse.gov.uk/pubns/indg136.pdf

PRACTICE QUESTIONS: health and safety

Design Engineering

1 Use one example of legislation to discuss the protection of workers/operators in an engineering workplace.

2 Give two examples of labels you would expect to find on a mechanical or electrical product.

3 Discuss the issues to be considered by an engineering company when implementing risk assessment procedures.

Fashion and Textiles

1 Use one example of legislation to discuss the protection of workers/operators in the textiles industry.

2 Give two examples of labels you would expect to find on a soft toy made of textile material.

3 Discuss the issues to be considered by textile manufacturers when implementing risk assessment procedures.

Product Design

1 Use one example of legislation to discuss the protection of workers/operators in the workshop environment.

2 Give two examples of labels you would expect to find on an electrical consumer product.

3 Discuss the issues to be considered by a manufacturer when implementing risk assessment procedures.

Photo credits

All photos that do not appear in the following list have been kindly supplied by the authors of this book.

p.2 © Babich Alexander/Shutterstock.com; p.7 © Trueffelpix/stock.adobe.com; **p.9** © Shaunwilkinson/stock.adobe.com; **p.10** © 111 Navy Chair; **p.14** © Claudio Divizia/stock.adobe.com; **p.16** (left) © bank_jay/stock.adobe.com, (right) © Aaron Bass/123RF; **p.18** © Krzysiek/stock.adobe.com; **p.19** (left) © Interfoto/Alamy Stock Photo, (right) © Syda Productions/stock.adobe.com; **p.28** (top) © Happy1e777/Shutterstock.com, (middle) © Ning2k/Shutterstock.com, (bottom) © Atstock Productions/Shutterstock.com; **p.29** © Believeinme33/123RF; **p.30** © Anelluk /123RF; **p. 36** © donaveh/123RF; **p.42** © Adrian Sherratt/Alamy Stock Photo; **p.43** © Scanrail/stock.adobe.com; **p.46** (top left) © Juergen Hanel/Alamy Stock Photo/Courtesy of Knoll, Inc., (bottom right) © Ratchapol yindeesuk/123RF; **p.47** (left) © Andrey_Arkusha/stock.adobe.com, (right) © Pavel Korotkov/stock.adobe.com; **p.49** © Charlie Pinatex Boot – Bourgeois Boheme – www.bboheme.com; **p.52** © imagedb.com/Shutterstock.com; **p.53** © Granger Historical Picture Archive / Alamy Stock Photo; **p.58** © Clari Massimiliano/Shutterstock.com; **p.72** © Günter Menzl/stock.adobe.com; **p.73** (top) © Massimo Cavallo/stock.adobe.com, (middle) © Kadmy/stock.adobe.com, (bottom) © REDPIXEL/stock.adobe.com; **p.75** © katukphoto/stock.adobe.com; **p.76** © Tomasz Zajda/stock.adobe.com; **p.78** (left) © Forest Stewardship Council®, (right) © Anthony Baggett/123RF; **p.79** (left) © Patryk Kosmider/stock.adobe.com, (right) © gkuna/123RF; **p.80** (left) © shooting88/stock.adobe.com, (right) © Fotos 593/stock.adobe.com; **p.81** © petovarga/stock.adobe.com; **p.82** © Egor/stock.adobe.com; **p.84** © EU Ecolabel; **p.89** (top) © rosinka79/stock.adobe.com, (bottom) © Taina Sohlman/123RF; **p.91** © Designua/Shutterstock.com; **p.96** © bonninturina/stock.adobe.com; **p.97** © Alfred Hofer/123RF; **p.107** © Soonthorn/stock.adobe.com; p.111 © Dzmitry Sukhavarau/stock.adobe.com; **p.113** (bottom right) © Iain Masterton/Alamy Stock Photo; **p.117** © Edwin Remsburg/VW Pics/Universal Images Group/Getty Images; **p.123** (right) © Petovarga/123RF; **p.125** (bottom) © folienfeuer/stock.adobe.com; **p.127** (right) © Fernando Batista/stock.adobe.com; **p.134** (top left) © John Hammond/The National Trust Photolibrary/Alamy Stock Photo, (top right) © Andreas von Einsiedel/Alamy Stock Photo, (bottom) © elenamas_86/stock.adobe.com, **p.136** (top) © Leo ho; (bottom) © Alexander Tolstykh/Alamy Stock Photo; **p.140** © vladimirzhoga/stock.adobe.com; **p.141** (top) © xMarshall/Alamy Stock Photo, (bottom) ©2004 Credit:TopFoto/ImageWorks; **p.142** (top) © dipling/stock.adobe.com, (bottom) © 2018 CARV; **p. 143** © OXO Good Grips Kitchen & Herb Scissors; **p.147** © trgowanlock/stock.adobe.com; **p.150** (right) © sergey0506/stock.adobe.com; **p.151** (right) © Jeanette Teare/123RF; **p.152** © Steven Urquhart/Shutterstock.com; **p.154** (left) © Christopher Marsh/Alamy Stock Photo, (right) © montego6/stock.adobe.com; **p.155** (left) © akulamatiau/stock.adobe.com, (right) © Monty Rakusen/Cultura Creative (RF)/Alamy Stock Photo; **p.157** (top) © Ian Kennedy/Shutterstock.com; **p.159** (bottom left) © James Dale/123RF, (right) © Destina/stock.adobe.com; **p.160** (bottom left) © ThamKC/stock.adobe.com; **p.161** (middle right) © Iakov Filimonov/Shutterstock.com; **p.162** © Kadmy/stock.adobe.com; **p.164** (right) © Studio KIVI/stock.adobe.com; **p.165** (top left) © praphab144/stock.adobe.com, (bottom right) © Jale Ibrak/stock.adobe.com; **p.168** (right) © Win Nondakowit/stock.adobe.com; **p.169** (top right) © beechtreebaby; **p.170** © Andrew Skinner/stock.adobe.com, **p.171** (bottom left) © Dusan Kostic/stock.adobe.com; **p.172** (left) © Monkey Business/stock.adobe.com; **p.173** © santypan/stock.adobe.com; **p.179** (right) © Sally Wallis/stock.adobe.com; **p.180** (left) © Tinxi/Shutterstock.com; **p.182** (left) © Dontree/Shutterstock.com, (right) © Tim Graham/Alamy Stock Photo; **p. 183** © Ruslan Kudrin/Alamy Stock Photo; **p.185** (left) © aapsky/stock.adobe.com, (right) © Kadmy/stock.adobe.com; **p.186** © bakicirkin/stock.adobe.com; **p.187** © eugenesergeev/stock.adobe.com; **p.188** (left) © Nor Gal/Shutterstock.com, (right) © H.S. Photos/Alamy Stock Photo; **p.189** (left) © frog/stock.adobe.com, (right) © nikolayn/stock.adobe.com; **p.190** (right) © Jonny White/Alamy Stock Photo; **p.191** (left) © Alexey Kamenskiy/Shutterstock.com, (right) © epitavi/stock.adobe.com; **p.196** © noxnorthy/123RF; **p.198** © magmac83/stock.adobe.com; **p.208** © Agefotostock/Art Collection/Alamy Stock Photo; **p.209** (top) © apugach/123RF, (bottom) © Stacy Barnett/123RF; **p.210** © Anna Baburkina/stock.adobe.com; **p.211** © Saklakova/stock.adobe.com; **p.212** © lubos K/stock.adobe.com; **p.215** © Fresh&Green/Alamy Stock Photo; **p.216** © travelib prime/Alamy Stock Photo; **p.217** (top left) © NaDi/stock.adobe.com, (bottom left) © Whiteaster/Alamy Stock

Index

Page locators in **bold** indicate key terms.